Massive MIMO for Future Wireless Communication Systems

IEEE Press
445 Hoes Lane
Piscataway, NJ 08854

Massive MIMO for Future Wireless Communication Systems

Technology and Applications

Edited by

Webert Montlouis
Johns Hopkins University, United States

Agbotiname Lucky Imoize
University of Lagos, Nigeria

IEEE Press
WILEY

Published by John Wiley & Sons, Inc., Hoboken, New Jersey.
Published simultaneously in Canada.

Library of Congress Cataloging-in-Publication Data applied for:

Hardback ISBN: 9781394228300

Cover Design: Wiley
Cover Image: © Yuichiro Chino/Getty Images

Set in 9.5/12.5pt STIXTwoText by Straive, Chennai, India

This book is dedicated with love and affection to my mother, thank you mom.

Webert Montlouis

In memory of my late grandma, Lucy Omoyeni Sado (Madam LS)

Agbotiname Lucky Imoize

Contents

About the Editors

Webert Montlouis (Fellow, IEEE) received his BS, MS, and PhD in electrical and computer engineering from Northeastern University, Boston, MA. He is currently at Johns Hopkins University, Baltimore, MD. He has served as chief scientist at the Applied Physics Laboratory (APL) and faculty in the electrical and computer engineering department. He has been the chair of the IEEE Massive MIMO Standard Development Working Group. He is also the co-chair of the Massive MIMO Working Group. He has served as general co-chair of the IEEE Massive MIMO Workshop and session chair for many IEEE conferences. He previously worked for the System Architecture, Design, and Integration Directorate, a division of the Raytheon Company in Boston, MA. During his tenure at Raytheon, he developed system architecture, operational concepts, algorithms, and performance analyses for phased array radar systems. He also worked as a technical consultant for firms specializing in developing ASIC and FPGA for communication systems. His research interests are Multi-Channel System Architecture, Sensing, Next Generation Radar Systems, Wireless Communications 5G and Beyond, Artificial Intelligence, Quantum Information Science, Digital Signal Processing, and Biomedical Signal Processing. He is a fellow of the IEEE and a member of the IEEE Signal Processing Society, the IEEE Communications Society, and the IEEE Aerospace and Electronic Systems Society.

Agbotiname Lucky Imoize (Senior Member, IEEE) received the BEng degree (Hons.) in electrical and electronics engineering from Ambrose Alli University, Nigeria, in 2008 and the MSc degree in electrical and electronics engineering from the University of Lagos, Nigeria, in 2012. He is a lecturer in the Department of Electrical and Electronics Engineering at the University of Lagos, Nigeria. Before joining the University of Lagos, he was a lecturer at Bells University of Technology, Nigeria. He was, until recently, a researcher at the Ruhr University Bochum, Germany, under the sponsorship of the Nigerian petroleum technology development fund (PTDF) and the German academic exchange service (DAAD) through the Nigerian-German postgraduate program. He was awarded the Fulbright fellowship as a visiting research scholar at the Wireless@VT Laboratory, Bradley Department of Electrical and Computer Engineering, Virginia Tech., Blacksburg, USA, where he worked under the supervision of Prof. R. Michael Buehrer from 2017 to 2018. He worked as a core network products manager at ZTE, Nigeria, and as a Network Switching Subsystem Engineer at Globacom, Nigeria. His research interests cover the fields of 6G wireless communication, wireless security systems, and artificial intelligence. He is the vice chair of the IEEE Communication Society Nigeria chapter. He is a registered engineer with the Council for the Regulation of Engineering in Nigeria and a member of the Nigerian Society of Engineers.

List of Contributors

Taufik Abrão
Electrical Engineering Department
Londrina State University
Brazil

Abdullateef Ola Adebayo
Department of Electrical and
Computer Engineering
Kwara State University
Malete
Nigeria

Anjana B. S.
Department of Communication
Engineering, School of Electronics
Engineering
Vellore Institute of Technology
Tamil Nadu, Vellore
India

Joseph Bamidele Awotunde
Department of Computer Science,
Faculty of Information and
Communication Sciences
University of Ilorin
Ilorin
Nigeria

Peace Oluwasijibomi Balogun
Department of Electrical and
Computer Engineering
Kwara State University
Malete
Nigeria

Nivetha Baskar
Department of Communication
Engineering, School of Electronics
Engineering
Vellore Institute of Technology
Tamil Nadu, Vellore
India

Emil Björnson
Department of Computer Science
KTH Royal Institute of Technology
Stockholm
Sweden

Yashuai Cao
Department of Electronic Engineering
Tsinghua University
Beijing China

and

Beijing National Research Center for Information Science and Technology (BNRist)
Beijing
China

Daniel Castanheira
Instituto de Telecomunicações (IT), and Departamento de Electrónica Telecomunicações e Informática (DETI)
Universidade de Aveiro
Aveiro
Portugal

Özlem Tuğfe Demir
Department of Electrical and Electronics Engineering
TOBB University of Economics and Technology
Ankara
Türkiye

Rui Dinis
Instituto de Telecomunicacões (IT), Faculdade de Ciências e Tecnologia
Universidade Nova de Lisboa
Lisboa
Portugal

Dipinkrishnan R
Department of Communication Engineering, School of Electronics Engineering
Vellore Institute of Technology
Tamil Nadu, Vellore
India

Unwana Macaulay Ekpe
Department of Electrical and Electronic Engineering, Faculty of Engineering
Akwa Ibom State University
Ikot Akpaden
Nigeria

Atílio Gameiro
Instituto de Telecomunicações (IT), and Departamento de Electrónica Telecomunicações e Informática (DETI)
Universidade de Aveiro
Aveiro
Portugal

Vishnu Vardhan Gudla
Department of Electronics and Communication Engineering
Aditya Engineering College
Andhra Pradesh
Surampalem
India

David William Marques Guerra
Electrical Engineering Department
Londrina State University
Brazil

Ekram Hossain
Department of Electrical and Computer Engineering
University of Manitoba
Canada

Agbotiname Lucky Imoize
Department of Electrical and Electronics Engineering, Faculty of Engineering
University of Lagos
Akoka, Lagos
Nigeria

and

Department of Electrical Engineering and Information Technology
Ruhr University
Bochum
Germany

Segun Jacob
Department of Electrical and Computer Engineering
Kwara State University
Malete
Nigeria

Thiruvengadam Sundarrajan Jayaraman
Department of Electronics and Communication Engineering
Thiagarajar College of Engineering
Tamil Nadu, Madurai
India

Helen Sheeba John Kennedy
Department of Communication Engineering, School of Electronics Engineering
Vellore Institute of Technology
Tamil Nadu, Vellore
India

Wilson Souza Jr.
Electrical Engineering Department
Londrina State University
Brazil

Debdatta Kandar
Department of Information Technology
North Eastern Hill University
Shillong, Meghalaya
India

Joumana Kassam
Instituto de Telecomunicações (IT), and Departamento de Electrónica Telecomunicações e Informática (DETI), Universidade de Aveiro
Aveiro
Portugal

Vinoth Babu Kumaravelu
Department of Communication Engineering, School of Electronics Engineering
Vellore Institute of Technology
Tamil Nadu, Vellore
India

Tiejun Lv
School of Information and Communication Engineering
Beijing University of Posts and Telecommunications (BUPT)
Beijing
China

José Carlos Marinello
Electrical Engineering Department
UTFPR
Brazil

Webert Montlouis
Applied Physics Laboratory (APL) and Electrical and Computer Engineering Department
Johns Hopkins University
Baltimore, MD
USA

Arthi Murugadass
School of Computer Science and Engineering
Vellore Institute of Technology
Chennai
Tamil Nadu
India

Abdulwaheed Musa
Department of Electrical and Computer Engineering
Kwara State University
Malete
Nigeria

and

Centre for Artificial Intelligence and Machine Learning Systems
Kwara State University
Malete
Nigeria

and

Institute for Intelligent Systems
University of Johannesburg
Johannesburg
South Africa

Wei Ni
Commonwealth Scientific and Industrial Research Organisation
Sydney, NSW
Australia

Vetriveeran Rajamani
Department of Micro and Nano Electronics, School of Electronics Engineering
Vellore Institute of Technology
Tamil Nadu, Vellore
India

Rajeshkumar V
Department of Communication Engineering, School of Electronics Engineering
Vellore Institute of Technology
Tamil Nadu, Vellore
India

Parisa Ramezani
Department of Computer Science
KTH Royal Institute of Technology
Stockholm
Sweden

Poongundran Selvaprabhu
Department of Communication Engineering, School of Electronics Engineering
Vellore Institute of Technology
Tamil Nadu, Vellore
India

Adão Silva
Instituto de Telecomunicações (IT), and Departamento de Electrónica Telecomunicações e Informática (DETI)
Universidade de Aveiro
Aveiro
Portugal

Arun Kumar Singh
Department of Electronics and Communication Engineering
Sikkim Manipal Institute of Technology
Sikkim Manipal University
Majitar, Sikkim
India

Velmurugan Periyakarupan Gurusamy Sivabalan
Department of Electronics and Communication Engineering
Thiagarajar College of Engineering
Tamil Nadu, Madurai
India

João Henrique Inacio de Souza
Electrical Engineering Department
Londrina State University
Brazil

Samarendra Nath Sur
Department of Electronics and Communication Engineering
Sikkim Manipal Institute of Technology
Sikkim Manipal University
Majitar, Sikkim
India

Anand Sreekantan Thampy
Centre for Nanotechnology Research
Vellore Institute of Technology
Tamil Nadu, Vellore
India

Hetong Wang
School of Information and Communication Engineering
Beijing University of Posts and Telecommunications (BUPT)
Beijing
China

Preface

The need to implement cutting-edge technologies to satisfy the demands of fast growing mobile usage by efficiently utilizing scarce spectrum resources is a critical requirement in modern wireless networks. Though several antenna configurations have been proposed to alleviate this problem, the existing techniques have not adequately addressed prevailing issues such as severe pathloss, limited network performance, interference, among others. A possible alternative is applying an advanced antenna technology called massive multiple-input multiple-output (MIMO). Massive MIMO is a key-enabling technology in ubiquitous 5G wireless systems, and it poses enormous prospects for the envisioned 6G and future wireless networks. Massive MIMO could integrate other cutting-edge technologies to improve rates, energy, and spectral efficiency, opening new frontiers in wireless communications.

In massive MIMO, a single base station (BS) can be exploited to eliminate inter-cell interference, employing directional beamforming. First, however, the problem of gaining an insightful understanding of the fundamental limits of massive MIMO persists. Second, the performance evaluation of massive MIMO under ideal and nonideal practical scenarios requires detailed investigation. Third, billions of massive devices communicate via open wireless channels, posing huge security risks to sensitive user data. Thus, robust security architectures for privacy preservation and security of critical information have become imperative in massive MIMO wireless communication networks.

This book presents applications and technology for massive MIMO in 6G and beyond. Specifically, the book presents a unified theoretical framework for analyzing the fundamental limits of massive MIMO, considering several practical network constraints. The book presents advanced signal processing algorithms to enable massive MIMO applications in realistic environments. Additionally, the book presents advanced mathematical tools to analyze multiuser dynamics in evolving wireless communication channels.

The key highlights of the book are outlined as follows:

- The book provides industry and academic researchers with new insights into the real-world deployment scenarios, design and implementation, application, technological trends, and associated benefits of massive MIMO in emerging wireless communication systems.
- The book addresses the need to design energy and spectral-efficient massive MIMO models to resolve several network issues, such as interference, pathloss, delay, traffic outage, and so on, in modern wireless communication systems.
- The book critically examines the fundamental limits of massive MIMO and proffer solutions to revamp the traditional MIMO architecture toward addressing the vast network issues, especially at the wireless edge.
- The book discusses critical security and privacy issues affecting all stakeholders in the wireless ecosystem, and provides practical and effective solutions to address these problems.

The book is structured into 13 chapters collected from industry experts and world-class academic researchers, resulting in diverse and high-quality work for the readers.

Chapter 1 provides introduction to massive MIMO for future wireless communication systems. Starting with the fundamentals of wireless communications and providing background information to facilitate understanding the book. The chapter examines how wireless communication has revolutionized the way we interact with technology, enabling seamless connectivity and communication between devices and networks without needing physical cables. As technology has advanced, each generation of wireless communication has brought significant improvements, leading us to the era of 5G and the promise of even more transformative advancements in the future with 6G and beyond. Wireless communication rely on electromagnetic waves to carry signals through the air or space. Understanding the fundamentals of wireless communication is essential to grasp how these systems work and their various applications. The key fundamentals are the electromagnetic spectrum, radio frequency, modulation, demodulation, antennas, propagation, signal-to-noise ratio (SNR), multiple access techniques, wireless standards, security, and latency, among others.

Chapter 2 examines the security and privacy of future wireless communication systems. The chapter summarizes the proliferating security vulnerability issues in wireless networks. Key challenges, including evolving security threats, massive MIMO security issues, and privacy risks from advanced data collection, are highlighted. Physical layer security methodologies, access control, monitoring tools, and privacy technologies are proposed as viable solutions. The impact of government policies, regulations, and emerging technologies like blockchain are highlighted. Last, open research issues are presented, emphasizing lightweight cryptography, new architectures, and aligning standards with the prevailing regulations.

Chapter 3 presents the applications of massive MIMO in wireless communication systems. The chapter presents a detailed capacity analysis, quantifying the immense throughput gains possible with large antenna arrays and wider bandwidths. Promising solutions drawing from existing literature are reviewed, including antenna architectures, estimation algorithms, signal processing advancements, network topologies, and cross-technology convergence that can help realize the potential of massive MIMO at mmWave/THz bands. The chapter identifies critical open-research challenges related to experimental validation, complexity, security, and standardization that need to be investigated to fully materialize the disruptive capabilities of massive MIMO for 6G and beyond. Last, the chapter provides a holistic perspective on harnessing the potential of massive MIMO in 6G by building on prior art while highlighting open problems, requiring further innovation tailored to wireless networks demands.

Chapter 4 considers cell-free massive MIMO technology and its applications in 6G. The chapter presents an overview of CF mMIMO systems, emphasizing the superiority of such systems over traditional cellular networks. The study covers the system architecture, namely centralized and distributed, along with recent research developments and beamforming algorithms to address the scalability issue of conventional CF systems. These techniques include the user-centric (UC) approach that seeks to build a realistic deployment, where each user equipment (UE) is only served by a small number of cooperative APs. In particular, the chapter considers the integration of CF with radio stripes (RSs) that improve robustness and reduce the high costs/complexity by employing a sequential topology, sharing the same cable for fronthaul and then connecting each RS to a single or multiple CPUs.

Chapter 5 focuses on localization in massive MIMO networks, considering the scenario from near-field to far-field. The chapter considers two subspace-based approaches; Multiple SIgnal Classification (MUSIC) and Estimation of Signal Parameters by Rotational Invariance Techniques (ESPRIT). Originally designed for far-field source localization, these methods have undergone several modifications to accommodate near-field scenarios. The chapter presents the foundations of MUSIC and ESPRIT algorithms and introduces some of their variations for both far-field and near-field localization by a single array of antennas. Finally, the chapter provides numerical examples to demonstrate the performance of the presented methods.

Chapter 6 considers energy-efficient (EE) uplink (UL) transmission in RIS-aided M-MIMO internet of things (IoT) systems. The chapter focuses on the EE UL transmission of M-MIMO IoT systems aided by an RIS. The work proposes and evaluates different schemes to minimize the total UL transmit power by optimizing the transmit power of IoT devices, the RIS phase-shift element, and the combining matrix at the BS. Particularly, special attention is given to manifold optimization techniques, which are well suited to the RIS phase-shifts optimization problems. Herein, the authors treat jointly via iterative alternating optimization (i-AO) approach, the three optimization variables: RIS phase-shift vector, BS combining matrix, and unit terminal (UT) power allocation vector. Extensive numerical results are provided and discussed, revealing that the proposed conjugate gradient (CG) method based on Riemannian manifold (RM) with the ZF combining achieves the highest power savings, being able to reduce the UL transmit power under typical operation conditions scenarios in comparison with conventional systems without RIS.

Chapter 7 deals with energy efficiency (EE) optimization of massive multiple-input multiple-output (M-MIMO) wireless communication systems aided by reconfigurable intelligent surfaces (RISs) and extralarge-scale M-MIMO (XL-MIMO) systems. The chapter presents a unified mathematical model for the wireless propagation channel and a set of optimization tools to optimize the EE in systems operating under realistic constraints. Particularly, two EE problem were formulated and solutions developed by considering different system configurations. In order to maximize the EE in RIS-aided M-MIMO systems, a sequential optimization method for joint RIS phase-shifts design, power allocation (PA), and optimization of the number of active BS antennas is proposed. Summarily, the chapter provides useful optimization tools and analytical frameworks, including techniques from convex optimization and evolutionary heuristic that can be applied to tackle a multitude of other EE-related problem formulations and feasible algorithmic solutions.

Chapter 8 presents NOMA-aided massive MIMO for next-generation networks. The system model of NOMA-aided mMIMO is presented and the sum rate expressions are derived for a two-user scenario. Results show that the proposed mMIMO-NOMA shows better performance than the existing mMIMO-OMA in terms of sum rate.

Chapter 9 proposes a low-complexity approach to achieving efficient hybrid precoding (HP) for millimeter-wave massive MIMO-NOMA systems. The chapter focuses on evaluating the performance of mmWave-mMIMO-NOMA systems in multiple user (MU) scenarios, with the overarching goal of designing a low-complexity HP scheme that elevates system performance. The chapter commences with an in-depth exploration of existing precoding algorithms pertinent to mMIMO-NOMA systems. Furthermore, the chapter proposes an innovative approach termed symmetric sequential over-relaxation (SSOR) complex regularized zero-forcing (CRZF) linear precoder. A pivotal aspect of the chapter involves comparing the performance of the proposed precoder against conventional linear precoders. The comparative analysis hones in on observable advantages such as improved spectral efficiency (SE), EE, and reduced computational complexity.

Chapter 10 investigates intelligent reflecting surfaces (IRS) and next-generation wireless systems. In particular, the chapter analyzes the slot-by-slot IRS reflection pattern design and two-timescale reflection pattern design schemes. For the slot-by-slot reflection optimization, an IRS to improve the propagation channel rank in mmWave massive MIMO systems without need to increase the transmit power budget was exploited. Then, the impact of the distributed IRS on the channel rank was analyzed. To further reduce the heavy overhead of channel training, channel state information (CSI) estimation, and feedback in time-varying MIMO channels, a two-timescale reflection optimization scheme is presented, where the IRS is configured relatively infrequently based on statistical CSI (S-CSI) and the active beamformers, and power allocation are updated based on quickly outdated instantaneous CSI (I-CSI) per slot. The achievable average sum-rate (AASR) of the system is maximized without excessive overhead of cascaded channel estimation. A recursive sampling particle swarm optimization (PSO) algorithm was developed to optimize the large-timescale IRS reflection pattern efficiently with reduced samplings of channel samples.

Chapter 11 conducts average bit error rate (ABER) performance evaluation of RIS-aided millimeter wave massive MIMO system under 3GPP 5G channels. The performance of the proposed system is evaluated in the indoor hotspot (InH)-indoor office and urban microcellular (UMi)-street canyon outdoor environments for 28 and 73 GHz mm Wave frequencies in 3GPP 5G channel models. The ABER of the proposed system is analyzed for indoor and outdoor scenarios by changing the number of reflector elements in RIS, and the number of BS antennas. The results showed that there is a significant SNR gain improvement while increasing the number of reflector elements and transmit antennas.

Chapter 12 explores massive MIMO for non-terrestrial wireless communication systems. The chapter offers a concise and comprehensive overview of the pivotal facets and formidable challenges inherent in non-terrestrial wireless communication networks, with a primary focus on signal processing for massive MIMO via satellites in the context of 5G and beyond. The themes presented reveal that orchestrating an intricate interplay among hundreds or even thousands of satellites and safeguarding a secure and dependable communications system will pose substantial challenges. Furthermore, and projecting forward, satellites are positioned as pivotal enablers for the forthcoming wave of wireless networks and will support a diverse array of applications, including massive machine-type communications, all underscored by the overarching need for secure connectivity. The chapter also presents ongoing issues related to spectrum allocation, interference management, and the need for security fortifications within the satellite communications sphere. As the demand for connectivity beyond the confines of earth continues to grow, the need for standardized and interoperable communication protocols becomes increasingly imperative. The chapter concludes by presenting recent solutions to problems in the ever-evolving landscape of non-terrestrial wireless communications and recommends a few important research pathways in this domain.

Chapter 13 integrates artificial intelligence and machine learning for channel estimation in massive MIMO wireless communication systems. Specifically, the chapter presents a comprehensive review of least squares (LS)- and minimum mean square error (MMSE)-based channel estimation in wireless communication systems, starting from single-input single-output (SISO) to massive MIMO in both flat and frequency-selective fading environments.

The book is an ideal reference to practitioners, industry and academic researchers, scientists, and engineers in the fields of wireless communications and networking, signal processing, 5G and 6G networks, massive MIMO standardization, antennas design, sensing and localization, channel modeling and measurement, artificial intelligence, machine learning, federated learning, terrestrial and non-terrestrial applications, security and privacy, and others. Also, the textbook is suitable for graduate and senior undergraduate courses in wireless communications and related fields.

The editors specially thank the reviewers of the original book proposal for their constructive suggestions. Special thanks to all authors of the chapters for their insightful contributions. Many thanks to the reviewers, and editorial assistants at Wiley and IEEE Press for their cooperation and support.

Agbotiname Lucky Imoize
Gelsenkirchen, North Rhine-Westphalia, Germany

Webert Montlouis
Baltimore, Maryland, United States

Acknowledgments

I extend my deepest gratitude to my family. Your unwavering support, patience, and love have been the foundation of all my endeavors, providing the strength and inspiration I needed to persevere. I am sincerely thankful to the editorial team at WILEY-IEEE Press for your exceptional leadership and guidance. Your expertise and dedication have been crucial in bringing this work to life. I am profoundly grateful to my friends for your persistent encouragement, moral support, and belief in my abilities. Your presence has been instrumental in my academic and professional growth. To my colleagues and students, thank you for continually inspiring me to explore new ideas and push the boundaries of knowledge. Your curiosity and enthusiasm have been a constant source of motivation and have enriched this journey immeasurably.

Webert Montlouis
Columbia, Maryland, United States

I sincerely express my profound gratitude to God for His divine wisdom and faithfulness in editing this book. The book would not have been possible without the support of the Department of Electrical Engineering and Information Technology, Institute of Digital Communication, Ruhr University Bochum, Germany, and the University of Lagos, Nigeria. I wholeheartedly acknowledge the sponsorship from the Nigerian Petroleum Technology Development Fund (PTDF) and the German Academic Exchange Service (DAAD) through the Nigerian-German Postgraduate Program. Additionally, I am indebted to the Deeper Life Bible Church, Essen Region, North Rhine-Westphalia, Germany, for their unwavering support. Special thanks to my beloved wife, Kelly, and our sons, Lucius, Luke, Lucas, and Luther. Last, I sincerely thank the editorial team at WILEY-IEEE Press for their support.

Agbotiname Lucky Imoize
Gelsenkirchen, North Rhine-Westphalia, Germany

1

Fundamentals of Wireless Communications: Massive MIMO Essentials for 6G and Beyond

Webert Montlouis[1] and Agbotiname Lucky Imoize[2]

[1] *Johns Hopkins University, Baltimore, MD, USA*
[2] *Department of Electrical and Electronics Engineering, Faculty of Engineering, University of Lagos, Akoka, Lagos, Nigeria*

1.1 Introduction

Wireless communication has revolutionized how we interact with technology, enabling seamless connectivity and communication between devices and networks without needing physical cables [1]. As technology has advanced, each generation of wireless communication has brought significant improvements, leading us to the era of 5G and the promise of even more transformative advancements in the future with 6G and beyond [2, 3]. Wireless communication relies on electromagnetic waves to carry signals through the air or space. Understanding the fundamentals of wireless communication is essential to grasp how these systems work and their various applications. The key fundamentals are the electromagnetic spectrum, radio frequency (RF), modulation, demodulation, antennas, propagation, signal-to-noise ratio (SNR), multiple access techniques, wireless standards, security, and latency.

The electromagnetic spectrum is the range of all possible frequencies of electromagnetic radiation. It includes radio waves, microwaves, infrared, visible light, ultraviolet, X-rays, and gamma rays. Different wireless communication technologies use different portions of this spectrum to transmit signals. In wireless communication, RF refers to the range of frequencies used to transmit and receive signals [4]. RF communication is most common and includes technologies like wireless fidelity (Wi-Fi), Bluetooth, cellular networks, and satellite communication.

Modulation is the process of impressing the information (voice, data, video, etc.) onto the carrier signal [5]. It allows the information to be transmitted efficiently over the wireless channel. Common modulation techniques include

Massive MIMO for Future Wireless Communication Systems: Technology and Applications,
First Edition. Edited by Webert Montlouis and Agbotiname Lucky Imoize.

amplitude modulation (AM), frequency modulation (FM), and phase modulation (PM). Demodulation is the reverse modulation process, extracting information from the received modulated signal. Antennas are essential components in wireless communication systems. They transmit and receive electromagnetic waves, allowing devices to communicate with each other.

Antennas can be directional or omnidirectional, depending on their design and application. Propagation refers to the way electromagnetic waves travel through the environment. The wireless signal can propagate through free space, atmosphere, buildings, and other obstacles. Propagation can be affected by factors like path loss, fading, reflection, diffraction, and interference. SNR measures the strength of the desired signal compared to the background noise in the communication channel. A higher SNR indicates a better quality signal and improved communication reliability.

Multiple access techniques allow multiple devices to share the same communication channel efficiently. Common techniques include frequency division multiple access (FDMA), time division multiple access (TDMA), code division multiple access (CDMA), and orthogonal FDMA (OFDMA). Various organizations define wireless communication standards to ensure compatibility and efficient use of the wireless spectrum. For example, the Institute of Electrical and Electronics Engineers (IEEE) sets standards for Wi-Fi (e.g., 802.11ac and 802.11ax), and the Third-Generation Partnership Project (3GPP) defines cellular standards like 4G LTE and 5G [6].

Wireless communication is susceptible to security threats like eavesdropping, data interception, and unauthorized access. Encryption and other security protocols protect the data and ensure secure communication. Latency is the time delay between the transmission and reception of data. Low latency is crucial in applications requiring real-time responsiveness (e.g., online gaming and video conferencing). Understanding these fundamental concepts of wireless communication helps engineers, researchers, and industry professionals design and optimize wireless systems, develop new technologies, and create innovative applications that continue to shape the world of wireless connectivity.

1.1.1 A Brief History of Wireless Communication

A brief history describing the evolution of wireless communication, the next phase in wireless communication, the beyond 5G – the future of wireless communications, and the applications landscape is presented as follows.

i) Evolution of Wireless Communication
   - 1G or First-generation mobile communication systems introduced in the 1980s, primarily based on analog technology.

- 2G or Second-generation mobile systems emerged in the early 1990s, utilizing digital technology to offer improved voice quality and the introduction of text messaging (SMS).
- 3G or Third-generation networks arrived in the early 2000s, providing faster data transfer, enabling mobile internet browsing, and supporting multimedia applications.
- 4G or Fourth-generation networks came into play around the mid-2000s, offering significantly faster data speeds, enhanced multimedia support, and the rise of app-based ecosystems.

ii) The Next Phase in Wireless Communication

- 5G or Fifth-generation wireless technology began to be deployed in the late 2010s and early 2020s. It represents a significant leap in performance, promising ultra-fast data speeds, lower latency, increased capacity, and improved reliability. Some key features of 5G include:
- Enhanced Mobile Broadband (eMBB): High-speed data transfer capable of delivering gigabit-per-second speeds, enabling seamless streaming, virtual reality (VR), and augmented reality (AR) experiences [7].
- Ultra-Reliable Low Latency Communications (URLLC): Reduced latency and increased reliability are critical for autonomous vehicles, remote surgeries, and industrial automation applications [8].
- Massive Machine-Type Communications (mMTC): 5G can connect a massive number of devices simultaneously, supporting the Internet of Things (IoT) and smart city applications [9, 10].

iii) Beyond 5G – The Future of Wireless Communications

As technology advances, researchers and industry experts are already exploring what comes after 5G. Several concepts and technologies are being considered for "Beyond 5G" or "6G":

- Terahertz (THz) Communication: Utilizing the higher-frequency spectrum (above 100 GHz) to achieve even faster data rates and greater capacity.
- Integrated Satellite Communication: Integrating terrestrial and satellite networks to provide seamless global coverage.
- Holographic Beamforming: Using advanced beamforming techniques to create highly focused, directional signals, improving efficiency and reducing interference.

iv) Applications Landscape

- 5G and beyond have the potential to revolutionize various industries and daily life. Some of the potential applications include:
- Enhanced virtual and augmented reality experiences.
- Smart cities with connected infrastructure and IoT devices.
- Advanced healthcare applications, such as remote surgeries and telemedicine.

- Self-driving cars and intelligent transportation systems.
- Improved real-time communication and collaboration tools.

Wireless communication has come a long way from its humble beginnings, and 5G represents a transformative step toward a more connected and efficient world. As technology continues to evolve, the possibilities of what future generations of wireless communication can achieve are boundless, promising a future of unprecedented connectivity and innovation.

1.1.2 Key Highlights of the Chapter

The following are the significant highlights of this chapter:

i) This chapter thoroughly explores the fundamental components of wireless communications.
ii) This chapter is an invaluable resource for researchers who may not have a prior foundation in wireless communication systems, offering them a rapid pathway to acquaint themselves with the subject and empowering them to make substantial contributions to the evolution of communication systems.
iii) Additionally, this chapter furnishes practical insights into the components of wireless communication and offers a rich set of references for those seeking to delve deeper into the subject.

1.1.3 Chapter Organization

The remaining part of this chapter is organized as follows. Section 1.2 covers digital and analog sources. Section 1.3 explores deterministic and random waveforms. Section 1.4 beams a laser focus on the propagation of electromagnetic waves. Section 1.5 examines the basics of information measures. Section 1.6 dissects the concept of wireless channel and information. Section 1.7 covers modulation and demodulation. Section 1.8 takes a closer look at the fundamentals of massive multiple input multiple output (MIMO) technology. Section 1.9 discusses critical security and privacy issues in emerging wireless communication systems. Finally, the conclusion to the chapter is drawn in Section 1.10.

1.2 Digital and Analog Sources

Digital and analog sources refer to different types of signals or data representations in electronic systems. These terms are commonly used in the context of various technologies, including signal processing, communication systems, and data storage. Digital signals are represented using discrete values, typically in the form of binary digits (bits). Each bit can take on one of the two values: 0 or 1.

Digital sources have well-defined levels and are less susceptible to noise and interference. They are precise and can be easily manipulated using digital processing techniques. Digital sources include digital audio signals, binary data in computers, digital images, and any information that can be represented using a series of discrete values.

Analog signals are represented using continuous values that vary smoothly over time. These signals can take any value within a range. Analog sources can represent a wide range of values with infinite resolution. However, they are more susceptible to noise and degradation over long distances. Analog sources include traditional audio signals, analog sensors (e.g., temperature sensors), analog video signals, and any continuously varying signal.

In digital communication systems, the conversion of analog signals to digital data is a crucial step for efficient processing. This transformation is accomplished through analog-to-digital conversion (ADC), a process that translates analog signals into their digital counterparts. Applications of ADC are diverse, ranging from digitizing audio signals for recording purposes to converting analog sensor readings into digital data for seamless integration with computer systems [11]. Conversely, digital-to-analog conversion (DAC) plays a pivotal role in various applications, such as audio playback. DAC is the process of converting digital signals back into analog form, allowing for the recreation of the original analog signal. In the sphere of audio playback, this is exemplified when digital audio files are transformed into analog signals for playback through speakers or headphones, ensuring a high-fidelity and immersive auditory experience.

Some communication systems still use analog signals, especially in certain audio and video broadcasting applications. However, analog communication is more susceptible to noise. Many modern communication systems use digital signals for transmission due to their robustness against noise. Digital communication allows for error detection and correction. The choice between digital and analog depends on the specific requirements of a given application. Digital technologies have become dominant in many areas due to their precision, reliability, and ease of processing advantages.

1.3 Deterministic and Random Waveforms

Deterministic signals are characterized by their accurate definition through mathematical functions or equations, affording them a precise and predictable behavior at any given instant. These signals are entirely governed by the mathematical constructs that describe them, leaving no room for ambiguity and ensuring their complete determination [12]. For instance, a deterministic signal can be explained using a mathematical formula detailing its amplitude, frequency, and phase as functions of time or another independent variable. The inherent predictability of deterministic signals originates from their mathematical representations, empowering us to ascertain their values with certainty at any specific point in time. These signals often exhibit a rhythmic and repetitive nature, underscoring their predictability and structured patterns. Familiar instances of deterministic signals include sine waves, square waves, triangular waves, and any other signal that lends itself to precise description through a mathematical equation. In fields as diverse as signal processing, control systems, and communication systems, deterministic signals assume a critical role due to their predictability and the ease with which they can be subjected to rigorous mathematical analysis. Their well-defined nature facilitates comprehensive understanding and positions them as indispensable tools in advancing scientific and technological pursuits.

Random signals, also known as stochastic signals, are signals that do not have a deterministic mathematical expression and exhibit an element of randomness or unpredictability. Unlike deterministic signals, a mathematical function cannot precisely define random signals. Instead, their behavior is characterized by statistical properties, and they may vary unpredictably over time. In contrast, random signals do not have a specific mathematical formula that precisely describes their values at any given point in time. Their behavior is inherently uncertain. These signals are characterized by statistical properties such as mean, variance, probability distribution, and correlation. These properties provide a way to describe the overall behavior of the signal. Common random signals include noise signals in communication systems, fluctuations in environmental conditions, or the output of a random number generator. Understanding random signals involves statistical analysis and probability theory to characterize and model their behavior. This is in contrast to deterministic signals, where precise mathematical expressions allow for deterministic analysis and predictability.

1.4 Propagation of Electromagnetic Waves

The phenomenon of electromagnetic wave propagation is a fascinating exploration into the dynamic journey of electromagnetic radiation as it traverses the vast region of space. These waves, intricate in their nature, unfold a mesmerizing interplay between electric and magnetic fields oscillating at right angles to each other while perpendicular to the direction of the wave's advancement. At the heart of understanding electromagnetic wave propagation lies a set of fundamental principles that guide their behavior. One such principle is the remarkable speed of light, which serves as a constant and defines the velocity at which these waves propagate. Wavelength and frequency, inseparable companions, determine electromagnetic waves' spatial and temporal characteristics, adding an extra layer of complexity to their intricate nature.

The medium through which electromagnetic waves travel is crucial in shaping their journey. Whether through the vacuum of space, a transparent medium like air, or a denser material, the properties of the medium influence the wave's behavior. This interaction with the medium leads to phenomena such as reflection, where the waves bounce off surfaces; refraction, where they change direction; and absorption and attenuation, where the waves are absorbed or lose energy as they propagate. These phenomena contribute to the rich tapestry of electromagnetic wave propagation, showcasing the intricate interplay between these waves and their surroundings. In essence, the propagation of electromagnetic waves unfolds as a work of physics, weaving together the threads of speed, wavelength, frequency, and medium interactions, creating a mesmerizing spectacle that extends our understanding of the universe.

A good grasp of the propagation of electromagnetic waves is crucial in various fields, including telecommunications, RF engineering, optics, and astronomy. The above principles govern how electromagnetic waves behave as they travel through different environments.

1.5 Information Measures

Information measures are quantitative metrics used to characterize the amount of information conveyed by a signal, message, or dataset [13]. These measures are fundamental in information theory, a branch of applied mathematics and electrical engineering that deals with the quantification of information.

Entropy measures the average uncertainty or randomness in a set of possible outcomes [14]. It is often used to quantify the amount of information in a probability distribution or a source of information. Higher entropy indicates higher uncertainty or randomness, while lower entropy indicates more predictability.

Shannon information measures the amount of information gained or conveyed by an event. It is calculated as the negative logarithm of the probability of the event as in (1.1):

$$I = -\log_2(P) \text{ bits} \tag{1.1}$$

where P is the probability of the event. This measure is used to quantify the surprise or unexpectedness of an outcome. The average information measure or entropy of a digital source is given by (1.2)

$$H = \sum_{j=1}^{m} P_j I_j = \sum_{j=1}^{m} P_j \log_2\left(\frac{1}{P_j}\right) \text{ bits} \tag{1.2}$$

where m is the number of possible messages and P_j is the probability of sending the jth message.

Mutual information measures the degree of dependence between two random variables, X and Y. It quantifies how much knowing the value of one variable reduces uncertainty about the other. High mutual information indicates strong dependence, while low mutual information suggests independence. Channel capacity represents the maximum rate at which information can be reliably transmitted over a communication channel. It depends on the bandwidth and SNR of the channel. The channel capacity theorem, developed by Claude Shannon, is a fundamental result in information theory.

1.6 Channel and Information

In information theory, a communication channel is a medium through which information is transmitted from a sender to a receiver. The study of communication channels involves understanding how information is encoded, transmitted, and received, considering the characteristics of the medium through which it travels. Some critical concepts related to communication channels and information are channel capacity, noisy channel, error-correction code, modulation, demonstration, bandwidth, data rate, and information theory [15].

The channel capacity is the maximum rate at which information can be reliably transmitted through a communication channel. It is influenced by the bandwidth of the channel and the SNR. Claude Shannon's channel capacity theorem limits the maximum data rate achievable without errors in a noisy channel [16]. The channel capacity is given by (1.3)

$$C = B\log_2\left(1 + \frac{S}{N}\right) \tag{1.3}$$

where B is the channel bandwidth in hertz (Hz), and S/N is the SNR.

In real-world communication systems, channels are often subject to noise, interference, and distortions that can introduce errors in the transmitted information. Noisy channels pose challenges in maintaining the integrity of the transmitted message, and error-correction techniques are employed to mitigate these effects.

Error-correction coding involves adding redundant information to the transmitted message to enable the receiver to detect and correct errors introduced during transmission [17]. Various error-correcting codes, such as Reed–Solomon codes and convolutional codes, are used to enhance the reliability of communication over noisy channels [18].

Modulation is the process of encoding information onto a carrier signal for transmission. Demodulation is the process of extracting the original information from the received modulated signal. Different modulation techniques, such as AM or FM, are used depending on the characteristics of the communication channel.

The bandwidth of a communication channel determines the range of frequencies it can accommodate. It is crucial in determining the maximum data rate that can be transmitted through the channel. The Nyquist–Shannon sampling theorem relates the bandwidth of a channel to the maximum achievable data rate [19].

Information theory, developed by Claude Shannon, provides a mathematical framework for quantifying information and communication. Shannon introduced concepts such as entropy, mutual information, and channel capacity. Information theory addresses fundamental questions about the limits of data compression, error correction, and reliable communication.

Understanding the characteristics of communication channels and employing appropriate coding and modulation techniques are essential in designing and optimizing communication systems, including wired and wireless networks. The goal is to maximize the reliable transmission of information despite the challenges posed by noise and other channel impairments.

1.7 Modulation and Demodulation

Modulation is a process used in communication systems to encode information onto a carrier signal [20]. The carrier signal is a high-frequency signal that serves as a medium for transmitting the information over a communication channel. The purpose of modulation is to enable the transmission of information over a distance without distortion and to make efficient use of the available bandwidth. There are different types of modulation techniques, and the choice of modulation depends on the specific requirements of the communication system. Some key aspects of modulation are the following: carrier signal, information signal, and types of modulation.

The carrier signal is a high-frequency waveform that can travel over long distances without significant attenuation. It provides the medium through which the information is transmitted. The information signal, also known as the baseband signal, contains the actual data or information that needs to be transmitted. This signal typically has a lower frequency compared to the carrier signal.

There are several types of modulation techniques, including:

- **Amplitude modulation (AM):** The amplitude of the carrier signal varies in proportion to the instantaneous amplitude of the information signal.
- **Frequency modulation (FM):** The frequency of the carrier signal is varied based on the instantaneous frequency of the information signal.
- **Phase modulation (PM):** The phase of the carrier signal is varied according to the instantaneous phase of the information signal.

The biggest advantage of modulation is the efficient use of bandwidth. Modulation allows for the transmission of information over a range of frequencies, making more efficient use of the available bandwidth. Modulated carrier signals are better suited for long-distance transmission compared to baseband signals.

Demodulation is the process of extracting the original information signal from the modulated carrier signal at the receiver. Demodulation is necessary to recover the transmitted information accurately. Modulation is widely used in various communication systems, including radio broadcasting, television, mobile communication, Wi-Fi, and satellite communication. Different modulation techniques are employed based on factors such as signal quality, bandwidth requirements, and the nature of the communication channel. Modulation is a key aspect of communication systems that allows information to be transmitted over a communication channel by encoding it onto a carrier signal. The choice of modulation technique depends on the specific requirements and characteristics of the communication system.

In 5G wireless networks, various modulation schemes are employed to transmit information over the airwaves. The choice of modulation scheme in 5G depends on factors such as data rate, signal quality, and spectral efficiency. Here are some modulation schemes commonly used in 5G:

1) **QPSK (quadrature phase shift keying):** QPSK is a digital modulation scheme that uses four different phase shifts to represent two bits per symbol. It provides better spectral efficiency than simpler modulation schemes like binary phase shift keying (BPSK) but is not as bandwidth-efficient as more complex schemes.
2) **16-QAM (quadrature amplitude modulation):** 16-QAM is a higher-order modulation scheme that uses a combination of both amplitude and phase changes to represent four bits per symbol. It offers higher data rates compared to QPSK but is more susceptible to noise and distortion.
3) **64-QAM:** 64-QAM further increases the number of symbols that can be represented in each modulation symbol, allowing for higher data rates. However, as the number of symbols increases, the scheme becomes more susceptible to noise and requires a higher SNR for reliable communication.
4) **256-QAM:** 256-QAM is an even higher-order modulation scheme, enabling even higher data rates. However, the increased complexity comes at the cost of higher sensitivity to noise and other impairments.
5) **256-QAM and Beyond:** In certain scenarios where the link quality is high and there is minimal interference, 256-QAM and even higher-order modulation schemes can be used to achieve very high data rates. These high-order schemes are often employed in conjunction with other advanced technologies, such as beamforming and massive MIMO, to enhance link reliability.

It is important to note that the choice of modulation in 5G is dynamic and can be adapted based on the channel conditions, interference levels, and quality of service requirements. Adaptive modulation and coding techniques are often used to dynamically adjust the modulation and coding schemes based on real-time channel conditions, optimizing data transmission performance. 5G networks use a variety of modulation schemes, including QPSK, 16-QAM, 64-QAM, and even higher-order schemes, to achieve different data rates and adapt to varying channel conditions. The flexibility and adaptability of modulation schemes contribute to the efficiency and reliability of 5G communication.

1.8 Massive MIMO

Massive MIMO utilizes a large number of antennas at the base station to improve spectral efficiency and capacity in wireless systems [21]. In comparison with the traditional MIMO systems that use a small number of antennas, massive MIMO employs a greater number of antennas to serve multiple users simultaneously, enabling efficient spatial multiplexing, beamforming, and interference management. The resultant benefits for the network providers and subscribers include but not limited to higher data rates, improved signal quality, and increased network capacity.

MIMO technology has evolved with time. The evolution to massive MIMO is critical to the fast-growing advancement in wireless communication systems, considering the case of 5G wireless networks [22]. In massive MIMO, a large number of antennas employed at base stations, enabling more parallel connections and better spatial multiplexing. Large or massive MIMO is a key component of ubiquitous 5G networks, helping satisfy the demand for increased mobile broadband, large machine-type connectivity, and ultra-reliable low-latency communication. A key driver of wireless communications evolution is massive MIMO [23]. The technology enables step-change improvements in spectral efficiency, link reliability, and network capacity [24].

A major leap forward in the development of MIMO technology is massive MIMO. The number of antennas in multiuser MIMO systems can be radically expanded, resulting in substantial gains in throughput and spectral efficiency in massive MIMO compared to the traditional small-scale MIMO systems [24]. Massive MIMO promises 10 times or greater capacity gains over the conventional multiuser MIMO system by scaling up antenna arrays to serve tens of users at the same time. According to the inventor of massive MIMO, Thomas Marzetta, massive MIMO systems could approach theoretical capacity limits in a simple and power-efficient manner by increasing the number of service antennas without bound [23].

Massive MIMO enables several other key 5G technologies to operate at scale [25]. This includes millimeter wave communication, massive machine-type connectivity, and ultra-dense small cell deployments. Thanks to the advanced interference management and spatial multiplexing capabilities of massive MIMO, making it a critical enabler for 5G and beyond. There is no denying the fact that massive MIMO has been a foundational technology for 5G wireless networks. However, new motivations, driving further innovation and advancement of massive MIMO capabilities for 6G and beyond, are envisioned.

1.8.1 Motivation for Massive MIMO in 6G and Beyond

Massive MIMO has been widely researched in 5G wireless networks, and it poses huge potential in the envisioned 6G wireless networks [26, 27]. The key motivations for projecting massive MIMO technology in future wireless systems such as 6G and beyond include but are not limited to the following: ubiquitous connectivity and enhanced coverage, exponential growth in data demand, ultra-Low latency and ultra-high reliability, higher-frequency bands and wider bandwidths, energy efficiency and sustainability, connection density and IoT scalability, diverse vertical applications, artificial intelligence (AI)-assisted massive MIMO optimization techniques, and integrated communication and sensing.

1) **Achieving ubiquitous connectivity and enhanced coverage**: It is envisioned that 6G will provide ubiquitous coverage, bridging urban and rural geographies. Specifically, massive distributed MIMO could integrate space-based and terrestrial systems to extend coverage to the remotest of areas not currently covered by traditional terrestrial networks.
2) **Exponential growth in data demand**: 6G is envisioned to cater to an exponential increase in data demand driven by emerging technologies like holographic communication, high-definition video streaming, and advanced virtual and augmented reality applications. Massive MIMO can handle a large number of spatial streams and efficiently utilize available spectrum in satisfying these high data rate requirements.
3) **Ultra-low latency and ultra-high reliability**: 6G aims to achieve ultra-low latency and ultra-high reliability to enable mission-critical applications like remote surgeries, autonomous vehicles, industrial automation, among others.
4) **Energy efficiency and sustainability**: As communication networks grow in scale, energy efficiency becomes paramount. Advanced massive MIMO techniques can help improve energy efficiency by optimizing transmission power, reducing interference, and enabling dynamic spectrum allocation.
5) **Higher-frequency bands and wider bandwidths**: 6G aims to explore higher-frequency bands, such as terahertz, to leverage wider bandwidths and achieve even higher data rates.

6) **Connection density and IoT scalability**: 6G envisions a massive number of connected devices, including the Internet of Things (IoT) and wireless sensor networks. Massive MIMO provides a scalable solution to manage the unprecedented connection densities, enabling seamless communication and resource allocation for a vast number of devices per unit area.
7) **Vertical applications**: It is envisioned that 6G will provide unique services to serve a wide range of vertical industries, each with unique communication requirements. 6G will drive applications like wireless sensing, robotics, and industrial automation. Interestingly, these applications will require highly reliable, low-latency communications.
8) **AI-assisted massive MIMO optimization techniques**: The integration of artificial intelligence and machine learning, including their variants such as deep and federated learning in massive MIMO systems, can be a game changer for MIMO-based wireless systems. AI and machine learning-based algorithms can help to dynamically optimize resource allocation, beamforming, and interference management. AI-driven MIMO optimization can facilitate gains in wireless network performance, and capacity enhancement in 6G networks.
9) **Integrated communication and sensing**: In the 6G era, communication and sensing could be integrated toward achieving more intelligent and context-aware wireless networks. Advanced massive MIMO techniques can enable joint communication and sensing, supporting real-time environmental monitoring and empowering other smart wireless applications.

1.8.2 The Potential of Massive MIMO Technology

Massive MIMO technology holds great potential for applications in future wireless communication systems [28]. A few of these potential applications of massive MIMO include the following: massive MIMO improving signal quality, massive MIMO improving spectral efficiency, massive MIMO supporting multiple users and dense devices, and massive MIMO enabling a wider coverage [29].

1) **Massive MIMO improving signal quality:** The application of multiple antennas in massive MIMO systems helps combat the effects of fading and interference, thereby improving signal quality. Massive MIMO systems support advanced signal processing techniques, such as beamforming and spatial multiplexing, enhancing signal quality, reducing interference, and improving overall network performance.
2) **Massive MIMO improving spectral efficiency:** In massive MIMO, a significant increase in spectral efficiency is achievable, enabling higher data rates and improved network capacity, due to the several number of antennas at the base station, allowing for the simultaneous transmission and reception of multiple data streams.

3) **Massive MIMO supporting multiple users and dense devices**: Massive MIMO systems can serve multiple users simultaneously, enabling efficient resource allocation, better user experience, and improved network capacity.
4) **Massive MIMO enabling a wider coverage:** The design of massive MIMO systems facilitates the focusing of beams, shaping the radiation pattern more accurately. Massive MIMO can help boost wireless network coverage in remote environments. By directing the signal toward specific users or locations, it improves signal strength and coverage, providing reliable communications.
5) **Massive MIMO for non-terrestrial wireless communication systems:** Recent works on non-terrestrial wireless communications are gaining grounds in the wireless research community [30]. Particularly, the use of large numbers of antennas and on-board satellites has attracted significant attention. Specifically, the deployment of Low Earth Orbit (LEO) satellite constellations and very high-throughput satellites in Geostationary Earth Orbit (GEO) has precipitated a transformative shift in global connectivity paradigms. Future wireless communication networks such as 6G and beyond, therefore, promise to deliver low-latency, high-speed data transfer capabilities, ushering in a new era of applications ranging from ubiquitous internet access to cutting-edge remote sensing. Satellites are positioned as pivotal enablers for the forthcoming wave of wireless networks and will support a diverse array of applications, including massive machine-type communications, all underscored by the overarching need for secure connectivity. As the demand for connectivity beyond the confines of the earth grows, the need for standardized and interoperable communication protocols becomes increasingly important. However, it should be noted that by orchestrating an intricate interplay among hundreds or even thousands of satellites and safeguarding a secure and dependable communications system will pose substantial challenges [31].
6) **Signal processing for non-terrestrial network-based massive MIMO systems:** Massive MIMO assumes that the number of transmit antennas, m on the non-terrestrial platform shall always be more than the number of single receive antennas, k, on the terrestrial/maritime/aeronautical-bound user equipment (UEs). In the envisioned fully converged future wireless communication systems, non-terrestrial networks will provide service to UEs in hard-to-reach areas characterized by sparse user population, low user density, and relatively high latency. The channel propagation characteristics of non-terrestrial networks are expected to have relatively high and variable Doppler shifts and spreads, and significant propagation delay [32]. These channel impediments need to be adequately compensated for through appropriate signal processing.

1.9 Security and Privacy of Wireless Systems

The recent advancement in wireless communication systems calls for overhauling the traditional security architecture to meet the growing demands for robust security and privacy preservation in modern wireless networks [33]. Currently, fast-growing security threats such as unauthorized access, malicious data breaches, and various network attacks in wireless systems need to be mitigated or resolved to protect critical user information [34]. The design of modern wireless networks should explore innovative privacy-enhancing technologies and protocols that empower users and ensure compliance with government policies and regulations. The security and privacy of wireless systems is considered under the following subheadings: identifying and resolving vulnerabilities and threats in wireless systems, and security and privacy issues in wireless systems.

1.9.1 Identifying and Resolving Vulnerabilities and Threats in Wireless Systems

In modern wireless networks design, robust security and privacy preservation mechanisms need to be put in place to protect critical user information from unlawful access and manipulations. Some critical security threats that need to be mitigated include but are not limited to the following: resolving man-in-the-middle attacks, identifying and addressing physical attacks, denial-of-service (DoS) attacks, tackling and resolving insider threats, risks associated with the integration of massive MIMO with IoTs, taming eavesdropping and unauthorized access, issues from network congestion and radio resource management, and vulnerabilities at the protocol level.

1) **Resolving man-in-the-middle attacks**: Nowadays, attackers are conducting sophisticated attacks on wireless nodes. They can reliably position themselves between the communication nodes, intercepting and altering the data exchanged between the nodes, thereby manipulating or intercepting sensitive user information, and compromising the integrity and confidentiality of the communication channel.
2) **Identifying and addressing physical attacks:** Malicious attackers can physically damage massive MIMO base stations and the associated equipment to compromise the functionality of the wireless network, leading to service disruption, unauthorized access, or information leakage.
3) **Denial-of-service (DoS) and jamming attacks:** Jamming and DoS attacks can be perpetrated by malicious personnel. They can launch jamming attacks by flooding the network with excessive traffic to disrupt the communication channel in massive MIMO systems, and this could potentially leads to service degradation or complete service outage.

4) **Tackling and resolving insider threats:** Insider threats pose significant risks to the security and privacy of massive MIMO systems. In several cases, trusted individuals within the network may misuse their privileges to gain unauthorized access, leak sensitive information, or deliberately disrupt the functionality of an operational system.
5) **Risks associated with the integration of massive MIMO with IoTs:** Undoubtedly, the integration of massive MIMO networks with IoT devices poses security concerns. Insecure IoT devices or gateways can serve as entry points for attackers, compromising the security and privacy of wireless networks.
6) **Taming eavesdropping and unauthorized access:** It is not surprising today that intruders or attackers can gain unlawful access to intercept and eavesdrop on sensitive information transmitted over unsecured wireless channels, hampering and compromising data confidentiality and system integrity.
7) **Issues from network congestion and radio resource management:** The design of massive MIMO systems needs to be scalable and sustainable. However, addressing the scalability and resource allocation challenges in massive MIMO systems still remains. These issues, if not properly addressed, can be exploited by attackers to degrade system performance or disrupt service availability in an operational wireless network.
8) **Vulnerabilities at the protocol level:** The design and implementation of wireless communication protocols in massive MIMO systems needs to be checked thoroughly for security vulnerabilities in the encryption algorithms, authentication mechanisms, or key exchange protocols. Such vulnerabilities can result in an unauthorized access to critical user data.

1.9.2 Security and Privacy Issues in Wireless Systems

Wireless communication systems transmit and receive user information via open wireless channels, posing huge security vulnerabilities such as cyber-attacks and data breaches [35]. Usually, malicious intruders can exploit these vulnerabilities in wireless networks to gain unauthorized access to sensitive data or disrupt operational wireless communication services. In the most cases, these attacks could have significant consequences, resulting in enormous financial losses, compromising user information, or total collapse of critical infrastructure. In order to mitigate these security risks, key physical layer security techniques need to be deployed appropriately.

1.9.2.1 Security Issues in Massive MIMO Systems

Massive MIMO technology holds great promise for enhancing the capacity and performance of wireless communication systems. However, the application of massive MIMO wireless communication systems could present some security issues that need to be addressed appropriately [36]. A few of these challenges are outlined as follows: user privacy protection, beamforming security, and channel estimation security, among others.

1) **User privacy protection:** The deployment of massive MIMO technology to wireless systems could present new privacy challenges. In massive MIMO, a large number of antennas and several network users could compound user privacy. In such a complex network configuration, adversaries can exploit the inherent characteristics of the network to access sensitive user information for unscrupulous gains.
2) **Beamforming security:** In massive MIMO, beamforming is critical to enhancing signal transmission and reception. However, the nature of the wireless channel makes it possible for beamed signals to be intercepted by intruders, making beamforming algorithms and protocols to be highly susceptible to various attacks such as eavesdropping and signal interception. Invariably, attackers can exploit vulnerabilities in the beamforming process to gain unauthorized access to transmitted information. This calls for strict security compliance in wireless networks designs.
3) **Channel estimation security:** Massive MIMO systems rely on accurate channel estimation for efficient data transmission. However, the vulnerabilities in the channel estimation procedures can be exploited by attackers. This implies that potential adversaries can launch attacks to manipulate channel estimates, leading to degraded system performance. Thus, the wireless channel needs to be protected appropriately at all times.

1.9.2.2 Probable Privacy Issues in Massive MIMO Systems

Massive MIMO systems have to contend with a myriad of security issues as outlined. In a similar fashion, there exists some possible privacy issues that need to be resolved in order to protect critical user information and ensure strict compliance with the prevailing privacy regulations [37]. These include but are not limited to the following: data leakage and information exposure, user profiling and tracking, and regulatory compliance issues, among others.

1) **Tackling data leakage and information exposure:** In massive MIMO systems, sensitive user data is being transmitted over unsecured wireless channels. In this case, the risk of data leakage and information exposure increases as the number of antennas and users in the system increases. Thus, the privacy of the transmitted information becomes very critical. In order to ensure the privacy of the data, the confidentiality and integrity of transmitted data must be guaranteed and protected at all costs.

2) **User profiling and tracking:** In massive MIMO systems, critical user data can be harvested, profiled, and tracked. Usually, the information is aggregated and analyzed. This raises serious privacy concerns since important user data can be accessed, tracked, and monitored by unauthorized users.
3) **Addressing regulatory compliance issues:** The prevailing privacy regulations specify strict requirements on the collection, storage, processing, and sharing of personal data. The transmission of user data in massive MIMO systems should comply strictly with these regulations to protect user privacy rights.

1.10 Conclusion

In this introductory chapter, the fundamentals of wireless communications with a keen focus on massive MIMO essentials for 6G and beyond are presented. Useful background information on digital and analog sources is provided, and deterministic and random waveforms are elaborated. In addition, the chapter considers the propagation of electromagnetic waves and examines the basics of information measures and wireless channels. Furthermore, the chapter discusses modulation and demodulation, and the critical features of massive MIMO technology. Last, the chapter examines security and privacy issues affecting all parties in the wireless ecosystem.

References

1 Banafaa, M., Shayea, I., Din, J. et al. (2023). 6G mobile communication technology: requirements, targets, applications, challenges, advantages, and opportunities. *Alexandria Engineering Journal* 64: 245–274.

2 Akyildiz, I.F., Kak, A., and Nie, S. (2020). 6G and beyond: the future of wireless communications systems. *IEEE Access* 8: 133995–134030.

3 Yeh, C., Do Jo, G., Ko, Y.J., and Chung, H.K. (2023). Perspectives on 6G wireless communications. *ICT Express* 9 (1): 82–91.

4 Baldini, G., Karanasios, S., Allen, D., and Vergari, F. (2013). Survey of wireless communication technologies for public safety. *IEEE Communications Surveys & Tutorials* 16 (2): 619–641.

5 Carson, J.R. (1922). Notes on the theory of modulation. *Proceedings of the institute of radio engineers* 10 (1): 57–64.

6 Imoize, A.L., Adedeji, O., Tandiya, N., and Shetty, S. (2021). 6G enabled smart infrastructure for sustainable society: opportunities, challenges, and research roadmap. *Sensors* 21 (5): 1709.

7 Gamage, H., Rajatheva, N., and Latva-Aho, M. (2017, June). Channel coding for enhanced mobile broadband communication in 5G systems. In: *2017 European Conference on Networks and Communications (EuCNC)*, 1–6. IEEE.

8 Pocovi, G., Shariatmadari, H., Berardinelli, G. et al. (2018). Achieving ultra-reliable low-latency communications: challenges and envisioned system enhancements. *IEEE Network* 32 (2): 8–15.

9 Qiao, L., Zhang, J., Gao, Z. et al. (2021). Massive access in media modulation based massive machine-type communications. *IEEE Transactions on Wireless Communications* 21 (1): 339–356.

10 Sharma, S.K. and Wang, X. (2019). Toward massive machine type communications in ultra-dense cellular IoT networks: current issues and machine learning-assisted solutions. *IEEE Communications Surveys & Tutorials* 22 (1): 426–471.

11 Palermo, S., Hoyos, S., Cai, S. et al. (2018). Analog-to-digital converter-based serial links: an overview. *IEEE Solid-State Circuits Magazine* 10 (3): 35–47.

12 Malanowski, M., Kulpa, K., Bączyk, M., and Maślikowski, Ł. (2020). Noise vs. deterministic waveform radar—possibilities and limitations. *IEEE Aerospace and Electronic Systems Magazine* 35 (10): 8–19.

13 Taneja, I.J. (1989). On generalized information measures and their applications. In: *Advances in Electronics and Electron Physics*, vol. 76, 327–413. Academic Press.

14 Rényi, A. (1961, January). On measures of entropy and information. In: *Proceedings of the Fourth Berkeley Symposium on Mathematical Statistics and Probability, Volume 1: Contributions to the Theory of Statistics*, vol. 4, 547–562. University of California Press.

15 Shannon, C.E. (2001). A mathematical theory of communication. *ACM SIGMOBILE Mobile Computing and Communications Review* 5 (1): 3–55.

16 Shannon, C.E. (1948). A mathematical theory of communication. *The Bell System Technical Journal* 27 (3): 379–423.

17 Zhang, Z. (2011). Theory and applications of network error correction coding. *Proceedings of the IEEE* 99 (3): 406–420.

18 Silva, D. and Kschischang, F.R. (2009). On metrics for error correction in network coding. *IEEE Transactions on Information Theory* 55 (12): 5479–5490.

19 Vitulyova, Y.S., Suleimenov, I.E., Matrassulova, D.K., and Bakirov, A.S. (2023). Discrete form of the Huygens-Fresnel principle: to the multi-dimensional analog of the Nyquist–Shannon sampling theorem. *International Journal of Information Technology* 15 (7): 3751–3759.

20 de Oliveira, L.G., Nuss, B., Alabd, M.B. et al. (2021). Joint radar-communication systems: modulation schemes and system design. *IEEE Transactions on Microwave Theory and Techniques* 70 (3): 1521–1551.

21 Khwandah, S.A., Cosmas, J.P., Lazaridis, P.I. et al. (2021). Massive MIMO systems for 5G communications. *Wireless Personal Communications* 120 (3): 2101–2115.

22 Albreem, M.A., Juntti, M., and Shahabuddin, S. (2019). Massive MIMO detection techniques: a survey. *IEEE Communications Surveys & Tutorials* 21 (4): 3109–3132.

23 Marzetta, T.L. (2015). Massive MIMO: an introduction. *Bell Labs Technical Journal* 20: 11–22.

24 Björnson, E., Hoydis, J., and Sanguinetti, L. (2017). Massive MIMO has unlimited capacity. *IEEE Transactions on Wireless Communications* 17 (1): 574–590.

25 Marzetta, T.L., Larsson, E.G., and Yang, H. (2016). *Fundamentals of Massive MIMO*. Cambridge University Press.

26 Saad, W. (2021). 6G wireless systems: Challenges and opportunities. In: Lin, X., Lee, N. (eds) *5G and Beyond: Fundamentals and Standards*. Springer, Cham. https://doi.org/10.1007/978-3-030-58197-8_7, 201-229.

27 Saad, W., Bennis, M., and Chen, M. (2019). A vision of 6G wireless systems: applications, trends, technologies, and open research problems. *IEEE Network* 34 (3): 134–142.

28 Lu, L., Li, G.Y., Swindlehurst, A.L. et al. (2014). An overview of massive MIMO: benefits and challenges. *IEEE journal of selected topics in signal processing* 8 (5): 742–758.

29 Kammoun, A., Debbah, M., and Alouini, M.S. (2017). Design of 5G full dimension massive MIMO systems. *IEEE Transactions on Communications* 66 (2): 726–740.

30 You, L., Li, K.X., Wang, J. et al. (2020). Massive MIMO transmission for LEO satellite communications. *IEEE Journal on Selected Areas in Communications* 38 (8): 1851–1865.

31 Zhang, Y., Wu, Y., Liu, A. et al. (2021). Deep learning-based channel prediction for LEO satellite massive MIMO communication system. *IEEE Wireless Communications Letters* 10 (8): 1835–1839.

32 Rinaldi, F., Maattanen, H.L., Torsner, J. et al. (2020). Non-terrestrial networks in 5G & beyond: a survey. *IEEE access* 8: 165178–165200.

33 Zhong, S., Zhong, H., Huang, X. et al. (2019). *Security and Privacy for Next-Generation Wireless Networks*. Springer International Publishing.

34 Abba Ari, A.A., Ngangmo, O.K., Titouna, C. et al. (2024). Enabling privacy and security in Cloud of Things: architecture, applications, security & privacy challenges. *Applied Computing and Informatics* 20 (1/2): 119–141.

35 Fang, D. and Qian, Y. (2020). 5G wireless security and privacy: architecture and flexible mechanisms. *IEEE vehicular technology magazine* 15 (2): 58–64.

36 Sikiru, I.A., Olawoyin, L.A., Faruk, N. et al. (2023). Physical layer security using boundary technique for emerging wireless communication systems. *Security and Privacy* e288.

37 Mathur, S., Reznik, A., Ye, C. et al. (2010). Exploiting the physical layer for enhanced security [security and privacy in emerging wireless networks]. *IEEE Wireless Communications* 17 (5): 63–70.

2

Security and Privacy of Future Wireless Communication Systems

Abdulwaheed Musa[1,2,3], Abdullateef Ola Adebayo[1], Peace Oluwasijibomi Balogun[1], and Segun Jacob[1]

[1] *Department of Electrical and Computer Engineering, Kwara State University, Malete, Nigeria*
[2] *Centre for Artificial Intelligence and Machine Learning Systems, Kwara State University, Malete, Nigeria*
[3] *Institute for Intelligent Systems, University of Johannesburg, Johannesburg, South Africa*

2.1 Introduction

Wireless communication has become an integral part of society, providing the infrastructure for various services from telephony to Internet access [1]. The evolution of wireless standards from 1G to 5G has enabled increasing data rates, reduced latency, and connectivity for massive numbers of devices. As we progress toward 6G networks and beyond, advanced technologies like massive multiple-input multiple-output (MIMO) will play a pivotal role by allowing simultaneous transmission to multiple users using large antenna arrays [2]. However, the increasing ubiquity of wireless technologies and the adoption of massive MIMO make security and privacy critical concerns [3].

Wireless communication relies on the transmission of signals over the air which are vulnerable to interception and manipulation. The open nature of the wireless medium allows attackers to eavesdrop on transmissions or inject malicious traffic without needing physical access [4]. Massive MIMO poses new security risks from pilot contamination attacks to compromise channel estimation and beamforming vulnerabilities to manipulate signal transmissions [5]. As networks densify and connect billions of devices, the attack surface grows exponentially. Such vulnerabilities can seriously impact the availability, integrity, and confidentiality of wireless services if robust security mechanisms tailored for massive MIMO are not implemented.

Along with security, user privacy is also a major concern with the pervasive collection of location and other personal data enabled by massive MIMO systems [6].

Massive MIMO for Future Wireless Communication Systems: Technology and Applications,
First Edition. Edited by Webert Montlouis and Agbotiname Lucky Imoize.

With the capability to simultaneously serve many users, MIMO networks generate fine-grained user information that can reveal behavior patterns and profiles if not anonymized properly. As regulations on data privacy tighten worldwide, disregarding user privacy can lead to loss of trust. Hence, privacy protection must be a priority in massive MIMO system design [7].

This chapter provides a timely survey on state-of-the-art security and privacy technologies proposed for safeguarding 6G wireless systems employing massive MIMO. Various physical layer security techniques leveraging the large antenna arrays are analyzed. Emerging cryptographic protocols, access control, authentication mechanisms, and intrusion detection systems specifically suitable for 6G and MIMO networks are examined. Additionally, privacy solutions like federated learning and differential privacy are highlighted. This comprehensive analysis aims to inspire industry and academia to jointly tackle the security and privacy challenges introduced by massive MIMO in 6G networks.

2.1.1 Motivation

As wireless networks continue to advance, ensuring robust security and preserving user privacy have become pressing needs. Wireless systems today face an escalating number of threats ranging from unauthorized access to malicious data breaches and network attacks. This proliferation of cybersecurity risks requires proactive measures tailored to emerging technologies and attack vectors to strengthen the protection of future wireless communication systems. At the same time, the pervasiveness of wireless networks and intensive data collection amplify concerns over personal data privacy. Users rightly worry about the protection of their personal information and potential misuse. There is a need to explore innovative privacy-enhancing technologies (PETs) and protocols that empower users and ensure compliance with evolving regulations. Looking ahead, the integration of security and privacy in a holistic manner will be crucial for establishing trustworthy communication environments in 6G networks and beyond. This integration comes with its own set of opportunities and challenges that need careful examination. Emerging technologies like blockchain and machine learning (ML) have the potential to contribute to this vision, but open research gaps persist. By diving into wireless security threats, privacy technologies, integration challenges, and research directions, this chapter aims to analyze the landscape and inspire future innovation to develop secure and privacy-enhanced wireless communication systems. The key contributions of this chapter are as follows:

- Provides an overview of vulnerabilities and threats in wireless networks, emphasizing robust security needs (Section 2.2).

- Discusses privacy challenges and examines PETs (Sections 2.4 and 2.6).
- Highlights the importance of security/privacy integration in 6G networks for trustworthy communication (Section 2.4).
- Proposes advanced security solutions like physical layer techniques, access control, and monitoring tools (Section 2.5).
- Analyzes the impact of emerging technologies on wireless security and identifies potential risks (Sections 2.7 and 2.8).
- Contributes insights and frameworks to help develop secure and privacy-enhanced wireless systems for researchers, industries, and regulators (overall analysis and frameworks).

The rest of the chapter is organized as follows: Section 2.2 provides an overview of massive MIMO systems and examines vulnerabilities and threats. Section 2.3 reviews related literature, while Section 2.4 details security and privacy challenges like evolving threats, the importance of security, and massive MIMO issues. Section 2.5 explores physical layer security techniques and technologies for massive MIMO. Privacy-enhancing technologies and their significance are discussed in Section 2.6. In Sections 2.7 and 2.8, emerging technologies and potential threats are, respectively, discussed. Section 2.9 identifies open research issues. Section 2.10 provides the limitations of the study. The chapter is finally concluded in Section 2.11.

2.2 Overview of the Current State of Massive MIMO Wireless Communication Systems

This section discusses the evolution of wireless networks, highlights key features and benefits of massive MIMO, examines its current applications and deployments, and analyzes vulnerabilities and threats present in massive MIMO systems.

Massive MIMO is a cutting-edge technology in wireless communication systems that utilizes a large number of antennas at the base station to improve spectral efficiency and capacity [8]. Unlike traditional MIMO systems that use a small number of antennas, massive MIMO employs dozens or even hundreds of antennas to serve multiple users simultaneously. This enables efficient spatial multiplexing, beamforming, and interference management, resulting in higher data rates, improved signal quality, and increased network capacity.

With the introduction of massive MIMO technology, wireless communication systems have witnessed a significant leap in performance and capacity. This breakthrough has played a pivotal role in shaping the evolution of wireless networks from 3G to the forthcoming 6G [9].

The evolution of wireless communication systems has witnessed significant advancements from 3G to the upcoming 6G networks. Each generation has introduced new technologies, features, and capabilities to meet the growing demands of mobile communication. Here's a brief overview of the evolution.

3G (Third Generation): Introduced in the early 2000s, marked a significant leap in mobile communication. It brought higher data rates, enabling services like video calling and mobile internet. 3G operated on frequency bands such as 850 MHz, 1900 MHz, and 2100 MHz, offering data rates of up to 2 Mbps [10, 11]. Although latency was relatively high at around 100 ms, 3G paved the way for the mobile data revolution. Key technologies of 3G included code division multiple access (CDMA), wideband code division multiple access (WCDMA), and universal mobile telecommunications system (UMTS).

4G (Fourth Generation): Emerged in the late 2000s, aiming to deliver faster speeds, improved reliability, and support for mobile gaming. Operating on frequency bands like 700 MHz, 1700 MHz, 1900 MHz, and 2600 MHz, 4G provided data rates of up to 100 Mbps, significantly enhancing the user experience [10, 11]. Latency was reduced to around 30 ms, making real-time applications more feasible. Notable technologies of 4G included LTE and WiMAX, enabling seamless wireless connectivity.

5G (Fifth Generation): The 5G has been a revolutionary leap in wireless communication. Introduced in the 2010s, 5G offers ultra-fast data rates, low latency, and the ability to connect massive numbers of IoT devices. Operating on a combination of sub-6 GHz and mmWave frequency bands, 5G can achieve data rates of up to 10 Gbps, enabling applications such as 4K video streaming and immersive virtual reality experiences [10]. Latency has been significantly reduced to below 1 ms, unlocking real-time communication possibilities. Key technologies driving 5G include mmWave, massive MIMO, and network slicing.

6G (Sixth Generation): The 6G is set to push the boundaries of wireless communication even further. 6G is expected to deliver higher speeds, near-instant connectivity, and integration of artificial intelligence (AI) and support for advanced applications. It is anticipated to operate on sub-terahertz and terahertz frequency bands, enabling data rates surpassing 100 Gbps. With latency in the sub-millisecond range, 6G will empower innovative applications such as augmented reality, holographic communications, and AI-driven networks [12]. Key technologies envisaged for 6G include terahertz communication and AI-driven network architectures [9, 13].

Table 2.1 summarizes the evolution of key parameters and technologies across wireless network generations from 3G to 5G and beyond, providing context on capabilities and motivation for future networks.

Table 2.1 Evolution of wireless communication systems.

Generation	Introduction	Main Advancements	Frequency bands	Data rates	Latency	Connection density	Key technologies
3G	Early 2000s	Higher data rates, video calling, mobile internet	850 MHz, 1900 MHz, 2100 MHz	Up to 2 Mbps	Around 100 ms	Moderate	CDMA, WCDMA, UMTS
4G	Late 2000s	Faster speeds, improved reliability, mobile gaming	700 MHz, 1700 MHz, 1900 MHz, 2600 MHz	Up to 100 Mbps	Around 30 ms	High	LTE, WiMAX
5G	2010s	Ultrafast data rates, low latency, massive IoT	Sub-6 GHz, mmWave	Up to 10 Gbps	Below 1 ms	Very high	mmWave, massive MIMO, network slicing
6G	Future	Higher speeds, near-instant connectivity, AI integration, advanced applications	Sub-terahertz, terahertz	Expected >100 Gbps	Sub-ms	Extremely high	Terahertz communication, AI-driven networks

Source: Adapted from Borgaonkar et al. [10].

2.2.1 Key Features and Benefits of Massive MIMO

Massive MIMO technology offers several key features and benefits that make it a promising solution for future wireless communication systems. Some of these include the following:

A) **Increased spectral efficiency:** Massive MIMO utilizes many antennas at the base station, allowing for the simultaneous transmission and reception of multiple data streams. This results in a significant increase in spectral efficiency, enabling higher data rates and improved network capacity [14, 15].
B) **Improved signal quality:** The use of multiple antennas in massive MIMO systems helps combat the effects of fading and interference. By employing advanced signal processing techniques, such as beamforming and spatial multiplexing, the system can enhance the signal quality, reduce interference, and improve overall network performance.
C) **Enhanced coverage:** With its ability to focus beams and shape the radiation pattern, massive MIMO can extend coverage to remote or challenging areas. By directing the signal toward specific users or locations, it improves signal strength and coverage, providing a more reliable and consistent connection.
D) **Support for multiple users and devices**: Massive MIMO systems have the capability to serve multiple users simultaneously. This enables efficient resource allocation, better user experience, and improved network capacity, particularly in densely populated areas with high user demand.

2.2.2 Current Applications and Deployments

Massive MIMO has found practical applications and deployments in various real-world scenarios. These include the following.

A) **Cellular networks:** Mobile network operators are increasingly deploying massive MIMO technology in their cellular networks to enhance capacity and coverage. By utilizing multiple antennas at base stations, they can serve a large number of users concurrently and improve the overall network performance.
B) **Internet of things (IoT):** Massive MIMO can support the connectivity requirements of IoT devices by providing efficient communication channels and handling a massive number of devices simultaneously. It enables seamless integration of IoT devices into wireless networks, supporting diverse IoT applications and services.
C) **Wireless broadband systems:** Massive MIMO has been adopted in wireless broadband systems, such as fixed wireless access (FWA) and broadband internet services. It enables high-speed connectivity to residential and business users, particularly in areas where wired infrastructure is limited or unavailable.

2.2.3 Vulnerabilities and Threats in Massive MIMO Wireless Communication Systems

A) **Eavesdropping and unauthorized access:** Attackers can attempt to intercept and eavesdrop on sensitive information transmitted over the wireless channel. They may also try to gain unauthorized access to the network or devices, compromising data confidentiality and system integrity [16].

B) **Jamming and denial-of-service (DoS) attacks:** Malicious actors can launch jamming attacks to disrupt the communication in massive MIMO systems. They can also launch DoS attacks, flooding the network with excessive traffic or overwhelming the system's resources, leading to service degradation or complete disruption.

C) **Man-in-the-middle attacks**: Attackers can position themselves between the communication nodes, intercepting and altering the data exchanged between them. This allows them to manipulate or steal sensitive information, compromising the integrity and confidentiality of the communication [17].

D) **Physical attacks:** Massive MIMO base stations and antennas can be physically tampered with, leading to service disruption, unauthorized access, or information leakage. Attackers may attempt to physically damage or tamper with the equipment to compromise the system's functionality [17].

E) **Insider threats:** Trusted individuals within the organization or network operators may misuse their privileges to gain unauthorized access, leak sensitive information, or disrupt the system intentionally. Insider threats pose significant risks to the security and privacy of massive MIMO networks [18].

F) **Protocol Vulnerabilities:** The design and implementation of communication protocols in massive MIMO systems may have vulnerabilities that can be exploited by attackers. Weaknesses in encryption algorithms, authentication mechanisms, or key exchange protocols can lead to unauthorized access or data breaches [19].

G) **IoT integration risks:** As massive MIMO networks integrate with IoT devices, additional vulnerabilities arise. Insecure IoT devices or compromised IoT gateways can serve as entry points for attackers, compromising the security and privacy of the overall system [20].

H) **Network congestion and resource management:** The scalability and resource allocation challenges in massive MIMO networks can lead to network congestion and inefficient resource management. These issues can be exploited by attackers to degrade system performance or disrupt service availability [21].

Figure 2.1 provides an overview of key vulnerabilities in massive MIMO systems across layers like physical, MAC, network, and applications that attackers can exploit to compromise confidentiality, integrity, and availability, emphasizing the

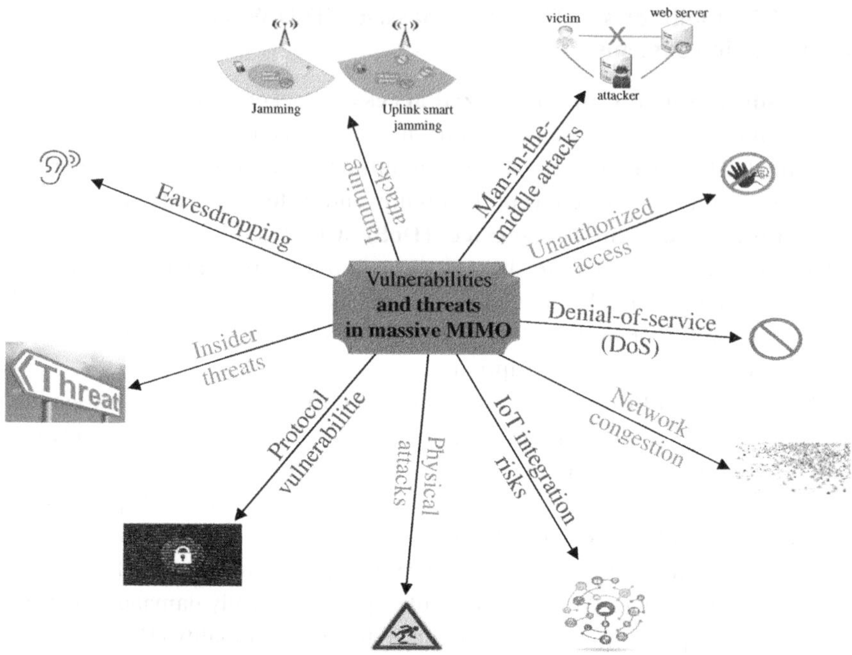

Figure 2.1 Vulnerabilities and threats in massive MIMO wireless communication systems.

critical need for robust, multilayered security mechanisms tailored to address the unique risks posed by massive MIMO technology as it gets adopted in 6G networks and beyond.

To provide a comprehensive overview, Table 2.2 highlights the vulnerabilities and threats in massive MIMO wireless communication systems along with their potential impact.

2.3 Related Reviews

This section summarizes and compares related literature on wireless security and privacy, analyzing the methodologies, strengths, weaknesses, and key findings of selected works. It provides insights into different research approaches applied to address security challenges.

The literature on wireless security and privacy employs a range of methodologies including qualitative research [22], forensic analysis [23], protocol design [24], security analysis [7], new technologies [25], and comprehensive surveys [26]. While the approaches differ, the works collectively highlight the increasing

Table 2.2 Vulnerabilities and threats in massive MIMO.

Vulnerability/threat	Description	Potential impact
Eavesdropping	Unauthorized interception of sensitive information	Data leakage, privacy breaches
Unauthorized access	Unauthorized access to the network or devices	Data compromise, system disruption
Jamming attacks	Deliberate disruption of wireless communication	Service degradation, connectivity issues
Denial-of-service (DoS)	Flooding the network to exhaust system resources	Service disruption, loss of connectivity
Man-in-the-middle attacks	Intercepting and altering communication between nodes	Data manipulation, unauthorized access
Physical attacks	Tampering with base stations or equipment	Service disruption, unauthorized access
Insider threats	Misuse of privileges by trusted individuals	Data breaches, system compromise
Protocol vulnerabilities	Weaknesses in communication protocols	Unauthorized access, data compromise
IoT integration risks	Insecurity of IoT devices or compromised gateways	System vulnerability, privacy breaches
Network congestion	Scalability and resource allocation challenges	Service degradation, inefficient resource usage

significance of threats like data breaches, vulnerabilities in systems/protocols, and emerging privacy risks associated with new technologies. The authors of [22, 23] demonstrate via interviews and forensic examination how even non-technical aspects like skills and design choices impact security. The authors of [7, 24] address identified weaknesses through novel protocols and cryptography enhancements. The authors of [25, 26] showcase how physical layer techniques and intelligent surfaces can strengthen wireless security. The literature emphasizes a need for continuous innovation, technology evolution, and systemic perspectives to ensure robust and holistic security solutions in the face of a dynamic threat landscape.

The works highlight the growing risks, demonstrate diverse problem-solving approaches, and establish the importance of continued security advancement to counter emerging challenges in wireless systems. Table 2.3 summarizes a few related literature by exploring each methodology, strengths, weaknesses, and results obtained in each.

Table 2.3 Review of related work.

References	Description of article	Methodology	Strength	Weaknesses	Results
[22]	Examines security and privacy challenges of big data adoption in the telecommunication industry	1. Qualitative research approach 2. Application of TOE framework (Technological, Organizational, and Environmental factors)	1. Focus on the telecommunication industry 2. In-depth insights and perspectives through qualitative research Comprehensive analysis from multiple dimensions	1. Sample size not mentioned 2. Lack of detailed description of data collection and analysis process No discussion of limitations or biases in the research design	Data management, data privacy, data compliance, and regulatory orchestration were identified as pressing challenges. Emphasizes addressing skill shortages in organizations adopting big data.
[23]	Examines privacy and security issues of the SimpliSafe security system on Android and iOS devices	1. Forensic analysis 2. Network traffic analysis 3. Examination of system and app interactions	1. Focus on a specific security system (SimpliSafe) 2. Utilizes forensic analysis to uncover vulnerabilities 3. Provides insights into security risks in home security systems	1. Sample size and participant characteristics not mentioned 2. Lack of detailed description of data collection and analysis process 3. No discussion of research design limitations or biases	1. Identifies potential vulnerabilities and privacy concerns in the SimpliSafe system 2. Emphasizes the need for developers to address these concerns. 3. Contributes to the understanding of privacy and security challenges in home security systems.

[24]	Proposes PI-MAKA protocol for secure and reliable wireless mobile communication	1. PI-MAKA protocol (multiple phases) 2. Comparative analysis of existing schemes 3. Experimental results	1. Addresses limitations of existing protocols 2. Reduces computational and communication costs 3. Highlights vulnerabilities of wireless mobile networks	1. No detailed description of data collection and analysis process 2. No discussion of research design limitations or biases	1. PI-MAKA achieves lower computational and communication costs than existing schemes. 2. Effectively addresses security and privacy challenges in wireless mobile communication.
[7]	The paper focuses on the security analysis of a privacy-preserving authentication scheme in VANETs. The authors propose modifications to enhance its security	1. The methodology involves concrete attack analysis to identify security vulnerabilities in the scheme 2. Modifications are proposed based on the vulnerabilities identified	1. Addresses security vulnerabilities in a privacy-preserving authentication scheme 2. Proposes modifications to enhance the security of the scheme.	Lacks a detailed description of the methodology used for the security analysis	The improved scheme achieves semantic security against forgery attacks in the random oracle model
[25]	The paper discusses the use of intelligent reflecting surfaces (IRSs) for enhancing wireless communication security and privacy	The methodology involves discussing the benefits and challenges of utilizing IRS for improving secrecy rates, communication covertness, and creating extra randomness in wireless propagations	1. Explores the potential of IRS in enhancing wireless communication security and privacy 2. Presents results on applying IRS to improve secrecy rates and communication covertness	Does not provide a detailed description of the methodology used for numerical studies	1. Application of IRS improves average secrecy rates in wiretap channels 2. Enables perfect communication covertness 3. Creates extra randomness in wireless propagations for hiding active wireless transmissions

(Continued)

Table 2.3 (Continued)

References	Description of article	Methodology	Strength	Weaknesses	Results
[26]	The paper provides a survey of physical layer security techniques for data transmission in future wireless networks	The methodology involves discussing various mobile technologies and techniques for physical layer security, including secure key generation, directional modulation (DM), spatial modulation (SM), covert communication, and intelligent reflecting surface (IRS)-aided communication	1. Provides a comprehensive overview of physical layer security techniques for data transmission in future wireless networks. 2. Highlights the importance of physical layer security and discusses various techniques and technologies to enhance security in wireless communication systems	Does not provide a detailed analysis or comparison of the discussed physical layer security techniques	1. Provides insights into various promising mobile technologies and techniques for physical layer security 2. Highlights the limitations of cryptography and advantages of physical security in protecting private information

2.4 Security and Privacy Challenges of Future Wireless Communication Systems

This section details the key security and privacy challenges arising from the evolving threat landscape, adoption of massive MIMO, and advanced data collection in future wireless systems. It emphasizes the critical importance of security and privacy in building trust and confidence.

The rapid advancement of wireless communication systems brings forth numerous opportunities for connectivity, convenience, and innovation. However, alongside these advancements, there is a growing concern regarding the security and privacy of these systems. Illustrated in Figure 2.2, future wireless communication systems face significant challenges in safeguarding sensitive information, protecting against cyberattacks, and addressing privacy concerns [27].

One of the primary challenges is the risk of cyberattacks and data breaches. Malicious actors can exploit vulnerabilities in network infrastructure, devices, and protocols to gain unauthorized access to sensitive data or disrupt communication services. These attacks can have severe consequences, ranging from financial losses to compromising personal information or critical infrastructure.

To mitigate these risks, physical layer security techniques play a crucial role. Technologies such as massive MIMO can enhance the security of data transmission by employing multiple antennas and beamforming to counter eavesdropping

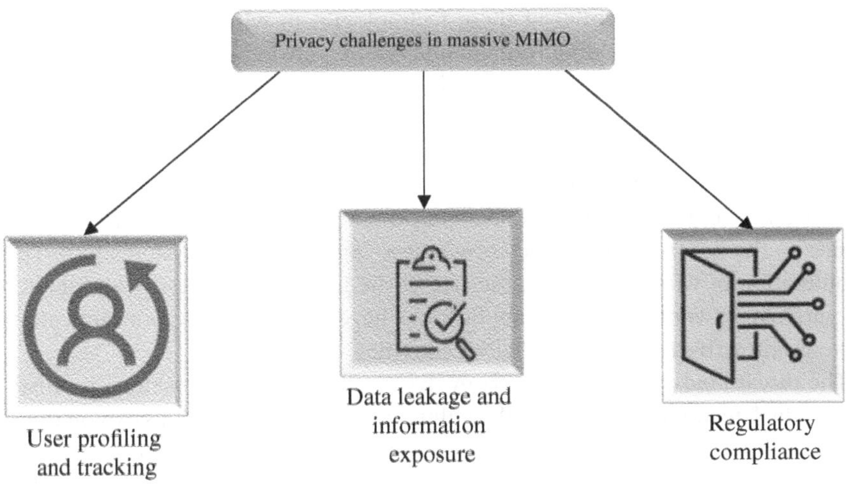

Figure 2.2 Privacy challenges in massive MIMO.

attempts [26]. By manipulating the wireless propagation environment, these techniques enable secure and reliable communication.

Implementing robust security mechanisms is vital to protect wireless communication systems. Encryption algorithms, authentication protocols, and intrusion detection systems are essential components of a comprehensive security framework. These mechanisms ensure the confidentiality, integrity, and availability of transmitted data and protect against unauthorized access or tampering.

In addition to security, privacy is a significant concern in future wireless communication systems. Users expect their personal information to be kept confidential and their communication activities to remain private. Privacy-enhancing technologies aim to address these concerns by anonymizing data, providing users with control over their personal information, and minimizing data collection and retention.

Compliance with regulations and standards is another crucial aspect of ensuring security and privacy in wireless communication systems [28]. Governments and regulatory bodies impose guidelines and requirements to protect user privacy and ensure the secure operation of networks. Adhering to these regulations helps establish trust among users and facilitates the secure deployment of wireless communication systems.

The emergence of new technologies, such as blockchain and ML, presents both opportunities and challenges for security and privacy in wireless communication systems. Blockchain offers decentralized and tamper-resistant transactional security, while ML can enhance anomaly detection and threat intelligence capabilities [29]. Exploring the potential of these technologies in wireless communication systems is crucial for building resilient and secure networks.

2.4.1 Evolving Threat Landscape

In recent years, the wireless communication landscape has witnessed a significant increase in security threats. The evolving threat landscape poses various challenges to the security and privacy of wireless communication systems [30]. The proliferation of wireless devices, the rise of the IoT, and the increasing reliance on wireless networks for critical applications have made wireless communication systems attractive targets for malicious activities.

The threat landscape is constantly evolving, with new attack vectors, sophisticated attack techniques, and growing vulnerabilities. Cybercriminals exploit these vulnerabilities to launch various types of attacks, such as eavesdropping, jamming, spoofing, and data manipulation. These attacks can have severe consequences, including unauthorized access to sensitive information, disruption of communication services, and compromise of user privacy.

2.4.2 Security Challenges in Massive MIMO

Massive MIMO technology has emerged as a promising solution to enhance the capacity and performance of wireless communication systems [31]. However, along with the benefits it offers, massive MIMO also introduces several security challenges that need to be addressed.

i) **Channel estimation security**: Massive MIMO systems rely on accurate channel estimation for efficient data transmission. However, the process of estimating the wireless channel introduces vulnerabilities that can be exploited by attackers [32]. Adversaries can launch attacks to manipulate channel estimates, leading to degraded system performance and compromised security.
ii) **Beamforming security**: Beamforming is a key technique in massive MIMO systems for enhancing signal transmission and reception. However, beamforming algorithms and protocols are susceptible to various attacks, including eavesdropping and signal interception. Attackers can exploit vulnerabilities in the beamforming process to gain unauthorized access to transmitted information.
iii) **User privacy protection**: The deployment of massive MIMO systems introduces new privacy challenges. The large number of antennas and users in massive MIMO networks makes it more difficult to ensure user privacy. Adversaries can exploit the inherent characteristics of massive MIMO to infer sensitive information about users, such as their location or behavior.

2.4.3 Privacy Challenges in Massive MIMO

In addition to security challenges, massive MIMO systems also face privacy challenges that need to be addressed to protect user information and ensure compliance with privacy regulations [4].

i) **User profiling and tracking**: The deployment of massive MIMO systems enables the collection of vast amounts of user data, which can be used for user profiling and tracking. The aggregation and analysis of this data raise concerns about user privacy and the potential for unauthorized tracking and monitoring.
ii) **Data leakage and information exposure**: Massive MIMO systems involve the transmission of large amounts of data over the wireless channel. The risk of data leakage and information exposure increases as the number of antennas and users in the system grows. Adequate measures must be implemented to protect the confidentiality and integrity of transmitted data.

iii) **Regulatory compliance**: Privacy regulations, such as the General Data Protection Regulation (GDPR), impose strict requirements on the collection, storage, processing, and sharing of personal data. Massive MIMO systems need to ensure compliance with these regulations and implement PETs to protect user privacy rights.

2.4.4 The Importance of Security and Privacy in Future Wireless Communication Systems

A) **Protection of sensitive information:** Wireless communication systems often transmit sensitive and personal information, including financial data, medical records, and personal communications. Without proper security measures, this information is vulnerable to interception, unauthorized access, and misuse. Robust security protocols and encryption mechanisms safeguard against data breaches and unauthorized disclosure, protecting individuals and organizations from potential harm.

B) **Trust and confidence:** Users rely on wireless communication systems for various purposes, such as online banking, e-commerce, and accessing personal accounts [11]. Maintaining trust and confidence in these systems is essential for their widespread adoption. By ensuring secure and private communication, users are more likely to have confidence in the system, leading to increased usage and customer satisfaction.

C) **Prevention of cyber-attacks:** Wireless communication systems are prime targets for cyberattacks due to their widespread usage and potential impact. Malicious actors can launch various attacks, such as eavesdropping, denial-of-service (DoS), or man-in-the-middle (MITM) attacks. Implementing robust security measures helps prevent these attacks, safeguarding the integrity and availability of wireless networks.

D) **Compliance with regulations:** Governments and regulatory bodies have established stringent regulations to protect user privacy and ensure secure communication. Compliance with these regulations is crucial for organizations operating wireless communication systems. By adhering to privacy and security guidelines, organizations not only avoid legal consequences but also demonstrate their commitment to protecting user data and privacy.

E) **Prevention of identity theft and fraud:** Wireless communication systems often involve the exchange of personal identification information and financial transactions. Inadequate security measures can lead to identity theft, where sensitive personal information is stolen and misused for fraudulent activities. By implementing strong authentication and encryption mechanisms, the risk of identity theft and fraud can be significantly mitigated.

F) **Protection of intellectual property:** Future wireless communication systems will likely be used for the transmission of intellectual property, proprietary information, and trade secrets. Without proper security measures, unauthorized access or interception of this data can lead to significant financial losses and competitive disadvantage. Strong security measures, including encryption and access controls, help protect intellectual property from theft and unauthorized disclosure.

G) **Safeguarding critical infrastructure:** Wireless communication systems play a vital role in critical infrastructure, including power grids, transportation systems, and healthcare facilities. Any compromise in the security or privacy of these systems can have severe consequences, affecting public safety, economic stability, and even human lives. Robust security measures are essential to protect critical infrastructure from cyberattacks and ensure their reliable operation.

2.5 Physical Layer Security and Technologies for Massive MIMO

This section explores physical layer security techniques like secure beamforming, artificial noise injection, and intelligent reflecting surfaces to enhance wireless security. It also discusses mechanisms beyond the physical layer like cryptography, access control, monitoring tools, and privacy technologies.

Physical layer security is a branch of cybersecurity that focuses on enhancing the security of wireless communication systems at the physical layer [33]. It employs various techniques and technologies to protect transmitted data from unauthorized access, interception, and manipulation [15, 34]. Unlike traditional cryptographic methods that primarily rely on encryption algorithms, physical layer security takes advantage of the unique characteristics of wireless communication channels to provide an additional layer of protection [15, 35].

The primary goal of physical layer security is to ensure secure and reliable communication between the transmitter and intended receiver while preventing eavesdroppers from decoding the transmitted information [36, 37]. It addresses the vulnerabilities and threats associated with wireless communication, such as eavesdropping, jamming, and unauthorized access. Here are some key aspects of physical layer security and technologies relevant to massive MIMO in 6G and beyond.

A) **Secure beamforming:** Beamforming is a fundamental technology in massive MIMO systems that directs the transmitted signals toward the intended receivers. Secure beamforming techniques focus on enhancing the confidentiality and integrity of the transmitted data by steering the beams toward the desired receivers and intelligently generating and transmitting artificial noise signals toward potential eavesdroppers to corrupt their reception [38].

B) **Artificial noise injection:** Artificial noise injection is another technique used in massive MIMO systems to enhance security. By intentionally introducing random noise in the transmission, artificial noise injection can mask the information-bearing signal, making it more challenging for eavesdroppers to decode the intended message [39].

C) **Channel randomness and exploitation**: Massive MIMO systems inherently exhibit high spatial diversity due to the presence of a large number of antennas at both the transmitter and the receiver. The spatial diversity creates randomness in the wireless channel, which can be exploited for enhancing security [40]. The variations in the channel conditions make it difficult for eavesdroppers to reliably intercept the transmitted signals.

D) **Cooperative beamforming and relaying:** Cooperative techniques, such as cooperative beamforming and relaying, can enhance the security of massive MIMO systems. By leveraging the cooperation among multiple antennas or relay nodes, secure transmission can be achieved through joint optimization of beamforming weights or relaying strategies.

E) **Intelligent reflecting surfaces (IRS):** IRSs, also known as reconfigurable intelligent surfaces, are a new technology that has the potential to significantly enhance the security of massive MIMO systems [41]. IRS consists of passive reflecting elements that can adaptively adjust the phase and amplitude of the reflected signals. By intelligently controlling the reflections, IRS can enhance the secrecy and confidentiality of the wireless communication links.

F) **Jamming and interference mitigation**: Massive MIMO systems can also utilize intentional jamming or interference techniques to disrupt unauthorized receivers. By injecting additional noise or interference signals, the communication can be protected against eavesdroppers or jammers [42].

Table 2.4 outlines key physical layer security techniques leveraging massive MIMO systems including secure beamforming, artificial noise injection, channel randomness exploitation, cooperative beamforming, intelligent reflecting surfaces, and jamming mitigation; comparing complexity, performance, and constraints to provide insights on integrating these mechanisms to enhance confidentiality and resilience against threats in wireless communication.

2.5.1 Other Security and Privacy Mechanisms

Along with physical layer security techniques, a holistic defense architecture needs to incorporate mechanisms at other layers as discussed below:

A) **Cryptographic techniques**: Cryptographic techniques play a fundamental role in securing wireless communication systems, including 6G and beyond.

Table 2.4 Physical layer security techniques for massive MIMO.

Technique	Description	Complexity	Performance	Constraints
Secure beamforming [38]	Directs signals toward intended receivers	Moderate	High secrecy rates	Channel state information (requires knowledge of eavesdropper location)
Artificial noise injection [39]	Introduces random noise to mask information-bearing signal	Low	Moderate secrecy capacity	Impact on system capacity (increases interference)
Channel randomness [40]	Exploits spatial diversity for enhanced security	Low	High secrecy rates	Dependence on channel conditions
Cooperative beamforming [38]	Leverages cooperation among multiple antennas	Moderate	High secrecy rates	Synchronization, overhead
Intelligent reflecting surfaces (IRS) [41]	Adjusts phase and amplitude of reflected signals	Moderate	Large secrecy rate gains	Complexity of IRS deployment (needs careful configuration)

These techniques ensure the confidentiality, integrity, and authenticity of data transmitted over the network. Some cryptographic mechanisms applicable to future wireless networks include the following:

i) Symmetric key encryption: Symmetric key encryption algorithms, such as advanced encryption standard (AES), ensure that the data transmitted between users is securely encrypted and can only be decrypted with the correct shared key.
ii) Asymmetric key encryption: Asymmetric key encryption, commonly known as public-key encryption, involves the use of a pair of keys: a public key for encryption and a private key for decryption. This technique enables secure key exchange, digital signatures, and secure communication between parties without the need for a pre-shared secret key.
iii) Hash functions: Hash functions are cryptographic algorithms that transform data into a fixed-size hash value. They are commonly used to verify the integrity of data by generating a hash value that can be compared before and after transmission to ensure data integrity.

B) **Access control:** Access control mechanisms are crucial for regulating user access to wireless networks. They help prevent unauthorized access and ensure that only authenticated and authorized users can connect to the network. Some access control mechanisms include the following:
   i) Authentication protocols: Authentication protocols, such as extensible authentication protocol (EAP), are used to verify the identity of users before granting access to the network. These protocols employ various techniques, including passwords, digital certificates, biometrics, or multifactor authentication, to establish user identity.
   ii) Authorization and role-based access control (RBAC): Authorization mechanisms define the permissions and privileges granted to authenticated users. RBAC is a widely used approach that assigns roles to users and grants access based on their assigned roles, ensuring that users have access only to the resources they need for their specific roles [43].

C) **Network monitoring and analysis**: Network monitoring and analysis tools are essential for detecting and mitigating security threats in real time. These tools help identify abnormal network behavior, intrusion attempts, and malicious activities. Some key monitoring and analysis mechanisms include the following:
   i) Intrusion detection and prevention systems (IDPS): IDPS monitors network traffic for signs of potential attacks or policy violations. They can automatically take action to block or mitigate suspicious activities, ensuring the network's security.
   ii) Security information and event management (SIEM): SIEM systems collect and analyze log data from various network devices, applications, and security appliances. They provide centralized monitoring, correlation, and analysis of security events, enabling proactive threat detection and response.

D) **Privacy-enhancing technologies**: Privacy-enhancing technologies are crucial for protecting the privacy of users' personal information and communications. These technologies aim to minimize the collection and retention of sensitive data, provide users with control over their information, and ensure data anonymity. Some PETs include the following:
   i) Anonymization techniques: Anonymization techniques, such as data anonymization and identity anonymization, help protect user privacy by removing or obfuscating personally identifiable information (PII) from datasets or communications [43].
   ii) Differential privacy: Differential privacy is a mathematical framework that provides a privacy guarantee by adding noise or perturbation to datasets while still allowing meaningful analysis. It allows statistical analysis while protecting the privacy of individual data contributors.

E) **Security standards and policies**: Security standards and policies establish guidelines and best practices to ensure the secure operation of wireless networks. These standards help organizations comply with security requirements and mitigate security risks. Some relevant security standards and policies for 6G and beyond include:
   i) Wireless security protocols: Standards like the Wi-Fi Protected Access (WPA) and WPA2/WPA3 ensure secure wireless communication by defining encryption methods, key management, and authentication protocols.
   ii) Data protection regulations: Privacy regulations, such as the GDPR and California Consumer Privacy Act (CCPA), impose obligations on organizations to protect users' personal data and ensure proper data handling practices.

By incorporating these security and privacy mechanisms, wireless communication systems in 6G and beyond can establish a comprehensive defense architecture. These mechanisms work in synergy to safeguard sensitive data, prevent unauthorized access, detect and mitigate threats, protect user privacy, and ensure compliance with security standards and regulations.

Figure 2.3 displays a range of security mechanisms, while Table 2.5 offers a comprehensive overview of diverse security and privacy mechanisms that extend beyond physical layer techniques. These mechanisms are vital components of a

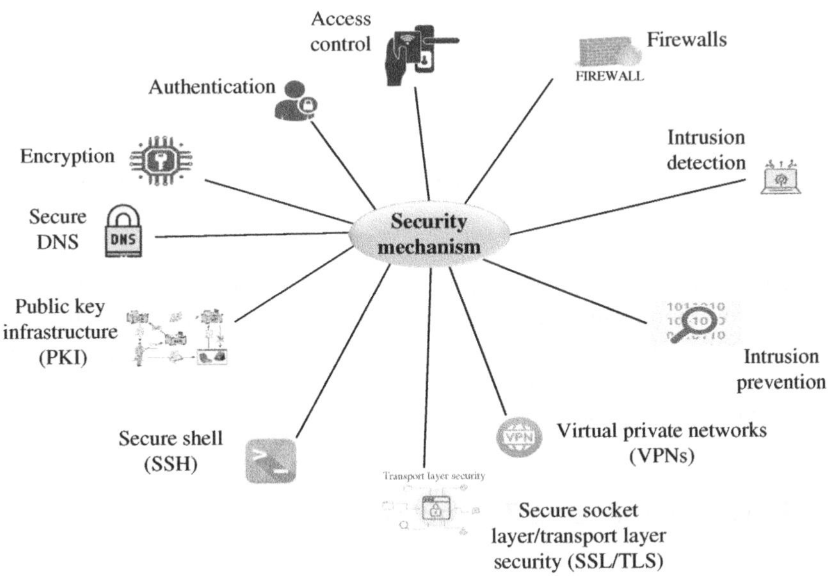

Figure 2.3 Security mechanisms.

Table 2.5 Security and privacy mechanisms beyond physical layer.

Security mechanism	Description	Key features
Encryption	Converts plaintext into ciphertext using algorithms	Confidentiality, integrity, data protection
Authentication	Verifies the identity of users or devices	User verification, access control, prevention of impersonation
Access control	Regulates user access to resources	User authorization, privilege management, resource protection
Firewalls	Monitors and filters network traffic	Network security, intrusion prevention
Intrusion detection	Identifies and alerts to potential attacks	Real-time monitoring, threat detection
Intrusion prevention	Blocks or mitigates attacks in real time	Attack prevention, network security
Virtual private networks (VPNs)	Securely connect remote networks or users	Confidentiality, secure communication
Secure socket layer/transport layer security (SSL/TLS)	Secures communication over networks	Encryption, data integrity, authentication
Secure shell (SSH)	Secure remote access and file transfer	Encryption, secure login, secure data transfer
Public key infrastructure (PKI)	Manages digital certificates and encryption keys	Authentication, secure communication, trust management
Secure DNS	Encrypts DNS traffic to prevent eavesdropping	Privacy, data protection, secure DNS resolution

robust defense architecture for wireless systems. They encompass encryption, access control, monitoring tools, privacy technologies, security standards, and policies. Together, they create a multilayered safeguarding framework for sensitive data, enabling threat detection, regulatory compliance, and the establishment of trustworthy communication.

2.6 Privacy-Enhancing Technologies

This section highlights the significance of PETs in future networks, examining their role in protecting personal data and complying with regulations. In the context of 6G networks and beyond, PETs play a crucial role in protecting the privacy of users' personal information and communications [44]. As wireless communication systems become more pervasive and interconnected, the need to address privacy concerns becomes increasingly significant. Here are some reasons highlighting the importance of PETs in the future:

i) **User control and consent:** Privacy-enhancing technologies empower users by giving them control over their personal information. They enable individuals to make informed decisions about the collection, use, and disclosure of their data. PETs facilitate mechanisms such as consent management tools and data subject rights to enhance transparency and user agency [45].
ii) **Data anonymization and pseudonymization:** PETs provide techniques for anonymizing or pseudonymizing data, reducing the risk of re-identification and protecting individuals' privacy. By removing or obfuscating PII, PETs help ensure that data cannot be traced back to specific individuals [43].
iii) **Secure data sharing and collaboration:** Privacy-enhancing technologies enable secure data sharing and collaboration while preserving privacy. Techniques such as secure multiparty computation and secure data federations allow different parties to collaborate and analyze data without sharing sensitive information.
iv) **Privacy by design:** PETs promote privacy by design principles, ensuring that privacy is considered from the initial stages of system design and development. By embedding privacy features and safeguards into the architecture, PETs help create privacy-preserving systems that minimize the risk of privacy breaches.

2.6.1 Impact of Regulations and Standards on Security and Privacy

Regulations and standards have a significant impact on the security and privacy aspects of 6G networks and beyond [46]. As technology evolves, it becomes crucial to establish guidelines and frameworks to ensure secure and privacy-respecting systems. Here are some ways in which regulations and standards influence security and privacy:

i) **Data protection and privacy regulations:** Regulations such as the European Union's GDPR and other data protection laws worldwide impose obligations on organizations regarding the collection, storage, processing, and sharing of personal data. Compliance with these regulations helps protect user privacy and establishes a framework for responsible data handling practices.
ii) **Security standards and best practices**: Standards bodies and organizations develop security standards and best practices that provide guidance on secure system design, encryption algorithms, authentication protocols, and

secure communication protocols. Adhering to these standards ensures that security measures are robustly implemented to protect against threats and vulnerabilities.

iii) **Privacy impact assessments**: Regulations often require organizations to conduct privacy impact assessments (PIAs) to evaluate the privacy risks associated with new technologies or data processing activities. PIAs help identify potential privacy concerns and recommend mitigation strategies to ensure compliance with privacy regulations.

iv) **Audits and certifications**: Regulations and standards may require organizations to undergo security audits or obtain certifications to demonstrate compliance with specific security and privacy requirements. These audits and certifications provide assurance to users and stakeholders that appropriate security measures are in place.

v) **Cross-border data transfers**: Regulations addressing cross-border data transfers, such as the EU-US Privacy Shield or Standard Contractual Clauses, aim to protect personal data when transferred to countries with different data protection regulations. These mechanisms ensure that data transferred across borders maintains an adequate level of protection.

2.7 Emerging Technologies

This section discusses emerging technologies like blockchain and ML that create new opportunities to strengthen wireless security and privacy. Emerging technologies like blockchain and ML present promising opportunities to enhance security and privacy in 6G networks.

i) **Blockchain:** Blockchain technology offers tamperproof decentralized ledgers that can be utilized to securely share data and validate transactions in 6G networks. Its distributed nature, cryptographic algorithms, and consensus mechanisms provide a high level of transparency, immutability, and trust in data transactions [47].

ii) **Machine learning:** Machine learning techniques can play a crucial role in enhancing security and privacy in 6G networks. By analyzing patterns and anomalies in network traffic, ML algorithms can detect and mitigate various types of attacks [48, 49]. They can also facilitate adaptive security mechanisms and enhance threat intelligence capabilities.

iii) **Federated learning:** Federated learning is a privacy-preserving ML approach that allows collaborative model training while keeping raw user data decentralized. This technique enables multiple devices to contribute to model updates without sharing their sensitive data, thereby improving privacy protection in 6G networks.

iv) **Homomorphic encryption:** Homomorphic encryption enables computations on encrypted data, reducing the exposure of sensitive user information in 6G networks. By performing operations directly on encrypted data, homomorphic encryption ensures privacy while allowing data processing and analysis without the need for decryption.

2.8 Potential Threats and Challenges of Future Wireless Communication Systems

This section identifies potential threats and challenges that need to be addressed proactively as wireless networks evolve, including risks from technologies like AI and increased virtualization. As 6G networks evolve, there are several potential threats and challenges that need to be considered for security and privacy. Some key points to note include the following:

i) **Increasing sophistication of cyberattacks with leverage of AI/ML and quantum computing:** With the advancement of technologies like AI/ML and quantum computing, cyberattacks are becoming more sophisticated and challenging to detect and mitigate [50]. Research is needed to develop proactive security measures and resilient defense mechanisms to counter these advanced threats.

ii) **New attack surfaces from holographic communication, extended reality, brain–computer interfaces:** The introduction of new technologies in 6G networks, such as holographic communication, extended reality, and brain–computer interfaces, creates new attack surfaces that can be exploited by adversaries. Understanding the security implications of these emerging technologies and developing appropriate security mechanisms is crucial to protect against potential attacks.

iii) **Privacy concerns around pervasive biometrics and human behavioral data collection:** 6G networks are expected to leverage pervasive biometrics and collect extensive human behavioral data for various applications. This raises privacy concerns regarding the collection, storage, and usage of personal data. Research is needed to develop privacy-preserving mechanisms that allow the utilization of such data while ensuring the protection of individuals' privacy rights.

iv) **Ethical risks of embedded and autonomous decision-making:** As 6G networks become more intelligent and autonomous, there are ethical risks associated with embedded decision-making capabilities. Ensuring that these systems adhere to ethical principles and values, such as fairness, transparency, and accountability, is critical. Research is needed to address the ethical implications of autonomous decision-making and develop frameworks for responsible and ethical 6G networks.

v) **Vulnerabilities from increased virtualization, softwarization, and dense edge networks:** The increasing virtualization, softwarization, and deployment of dense edge networks in 6G introduce new vulnerabilities that can be exploited by attackers. Research is needed to identify and address these vulnerabilities, develop robust security mechanisms for virtualized environments, and ensure secure communication and computation at the network edge [50].

vi) **Lack of security standardization among proprietary semiautonomous network implementations:** With the introduction of new technologies and proprietary network implementations, there is a risk of fragmented security measures and the lack of standardized security protocols. Research is needed to establish industry-wide security standards and promote interoperability among different 6G network implementations, ensuring a consistent and robust security posture across the ecosystem.

vii) **Insufficient transparency into complex AI systems and decentralized data flows:** The integration of AI systems in 6G networks introduces complexities in data flows and decision-making processes. Ensuring transparency and accountability in AI systems is crucial for maintaining user trust and detecting potential biases or malicious behaviors. Research is needed to develop explainable AI techniques and mechanisms for auditing decentralized data flows to ensure transparency and accountability.

2.9 Open Research Issues and the Future of Security and Privacy

This section presents open research issues and future directions for advancing security and privacy in beyond 5G and 6G networks. While extensive research has been done on wireless security and privacy, several open issues remain to be addressed for 6G networks. These open research issues include the following:

i) **Developing lightweight cryptography and physical layer security suitable for resource-constrained 6G devices:** As 6G networks aim to connect a vast number of devices, including IoT devices with limited resources, there is a need for efficient cryptographic algorithms and physical layer security mechanisms that can provide robust protection while minimizing computational and power requirements [51].

ii) **Adapting existing solutions to new 6G architectures like cell-free massive MIMO and intelligent surfaces:** 6G networks are expected to incorporate new architectural paradigms, such as cell-free massive MIMO and intelligent reflecting surfaces. Research is needed to adapt existing security and privacy solutions to these novel architectures and explore their implications on network security.

iii) **Validating theoretical gains of physical layer security under practical constraints:** While physical layer security has shown promising results in theory, its practical implementation and performance in real-world scenarios need to be validated. Research is required to investigate the limitations and practical considerations of physical layer security techniques and ensure their effectiveness in diverse deployment scenarios.

iv) **Designing holistic cross-layer security architectures spanning the physical to application layer:** 6G networks will require comprehensive security architectures that consider vulnerabilities and threats across multiple network layers [51]. Developing cross-layer security mechanisms that can seamlessly integrate security measures from the physical layer to the application layer is crucial for ensuring end-to-end security and privacy.

v) **Leveraging technologies like blockchain, quantum communications, and machine learning:** Emerging technologies such as blockchain, quantum communications, and ML offer new possibilities for enhancing security and privacy in 6G networks. Research is needed to explore the integration of these technologies into the network infrastructure and develop novel security and privacy mechanisms leveraging their capabilities [52].

vi) **Aligning technical standards with emerging privacy regulations worldwide:** With the increasing focus on privacy regulations worldwide, it is important to align technical standards and protocols with these regulations. Research is needed to ensure that security and privacy mechanisms implemented in 6G networks comply with privacy regulations while providing robust protection against emerging threats.

vii) **Analyzing information-theoretic secrecy capacity limits under complex deployment scenarios:** Information-theoretic secrecy capacity is a fundamental measure of the security level achievable in wireless communication. Research is required to analyze the secrecy capacity limits in complex deployment scenarios of 6G networks, taking into account factors such as channel characteristics, network topology, and interference.

viii) **Prototyping and experimental analysis of sophisticated cross-layer wireless attacks**: To develop effective countermeasures against evolving wireless attacks, it is essential to conduct prototyping and experimental analysis of sophisticated cross-layer attacks [53]. This research will enable the identification of vulnerabilities and the development of mitigation strategies to enhance the security of 6G networks.

Advancing research in these directions will be crucial for building secure and privacy-preserving 6G wireless networks.

2.10 Limitations of the Study

While this study provides a comprehensive survey and discussions of security and privacy in wireless communication systems, it has certain limitations that need to be considered:

i) The literature review primarily focused on technical research papers and excluded perspectives from law, policy, ethics, and society-related disciplines.
ii) The scope was limited to emerging 5G, beyond 5G, and 6G technologies like massive MIMO. Legacy wireless technologies still in use were not discussed.
iii) Only a subset of physical layer security techniques was covered in detail given the broad nature of the survey. An exhaustive analysis of all techniques was out of scope.
iv) Specific implementation details and experimental results demonstrating the performance of various security and PETs were not presented.
v) The treatment of PETs was limited in breadth and depth.
vi) Quantitative security and privacy modeling techniques were not included in the study.
vii) Potential solutions were presented at a high conceptual level without rigorous technical validation through simulations, testbeds, etc.
viii) References from industry technology standards and roadmap documents were limited.
ix) The evolving nature of threats and rapid pace of technology changes may render some of the discussed content outdated.

Further research can address these limitations by incorporating more diverse interdisciplinary perspectives, quantitative analysis, prototype implementation results, insights from industry standards activities, and updating the contents with the latest advancements.

2.11 Lessons Learned

This section summarizes the key lessons learned from the comprehensive analysis of security and privacy in future wireless communication systems presented in this chapter.

i) **Adapting to an evolving threat landscape:** As wireless networks evolve, they face an increasingly complex and growing threat landscape. To address this, proactive security measures tailored to new technologies and attack vectors are essential.

ii) **Securing massive MIMO technology:** While massive MIMO enables high throughput and connectivity, it also introduces new vulnerabilities. To counter these risks, robust physical layer techniques and defense-in-depth mechanisms across layers must be implemented.
iii) **Amplified privacy risks:** Advanced data collection and analytics amplify privacy risks. It is crucial to incorporate PETs and adopt responsible data practices.
iv) **Multilayered security approach:** There is no "one-size-fits-all" solution to security. A multilayered approach involving security protocols, access controls, monitoring, standards, and policies is necessary.
v) **Integrating security and privacy from the ground up:** Security and privacy should be foundational pillars in the design of future networks, not afterthoughts. A holistic approach spanning from architecture to implementation is key.
vi) **Leveraging emerging technologies for security:** The strategic use of emerging technologies like AI, blockchain, and quantum communications can strengthen security. However, it also presents new challenges that need to be carefully considered.
vii) **Aligning research with standards and regulations:** Academic research should align with and inform the development of technical standards and privacy regulations worldwide to ensure cohesive and effective security practices.
viii) **Real-world validation and testing:** Prototyping, testbeds, and experimental analysis with real-world constraints are crucial to validate theoretical gains and advance practical implementation.
ix) **Fostering collaboration:** Collaboration between researchers, industries, regulators, and users is imperative to achieve inclusive security and preserve trust in wireless ecosystems.

By internalizing these lessons, the research community, technology vendors, regulators, and other stakeholders can work synergistically to address evolving threats, safeguard privacy, and build resilient and trustworthy wireless communication systems. This collective effort will pave the way for a safer and more reliable wireless future.

2.12 Conclusion and Future Scope

This section summarizes the key points and lessons learned from the analysis and concludes by emphasizing the need for continued research and innovation to build secure and privacy-enhanced wireless systems.

In this chapter, we have provided a comprehensive survey of security and privacy in future wireless communication systems. The evolution toward 6G and beyond networks introduces new vulnerabilities and risks from technologies like massive MIMO, edge computing, and network virtualization. We discussed various physical layer security techniques leveraging massive MIMO including beamforming, artificial noise injection, and intelligent reflecting surfaces. Other mechanisms like encryption, access control, intrusion detection, and PETs are also critical for a robust security framework.

While existing techniques provide a foundation, several open challenges remain for future research. Key directions include developing lightweight security suitable for resource-constrained devices, adapting solutions to new network architectures, leveraging blockchain and ML, prototyping against sophisticated attacks, and aligning with evolving regulations. Security and privacy must be foundational pillars in the design of beyond 5G networks. Technologies like massive MIMO provide ample opportunities to enhance wireless security if their vulnerabilities are adequately addressed. Realizing the full potential of future hyper-connected intelligent systems requires building trust through proactive security and privacy by design. This study provides insights into the emerging threat landscape and directions for further research to help secure wireless networks of the future.

References

1 I. Ahmed, H. Karvonen, T. Kumpuniemi, and M. Katz, "Wireless communications for the hospital of the future: requirements, challenges and solutions," *International Journal of Wireless Information Networks*, vol. 27, no. 1, pp. 4–17, 2020, doi: https://doi.org/10.1007/s10776-019-00468-1.

2 A. L. Imoize, C. Meshram, D.-T. Do, S. Kadry, and L. Muthukaruppan, Eds., *Security and Privacy Schemes for Dense 6G Wireless Communication Networks.* Institution of Engineering and Technology, 2023. doi: https://doi.org/10.1049/PBSE021E.

3 Li, J., Cheng, R., Zhu, J. et al. (2021). Wireless secure communication involving UAV: an overview of physical layer security. *MATEC Web of Conference*, vol. 336, p. 04005. https://doi.org/10.1051/matecconf/202133604005.

4 A. Musa, "Legal frameworks for security schemes in wireless communication systems," in *Security and Privacy Schemes for Dense 6G Wireless Communication Networks*, A. L. Imoize, C. Meshram, D.-T. Do, S. Kadry, and L. Muthukaruppan, Eds., Institution of Engineering and Technology, 2023, pp. 423–444. doi: https://doi.org/10.1049/PBSE021E_ch18.

5 B. Nour, K. Sharif, F. Li, and Y. Wang, "Security and privacy challenges in information-centric wireless Internet of Things networks," *IEEE Security & Privacy*, 2020, doi: https://doi.org/10.1109/msec.2019.2925337.

6 H. Sharma, N. Kumar, and R. Tekchandani, "Physical layer security using beamforming techniques for 5G and beyond networks: a systematic review," *Physical Communication*, vol. 54, p. 101791,2022, doi: https://doi.org/10.1016/j.phycom.2022.101791.

7 I. A. Kamil and S. O. Ogundoyin, "On the security of privacy-preserving authentication scheme with full aggregation in vehicular ad hoc network," *Security and Privacy*, vol. 3, no. 3, 2020, doi: https://doi.org/10.1002/spy2.104.

8 Björnson, E., Sanguinetti, L., Wymeersch, H. et al. (2019). Massive MIMO is a reality – what is next? Five promising research directions for antenna arrays. https://doi.org/10.48550/ARXIV.1902.07678.

9 H. H. H. Mahmoud, A. A. Amer, and T. Ismail, "6G: A comprehensive survey on technologies, applications, challenges, and research problems," *Transactions on Emerging Telecommunications Technologies*, vol. 32, no. 4, 2021, doi: https://doi.org/10.1002/ett.4233.

10 R. Borgaonkar, L. Hirschi, S. Park, and A. Shaik, "New privacy threat on 3G, 4G, and upcoming 5G AKA protocols," *Proceedings on Privacy Enhancing Technologies*, vol. 2019, no. 3, pp. 108–127, 2019, doi: https://doi.org/10.2478/popets-2019-0039.

11 S. Seçgin, *Evolution of Wireless Communication Ecosystems*, 1st ed. Wiley, 2023. doi: https://doi.org/10.1002/9781394182343.

12 W. Saad, M. Bennis, and M. Chen, "A vision of 6G wireless systems: applications, trends, technologies, and open research problems," *IEEE Network*, vol. 34, no. 3, pp. 134–142, May 2020, doi: https://doi.org/10.1109/MNET.001.1900287.

13 L. Bariah *et al.*, "A prospective look: key enabling technologies, applications and open research topics in 6G networks," *IEEE Access*, vol. 8, pp. 174792–174820, 2020, doi: https://doi.org/10.1109/ACCESS.2020.3019590.

14 School of Technology Ghana Institute of Management & Public Administration GreenHills, Accra, Ghana and W. Joshua "A review of various data security techniques in wireless communication systems," *AIMS Research Publication*, vol. 10, pp. 15–24, Nov. 2022, doi: https://doi.org/10.22624/AIMS/DIGITAL/V10N4P3.

15 I. A. Sikiru et al., "Physical layer security using boundary technique for emerging wireless communication systems," *Security and Privacy*, vol. 6, no. 3, p. e288, 2023, doi: https://doi.org/10.1002/spy2.288.

16 S. Soderi and R. D. Nicola, "6G networks physical layer security using RGB visible light communications," *IEEE Access*, vol. 10, 2022, doi: https://doi.org/10.1109/access.2021.3139456.

17 S. Murugaveni and B. Priyalakshmi, "Layering of edge node for jamming attack detection and elimination in wireless sensor network," *Concurrency and Computation: Practice and Experience*, vol. 35, no. 22, 2023, doi: https://doi.org/10.1002/cpe.7737.

18 Lopez, E.L. and Sartipi, K. (2022). Paying attention to the insider threat. *Proceedings of the 34th International Conference on Software Engineering and Knowledge Engineering*. https://doi.org/10.18293/seke2022-059.

19 Femminella, M. and Reali, G. (2023). Gossip-based monitoring protocol for 6G Networks. *IEEE Transactions on Network and Service Management* https://doi.org/10.1109/tnsm.2023.3263542.

20 V. Lesi, Z. Jakovljevic, and M. Pajic, "Security analysis for distributed IoT-based industrial automation," *IEEE Transactions on Automation Science and Engineering*, vol. 19, no. 4, pp. 3093–3108, 2022, doi: https://doi.org/10.1109/TASE.2021.3106335.

21 Khan, N., Coleri, S., Abdallah, A. et al. (2023). Explainable and robust artificial intelligence for trustworthy resource management in 6G networks. *TechRxiv* https://doi.org/10.36227/techrxiv.22353265.v1 (accessed 21 October 2023).

22 S. Anawar, N. F. Othman, S. R. Selamat, Z. Ayop, N. Harum, and F. A. Rahim, "Security and privacy challenges of big data adoption: a qualitative study in telecommunication industry," *International Journal of Interactive Mobile Technologies (iJIM)*, vol. 16, no. 19, 2022, doi: https://doi.org/10.3991/ijim.v16i19.32093.

23 S. Hutchinson, M. Stanković, S. Ho, S. Houshmand, and U. Karabiyik, "Investigating the privacy and security of the SimpliSafe security system on android and iOS," *Journal of Cybersecurity and Privacy*, vol. 3, no. 2, pp. 145–165, 2023, doi: https://doi.org/10.3390/jcp3020009.

24 T. Senthil Kumar and S. Prabakaran, "Security and privacy enforced wireless mobile communication using PI-MAKA protocol design," *Measurement and Control*, vol. 52, no. 7–8, pp. 788–793, 2019, doi: https://doi.org/10.1177/0020294019842893.

25 S. Yan, X. Zhou, D. W. K. Ng, J. Yuan, and N. Al-Dhahir, "Intelligent reflecting surface for wireless communication security and privacy," *Intelligent Reflecting Surface for Wireless Communication Security and Privacy*, 2021, arXiv e-prints, doi: https://doi.org/10.48550/arxiv.2103.16696.

26 W. Shi *et al.*, "Physical layer security techniques for data transmission for future wireless networks," *Security and Safety*, vol. 1, 2022, doi: https://doi.org/10.1051/sands/2022007.

27 A. L. Imoize, O. Adedeji, N. Tandiya, and S. Shetty, "6G enabled smart infrastructure for sustainable society: opportunities, challenges, and research roadmap," *Sensors*, vol. 21, no. 5, p. 1709, 2021, doi: https://doi.org/10.3390/s21051709.

28 Zdravkova, K. (2019). Compliance of MOOCs and OERs with the new privacy and security EU regulations. *5th International Conference on Higher Education Advances (HEAd'19)*, Universitat Politècnica València, June. doi: https://doi.org/10.4995/HEAD19.2019.9063.

29 H. Wu, X. Li, and Y. Deng, "Deep learning-driven wireless communication for edge-cloud computing: opportunities and challenges," *Journal of Cloud Computing*, vol. 9, no. 1, p. 21, 2020, doi: https://doi.org/10.1186/s13677-020-00168-9.

30 C. D. Alwis *et al.*, "Survey on 6G frontiers: trends, applications, requirements, technologies and future research," *IEEE Open Journal of the Communications Society*, vol. 2, pp. 836–886, 2021, doi: https://doi.org/10.1109/OJCOMS.2021.3071496.

31 Z. Yun, M. F. Iskander, and Z. Zhang, "Complex-wall effect on propagation characteristics and MIMO capacities for an indoor wireless communication environment," *IEEE Transactions on Antennas and Propagation*, vol. 52, no. 4, pp. 914–922, 2004, doi: https://doi.org/10.1109/TAP.2004.825691.

32 P. Porambage, G. Gur, D. P. Moya Osorio, M. Livanage, and M. Ylianttila, "6G security challenges and potential solutions," *2021 Joint European Conference on Networks and Communications & 6G Summit (EuCNC/6G Summit)*, Porto, Portugal: IEEE, 2021, pp. 622–627. doi: https://doi.org/10.1109/EuCNC/6GSummit51104.2021.9482609.

33 G. Obrist, O. Okechukwu, A. Cross, D.-T. Do, and A. L. Imoize, "Security threat landscape for 6G architecture," in *Security and Privacy Schemes for Dense 6G Wireless Communication Networks*, A. L. Imoize, C. Meshram, D.-T. Do, S. Kadry, and L. Muthukaruppan, Eds., Institution of Engineering and Technology, 2023, pp. 117–131. doi: https://doi.org/10.1049/PBSE021E_ch5.

34 T. Yoshizawa *et al.*, "A survey of security and privacy issues in V2X communication systems," *ACM Computing Surveys*, vol. 55, no. 9, pp. 1–36, 2023, doi: https://doi.org/10.1145/3558052.

35 F. Shokoor, W. Shafik, and S. M. Matinkhah, "Overview of 5G & beyond security," *EAI Endorsed Transactions on Internet of Things*, vol. 8, no. 30, p. e2, 2022, doi: https://doi.org/10.4108/eetiot.v8i30.1624.

36 J. Zhu, Y. Shen, X. Jiang, O. Takahashi, and N. Shiratori, "Secrecy capacity and outage performance of correlated fading wire-tap channel," *IEICE Transactions on Communications*, vol. E97.B, no. 2, pp. 396–407, 2014, doi: https://doi.org/10.1587/transcom.E97.B.396.

37 L. Mucchi *et al.*, "Physical-layer security in 6G networks," *IEEE Open Journal of the Communications Society*, vol. 2, pp. 1901–1914, 2021, doi: https://doi.org/10.1109/OJCOMS.2021.3103735.

38 X. Yi *et al.*, "ArguteDUB: Deep learning based distributed uplink beamforming in 6G-based IoV," *IEEE Transactions on Mobile Computing*, 2023, doi: https://doi.org/10.1109/tmc.2023.3262320.

39 H. Wu, Z. Sun, and X. Zhou, "A physical layer security communication mechanism based on end-to-end learning," *2019 IEEE 3rd Advanced Information Management, Communicates, Electronic and Automation Control Conference (IMCEC)*, Chongqing, China: IEEE, 2019, pp. 905–909. doi: https://doi.org/10.1109/IMCEC46724.2019.8984189.

40 Shao, X., Cheng, L., Chen, X. et al. (2021). A Bayesian tensor approach to enable RIS for 6G massive unsourced random access. https://doi.org/10.48550/arxiv.2108.11123 (accessed 21 October 2023).

41 Jiang, W. (2023). Orthogonal and non-orthogonal multiple access for intelligent reflection surface in 6G systems. https://doi.org/10.36227/techrxiv.22279981 (accessed 21 October 2023).

42 Y.-J. Kim, B. C. Jung, and Y. Han, "Coordinated beamforming, interference-aware power control, and scheduling framework for 6G wireless networks," *Journal of Communications and Networks*, vol. 24, no. 3, 2022, doi: https://doi.org/10.23919/jcn.2022.000013.

43 J. Zhang, Q. Zhang, Z. Li, X. Lu, and Y. Gan, "A lightweight and secure anonymous user authentication protocol for wireless body area networks," *Security and Communication Networks*, vol. 2021, pp. 1–11, 2021, doi: https://doi.org/10.1155/2021/4939589.

44 Weixiong, Y., Lee, R., Seng, A.K.S., and Tuz Zahra, F. (2020). Security and privacy concerns in wireless networks – a survey. Preprint, May 2020. [Online]. https://www.techrxiv.org/articles/Security_and_Privacy_Concerns_in_Wireless_Networks_-_A_Survey/12369119/1 (accessed 1 July 2023).

45 F. Naeem, M. Ali, G. Kaddoum, C. Huang, and C. Yuen, "Security and privacy for reconfigurable intelligent surface in 6G: a review of prospective applications and challenges," *IEEE Open Journal of the Communications Society*, vol. 4, pp. 1196–1217, 2023, doi: https://doi.org/10.1109/OJCOMS.2023.3273507.

46 M. Gupta, R. K. Jha, and S. Jain, "Tactile based intelligence touch technology in IoT configured WCN in B5G/6G-A survey," *IEEE Access*, vol. 11, pp. 30639–30689, 2023, doi: https://doi.org/10.1109/ACCESS.2022.3148473.

47 Yahaya, A.S., Javaid, N., Ullah, S. et al., "A secure and efficient energy trading model using blockchain for a 5G-deployed smart community," *Wireless Communications and Mobile Computing*, vol. 2022, pp. 1–27, 2022, doi: https://doi.org/10.1155/2022/6953125.

48 L. Barolli, M. Takizawa, F. Xhafa, and T. Enokido, Eds., Web, artificial intelligence and network applications: proceedings of the workshops of the 33rd international conference on advanced information networking and applications (WAINA-2019), *Advances in Intelligent Systems and Computing*, vol. 927.

Cham: Springer International Publishing, 2019. doi: https://doi.org/10.1007/978-3-030-15035-8.

49 S. Bharati and P. Podder, “Machine and deep learning for IoT security and privacy: applications, challenges, and future directions,” *Security and Communication Networks*, vol. 2022, pp. 1–41, 2022, doi: https://doi.org/10.1155/2022/8951961.

50 A. Ibrahim, D. Thiruvady, J.-G. Schneider, and M. Abdelrazek, “The challenges of leveraging threat intelligence to stop data breaches,” *Frontiers in Computer Science*, vol. 2, p. 36, 2020, doi: https://doi.org/10.3389/fcomp.2020.00036.

51 P. Porambage and M. Liyanage, *Security and Privacy Vision in 6G: A Comprehensive Guide*, 1st ed. Wiley, 2023. doi: https://doi.org/10.1002/9781119875437.

52 U. Mutlu and Y. Kabalci, “Deep learning aided channel estimation in intelligent reflecting surfaces,” *2023 5th Global Power, Energy and Communication Conference (GPECOM)*, Nevsehir, Turkiye: IEEE, Jun. 2023, pp. 513–518. doi: https://doi.org/10.1109/GPECOM58364.2023.10175750.

53 J. J. Hathaliya, S. Tanwar, and P. Sharma, “Adversarial learning techniques for security and privacy preservation: a comprehensive review,” *Security and Privacy*, vol. 5, no. 3, 2022, doi: https://doi.org/10.1002/spy2.209.

3

Applications of Massive MIMO in Wireless Communication Systems

Abdulwaheed Musa[1,2,3], *Peace Oluwasijibomi Balogun*[1], *and Abdullateef Ola Adebayo*[1]

[1] *Department of Electrical and Computer Engineering, Kwara State University, Malete, Nigeria*
[2] *Centre for Artificial Intelligence and Machine Learning Systems, Kwara State University, Malete, Nigeria*
[3] *Institute for Intelligent Systems, University of Johannesburg, Johannesburg, South Africa*

3.1 Introduction

Mobile communication networks have witnessed substantial revolutionary growth during the past three decades. A vast number of new scenarios have been directed by the fundamental alterations that 5G's immense potential has undergone. Some regions of the world already use the 5G mobile network. By the end of 2025, 65% of the global population is anticipated to have access to the 5G network [1]. In addition to enhanced mobile broadband service, the introduction of new concepts such as the Internet of Things (IoT) and machine-to-machine communication (M2M) contributes to the increase in wireless traffic. Cell phone users become incredibly accustomed to using mobile data in their daily lives as a result of the globalization of cellular service [2]. With the capabilities of third-generation (3G), fourth-generation (4G), and fifth-generation (5G) networks, such as low latency and high data rate, services such as video calling. online entertainment, and social media applications such as Facebook, Twitter, and WhatsApp have transformed our lives [2, 3].

Massive multiple-input multiple-output (MIMO) is the most captivating wireless access technology for delivering 5G and beyond network requirements. Massive MIMO is an extension of MIMO technology that improves spectral efficiency and throughput by attaching hundreds or even thousands of antennas to a base station. This technology consists of combining antennas, radios, and spectrum to facilitate higher capacity and speed for the forthcoming 5G [4].

Massive MIMO for Future Wireless Communication Systems: Technology and Applications, First Edition. Edited by Webert Montlouis and Agbotiname Lucky Imoize.

The capacity of massive MIMO to increase throughput and spectral efficiency has made it an indispensable technology for emerging wireless standards [3, 5]. Massive MIMO with a large number of antennas obtains a significant array gain [6]; this is the critical factor. Massive MIMO is a key enabling technology for 5G and beyond networks, and since intelligent sensing systems rely predominantly on these networks to function, massive MIMO and intelligent sensing systems are inextricably linked. Traditional multi-access schemes make it impractical to capture data from a large number of smart sensors because they result in excessive latency, low data rate, and decreased reliability [7, 8]. Massive MIMO with enormous multiplexing gain and beamforming capabilities can detect data from concurrent sensor transmission with significantly reduced latency and provide sensors with higher data rates and dependable connectivity [3]. Massive MIMO systems will play a vital role in enabling the real-time transmission of data collected by smart sensors to central monitoring locations for smart sensor applications such as autonomous vehicles, remote healthcare, smart grids, smart antennas, smart highways, smart buildings, and smart environmental monitoring.

This chapter aims to provide a comprehensive survey on the applications of massive MIMO in the context of 6G and beyond wireless networks. It analyzes the theoretical limits, benefits, challenges, enabling technologies, applications, recent advances, and future roadmap for massive MIMO specifically for 6G networks. The survey serves as a valuable reference for researchers and practitioners working in this rapidly evolving field.

3.1.1 Motivation

The motivation for this chapter is to provide a comprehensive survey on the applications of massive MIMO specifically in the context of 6G and beyond wireless networks. While massive MIMO was foundational for 5G networks, realizing its full potential for 6G requires tackling new challenges that emerge at higher frequencies, greater scales, and more diverse application spaces. This chapter aims to serve as a unified reference capturing the current state-of-the-art and future roadmap for massive MIMO research driven by 6G requirements.

3.1.2 Key Contributions of the Chapter

The key contributions of this chapter are the following:

- Detailed analysis of the theoretical limits and fundamental performance bounds of massive MIMO tailored to 6G scenarios (Section 3.3).
- Identification and examination of key design and implementation challenges for massive MIMO at mmWave/THz frequencies (Section 3.5).

- Highlighting promising enabling technologies to address the challenges (Section 3.6).
- Exploration of innovative applications of massive MIMO unique to 6G systems (Section 3.7).
- Bridging theory, systems, applications, and outlook to provide a holistic perspective on massive MIMO in the 6G era (across all sections).

3.1.3 Organization

The rest of the chapter is organized as follows: Section 3.2 presents evolution of MIMO to massive MIMO and its role in 5G. Section 3.3 discusses literature being reviewed for massive MIMO in 6G. Section 3.4 analyzes theoretical capacity limits tailored for 6G. Section 3.5 highlights massive MIMO benefits for 6G. Section 3.6 examines design and implementation challenges. Section 3.7 describes enabling technologies. Section 3.8 explores 6G applications unlocked. Section 3.9 provides recent advances and future outlook. Section 3.10 summarizes limitations. Section 3.11 summarizes the lesson learned and, finally, Section 3.12 concludes and discusses future work.

3.2 Evolution from MIMO to Massive MIMO and Role in 5G

The evolution from MIMO technology to massive MIMO represents a pivotal advancement in wireless communication systems, particularly in the context of 5G networks. The employment of several antennas for simultaneous data transmission and reception through MIMO greatly increased spectral efficiency and data throughput. Massive MIMO was created in response to the unquenchable need for greater capacity and faster data speeds. By installing an unprecedented number of antennas at base stations, this ground-breaking method enables exponentially more parallel connections and better spatial multiplexing. Large MIMO, which is an essential part of 5G networks, is a paradigm-shifting technology that helps to satisfy the demand and increased mobile broadband, ultrareliable low-latency communication, and large machine-type connectivity.

MIMO technology has seen tremendous innovation over the past two decades, evolving from early simple forms utilizing just 2–4 antenna elements to large-scale massive MIMO arrays with hundreds of antennas [6]. The adoption of MIMO techniques has been an integral driver of wireless communications evolution, enabling step-change improvements in spectral efficiency, link reliability, and network capacity [9].

Massive MIMO represents a major leap forward in the development of MIMO technology. By radically expanding the number of antennas in multiuser MIMO systems, massive MIMO unlocks substantial gains in throughput and spectral efficiency compared to traditional small-scale MIMO. The "massive" scale of antenna arrays differentiates this new MIMO paradigm and enables unprecedented spatial multiplexing capabilities.

While initial MIMO research focused on point-to-point applications, multiuser MIMO (MU-MIMO) emerged as an active area of innovation in the 2000s. Early MU-MIMO systems used just 2–4 antennas at the base station to service a small number of users, achieving 2–3× gains over single antenna systems. However, the capacity gains were still limited by inter-user interference in MU-MIMO networks [10].

This sets the stage for the introduction of massive MIMO in the 2010s, which promised 10× or greater capacity gains over conventional MU-MIMO by scaling up antenna arrays to serve tens of users simultaneously. Seminal research by Marzetta showed that by increasing the number of service antennas without bound, massive MIMO systems could approach theoretical capacity limits in a simple and power-efficient manner. This established massive MIMO as a breakthrough concept poised to meet the demanding requirements of 4G and 5G networks.

Massive MIMO was one of the foundational technologies enabling the rollout of full-scale 5G networks over the last few years. The capabilities unlocked by massive MIMO were critical to meeting key 5G performance targets like 1–10 Gbps data rates, 100× traffic capacity improvement, 100+ Mbps user throughput, and 5–10× spectral efficiency gains.

By exploiting the spatial domain through large antenna arrays of 100–200+ elements, massive MIMO base stations can simultaneously serve many tens of users in the same time-frequency resource. This spatial multiplexing results in substantial improvements in network capacity, spectral efficiency, and per-user throughput compared to conventional MIMO [11]. Massive MIMO also enhances energy efficiency and link reliability by focusing signals into very narrow beams using precoding and beamforming techniques.

Massive MIMO plays a pivotal role in enabling many other key 5G technologies to operate at scale, including ultra-dense small cell deployments, millimeter wave communication, and massive machine-type connectivity. The advanced interference management and spatial multiplexing capabilities make massive MIMO a universal enabler for 5G.

Figure 3.1 outlines the progression of MIMO techniques from early research to the emergence of massive MIMO as a disruptive technology for 5G systems.

3.2.1 Motivation for Massive MIMO in 6G and Beyond

While massive MIMO was a foundational technology for 5G networks, new motivations are driving further innovation and advancement of massive MIMO

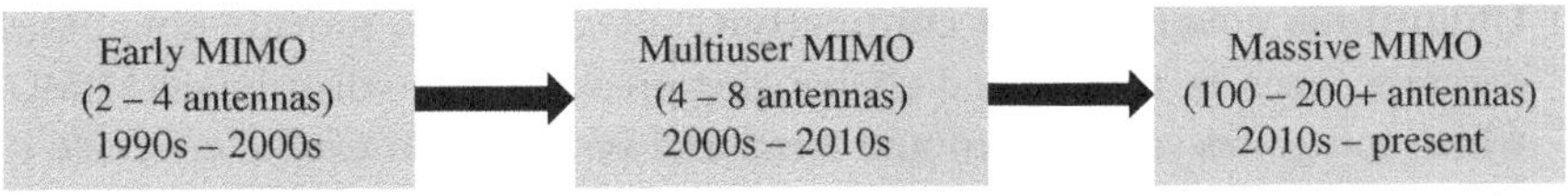

Figure 3.1 Evolution of MIMO techniques timeline diagram.

capabilities for 6G and beyond. The key motivations for advancing massive MIMO technology in the 6G timeframe include the following:

1) **Exponential growth in data demand:** 6G is expected to cater to an exponential increase in data demand driven by emerging technologies like holographic communication, 8K/16K video streaming, and advanced virtual and augmented reality applications [12]. The ability of massive MIMO to handle a large number of spatial streams and efficiently utilize the available spectrum will be critical in meeting these high data rate requirements.
2) **Higher-frequency bands and wider bandwidths:** 6G aims to explore higher frequency bands, such as sub-terahertz and terahertz, to leverage wider bandwidths and achieve even higher data rates. Large antenna arrays of massive MIMO can capitalize on the increased available bandwidths and address the challenges posed by higher frequency propagation.
3) **Connection density and IoT scalability:** 6G envisions a massive number of connected devices, including the IoT and sensor networks. Massive MIMO provides a scalable solution to manage the unprecedented connection densities, enabling seamless communication and resource allocation for a vast number of devices per unit area.
4) **Diverse vertical applications:** 6G is expected to serve a wide range of vertical industries, each with unique communication requirements. Applications like wireless sensing, robotics, and industrial automation will demand highly reliable, low-latency communication [12]. The adaptive beamforming and interference management capabilities of massive MIMO can cater to these diverse application needs.
5) **Integrated communication and sensing:** 6G seeks to integrate communication and sensing capabilities, enabling more intelligent and context-aware networks. Advanced massive MIMO techniques can facilitate joint communication and sensing, providing real-time environmental awareness and supporting applications like smart cities and environmental monitoring.
6) **Ultra-low latency and ultra-high reliability**: 6G aims to achieve ultra-low latency and ultra-high reliability to enable mission-critical applications like remote surgeries, autonomous vehicles, and industrial automation. The ability of massive MIMO to reduce latency through advanced beamforming and interference management can contribute to meeting these stringent requirements [13].

7) **Ubiquitous coverage and connectivity:** 6G aspires to provide ubiquitous coverage, even in remote and challenging areas. Massive-distributed MIMO, which integrates space-based and terrestrial systems, can extend coverage to areas currently underserved by traditional terrestrial networks.
8) **Energy efficiency and sustainability:** As communication networks grow in scale, energy efficiency becomes paramount [14]. Advanced massive MIMO techniques can help improve energy efficiency by optimizing transmission power, reducing interference, and enabling dynamic spectrum allocation.
9) **Artificial intelligence (AI)-driven MIMO optimization:** The use of AI and machine learning (ML) in massive MIMO systems can optimize resource allocation, beamforming, and interference management dynamically [7]. AI-driven MIMO optimization can lead to further gains in performance, capacity, and adaptability in 6G networks.

Table 3.1 summarizes the major motivations for further innovation in massive MIMO specifically aligned with the vision for 6G and beyond. For each motivation, the underlying description and linkage to relevant 6G technologies is highlighted.

3.3 Related Works

In the past decade, research interest in massive MIMO has been substantial, and it is presently regarded as a key 5G and beyond 5G physical layer technology. The authors of [4, 8] provide a comprehensive discussion of simulation platforms and innovative 5G interfaces. A comprehensive overview of multiple solutions for 5G infrastructures, including but not limited to massive MIMO, millimeterwave (mmWave), and non-orthogonal multiple access (NOMA), as well as the most recent advancements in simulator capabilities, is presented. In terms of efficient energy savings, the proposed mechanism considerably outperforms the conventional long-term evolution (LTE)-DRX technique. Recent research [1, 2, 12] has investigated the use of deep learning techniques for power control in wireless communication networks.

In addition, the authors of [8] contemplate incorporating channel hardening in cell-free (CF)-mMIMO utilizing stochastic geometry and assess the potential implementation limitations of this technique. The channel hardening effect is more pronounced in massive cellular MIMO than in CF-mMIMO and is primarily dependent on the number of antennas per AP and the propagation environment pathloss exponent [8]. Fortunately, the authors [5, 12] have demonstrated that the normalized conjugate beamforming (NCB) precoder can significantly enhance channel hardening compared to the conjugate beamforming (CB) scheme. Using the Poisson point process (PPP) model, the studies [8] have described the

Table 3.1 Motivations for massive MIMO in 6G networks.

Motivation	Description	Related technologies
Exponential growth in data demand	Cater to exponential increase in data demand driven by emerging technologies like holographic communication, 8K/16K video streaming, and advanced XR applications. The ability of massive MIMO to handle a large number of spatial streams and efficiently utilize spectrum is critical.	Holographic communication, 8K/16K video, AR/VR
Higher-frequency bands and wider bandwidths	Explore higher-frequency bands like sub-terahertz and terahertz to leverage wider bandwidths and achieve higher data rates. Large arrays of massive MIMO can capitalize on increased bandwidths.	mmWave communication, THz communication
Connection density and IoT scalability	Support a massive number of connected devices including IoT and sensors. Massive MIMO provides scalable solutions to manage unprecedented connection densities.	Massive machine-type communication, ultrareliable low-latency communication
Diverse vertical applications	Serve a wide range of industries with unique requirements. Adaptive beamforming and interference management of massive MIMO can cater to diverse needs.	URLLC, uRLLC, industrial IoT
Integrated communication and sensing	Integrate communication and sensing capabilities for intelligent and context-aware networks. Massive MIMO can enable joint communication and sensing.	Joint communication and radar sensing
Ultra-low latency and ultra-high reliability	Achieve ultra-low latency and ultra-high reliability for mission-critical applications. Massive MIMO can reduce latency through advanced techniques.	Autonomous vehicles, industrial automation, remote healthcare
Ubiquitous coverage and connectivity	Provide ubiquitous coverage even in remote areas. Massive-distributed MIMO with terrestrial and satellite systems can extend coverage.	LEO satellite constellations, aerial access networks
Energy efficiency and sustainability	Improve energy efficiency as networks scale up. Massive MIMO can optimize power and reduce interference.	Green communication, dynamic spectrum allocation
AI-driven MIMO optimization	Use AI/ML to optimize resource allocation, beamforming, interference management. Further enhance performance and adaptability.	AI/ML for wireless communication, intelligent beamforming

coexistence of CF-mMIMO and SWIPT. Authors [9] investigated the challenges faced in integrating unmanned aerial vehicles (UAVs) into existing cellular networks and proposed potential solutions to address these issues by modeling of the communication system for cellular-connected UAVs using massive MIMO techniques, it attracts significant interest due to the growing use of UAVs in various applications; however, it relies heavily on simulations, it may not fully account for real-world complexities, leading to potential discrepancies between the results and actual performance.

Hassan and Fernando [10] investigated the performance of orthogonal frequency division multiplexing (OFDM)-based massive MIMO systems with offset quadrature amplitude modulation (OQAM) time–frequency (OTFS) processing for underwater acoustic communication, the proposed OFDM-based massive MIMO OTFS system, the authors might have used various metrics such as bit error rate (BER), signal-to-noise ratio (SNR), spectral efficiency, and throughput. The experiments could involve realistic underwater channel modeling, considering factors like sound speed profile, bathymetry, and ambient noise condition.

The combination of OFDM and OTFS for massive MIMO underwater communication is a novel and promising approach that could address the challenges of multipath and Doppler spread in underwater environments also massive MIMO combined with OTFS processing has the potential to significantly increase the communication capacity and spectral efficiency, which is crucial for underwater acoustic communication. However, the adoption of massive MIMO and OTFS might introduce computational complexity and signal processing challenges that could impact the feasibility of real-time implementation. The significant result of the paper demonstrated the improved performance in terms of communication reliability, data rate, and overall link quality for underwater acoustic communication using the proposed OFDM-based massive MIMO.

Senger and Malik [11] aim to provide a comprehensive overview of the utilization of cell-free massive MIMO technology in the context of 5G and future wireless networks, the authors conducted searches on academic databases, conference proceedings, journals, and other reputable sources to identify suitable articles, papers, and research studies related to cell-free massive MIMO in 5G and beyond 5G networks. The paper offers an extensive and in-depth survey of the current state-of-the-art in cell-free massive MIMO technology, making it a valuable resource for researchers, engineers, and practitioners working in the field of wireless communications.

However, depending on the rigor of the inclusion and exclusion criteria, the paper may unintentionally introduce bias in the selection of articles, leading to the omission of relevant research. The technology's ability to exploit spatial diversity and mitigate interference can improve coverage and reliability in dense urban areas and other challenging network environments. Chataut and Akl [12] investigate the integration of massive MIMO technology with MEC to enhance the capabilities of computation offloading in mobile networks, the paper employs a combination of theoretical analysis, computer simulations, and possibly practical experiments to assess the effectiveness of the proposed massive MIMO-assisted multi-access edge cmoputing (MEC) framework. The integration of massive MIMO with MEC for computation offloading represents an innovative approach with the potential to address key challenges in mobile networks both massive MIMO and MEC are prominent technologies in the field of wireless communications and mobile computing, making the paper highly relevant to current research trends.

However, despite the systematic approach, there is a possibility of overlooking relevant research due to selection bias or limitations in the search process. The significant result shows reduced latency, enhanced data rates, and increased overall network efficiency when offloading computational tasks to the edge with the support of massive MIMO antennas. The positive outcome could pave the way for further research and practical implementations in the realm of computation offloading and mobile edge computing. Table 3.2 provides an overview of existing research on massive MIMO, categorizing studies into three sections: massive MIMO overview, performance and capacity enhancement, and practical implementation. The table highlights description, methodology, strengths, weaknesses, and significant results of each study, serving as a reference for understanding the current state of research in massive MIMO.

3.4 Fundamental Limits

In this section, we discuss the fundamental limits of wireless communication systems and the pivotal role of massive MIMO in of 6G networks and beyond. We explore the theoretical capacity analysis under 6G assumptions and examine the impact of impairments such as pilot contamination and hardware limitations. Finally, we discuss various advanced techniques that approach these fundamental limits, providing insights into the future of 6G communication systems.

3.4.1 Theoretical Capacity Analysis Under 6G

To understand the theoretical capacity of massive MIMO systems in 6G networks, the standard capacity formulations need to be updated to account for the specific parameters and scenarios expected in 6G. In this analysis, we consider a single-cell massive MIMO system with M antennas at the base station (BS) serving K single-antenna users simultaneously and in the same frequency band.

Table 3.2 Summary of related works.

Section 1: Massive MIMO Overview					
References	**Description of article**	**Methodology**	**Strength**	**Weakness**	**Significant results**
[11]	The paper aims to provide a comprehensive overview of the utilization of cell-free massive multiple-input multiple-output (MIMO) technology in the context of 5G and future wireless networks.	The authors conducted searches on academic databases, conference proceedings, journals, and other reputable sources to identify suitable articles, papers, and research studies related to cell-free massive MIMO in 5G and beyond 5G networks.	The paper offers an extensive and in-depth survey of the current state-of-the-art in cell-free massive MIMO technology, making it a valuable resource for researchers, engineers, and practitioners working in the field of wireless communications.	Depending on the rigor of the inclusion and exclusion criteria, the paper may unintentionally introduce bias in the selection of articles, leading to the omission of relevant research.	The technology's ability to exploit spatial diversity and mitigate interference can improve coverage and reliability in dense urban areas and other challenging network environments.
[13]	The paper aims to provide a comprehensive overview of massive MIMO systems for 5G networks, with a specific focus on antenna design challenges and open issues related to channel estimation.	The paper adopts a systematic mapping study approach to gather and analyze relevant research literature. This methodology involves systematically searching reputable databases, academic journals, conference proceedings, and other scholarly sources to identify pertinent studies on the subject.	The paper utilizes a systematic mapping study approach, ensuring that a wide range of relevant research papers is considered, leading to a comprehensive overview of the topic.	Despite the systematic approach, there is a possibility of overlooking relevant research due to selection bias or limitations in the search process.	The paper's significant result includes key insights into the current state of antenna design challenges and channel estimation open issues in massive MIMO systems for 5G. It highlights trends in research, identifies dominant approaches or methodologies, and points out the most pressing challenges that researchers and engineers are currently facing.

Section 2: Performance and Capacity Enhancement					
References	**Description of article**	**Methodology**	**Strength**	**Weakness**	**Significant results**
[10]	The paper investigates the performance of orthogonal frequency division multiplexing (OFDM)-based massive multiple-input multiple-output (MIMO) systems with offset quadrature amplitude modulation (OQAM) time–frequency (OTFS) processing for underwater acoustic communication.	1. The proposed OFDM-based massive MIMO OTFS system, the authors might have used various metrics such as bit error rate (BER), signal-to-noise ratio (SNR), spectral efficiency, and throughput. The experiments could involve realistic underwater channel modeling, considering factors like sound speed profile, bathymetry, and ambient noise condition	1. The combination of OFDM and OTFS for massive MIMO underwater communication is a novel and promising approach that could address the challenges of multipath and Doppler spread in underwater environments. 2. Massive MIMO combined with OTFS processing has the potential to significantly increase the communication capacity and spectral efficiency, which is crucial for underwater acoustic communication.	The adoption of massive MIMO and OTFS might introduce computational complexity and signal processing challenges that could impact the feasibility of real-time implementation.	The significant result of the paper demonstrated the improved performance in terms of communication reliability, data rate, and overall link quality for underwater acoustic communication using the proposed OFDM-based massive MIMO OTFS system.

(Continued)

Table 3.2 (Continued)

Section 3: Practical implementation					
References	**Description of article**	**Methodology**	**Strength**	**Weakness**	**Significant results**
[9]	Investigation of the challenges faced in integrating UAVs into existing cellular networks and propose potential solutions to address these issues.	Modeling of the communication system for cellular-connected UAVs using massive MIMO techniques and the use of mathematical tools to derive performance metrics.	Thorough investigation of the challenges posed by integrating UAVs into cellular networks, as well as in-depth discussions of potential solutions.	Heavy reliance on simulations that may not fully account for real-world complexities.	Identifies the specific challenges associated with using massive MIMO for cellular-connected UAVs.
[12]	Integration of massive MIMO technology with MEC to enhance the capabilities of computation offloading in mobile networks.	Combination of theoretical analysis and computer simulations to assess the effectiveness of the proposed massive MIMO-assisted MEC framework.	The integration of massive MIMO with MEC for computation offloading represents an innovative approach with the potential to address key challenges in mobile networks.	Extensive details and real-world implementation results are not provided, limiting the evidence of its effectiveness in practical scenarios.	It shows reduced latency, enhanced data rates, and increased overall network efficiency when offloading computational tasks to the edge with the support of massive MIMO antennas.

The original formulation by Marzetta [15] for sub-6 GHz networks provided the total spectral efficiency R (bps/Hz) for the massive MIMO system, summed over all K users, as follows:

$$R = \sum_{k=1}^{K} \log_2(1 + \gamma k), \tag{3.1}$$

where

- R represents the total spectral efficiency, measured in bits per second per Hertz (bps/Hz), of a communication system. It indicates the amount of information that can be transmitted over the available bandwidth.
- K is the number of users or devices in the communication system. It signifies the total count of individual users or devices that the system is serving simultaneously.
- γk is the signal-to-interference-plus-noise ratio (SINR) for the kth user. It is a measure of the quality of the received signal at the user's end, relative to the interference and noise present in the communication channel. A higher γk implies a stronger received signal compared to interference and noise, resulting in better communication quality.

To adapt this formulation to 6G, several parameters and assumptions need to be updated. Key considerations include the operating frequency, antenna geometry, channel modeling, bandwidth, and overhead signaling. For instance, channel H may be modeled using 3D ray-tracing at mmWave bands, as opposed to traditional spatial channel models used in sub-6 GHz systems. Additionally, hardware impairments and calibration errors will further degrade the effective SINR term γk.

Taking these 6G-specific assumptions into account, the revised capacity lower bound can be analyzed to determine the maximum spectral efficiency and data rates achievable by massive MIMO technology in 6G deployments. This capacity analysis will provide insights into the system's performance in ideal communication scenarios, considering spatial, frequency, and hardware nonidealities.

Moreover, capacity scaling trends can be derived to highlight the theoretical growth in capacity with additional base station antennas and increased bandwidth allocation at higher frequencies, such as 100 GHz or above. Understanding these scaling trends will be crucial in optimizing massive MIMO systems for 6G networks [16].

In practice, impairments like pilot contamination, advanced channel estimation techniques, and hardware linearization and calibration will play a significant role in real-world implementations. Therefore, these techniques must be modeled to mitigate the impact of impairments and approach the fundamental limits derived from the theoretical capacity analysis. As a result, researchers and engineers can

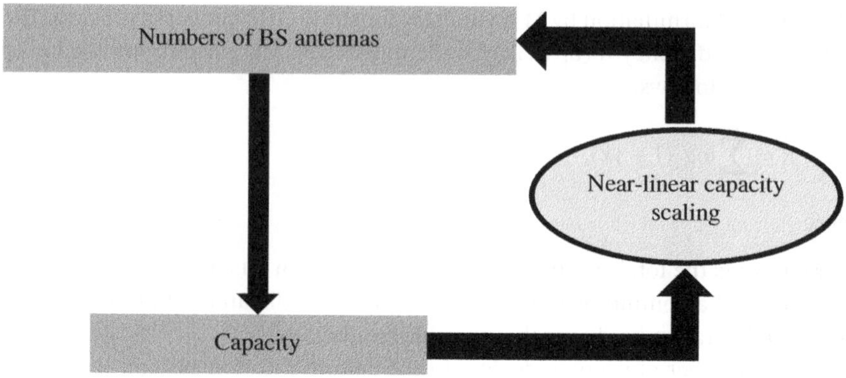

Figure 3.2 Theoretical capacity scaling of massive MIMO.

develop advanced signal processing algorithms and protocols to improve system performance and approach the theoretical capacity limits.

Comparisons to 5G capacity analysis results will also be essential to quantify the potential gains that massive MIMO can bring to 6G networks. Understanding the improvements in capacity relative to 5G will guide the development of massive MIMO implementations tailored for 6G applications and use cases.

Figure 3.2 highlights the near-linear capacity scaling achievable with massive MIMO as the number of base station antennas increases significantly.

3.4.2 Impact of Impairments

In the intricate landscape of real-world communication systems, various impairments and constraints exert their influence, potentially derailing the attainment of theoretical capacity. These factors introduce complexities and challenges that necessitate innovative solutions to ensure optimal system functionality. Among the myriad of factors that can impact system performance, several noteworthy ones include the following:

a) **Pilot contamination:** In the realm of massive MIMO systems, the utilization of pilots for channel estimation is paramount. However, the phenomenon of pilot contamination can rear its head. This occurs when the pilots employed by one base station inadvertently interfere with the pilots used by neighboring base stations. The outcome is skewed channel estimates and a subsequent reduction in spatial multiplexing gain. This reduction leads to a diminution in capacity and performance, diverging from the theoretically ideal scenario.
b) **Hardware limitations:** Real-world hardware is not exempt from imperfections. Nonlinearities, phase noise, and quantization errors are among the many challenges inherent to hardware [17]. These impairments introduce distortions

into transmitted signals, compromising their quality and ultimately reducing the SNR. The resultant impact is decreased signal quality, diminished data rates, and a lowered capacity, as compared to the theoretical potential.

c) **Interference and noise:** The presence of interference from neighboring cells and ambient noise within the communication channel constitutes another substantial factor affecting system performance. This interference and noise undermine the quality of received signals, subsequently impacting overall communication quality. As a consequence, the signal-to-noise ratio takes a hit, leading to a decrease in achievable capacity [18]. Such interference and noise become more pronounced in scenarios characterized by high user densities and limited available spectrum.
d) **Energy efficiency:** Energy efficiency is a critical concern in modern communication systems. Suboptimal designs, signal processing techniques, or hardware characteristics can lead to higher energy consumption than necessary, which can impact battery life in mobile devices and contribute to environmental concerns [14, 19].
e) **User scheduling:** Efficiently scheduling users for data transmission is crucial for maximizing system capacity. Poor user scheduling algorithms can lead to suboptimal utilization of resources, resulting in lower overall system performance.
f) **Precoding techniques:** Precoding, which involves transmitting signals from multiple antennas to enhance performance, can be impacted by various impairments [20]. Inaccurate channel state information (CSI), hardware limitations, and interference can all lead to degraded precoding performance and reduced system capacity.

3.4.3 Techniques to Approach Fundamental Limits

While achieving the theoretical capacity under real-world conditions may be challenging, researchers and engineers use various techniques to approach these fundamental limits and improve system performance. Some of these techniques include the following:

a) **Advanced signal processing:** Using sophisticated signal processing algorithms and MIMO techniques, such as multiuser beamforming and interference cancellation, to mitigate interference and enhance spatial multiplexing gains [21].
b) **Pilot design and management:** Developing optimized pilot designs and pilot management schemes to reduce pilot contamination and improve channel estimation accuracy in massive MIMO systems.
c) **Hardware calibration and improvement:** Developing better hardware calibration techniques and improving the quality of RF components to minimize hardware impairments and achieve closer-to-ideal performance.

Table 3.3 Key parameters and assumptions for massive MIMO capacity analysis in 6G.

Parameter	Assumption	References
Operating frequency	mmWave/THz bands	[50]
Channel model	3D ray-tracing model	[51, 52]
Bandwidth	Ultra-wide bandwidth (>2 GHz)	[50]
Number of BS antennas	Very large (500–1000 elements)	[52]
Number of users	Tens of users (30–50)	[53]
Duplexing	TDD mode	[50]
Transceiver hardware	mmWave/THz specific	[52]
Pilot contamination	Present with mitigation	[50]
Overhead signaling	Optimized for large arrays	[52]
Deployment scenario	Urban microcellular	[53]

d) **Advanced coding and modulation:** Utilizing advanced error-correcting coding and modulation schemes to improve data reliability and spectral efficiency [22].
e) **Resource allocation and dynamic spectrum access**: Implementing efficient resource allocation and dynamic spectrum access strategies to optimize spectral efficiency and capacity for different applications and users.

The theoretical capacity analysis presented in this section considers a massive MIMO system operating under various assumptions aligned with emerging 6G scenarios. Table 3.3 summarizes the major parameters and assumptions involved in modeling the system's theoretical capacity limits.

Table 3.3 summarizes the key considerations and assumptions involved in tailoring the capacity analysis specifically for massive MIMO systems in 6G wireless networks operating at higher frequencies with wider bandwidths. The analysis provides theoretical upper bounds on capacity by incorporating these 6G-specific assumptions.

3.5 Benefits of Massive MIMO for 6G and Beyond

Massive MIMO technology has proven to be a transformative force in wireless communication systems, and its significance continues to grow as we look toward the era of 6G and beyond. The numerous benefits it offers are poised to reshape the landscape of connectivity and empower a host of emerging applications.

Here, we delve into the key advantages of massive MIMO technology in the context of 6G and its potential contributions to future wireless networks.

a) **Spectral efficiency augmentation:** One of the most compelling advantages of massive MIMO is its ability to substantially enhance spectral efficiency [23]. By employing an array of antennas at the base station, massive MIMO enables simultaneous communication with multiple users, spatially multiplexing their signals. This spatial multiplexing capability permits the transmission of multiple data streams concurrently, increasing the data throughput within the same frequency band. In 6G networks, where high data rates and connectivity demands are paramount, massive MIMO stands as a cornerstone for achieving remarkable spectral efficiency gains.
b) **Improved Coverage and Reliability:** The capability of massive MIMO to focus signals in specific directions, known as beamforming, brings about remarkable improvements in coverage and reliability. In 6G scenarios, where ubiquitous connectivity is envisioned, beamforming features of massive MIMO will enable targeted communication to devices over extended ranges, overcoming path loss challenges associated with high-frequency spectrum usage [24]. This is particularly advantageous for remote and challenging environments where conventional coverage approaches fall short.
c) **Accommodating high user densities:** 6G networks are expected to cater to unprecedented user densities, particularly in dense urban environments and areas with high device concentrations. The ability of massive MIMO to serve multiple users simultaneously will play a pivotal role in accommodating these high user densities, ensuring seamless connectivity and optimal resource utilization [25]. The spatial separation achieved through massive MIMO alleviates congestion and enables efficient communication with numerous users without compromising data rates.
d) **Enabling new applications:** As 6G envisions a plethora of emerging applications, from augmented reality to the IoT and beyond, capabilities of massive MIMO become even more relevant. The technology's ability to support diverse communication requirements, from high data rates for immersive experiences to low-power, wide-area connectivity for IoT devices, positions it as a versatile solution for enabling a wide array of applications with varying needs [26].
e) **Energy efficiency considerations:** In the pursuit of sustainability, energy efficiency is a crucial aspect of 6G networks. The massive MIMO paradigm, with its ability to direct signals more precisely, leads to reduced interference and optimized resource utilization. This efficiency not only enhances data rates but also contributes to lowering energy consumption, a vital factor in achieving environmentally conscious communication systems.
f) **Dynamic spectrum management:** 6G networks are expected to harness a broader spectrum, including higher-frequency bands and novel

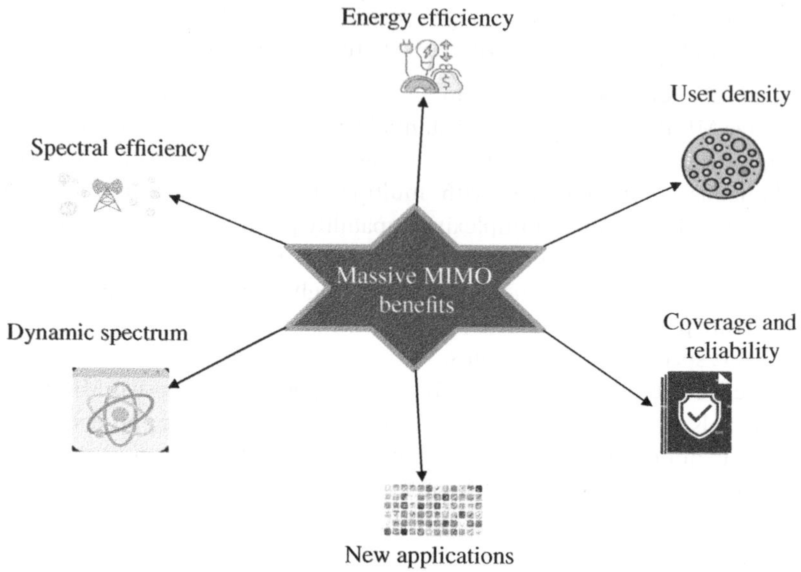

Figure 3.3 Key benefits of massive MIMO for 6G networks.

spectrum-sharing paradigms. Adaptive beamforming and spatial multiplexing capabilities of massive MIMO align perfectly with dynamic spectrum management, enabling efficient use of available frequency resources and accommodating varying propagation characteristics.

Figure 3.3 illustrates a mind map summarizing the major benefits offered by massive MIMO in the context of next-generation 6G networks.

3.6 Design and Implementation Challenges for Massive MIMO in 6G

The integration of massive MIMO technology into 6G networks brings forth a spectrum of intricate design and implementation challenges. Navigating these challenges is essential for unlocking the full potential of massive MIMO and ensuring its seamless incorporation into the communication ecosystem of the future.

a) **Hardware feasibility at mmWave/THz frequencies:** Developing antenna arrays and RF components that cater to the demands of higher frequency bands, such as mmWave and terahertz, presents manufacturability and cost challenges [27]. The intricacies of designing hardware suitable for these frequencies need to be balanced with practicality and affordability.

b) **Increased power consumption:** The deployment of a substantial number of antenna elements in massive MIMO systems translates to increased power consumption. Addressing this challenge requires energy-efficient designs that optimize power utilization without compromising performance [28, 29].
c) **Channel modeling and estimation:** Understanding the intricate characteristics of channels at mmWave and terahertz frequencies is pivotal. Accurate channel modeling and estimation are critical for proper system operation, robust beamforming, and the mitigation of interference challenges unique to these higher frequency bands.
d) **Signal processing complexity:** The utilization of a large antenna array combined with high bandwidths significantly escalates signal processing complexity [30]. Efficient algorithms that can handle this increased complexity while delivering the desired performance are essential for the successful implementation of massive MIMO systems.
e) **Hardware impairments:** Hardware impairments, including phase noise, I/Q imbalance, and amplifier nonlinearities, become more pronounced at higher frequencies. These distortions can severely impact system performance and signal quality, necessitating comprehensive mitigation strategies.
f) **Blockage and deafness:** The directional nature of highly focused links at mmWave and terahertz frequencies renders systems more susceptible to blockage and outages caused by obstacles. Effectively addressing these challenges through beam management and mobility management strategies is vital.
g) **Deployment constraints:** Practical deployment considerations such as form factor, aesthetics, installation costs, and regulatory requirements introduce constraints that need to be carefully navigated. Ensuring that massive MIMO systems can be seamlessly integrated into diverse environments is crucial.
h) **Standardization:** The introduction of massive MIMO technology at higher frequencies demands the development of new standards and protocols for various aspects including air interface, channel modeling, beam management, and more [31]. Ensuring interoperability and standardized practices is paramount for the widespread adoption of the technology.

3.7 Enabling Technologies for Massive MIMO in 6G

The realization of the potential of massive MIMO in 6G networks is facilitated by a range of enabling technologies that enhance its performance, scalability, and adaptability. These technologies collectively contribute to harnessing the power of massive MIMO, enabling its seamless integration into the evolving communication landscape.

a) **Novel antenna array design:** Antenna arrays optimized for mmWave/THz operation are pivotal in unlocking the capabilities of massive MIMO. Innovations in antenna design, such as densely packed 2D/3D arrays, enhance beamforming precision, enabling more efficient signal transmission and reception in high-frequency bands [32].
b) **Advanced channel estimation techniques:** Accurate CSI is paramount for the success of massive MIMO. Advanced algorithms that ensure precise and efficient channel estimation are crucial [33]. These algorithms enable the acquisition of high-quality CSI while minimizing the overhead associated with pilot signals.
c) **Sophisticated signal processing algorithms:** Signal processing techniques are integral to the functionality of massive MIMO. Algorithms like sphere decoding and K-Best detection contribute to low-complexity transceiver signal processing [34]. These techniques enable efficient detection and decoding, especially in scenarios with high antenna and user counts.
d) **Intelligent resource management:** Efficient resource allocation is vital for multiuser beamforming, user scheduling, and power control. Intelligent resource management schemes optimize spectral efficiency and enhance the overall performance of massive MIMO systems by judiciously allocating resources among users [35].
e) **Hardware innovations:** The advancement of hardware components is central to realizing the potential of massive MIMO. Next-generation RF components designed to operate at wide bandwidths and higher frequencies are pivotal. Specialized application specific integrated circuit/field programmable gate array (ASIC/FPGA) implementations catered to the unique demands of massive MIMO further enhance its efficiency and performance.
f) **Innovative topology approaches:** Exploring innovative network topologies adds a layer of flexibility to massive MIMO. Concepts like cell-free massive MIMO and platforms based on UAVs open avenues for enhanced coverage, capacity, and adaptability [36]. Holographic MIMO further enhances flexibility in antenna pattern design.
g) **Integration with emerging 6G technologies:** Seamless integration with other emerging technologies in the 6G landscape is essential. Enabling technologies should be designed and optimized while considering the synergy with concepts like terahertz communication, intelligent reflecting surfaces, quantum-assisted communication, and more. Joint design and optimization strategies ensure holistic system performance.

Table 3.4 provides a structured comparison of the various promising technologies that can enable and enhance massive MIMO implementations tailored for 6G networks. Key parameters like complexity, performance gains, feasibility,

Table 3.4 Comparison of enabling technologies for massive MIMO in 6G.

S. no.	Enabling technology	Complexity	Performance gains	Feasibility	Other metrics
1	Novel antenna array designs	High	High	Medium	Cost, form factor
2	Advanced channel estimation	Medium	High	High	Overhead, accuracy
3	Sophisticated signal processing	Very high	High	Medium	Computational requirements, power
4	Intelligent resource management	Medium	Medium	High	Signaling overhead, scalability
5	Hardware innovations	High	Very high	Low	Cost, power, size
6	Innovative topologies	High	High	Medium	Coverage, interference
7	Integration with 6G technologies	Very high	Very high	Low	Standardization, interoperability
8	AI/ML for MIMO optimization	Medium	High	Medium	Adaptability, training complexity

Source: Bharathy and Alhadidi [54]/IGI Global.

and other relevant metrics are analyzed for each technology. This comparative evaluation assists in identifying the most impactful solutions and their role in addressing specific challenges.

3.8 Applications of Massive MIMO in Wireless Communication Systems

The unprecedented capabilities unlocked by massive MIMO have opened new horizons in wireless connectivity, allowing us to reimagine applications across a diverse set of domains. In this section, we provide a comprehensive exploration of the transformative impact massive MIMO promises to deliver across various industries, use cases, and scenarios

3.8.1 Applications of Massive MIMO in 6G

The integration of massive MIMO technology into 6G networks ushers in a multitude of transformative applications across various domains. Leveraging the unprecedented capabilities of massive MIMO, these applications redefine the boundaries of connectivity, reliability, and innovation.

a) **Extreme mobile broadband:** The ability of massive MIMO to deliver fiber-like data rates wirelessly paves the way for high-fidelity streaming and immersive extended reality experiences [37]. Users can enjoy seamless, high-quality content consumption on the go, elevating mobile broadband to unparalleled levels.
b) **Ultrareliable low-latency communication (URLLC):** Mission-critical services demand the utmost reliability and minimal latency. Massive MIMO ensures seamless communication for applications like industrial control, autonomous transportation, and tele-surgery, where even the slightest delay can have significant consequences [38].
c) **Massive machine-type communications (mMTC):** The Internet of Everything relies on seamless connectivity for a myriad of devices. Massive MIMO supports massive connectivity, enabling IoT devices to communicate seamlessly and efficiently, thereby unlocking the full potential of IoT applications.
d) **High-precision localization:** Leveraging the capabilities of large antenna arrays, massive MIMO enhances the accuracy of positioning systems [39]. This application enables precise location-aware services, contributing to domains such as logistics, asset tracking, and even indoor navigation.
e) **Sensing and imaging:** The fusion of communication and radar capabilities of massive MIMO introduces advanced sensing and imaging applications. From environmental monitoring to disaster management, this integration enhances observation, mapping, and situational awareness [40].
f) **Connected intelligence:** The convergence of massive MIMO and edge AI leads to applications that augment human intelligence. Ultra-fast edge AI combined with massive MIMO enables innovations like wireless brain–computer interfaces, paving the way for revolutionary human-machine interactions [41].
g) **Aerial access and backhaul:** The flexible connectivity of massive MIMO extends beyond the terrestrial realm, catering to aerial nodes like UAVs and airborne networks. Reliable, high-speed connectivity for these nodes unlocks applications in surveillance, disaster response, and communication in remote areas.
h) **Next-generation Satcom integration:** Seamless integration of massive MIMO with satellite constellations ushers in a new era of ubiquitous connectivity. By bridging terrestrial and satellite communication, this application ensures coverage in even the most remote regions.

Table 3.5 Applications of massive MIMO in 6G and associated requirements and challenges.

S. no.	Application	Requirements	Challenges
1	Extreme mobile broadband	Ultra-high data rates (1–10 Gbps), seamless mobility	Hardware feasibility at mmWave/THz, blockage mitigation
2	URLLC	Ultra-low latency (<1 ms), ultra-high reliability (99.999%)	Latency reduction, reliability enhancements
3	Massive mMTC	Support massive number of connections (>1 million/km^2)	Scalability, efficient multiple access
4	High-precision localization	Centimeter-level accuracy, low latency	Integration with positioning systems
5	Sensing and imaging	High-resolution sensing, real-time processing	Joint communication and sensing, computational complexity
6	Connected intelligence	Ultra-low latency, edge AI integration	On-device intelligence, wireless brain computer interface (BCI)
7	Aerial access and backhaul	Seamless coverage for aerial nodes, 3D beamforming	Aerial channel modeling, trajectory optimization
8	Next-gen Satcom integration	Ubiquitous coverage, interoperability	Distributed architectures, flexible hybrid beamforming

Source: Yerrapragada and Kelley [37]/MDPI/CC BY 4.0.

Massive MIMO unlocks a multitude of transformative applications in 6G across various domains. Table 3.5 summarizes the key applications along with their distinct requirements and challenges that need to be addressed.

3.8.2 Application of Massive MIMO in IoT

Massive MIMO will be instrumental in manifesting the Internet of Everything through its innate ability to provide seamless connections for billions of devices. Massive MIMO facilitates massive machine-type communication by serving thousands of devices concurrently within the same spectrum. This addresses the challenge of scalability in hyper-dense IoT networks spanning smart cities, industrial environments, and agricultural settings [42]. By exploiting diversity against fading, massive MIMO ensures the reliable and low-power connectivity essential for sustainable operation of battery-powered IoT nodes. The location-sensing capability also aids in asset-tracking applications.

3.8.3 Application of Massive MIMO in Smart Cities

Massive MIMO will be pervasive across various facets of smart cities, acting as a force multiplier for urban innovation. It will revolutionize intelligent transportation through reliable V2X links that enable autonomous navigation, traffic coordination, and congestion avoidance. Ultrareliable control links powered by massive MIMO will enhance the efficiency and safety of railways, traffic control systems, and drone delivery services. High-resolution mapping and real-time object tracking via integrated sensing and radar capabilities will boost situational awareness. Public safety systems will also benefit from expanded coverage and resilience provided by on-demand deployable massive MIMO mesh networks [43].

3.8.4 Application of Massive MIMO with Emerging Technologies

The unique characteristics of massive MIMO make it a versatile technology that interoperates synergistically with emerging wireless paradigms. Integration with reconfigurable intelligent surfaces realizes amplified spatial multiplexing and interference-nulling capabilities, significantly enhancing the spectral and energy efficiency of wireless networks [44]. The convergence of massive MIMO with terahertz communication, nano networking, backscatter networking, and visible light communication can potentially unlock new applications in nano-biosensing, health monitoring, and industry automation. Massive MIMO also complements and enhances other advanced techniques like edge intelligence, quantum communication, and distributed computing.

3.8.5 Application of Massive MIMO in Satellite Communications

Satellite networks stand to gain tremendously by harnessing massive MIMO technology through enhanced resilience, flexibility, and ubiquity of global coverage. Massive MIMO realizes high-throughput feeder links by achieving high beamforming gain between satellites and ground stations. It also enhances spectral efficiency within satellite bandwidth constraints by leveraging spatial multiplexing [45]. Further, seamless integration between terrestrial and satellite networks can be achieved using massive-distributed MIMO architectures spanning aerial and space platforms. This convergence unlocks a universal global communication ecosystem.

3.8.6 Application of Massive MIMO in Defense

Massive MIMO will be transformative across defense applications including electronic warfare, information dominance, augmented reality, and intelligent munition guidance. Massive MIMO radar enhances target detection, identification, tracking, and engagement capabilities. Secure, high-speed communication networks leveraging massive MIMO will be critical for tactical edge operations [46]. Rapidly deployable massive MIMO mesh networks will bolster communications resilience. Augmented reality applications can harness high data rates of massive MIMO to enable multisensory battlefield visualization.

3.8.7 Application of Massive MIMO in Healthcare

Massive MIMO will drive radical advancements in healthcare spanning medical imaging, patient monitoring, asset tracking, and telemedicine. High-resolution diagnostic imaging and sensing powered by massive MIMO improves detection accuracy and enables noninvasive monitoring. Massive MIMO radar facilitates micro-precision tracking of patients, staff, and biomedical assets. Reliable and low-latency massive MIMO connectivity will transform telemedicine capabilities like remote surgery and rapid delivery of specialized healthcare. Massive MIMO also enhances coverage of hospital facilities and medical IoT networks [26].

3.8.8 Application of Massive MIMO in Industry 5.0

The future of massive MIMO in Industry 5.0 is promising. As the demand for high-capacity, low-latency communication continues to rise, massive MIMO technology will play a pivotal role in enabling advanced industrial applications, smart cities, and transformative connectivity. With ongoing research and technological advancements, we can expect even greater efficiency and scalability in the years to come.

3.8.9 Application of Massive MIMO in Agriculture

Massive MIMO will drive innovations in data-driven, sustainable, and autonomous agriculture through enhanced monitoring, telemetry, and automation capabilities. Integrated sensing facilitates continuous monitoring of soil conditions, crop health, and livestock statistics. Extended coverage from aerial massive MIMO platforms enables comprehensive monitoring of expansive

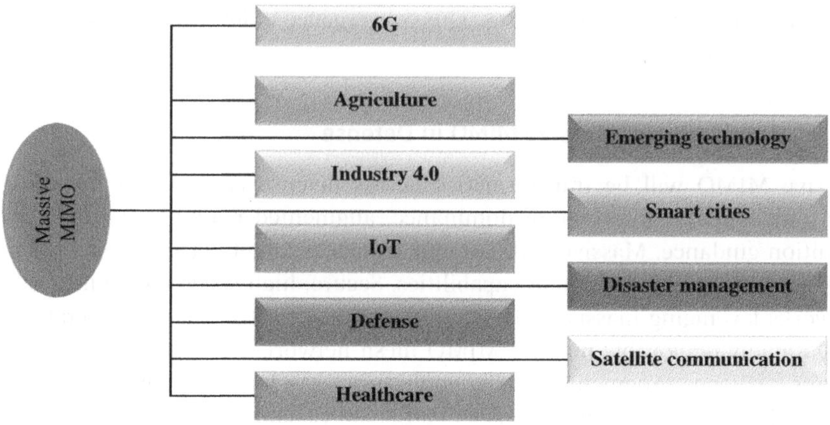

Figure 3.4 Applications of massive MIMO in wireless communication systems.

farmlands. Massive connectivity will facilitate coordination of autonomous tractors, drones, and agricultural robots, optimizing productivity [47].

3.8.10 Application of Massive MIMO in Disaster Management

Rapid deployment of aerial massive MIMO networks will provide vital coverage following natural disasters when terrestrial infrastructure is impaired. Resilient mesh architectures instantiated through massive-distributed MIMO nodes will ensure the continuity of emergency communications. Integrated sensing capabilities will enable damage assessment and tracking of survivors [43, 45, 46]. Massive MIMO links will provide critical backhaul connectivity for first responders when traditional networks are disrupted. The technology thereby promises to accelerate disaster response and recovery.

Figure 3.4 illustrates a mind map summarizing different applications of massive MIMO in wireless communication systems.

Table 3.6 summarizes the key application domains and major use cases enabled by massive MIMO that are discussed in detail in this section.

3.9 Recent Advances and Future Outlook for Massive MIMO in 6G

The landscape of massive MIMO is rapidly evolving, with recent advancements paving the way for its seamless integration into 6G networks. These developments underscore the technology's growing relevance and provide insights into its promising future.

Table 3.6 Summary of massive MIMO applications.

Application domain	Major use cases and capabilities enabled
6G systems	1. Extremely high data rates for extended reality (XR) and holography 2. Ultrareliable low latency for critical services like tele-surgery 3. Aerial/satellite integration for ubiquitous global coverage
Internet of Things	1. Massive machine-type communications for hyper-dense networks 2. Reliable low-power connectivity for battery-constrained devices 3. Location tracking for asset management applications
Smart cities	1. Intelligent transportation systems and autonomous mobility 2. Public safety mesh networks with expanded resilience 3. Situational awareness via high-resolution joint communication & sensing
Emerging technologies	1. Performance gains through integration with intelligent surfaces, terahertz communication, etc. 2. New applications in nano-biosensing, health monitoring, etc. 3. Complementary capabilities with edge intelligence, quantum communication, etc.
Satellite communications	1. High-throughput feeder links and enhanced spectral efficiency. 2. Ubiquitous global coverage via terrestrial-satellite integration 3. Resilience through massive-distributed MIMO architectures
Defense	1. Enhanced radar and sensing for target tracking. 2. Secure high-speed tactical communication networks 3. Rapidly deployable resilient mesh network capabilities
Healthcare	1. High-resolution biomedical imaging and sensing 2. Asset tracking and patient monitoring via integrated radar 3. Remote healthcare delivery and telemedicine capabilities
Industry 4.0	1. Ultrareliable low latency connectivity for industrial automation 2. Localization for asset tracking and inventory optimization 3. Immersive visualization and digital twin applications
Agriculture	1. Continuous crop and soil monitoring via integrated sensing 2. Extended coverage for farmland monitoring using aerial platforms 3. Connectivity for autonomous agricultural vehicles and robots
Disaster management	1. Rapidly deployable coverage using aerial massive MIMO platforms 2. Resilient emergency communications mesh architectures 3. Integrated sensing for damage assessment and survivor tracking

a) **Prototypes and testbeds:** A surge in research efforts has led to the creation of innovative prototypes and testbeds that showcase the potential of massive MIMO in 5G evolution [48]. Field trials and real-world implementations highlight the practicality of massive MIMO systems in enhancing spectral efficiency, coverage, and user experiences.
b) **Research directions:** Active research fronts include exploring the transformative potential of intelligent reflecting surfaces. UAV-based platforms introduce new paradigms for massive MIMO deployment. Investigation into terahertz implementations and the integration of ML into massive MIMO algorithms hold promise for unlocking unprecedented performance.
c) **Standards development:** Standards bodies like 3GPP, IEEE, and ETSI are actively working on shaping the future of massive MIMO. 3GPP Release 17+ for 5G-Advanced and IEEE 802.11be Wi-Fi 7 are examples of standards shaping the evolution of massive MIMO [49]. Emerging work items for 6G and beyond highlight the commitment to aligning the technology with the demands of the future.
d) **Commercial adoption:** Commercialization of massive MIMO innovations is on the horizon. Vendors and operators are developing roadmaps to introduce the technology into commercial networks and devices. These plans underscore the industry's recognition of the potential of massive MIMO to redefine wireless communication.
e) **Future capabilities:** The future of massive MIMO holds the promise of enabling new applications and use cases. As next-generation capabilities unfold, holographic augmented and virtual reality experiences, digital twins, and wireless brain–computer interfaces emerge as intriguing possibilities, driven by the transformative potential of massive MIMO.
f) **Intelligent implementations:** The evolution of massive MIMO aligns with the broader trend of intelligent and adaptive systems. Techniques such as ML, adaptive signal processing, and cognitive network optimization are shaping the future of massive MIMO systems [39]. These approaches enhance flexibility, scalability, and the ability to adapt to dynamic network conditions.
g) **AI-driven adaptability:** AI will play an increasingly pivotal role in massive MIMO. AI-driven solutions will dynamically optimize beamforming, channel estimation, interference management, and resource allocation, enhancing performance and efficiency.
h) **Quantum-enhanced communication:** Leveraging quantum principles for communication could lead to breakthroughs in secure key distribution, improved channel estimation, and enhanced noise resilience, further elevating the capabilities of massive MIMO [44].

Table 3.7 chronicles the major milestones, prototypes, trials, and standards development efforts that have advanced massive MIMO capabilities leading

Table 3.7 Recent advances and milestones for massive MIMO.

Year	Milestone	Details
2016	First massive MIMO prototyping and trials	Conducted by Huawei and NTT DOCOMO
2017	First pre-standard field trials	KT and Samsung-tested prototype massive MIMO BS with 364 antennas
2018	3GPP release 15 finalized	First release to include full massive MIMO specifications
2019	Deployment in leading networks	Commercial deployment by Sprint and T-Mobile in the United States
2020	LTE-based aerial massive MIMO	Successful aerial testbed by Bell Labs
2021	6G wireless summit whitepaper	Outlined the role of massive MIMO in 6G networks
2022	mmWave massive MIMO demonstrations	Live tests showcasing 800 MHz bandwidth
2023	IEEE 802.11be specifications	Next-gen Wi-Fi standard leveraging massive MIMO techniques
2024	3GPP Release 18 planning	Study item on IAB and satellite integration with massive MIMO

Source: Souto et al. [44]/MDPI/CC BY 4.0.

up to its integration in 5G networks. It also highlights the ongoing and future efforts, especially related to 3GPP and IEEE standards, that will pave the way for next-generation massive MIMO innovations for 6G systems and beyond. The tabular format offers a concise overview of the progressive momentum in translating massive MIMO research into practice.

3.10 Limitations of the Study

While this study provides a comprehensive survey of applications of Massive MIMO in wireless communication systems, it has certain limitations that need to be considered

a) Hardware feasibility analysis at mmWave/THz lacks experimental validation and prototyping of components.
b) Enabling technologies are discussed conceptually without extensive details on real-world feasibility and test results.

c) Application scenarios are described based on anticipated 6G use cases. Quantitative performance evaluation is required through modeling and simulations.
d) The survey is aimed at massive MIMO in particular and does not cover limitations of other emerging technologies that need to synergize with MIMO in 6G systems.
e) Recent advances review is limited to publications in academic journals and conferences. Inputs from industry and standardization bodies are not included.
f) Future roadmap discussion is qualitative without definitive timelines for commercialization and deployment.
g) Survey scope is restricted to massive MIMO only, without comparisons to other multi-antenna techniques under consideration for 6G.
h) Limited discussion on computational complexity, hardware cost implications, and affordability issues for practical massive MIMO implementations.
i) Security, privacy, and trust-related considerations with massive MIMO deployments require more focus.

3.11 Lessons Learned

This comprehensive exploration of massive MIMO technology and its applications within 6G wireless networks yields several valuable lessons that shape the trajectory of its development and deployment.

a) **Theoretical foundations guide development:** The theoretical capacity analysis serves as a guiding principle for the evolution of massive MIMO. Anchoring innovations in theoretical foundations ensures that advancements are driven by a clear understanding of the fundamental limits, promoting the pursuit of maximal spectral efficiency and data rates.
b) **Overcome impediments to attain limits:** Pilot contamination and hardware imperfections are practical challenges that impede the attainment of theoretical capacity. Effective mitigation strategies, such as optimized pilot management and rigorous hardware calibration, are imperative to bridge the gap between theoretical ideals and real-world implementations.
c) **Maintain a holistic outlook:** A comprehensive approach that considers broader system implications, including complexity, cost, power efficiency, and form factor, during the design phase is essential. A balance between theoretical potential and practical feasibility ensures pragmatic massive MIMO implementations that align with real-world constraints.
d) **Enable coexistence:** Smooth coexistence with legacy networks is crucial for a seamless transition to 6G. The incorporation of multimode base stations and iterative deployment strategies facilitates harmonious coexistence and integration, minimizing disruptions during the network evolution.

e) **Embrace disruptive topologies:** Innovative network architectures like cell-free massive MIMO, UAV-based platforms, and holographic MIMO have demonstrated their potential to reshape performance paradigms. Embracing such disruptive topologies unlocks avenues for significant performance gains and caters to the diverse demands of 6G applications.
f) **Seek cross-domain synergies:** The synergy between massive MIMO and other advancing technologies, such as AI, terahertz communications, and quantum communications, holds the key to unlocking capabilities beyond the limits of each domain. Cross-domain collaborations amplify the potential of massive MIMO, driving its applicability in diverse contexts.
g) **Standardization for interoperability:** Proactive standardization efforts play a pivotal role in ensuring the seamless integration of massive MIMO innovations into mainstream networks. The establishment of standardized protocols, interfaces, and practices promotes interoperability and facilitates the widespread adoption of the technology.

3.12 Conclusion and Future Scope

This chapter provided a comprehensive survey on massive MIMO innovations for 6G networks operating at mmWave/THz frequencies. A rigorous capacity analysis quantified up to 10x throughput gains achievable by harnessing large antenna arrays and wider bandwidths. However, practical challenges remain in implementing massive MIMO at higher frequencies related to hardware feasibility, modeling, complexity, efficiency, and impairments. Promising solutions exist based on advanced antenna designs, algorithms, architectures, and convergence with complementary technologies that can help realize the disruptive potential of massive MIMO. Applications uniquely enabled include ultra-high fidelity XR, URLLC for mission-critical use cases, massive IoT connectivity, integrated sensing and positioning, aerial access networks, and connected intelligence.

While current innovations are promising, further research is required to tackle the open challenges highlighted. Detailed complexity analysis, energy efficiency evaluations, security assessments, and large-scale experimental validation through trials and testbeds will be crucial next steps. Investigating optimal convergence strategies with complementary 6G technologies using co-design approaches also merits attention. The development of flexible and scalable architectures for seamless integration into 5G infrastructure is vital for smooth evolution to 6G. Finally, standardization efforts and quantified techno-economic studies are imperative to catalyze the adoption of massive MIMO in future 6G networks and enable the full realization of its disruptive capabilities.

References

1 M. Banafaa, Shayea, I., Din, J. et al., "6G mobile communication technology: requirements, targets, applications, challenges, advantages, and opportunities," *Alexandria Engineering Journal*, vol. 64, pp. 245–274, 2023, doi: https://doi.org/10.1016/j.aej.2022.08.017.

2 M. A. S. Sejan, M. H. Rahman, M. A. Aziz, D. S. Kim, Y. H. You, and H. K. Song, "A comprehensive survey on MIMO visible light communication: current research, machine learning and future trends," *Sensors*, vol. 23, no. 2, 2023, doi: https://doi.org/10.3390/s23020739.

3 D. Borges, P. Montezuma, R. Dinis, and M. Beko, "Massive MIMO techniques for 5G and beyond—opportunities and challenges," *Electronics (Switzerland)*, vol. 10, no. 14, pp. 1–29, 2021, doi: https://doi.org/10.3390/electronics10141667.

4 Y. Liu, "Analysis of massive MIMO&NOMA of 5G and future prospects," *Journal of Physics: Conference Series*, vol. 1738, no. 1, 2021, doi: https://doi.org/10.1088/1742-6596/1738/1/012004.

5 S. A. Khwandah, J. P. Cosmas, P. I. Lazaridis, Z. D. Zaharis, and I. P. Chochliouros, "Massive MIMO systems for 5G communications," *Wireless Personal Communications*, vol. 120, no. 3, pp. 2101–2115, 2021, doi: https://doi.org/10.1007/s11277-021-08550-9.

6 Bhatti, S.(2019). Massive MIMO, mm Wave and 5G technology insights and challenges. *3C Tecnología_Glosas De Innovación Aplicadas a La Pyme*. doi: 10.17993/3ctecno.2019.specialissue2.498-517.

7 F. A. P. De Figueiredo, F. A. C. M. Cardoso, I. Moerman, and G. Fraidenraich, "On the application of massive MIMO systems to machine type communications," *IEEE Access*, vol. 7, no. December, pp. 2589–2611, 2019, doi: https://doi.org/10.1109/ACCESS.2018.2886030.

8 H. I. Obakhena, A. L. Imoize, F. I. Anyasi, and K. V. N. Kavitha, Application of cell-free massive MIMO in 5G and beyond 5G wireless networks: a survey, vol. 68, no. 1. *Journal of Engineering and Applied Science*, 2021. doi: https://doi.org/10.1186/s44147-021-00014-y.

9 P. Händel and D. Rönnow, "MIMO and massive MIMO transmitter crosstalk," *IEEE Transactions on Wireless Communications*, vol. 19, no. 3, 2020, doi: https://doi.org/10.1109/twc.2019.2959534.

10 N. M. Hassan and X. Fernando, "Massive MIMO wireless networks: an overview," *Electronics*, vol. 6, no. 3, 2017, doi: https://doi.org/10.3390/electronics6030063.

11 S. Senger and P. K. Malik, "A comprehensive survey of massive- MIMO based on 5G antennas," *International Journal of RF and Microwave Computer-Aided Engineering*, vol. 32, no. 12, 2022, doi: https://doi.org/10.1002/mmce.23496.

12 R. Chataut and R. Akl, "Massive MIMO systems for 5G and beyond networks—overview, recent trends, challenges, and future research direction," *Sensors*, vol. 20, no. 10, 2020, doi: https://doi.org/10.3390/s20102753.

13 Y. Su, H. Gao, and S. Zhang, "Secure massive MIMO system with two-way relay cooperative transmission in 6G networks," *EURASIP Journal on Wireless Communications and Networking*, vol. 2023, no. 1, 2023, doi: https://doi.org/10.1186/s13638-023-02280-1.

14 Y. Su, H. Gao, and S. Zhang, "Energy-efficient resource management for CCFD massive MIMO systems in 6G networks," *Journal of Systems Engineering and Electronics*, vol. 33, no. 4, 2022, doi: https://doi.org/10.23919/jsee.2022.000085.

15 T. L. Marzetta, "Noncooperative cellular wireless with unlimited numbers of base station antennas," *IEEE Transactions on Wireless Communications*, vol. 9, no. 11, 2010, doi: https://doi.org/10.1109/twc.2010.092810.091092.

16 Ngo, H.Q. (2015). Massive MIMO: fundamentals and system designs. *scite.ai* https://doi.org/10.3384/lic.diva-112780 (accessed 5 August 2023.

17 Gustavsson, U., Sanchéz-Perez, C., Eriksson, T. et al. (2014). On the impact of hardware impairments on massive MIMO. *IEEE Globecom Workshops (GC Wkshps)*. https://doi.org/10.1109/glocomw.2014.7063447.

18 Ngo, H.Q., Matthaiou, M., Smith, P. et al. (2019). Massive MIMO with a generalized channel model: fundamental aspects. *IEEE 20th International Workshop on Signal Processing Advances in Wireless Communications (SPAWC)*. https://doi.org/10.1109/spawc.2019.8815518.

19 Zhang, J., Wei, Y., Björnson, E. et al. (2017). Spectral and energy efficiency of cell-free massive MIMO systems with hardware impairments. *9th International Conference on Wireless Communications and Signal Processing (WCSP)*. https://doi.org/10.1109/wcsp.2017.8171057.

20 Athley, F., Durisi, G., and Gustavsson, U. (2015). Analysis of massive MIMO with hardware impairments and different channel models. https://doi.org/10.48550/arxiv.1501.04200 (accessed 5 August 2023).

21 Papazafeiropoulos, A.K., Sharma, S.K., and Chatzinotas, S. (2015). Impact of transceiver impairments on the capacity of dual-hop relay massive MIMO systems. *IEEE Globecom Workshops (GC Wkshps)*. https://doi.org/10.1109/glocomw.2015.7414137.

22 Belaoura, W., Shakir, M.Z., and Qaraqe, K.A. (2020). Impact of hardware impairments on the performance of millimeter-wave massive MU-MIMO systems with distributed antennas. *IEEE Eighth International Conference on Communications and Networking (ComNet)*. https://doi.org/10.1109/comnet47917.2020.9306107.

23 T. E. Bogale and L. B. Le, "Massive MIMO and mmWave for 5G wireless HetNet: potential benefits and challenges," *IEEE Vehicular Technology Magazine*, vol. 11, no. 1, 2016, doi: https://doi.org/10.1109/mvt.2015.2496240.

24 Becirovic, E. (2022). Signal processing aspects of massive MIMO. Dissertations. Linköping Studies in Science and Technology. https://doi.org/10.3384/9789179294540.

25 Le, M.T.P., Sanguinetti, L., Björnson, E., and Benedetto, M.-G.D. (2019). What is the benefit of code-domain NOMA in massive MIMO? *IEEE 30th Annual International Symposium on Personal, Indoor and Mobile Radio Communications (PIMRC)*. https://doi.org/10.1109/pimrc.2019.8904268.

26 Becirovic, E. (2020). On massive MIMO for massive machine-type communications. Licentiate Thesis. Linköping Studies in Science and Technology.https://doi.org/10.3384/lic.diva-162586.

27 Yang, X., Lu, W., Wang, N. et al., "Design and implementation of a TDD-based 128-antenna massive MIMO prototype system," *China Communications*, vol. 14, no. 12, 2017, doi: https://doi.org/10.1109/cc.2017.8246333.

28 Ou, Q., Wang, Y., Bao, L. et al. "Power allocation in smart grid aided cell free massive MIMO systems for URLLC applications," *Journal of Physics: Conference Series*, vol. 2310, no. 1, 2022, doi: https://doi.org/10.1088/1742-6596/2310/1/012069.

29 Choi, J.-S., Park, J.-H., and Lee, N. (2021). Energy efficiency maximization precoding for quantized massive MIMO systems. https://doi.org/10.48550/arxiv.2108.03048 (accessed 5 August 2023).

30 N. S. Ahmed, N. S. M. A. Hasan, and N. A. K. M. N. Islam, "Design of an optimum massive MIMO FSO system and analysis of its performance in different weather conditions," *Mist International Journal of Science and Technology*, vol. 10, 2022, doi: https://doi.org/10.47981/j.mijst.10(03)2022.335(43-52).

31 N. Shlezinger and Y. C. Eldar, "On the spectral efficiency of noncooperative uplink massive MIMO systems," *IEEE Transactions on Communications*, vol. 67, no. 3, 2019, doi: https://doi.org/10.1109/tcomm.2018.2884008.

32 Koundal, P., Singh, S., and Kaur, R. (2023). Massive MIMO antenna array at terahertz band for short-range communications. https://doi.org/10.21203/rs.3.rs-2785864/v1 (accessed 5 August 2023).

33 Al-Salihi, H., Nakhai, M.R., and Le, T.A. (2018). DFT-based channel estimation techniques for massive MIMO systems. *25th International Conference on Telecommunications (ICT)*. https://doi.org/10.1109/ict.2018.8464829.

34 J. Sánchez, F. Rusek, O. Edfors, M. Sarajlic, and L. Liu, "Decentralized massive MIMO processing exploring daisy-chain architecture and recursive algorithms," *IEEE Transactions on Signal Processing*, vol. 68, 2020, doi: https://doi.org/10.1109/tsp.2020.2964496.

35 Liu, L., Wang, Z., and Cui, S. (2020). Intelligent reflecting surface assisted massive MIMO communications. *IEEE 21st International Workshop on Signal Processing Advances in Wireless Communications (SPAWC)*. https://doi.org/10.1109/spawc48557.2020.9154291.

36 Neil, C.T., Smith, P., and Dmochowski, P.A. (2015). On the impact of antenna topologies for massive MIMO systems. *IEEE International Conference on Communications (ICC)*. https://doi.org/10.1109/icc.2015.7248624.

37 A. K. Yerrapragada and B. Kelley, "On the application of K-user MIMO for 6G enhanced mobile broadband," *Sensors*, vol. 20, no. 21, 2020, doi: https://doi.org/10.3390/s20216252.

38 T. N. Vu, C.-F. Liu, M. Bennis, M. Debbah, M. Latva-aho, and C. S. Hong, "Ultra-reliable and low latency communication in mmWave-enabled massive MIMO networks," *IEEE Communications Letters*, vol. 21, no. 9, 2017, doi: https://doi.org/10.1109/lcomm.2017.2705148.

39 M. Sharma, S. Verma, and S. Verma, "Optimization of cell-free massive MIMO system," *Journal of Physics: Conference Series*, vol. 2327, no. 1, 2022, doi: https://doi.org/10.1088/1742-6596/2327/1/012056.

40 A. Waseem, A. Naveed, M. Arshad, H. Anis, and I. M. Qureshi, "Compressive sensing based channel estimation for massive MIMO communication systems," *Wireless Communications and Mobile Computing*, vol. 2019, 2019, doi: https://doi.org/10.1155/2019/6374764.

41 L. Iliadis, Z. D. Zaharis, S. P. Sotiroudis, P. Sarigiannidis, G. K. Karagiannidis, and S. K. Goudos, "The road to 6G: a comprehensive survey of deep learning applications in cell-free massive MIMO communications systems," *EURASIP Journal on Wireless Communications and Networking*, vol. 2022, no. 1, 2022, doi: https://doi.org/10.1186/s13638-022-02153-z.

42 A.-S. Bana, de Carvalho, E., Soret, B. et al., "Massive MIMO for Internet of Things (IoT) connectivity," *Physical Communication*, vol. 37, 2019, doi: https://doi.org/10.1016/j.phycom.2019.100859.

43 N. Komninos, C. Bratsas, C. Kakderi, and P. Tsarchopoulos, "Smart city ontologies: improving the effectiveness of smart city applications," *Journal of Smart Cities*, vol. 1, no. 1, 2016, doi: https://doi.org/10.18063/jsc.2015.01.001.

44 V. D. P. Souto, Dester, P.S., Facina, M.S.P. et al., "Emerging MIMO technologies for 6G networks," *Sensors*, vol. 23, no. 4, 2023, doi: https://doi.org/10.3390/s23041921.

45 L. You, K. Li, J. Wang, X. Gao, X.-G. Xia, and B. Ottersten, "Massive MIMO transmission for LEO satellite communications," *IEEE Journal on Selected Areas in Communications*, vol. 38, no. 8, 2020, doi: https://doi.org/10.1109/jsac.2020.3000803.

46 P. Chandhar and E. G. Larsson, "Massive MIMO for connectivity with drones: case studies and future directions," *IEEE Access*, vol. 7, 2019, doi: https://doi.org/10.1109/access.2019.2928764.

47 Haghighatshoar, S., Khalilsarai, M.B., and Caire, G. (2018). Multi-band covariance interpolation with applications in massive MIMO. *IEEE International*

Symposium on Information Theory (ISIT). https://doi.org/10.1109/isit.2018.8437890.

48 Alexandropoulos, G.C., Islam, A., and Smida, B. (2022).Full duplex massive MIMO architectures: recent advances, applications, and future directions. https://doi.org/10.48550/arxiv.2205.08393 (accessed 5 August 2023).

49 F. Qamar, M. U. A. Siddiqui, M. N. Hindia, R. Hassan, and Q. V. Nguyen, "Issues, challenges, and research trends in spectrum management: a comprehensive overview and new vision for designing 6G networks," *Electronics*, vol. 9, no. 9, 2020, doi: https://doi.org/10.3390/electronics9091416.

50 Y. Hu, M. Yin, W. Xia, S. Rangan, and M. Mezzavilla, "Multi-frequency channel modeling for millimeter wave and THz wireless communication via generative adversarial networks," in *IEEE Journal on Selected Areas in Communications*, vol. 39, no. 7, pp. 2061–2074, 2021, doi: https://doi.org/10.1109/JSAC.2021.3082469.

51 Salous, S., Gaillot, D., Molina-Garcia-Pardo, J.-M. et al. (2021). IRACON channel measurements and models. In: *Inclusive Radio Communications for 5G and Beyond* (ed. C. Oestges and F. Quitin), 49–105. Academic Press ISBN: 9780128205815, doi: https://doi.org/10.1016/B978-0-12-820581-5.00009-2.

52 Khan, M.M., Hossain, S., Mozumdar, P. et al. (2022). A review on machine learning and deep learning for various antenna design applications. *Heliyon* 8 (4): e09317. ISSN: 2405-8440, doi: https://doi.org/10.1016/j.heliyon.2022.e09317.

53 Zöchmann, E., Guan, K., and Rupp, M. (2017). Two-ray models in mmWave communications. *2017 IEEE 18th International Workshop on Signal Processing Advances in Wireless Communications (SPAWC)*, Sapporo, Japan, pp. 1–5, doi: https://doi.org/10.1109/SPAWC.2017.8227646.

54 Bharathy, A.M.V., Alhadidi, B., and (2022). *Challenges and Risks Involved in Deploying 6G and NextGen Networks*. Hershey, PA: Information Science Reference, an imprint of IGI Global.

4

Cell-Free Massive MIMO Technology and Applications in 6G

Joumana Kassam[1], Daniel Castanheira[1], Adão Silva[1], Rui Dinis[2], and Atílio Gameiro[1]

[1]*Instituto de Telecomunicações (IT), and Departamento de Electrónica, Telecomunicações e Informática (DETI), Universidade de Aveiro, Aveiro, Portugal*
[2]*Instituto de Telecomunicacões (IT), Faculdade de Ciências e Tecnologia, Universidade Nova de Lisboa, Lisboa, Portugal*

4.1 Introduction

The rise of 5G and beyond networks marks a new era known as the Internet of Intelligence, connecting people and things seamlessly. This requires networks to meet demanding requirements such as high efficiency, low-latency, reliability, and robust connections for real-time applications. Overcoming these challenges is crucial for unlocking the full potential of a connected world, enabling innovative applications and transforming industries globally. Multiple-input multiple-output (MIMO) and millimeter-wave (mmWave) are among the efficient technologies employed in 5G and future networks.

MIMO is a wireless communication technology that enhances the performance of data transmission by using multiple antennas at both ends. It enables simultaneous transmission of multiple data streams. Massive MIMO (mMIMO) takes this concept further by using a significantly higher number of antennas at the terminals. With potentially hundreds or thousands of antennas, mMIMO offers substantial gains in capacity, coverage, and energy efficiency (EE). It needs advanced signal processing techniques like beamforming and spatial multiplexing to serve multiple users simultaneously, improving spectral efficiency (SE) and user experiences, particularly in densely populated areas.

MmWave frequencies, ranging from 30 to 300 GHz, provide high bandwidth for transmitting large amounts of data at extremely high speeds. These frequencies are being researched for 5G wireless systems due to their potential to meet the

Massive MIMO for Future Wireless Communication Systems: Technology and Applications,
First Edition. Edited by Webert Montlouis and Agbotiname Lucky Imoize.

growing demand for high capacity and low latency. However, mmWave signals have a shorter range and are susceptible to absorption and blockage, necessitating advanced techniques like beamforming and mMIMO to address these challenges.

Therefore, this section focuses on these advanced technologies, aiming to provide a comprehensive understanding of their capabilities and applications in current and future wireless communication networks.

4.1.1 MIMO Technologies

In recent years, MIMO concept has emerged as a disruptive technology [1]. By employing multiple antennas at the transmitter and/or receiver, MIMO systems unlocked the benefits of spatial multiplexing and diversity gains, leading to significant improvements in SE without necessitating additional frequency, power, or time resources. Building upon the MIMO framework, the concept of mMIMO has drawn a lot of interest as a crucial technology for 5G and beyond networks [2]. It can leverage a large number of antennas to concurrently serve a limited number of user equipment's (UEs) throughout each time–frequency resource slot. This approach effectively reduces transmission energy requirements and mitigates inter-cell interference [3], utilizing either centralized architectures with a large-scale antenna array at the base station or distributed design with antennas spread throughout the cell [1, 3]. To meet the diverse requirements of modern networks, researchers have extensively investigated various paradigms of mMIMO, as depicted in Figure 4.1, including co-located [4], distributed [5], and cell-free (CF) mMIMO [6] configurations. These mMIMO systems hold substantial promise in meeting the demands of 5G and beyond, delivering high data rates, reliable coverage, and enhanced quality-of-service (QoS) for users, particularly when deployed in distributed or CF arrangements [6, 7].

Co-located mMIMO employs a large number of antennas on macro base stations (BSs) to communicate, simultaneously, with a smaller number of UEs

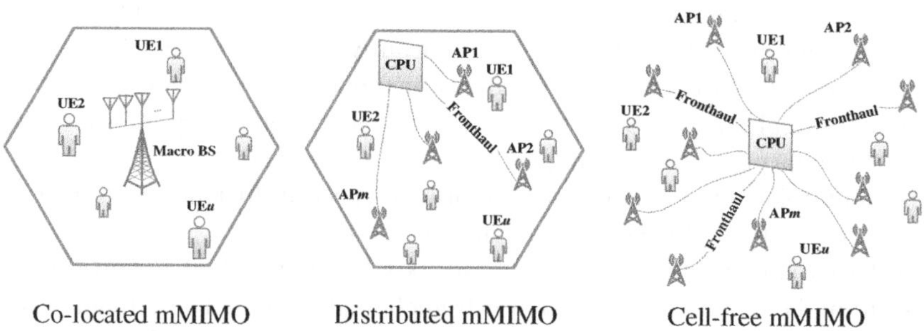

Figure 4.1 Massive MIMO architectures.

using beamforming techniques. This setup increases data rates and improves network coverage and EE. However, co-located mMIMO systems may experience performance degradation for cell-edge users due to the reduced channel gains from the serving cell and interference inherited from nearby cells [8]. Therefore, distributed mMIMO has been suggested as an alternative solution. It can operate by deploying numerous geographically distributed access points (APs), connected to a central processing unit (CPU) through fronthaul links. These systems have been addressed in single and multicell scenarios, where they are similar to network MIMO [9], coordinated multipoint (COMP) [10], and distributed antenna systems (DASs) [11]. Each cell cooperatively serves users within its range of influence. Due to the diversity benefits obtained from different large-scale fading (LSF) components received by each user from multiple APs, distributed mMIMO systems provide better data rates than co-located mMIMO counterpart. Furthermore, it can provide better QoS to cell-edge users due to the presence of more APs in their proximity [12]. However, the implementation of distributed large MIMO systems is constrained by practical issues such as network synchronization, low-latency, and high-rate fronthaul requirements.

CF MIMO is an emerging embodiment of distributed mMIMO systems that, unlike the previous mMIMO systems, do not use the cellular or cell boundary concepts [13]. As depicted in Figure 4.2, it employs time-division duplex (TDD) operation, in which several APs, with single or multiple antennas, serve all UEs on the same time–frequency resources [14] by leveraging local channel state information (CSI) and conducting joint transmission. The CPU sends downlink (DL) data and power control coefficients to the APs, while the APs provide feedback

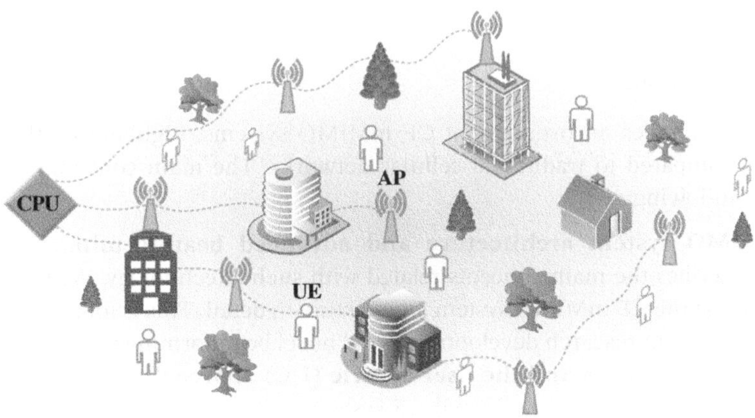

Figure 4.2 Conventional cell-free mMIMO architecture.

to the CPU through the fronthaul connection on the uplink (UL) data received from UEs [15]. In such systems, only channel statistics are used at the CPU for joint detection, resulting in high performance. These systems are especially interesting for indoor and hot-spot coverage scenarios such as smart factories, railway stations, malls, and stadiums, since they offer increased system performance and enhanced SE. They also provide seamless mobility assistance without the need for handovers, providing consistent QoS in future vehicular scenarios [16].

4.1.2 Bandwidth Expansion and Beamforming

In response to the high congestion in the microwave radio spectrum, as well as the insufficient capacity of massive antenna systems and network densification, an alternative physical layer solution emerged to meet the QoS demands of 5G wireless communication networks [17]. This solution aimed to expand the available spectrum (sub-6 GHz) by exploring less congested frequency bands, such as millimeter wave (mmWave), which exhibit great potential in 5G and beyond networks [18]. Unlike microwave bands, mmWave frequencies offer a significantly larger spectrum, and thus the advancements in low-power radio frequency (RF) miniaturization have made it feasible to integrate numerous antenna elements into compact form factors [19]. By leveraging large antenna arrays, mmWave mMIMO schemes can be effectively implemented. These schemes employ appropriate hybrid analog–digital beamforming techniques, which involve a smaller number of RF chains compared to antennas, enabling them to overcome the substantial increase in free-space path loss (PL) caused by higher frequencies [19, 20]. In contrast, sub-6 GHz systems typically rely on full-digital baseband beamforming, connecting each antenna element to a dedicated RF chain [21].

4.1.3 Contributions

This chapter provides an overview of CF mMIMO systems, highlighting their advantages compared to traditional cellular networks. The main contributions include the following:

- **CF mMIMO system architecture and advanced beamforming:** This section describes the main concepts related with such a technology. We begin by examining the CF mMIMO system architecture in detail. This is followed by some of the recent research developments and novel beamforming algorithms.
- **Scalability challenges and the user-centric (UC) approach:** The scalability challenges inherent to traditional CF systems are analyzed. In particular, it focuses on the UC approach, which is a technology developed for practical deployment scenarios where the number of cooperative APs serving a UE is limited.

- **Integration of CF with radio stripes (RSs):** This strategic integration enhances system resilience while simultaneously reducing the high cost and complexity associated with a sequential topology. This involves shared fronthaul cabling and the connection of each RS to one or more CPUs, marking a transformative step toward a more efficient and cost-effective network design paradigm.

4.1.4 Chapter Organization

The remainder of this chapter is organized as follows: Section 4.2 presents in detail the conventional CF mMIMO systems in terms of the literature review, the system model architecture, design overview, and the numerical results. Then, Section 4.3 addresses some of the recent advancements that have been used to build scalable CF mMIMO systems, namely, the UC approach. Section 4.4 presents one of the promising technologies for B5G networks, known as RSs, highlighting its integration with CF mMIMO systems. Finally, the main conclusions and future work are drawn in Section 4.5.

4.2 Conventional Cell-Free mMIMO Systems

Each UE in conventional CF mMIMO systems has the capability to be served by all APs. Building upon this assumption, this section provides an overview of recent studies that have been proposed in the literature. Then follows a description of the system model design and some numerical results.

4.2.1 Literature Review

Recently, there has been a significant and extensive focus on studying the performance of conventional CF mMIMO systems. Researchers have dedicated considerable efforts to investigate different aspects of such systems, considering sub-6 GHz frequency bands with power allocation [14, 22–25], or with the impact of employing a full-digital beamforming [26–29]. On the other hand, other authors concentrated on analyzing the performance of CF systems at mmWave frequencies for both narrowband-based [30–32] and wideband-based [33, 34] systems. These studies aimed to explore the full potential of CF mMIMO and optimize its performance in terms of capacity, coverage, SE, EE, and overall system performance.

4.2.1.1 CF mMIMO at Sub-6 GHz with Power Allocation

The comparative analysis of [14] focused on the performance of CF mMIMO and small-cells (SCs) systems in terms of achievable data rates for both DL and UL transmissions. The results demonstrated that the CF mMIMO system outperformed the fully distributed SCs one, particularly in terms of the 95%-likely per user throughput. Moreover, by implementing max–min power control, CF mMIMO also guaranteed consistent and satisfactory service for all UEs across the coverage area. In [22], the authors in their study on enhancing the performance of the CF system, proposed a novel approach to address the UL max–min signal-to-interference-plus-noise ratio (SINR) problem, dividing the optimization problem into two sub-problems: receiver filter coefficient design and power allocation. An iterative technique was used to solve these sub-problems, resulting in a globally optimum solution. They also presented a sub-optimal max–min SINR scheme based on standard geometric programming, eliminating the need for any iterative approach. The authors in [23] investigated the application of an expanded compute-and-forward (ECF) framework, aiming to improve the achievable sum rate. They proposed several strategies, including a quadratic programming relaxation-based coefficient vector selection, LSF-based low-complexity AP selection, and power control algorithms to enhance the system's performance. Then, they derived a sub-optimal decoding order of combinations and developed three assignment algorithms to obtain a sub-optimal decoding order of users. The work of [24] proposed a novel algorithm for joint channel estimation and data detection, enabling reliable transmission to a large number of UEs with short packets. This approach aimed to reduce the channel training overhead imposed by several UEs broadcasting at the same time, by exploiting the sparsity of channels and the boundedness of quadrature amplitude modulation (QAM) constellations. Then, different methods were developed to permute the APs and UE indices, creating virtual cells, and enhancing initial solutions and convergence. The main objective of [25] was to develop a power allocation algorithm that maximizes EE while adhering to both the sum power constraint at each AP and the individual QoS constraint at each UE. The authors proposed a low-complexity iterative power control algorithm based on the accelerated projected gradient (APG) method. The algorithm's iterations were carried out using simple closed-form expressions, resulting in significantly lower computational complexity compared to sequential second-order cone programs (SOCPs)-based techniques while achieving the same performance.

4.2.1.2 CF mMIMO at Sub-6 GHz with Full-Digital Beamforming

Extensive research has been conducted on the performance of conventional CF mMIMO systems using sub-6 GHz frequency bands and full-digital beamforming at APs/CPU [26–29]. The authors of [26] developed low-complexity power optimization algorithms for DL-distributed systems. These algorithms, based on the max–min criteria, were designed to achieve optimal and sub-optimal solutions, considering the use of conjugate beamforming and zero-forcing (ZF) precoder schemes. The study of [27] examined the UL SE of various coordinated beamforming implementations (from fully distributed to fully centralized). The analysis considered both global and local minimum mean squared error (MSE) criteria. Then, the closed-form SE expressions for two-layer decoding processing were derived using a maximum-ratio combining (MRC) scheme. In [28], the work concentrated on the exact closed-form expression for the achievable SE of a centralized DL system, called enhanced normalized conjugate beamforming system. Then, they used LSF quantities to develop an optimal max–min fairness power control algorithm. In order to prevent direct matrix inversion in a DL-distributed ZF-based precoder system, several iterative matrix inversion methods were proposed in [29], aiming to reduce computational complexity while achieving a near-optimal performance.

However, it is crucial to note that the assumptions, algorithms, and analytical expressions used in previous studies may not be directly applicable when considering mmWave frequency bands. Consequently, several researchers have focused on analyzing the performance of CF mMIMO systems at mmWave frequencies [30–34], emphasizing the use of hybrid analog-digital beamforming techniques to overcome the challenges associated with the higher-frequency range.

4.2.1.3 CF mMIMO at mmWave for Narrowband Systems

Starting with the literature that concentrated on narrowband mmWave-based systems. First, the work of [30, 31] considered semi-centralized coordinated beamforming architecture to evaluate their systems. Specifically in [30], sophisticated algorithms for both UL and DL low-complexity multiuser hybrid beamforming were designed, taking into account several limitations, such as imperfect CSI estimation, the capacity limitations of fronthaul connections, power allocation, and non-orthogonal pilot sequences. The proposed approach leveraged large-scale second-order CSI for the analog part of the precoder/decoder, while the digital part was implemented using a ZF technique based on small-scale instantaneous CSI. In [31], the authors, besides the semi-centralized, investigated an UL-distributed architecture, using low-complexity algorithms to dynamically enable/disable the RF chains, as well as the analog-to-digital converters (ADCs), and phase shifters at the APs. The objective was to minimize power consumption, maintain the total achievable rate, and simultaneously enhance the system EE. Three distributed analog precoding schemes were proposed together with ZF digital precoding in [32]. The researchers examined the impact of correlated shadowing and pilot assignment on the average SE performance of the designed DL precoding algorithms under both indoor and outdoor scenarios.

4.2.1.4 CF mMIMO at mmWave for Wideband Systems

The works of [33, 34] considered wideband mmWave channel. More precisely, in [33], the authors presented a centralized hybrid beamforming architecture specifically designed for orthogonal frequency-division multiplexing (OFDM) systems. The architecture employed a beam squint-aware channel covariance to improve its performance. They introduced UL/DL analog precoder/combiner algorithms based on high-dimensional single- and double-phase shifters, while considering various assumptions regarding beam squint awareness. In [34], an UL low-complexity analog precoder based on angle of departure (AoD) was developed together with a joint design approach to optimization of the distributed hybrid equalizer at the AP and the combiner at the CPU, considering discrete Fourier transform spread-OFDM (DFT-S-OFDM) as an access technique. The objective of this joint design was to minimize the average MSE between the transmitted signals and the signals estimated at the CPU. To improve the overall detection performance, a novel iterative cyclic minimization approach was developed, allowing to optimize the AP hybrid equalizer and CPU combiner based on a set of error variances.

4.2.2 System Model and Design

This section presents the system model for conventional CF mMIMO systems, emphasizing the design of the UE/APs terminals, and the APs hybrid analog–digital equalizer as well as the CPU data combiner.

Let us consider a multiuser UL CF mMIMO system that uses DFT-S-OFDM as an access technique. The system consists of M APs and U UEs. The APs process the signals of all UEs, and these resulting signals are transmitted to a CPU through fronthaul connections. The CPU combines the received signals and performs data detection for each user. Both the UEs and APs are equipped with multiple antennas. The terminals are geographically distributed within a defined area. The geographical distribution of the terminals is assumed to be random following a uniform distribution.

As depicted in Figure 4.3, at the transmitter side, the bits are first modulated into a Q-ary QAM (Q-QAM) signal. This signal is subsequently mapped into a single DFT-S-OFDM symbol. The resulting digital signal is then converted into analog format to pass through the analog part of a hybrid analog–digital precoder. Therefore, the transmitted signal for the uth UE at subcarrier k is given by

$$\mathbf{x}_{u,k} = \mathbf{w}_{a,u} s_{u,k}, \tag{4.1}$$

where $s_{u,k}$ represents the output of the DFT-S-OFDM block and $\mathbf{w}_{a,u}$ is the analog precoder that is composed of a number of phase shifters connecting the dedicated RF chain to all transmitting antennas. Notice that the design of the analog precoder relies only on the knowledge of the average AoDs between the UE and APs [34]. Then, the UE transmits only one data stream of information per subcarrier

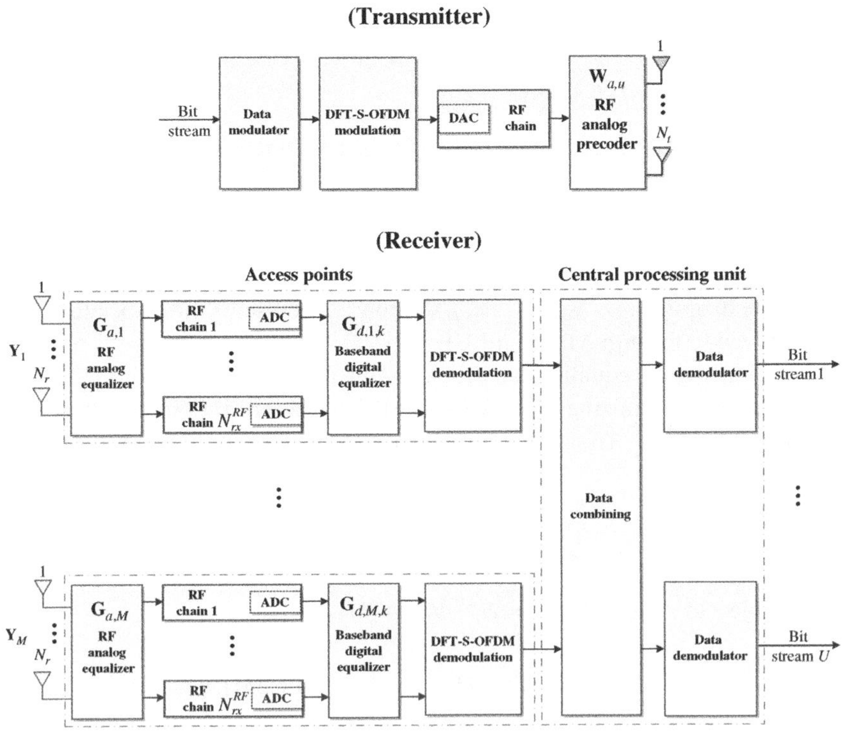

Figure 4.3 Cell-free mMIMO transceiver system model.

to the receiver which passes through the channel commonly modeled using a clustered wideband channel model.

At the receiver side, a distributed architecture is employed, dividing the signal processing tasks between the APs and CPU. At each AP, the N_r received signals are combined using a hybrid equalizer. The equalized signal is then sent through the fronthaul connections to the CPU for further combining. The received signal at the mth AP at subcarrier k is

$$\mathbf{y}_{m,k} = \sum_{u=1}^{U} \mathbf{H}_{m,u,k}\mathbf{x}_{u,k} + \mathbf{n}_{m,k}, \tag{4.2}$$

where $\mathbf{H}_{m,u,k}$ denotes the channel between the mth AP and uth UE at subcarrier k and $\mathbf{n}_{m,k}$ denotes the additive white Gaussian noise (AWGN) with variance σ_n^2.

At the mth AP, the signal passes through the analog part of the hybrid equalizer, denoted as $\mathbf{G}_{a,m}$, composed only by phase shifters, as $|\mathbf{G}_{a,m}[i,j]|^2 = N_r^{-1}$, then the output is further processed by N_{rx}^{RF} receiving RF chains and by the digital part $\mathbf{G}_{d,m,k}$. Consequently, the estimated signal after applying the hybrid analog–digital equalizer at the kth subcarrier can be expressed as

$$\hat{\mathbf{s}}_{hyb,m,k} = \mathbf{G}_{d,m,k}^{H} \mathbf{G}_{a,m}^{H} \mathbf{y}_{m,k}. \tag{4.3}$$

The obtained estimates are sent to the CPU to be combined for final data estimate for the UEs at the kth subcarrier as

$$\hat{\mathbf{s}}_k = \sum_{m=1}^{M} \mathbf{V}_m \hat{\mathbf{s}}_{hyb,m,k}, \tag{4.4}$$

where $\mathbf{V}_m = \text{diag}(v_{m,1}, \dots, v_{m,u}, \dots, v_{m,U})$, with $v_{m,u}$ represents the associated CPU combiner weight for mth AP and $u$thUE.

The analog–digital equalizer at the APs and the combiner at the CPU can be derived by minimizing the average MSE between the transmitted and CPU-received signals. Thus, the optimization problem can be mathematically formulated as

$$\begin{aligned} &\underset{\{\mathbf{V}_m, \mathbf{G}_{d,m,k}, \mathbf{G}_{a,m}\}_m}{\text{minimize}} \quad N_c^{-1} \sum_{k=1}^{N_c} \mathbb{E}\left[\|\hat{\mathbf{s}}_k - \mathbf{s}_k\|^2\right] \\ &\text{subject to} \quad |\mathbf{G}_{a,m}[i,j]|^2 = N_r^{-1}, \forall m \in [M], i \in [N_r], j \in \left[N_{\text{rx}}^{\text{RF}}\right] \\ &\qquad N_c^{-1} \sum_{k=1}^{N_c} \mathbf{e}_u^H \mathbf{G}_{d,m,k}^H \mathbf{H}_{a,m,k} \mathbf{e}_u = 1, \forall m \in [M], u \in [U] \\ &\qquad \sum_{m=1}^{M} \mathbf{V}_m = \mathbf{I}, \forall m \in [M], \end{aligned} \tag{4.5}$$

where the first constraint enforces that all phase shifters have an equal magnitude, the second ensures zero estimation bias at each AP, while the third enforces zero bias at the CPU.

One option to solve the MSE problem (4.5) is the cyclic minimization approach presented in [34]. The weights of the CPU combiner $v_{m,u}$ and the coefficients of the APs hybrid analog–digital equalizer are iteratively optimized in this manner. During the optimization process, the CPU combiner/APs hybrid equalizer employs a priori information sent by the APs/CPU to weight the MSE of each UE.

4.2.3 Numerical Results

In this section, we present some results for conventional CF mMIMO mmWave systems, where they are obtained through Monte Carlo simulation. Figure 4.4 compares the BER performance between the work proposed in [31, 34] that basically solves (4.5), considering wideband distributed CF mMIMO systems, in which

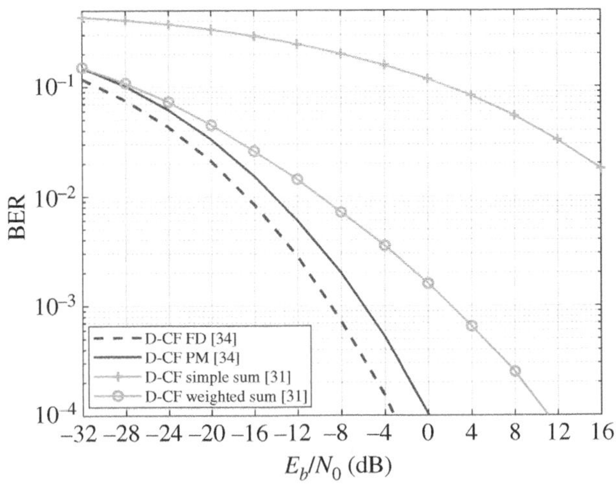

Figure 4.4 BER performance comparison between the CF work proposed in [31, 34].

the processing is divided between the APs and the CPU. As shown in Table 4.1, the experiment analysis was obtained using four APs and two UEs, where the number of receive RF chains is equal to the number of UEs, considering the same number of transmit and receive antennas $N_{tx} = N_{rx} = 16$. For a fair comparison, all schemes in this figure consider the AoD analog-based precoder at the UEs [34]. We can find that the performance of the distributed method (D-CF PM) presented in [34] achieves a performance close to the distributed full-digital curve (D-CF FD) with a penalty of approximately 3 dB for a BER target of 10^{-3}. This gap is a result of using

Table 4.1 Conventional CF mMIMO simulation parameters.

Parameters	Values
Modulation technique	DFT-S-OFDM
Number of subcarriers	512
DFT data blocks	4
Number of UEs	2
Number of APs	4
Number of received RF chains	2
Number of transmit antennas per UE	16
Number of APs antennas	16
Coverage radius of each AP	500 m

a hybrid analog-digital architecture instead of a full-digital one. Then, we present results for [31] under two different cases:

1) The soft estimates obtained at each AP are simply added at the CPU, labeled as "D-CF simple sum [31]," as considered in [31].
2) The soft estimates are weighted-summed at the CPU, labeled as "D-CF weighted sum [31]" as proposed in [34].

From the obtained results, it is evident that the performance of original D-CF simple sum method [31] is quite far from the one obtained with the D-CF PM [34] and D-CF weighted sum method [31]. This is due to the different signal attenuation suffered by the signals received at each AP. Consequently, the estimates obtained at each AP may be optimally combined considering these differences. Comparing D-CF weighted sum [31] with D-CF PM [34], we note that the performance is relatively close for the low SNR regime but degrades in the medium-to-high SNR regimes. This degradation can be attributed to the fact that the hybrid equalizers designed in [31] do not effectively leverage the inherent space–frequency diversity offered by the considered DFT-S-OFDM systems, in contrast to [34] that explicitly focused on removing interference by efficiently exploiting the available space–frequency diversity.

Figure 4.5 compares the BER performance of the distributed conventional CF mMIMO systems adopted in [34] with SCs counterpart. In this figure, we consider the same parameters as in Figure 4.4, notice that in SCs system, each user is served by one AP. We can observe that the performance of the hybrid method proposed in [34], labeled as "PM" in both CF and SC is relatively close to that of the full-digital

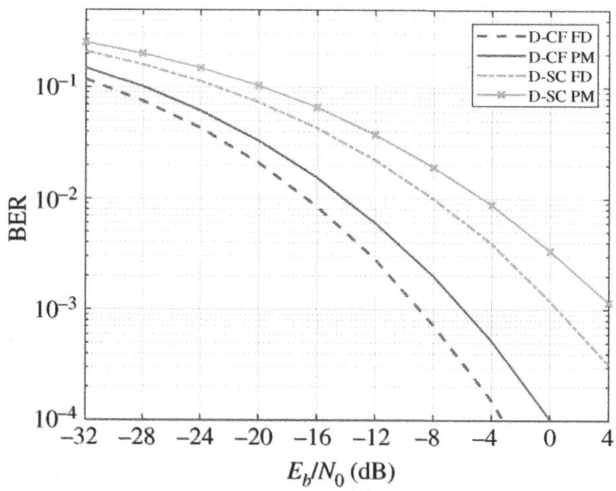

Figure 4.5 BER performance comparison between CF and SCs systems.

"FD" approach, with a penalty of less than 4 dB for a BER target of 10^{-3}. However, when comparing the results of SCs with CF systems, it becomes evident that CF outperforms SCs. Due to the fact that in a CF system, two UEs transmit data to four APs, resulting in increased gain as the signals from multiple APs are combined at the CPU to recover the data for a specific user.

Although the outcomes of the conventional CF mMIMO studies, in which each UE can be served by all APs, have provided valuable insights, the practical implementation of such systems poses challenges, such as the need for massive fronthaul signaling, high computational complexity, and increased power consumption. Consequently, researchers have shifted their focus toward evaluating alternative solutions that enable the construction of scalable CF mMIMO systems with less fronthaul requirements. These methods allow for more practical implementation and feasible deployments in real-world scenarios since each UE can be served by a group of APs. More details related to these challenges will be discussed in the next sections.

4.3 Scalable User-Centric Cell-Free mMIMO Systems

This section aims to shed light on the diverse algorithms that can be employed to address the challenges associated with conventional CF systems. The focus begins by highlighting the shift from unscalable CF systems to scalable ones through the adoption of a UC approach. Then follows a literature review about the advancements made in CF mMIMO systems with limited fronthaul capacity. It provides an overview of the state-of-the-art techniques and algorithms proposed in the academia. These advancements aim to optimize fronthaul utilization, manage computational complexity, and enhance the performance of CF mMIMO systems.

4.3.1 Literature Review

Early studies on network MIMO assumed that each UE in a wireless access network should be served by many APs within their coverage area [35]. However, these assumptions are impractical due to the high computational complexity and immense fronthaul signaling for CSI and data sharing [36]. Later evidence showed that network MIMO can operate without CSI sharing, sacrificing interference cancellation [37]. Initially, a network-centric approach was considered, combining the APs into nonoverlapping cooperative clusters. In these clusters, the APs can share data, and possibly CSI, to only serve the UEs located within the coverage region [9]. This in fact yielded limited gains in practice since UEs at cluster edges experienced significant intercluster interference [38]. As an

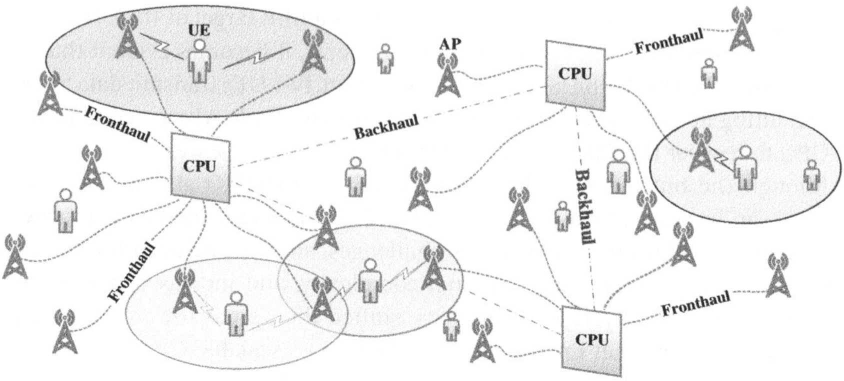

Figure 4.6 Scalable user-centric cell-free mMIMO system.

alternative, a UC approach has been proposed where each UE can be served by the best AP subset based on channel conditions. This eliminates nonoverlapping cooperation clusters, requiring APs to cooperate with different APs for different UEs sharing the same resources [39].

As stated, the computational complexity and fronthaul capacity needed for each AP to process and exchange the data signals relevant to all UEs rendered the initial form of CF mMIMO unscalable [40]. To overcome these scalability challenges and establish a more practical deployment framework, the UC approach was reintroduced again to build a scalable CF system [41]. As depicted in Figure 4.6, each UE is served by only a small subset of cooperative APs, effectively reducing the amount of information that needs to be sent to the CPU. Table 4.2 presents an overview of recent research [42–51] that has been focused on UC CF mMIMO systems. These studies proposed different APs–UEs association techniques to achieve their specific objectives in building scalable systems. It is of paramount importance to notice that UC CF mMIMO topologies play a crucial role in ensuring consistent coverage and performance for all users across the network area. By intelligently allocating resources and limiting the number of APs serving each UE, the UC approach successfully addresses the scalability concerns of CF mMIMO systems and paves the way for efficient and reliable communication in future networks [52].

On the other hand, the primary challenge in the conventional CF mMIMO systems arises from the constrained capacity of the links connecting the APs to the CPU, as shown in Figure 4.7. In practice, the assumption of having unlimited fronthaul capacity is unfeasible. Consequently, the consideration of CF mMIMO networks with capacity-limited fronthaul links has recently been undertaken [30, 53–59], subject to specific scenario conditions to improve the achievable user rates.

Table 4.2 Literature comparison for user-centric cell-free mMIMO systems.

References	Data transmission	Channel model	Performance analysis	APs–UEs selection-based approach
[42]	DL	Block fading	Average sum rate	Effective channel gain
[43]	DL	Rayleigh fading	Achievable rate per user	LSF
[44]			Average secrecy rate	
[45]			EE	1. LSF 2. Received power
[46]			Accuracy	Nearest AP distance
[47]	UL		SE + EE	
[48]	UL/DL	mmWave	Achievable rate per user + EE	Largest Frobenius norm
[49]	DL		BER + Average sum rate	Threshold received SNR
[50]			Average sum rate	LSF
[51]	UL		BER	SINR

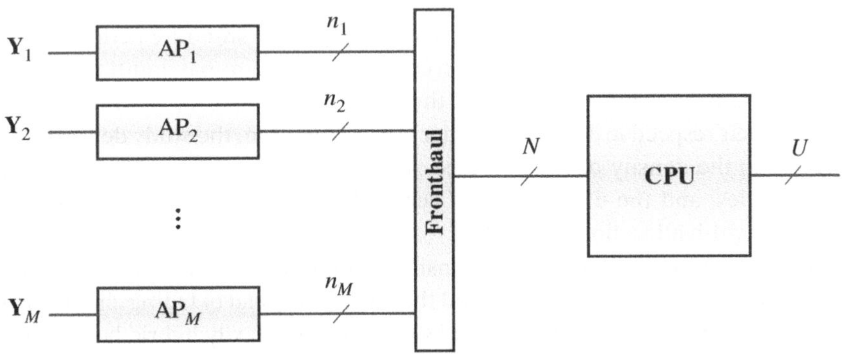

Figure 4.7 Cell-free mMIMO with limited fronthaul capacity links.

Specifically, in [53], three different UL transmission techniques at the APs from different perspectives, such as SE, EE, the effects of pilot reuse, and multiple-antenna APs were evaluated. The authors proposed an efficient and low-complexity algorithm for fronthaul rate allocations, enabling the optimal sharing of fronthaul connection capacity for transmitting CSI and data signals from the APs to the CPU. Two power allocation schemes were discussed:

one provided an approximate solution based on geometric programming and the other employed a general iterative algorithm. The authors of [54] designed a practical framework to facilitate a fair comparison between different CPU-AP functional splits. The framework focused on quantizing the signals shared among the APs and CPU during both the UL and DL transmission phases. They employed low-resolution ADCs to compute closed-form solutions for the max–min power allocation and fronthaul quantization problems. Bashar et al. [55] examined how quantization distortion correlation affected the system's capacity. The work specifically used local MSE receivers and investigated two distinct scenarios under the assumption of uncorrelated quantization distortion: the exact UL per-user rate and the UL per-user rate without taking the correlation between the inputs of the quantizers into account. The research conducted by Bashar et al. [56] examined the impact of fronthaul quantization and imperfect channel acquisition on system performance. Three scenarios were considered to analyze the total power consumption of the system: estimate and quantize, quantize and estimate, and decentralized cases. The standard linear receivers (MRC, ZF, and MMSE) were evaluated, taking into account the limitation of fronthaul links, quantization errors, and pilot contamination. Then, they proposed an efficient user assignment algorithm to enhance the decentralized system's performance. In [57], the authors proposed an OFDM-based system that takes into consideration two constraints: the transmit power and fronthaul capacity of each AP in the DL transmission. They adopted two quadratic transform techniques to optimize the minimum rate of different users and enhance the system sum rate performance. In [58], it was assumed that the proposed UL architecture has multiple CPUs. The APs are connected to each CPU through wired mmWave-fronthaul links based on their respective distances. Under this assumption, the study derived two expressions: the density of the activated APs as a function of the blockage and CPU densities, and the distribution of achievable fronthaul capacity, assuming equal bandwidth allocation to all APs.

Unlike [53–58], the studies [30, 59] considered mmWave communications in the access links between the transmitter and the receiver instead of fading channels to evaluate their systems performance. In [30], a low-complexity hybrid beamforming implementation was proposed based on large-scale second-order statistics of the propagation channel. Along with the fronthaul capacity constraints, the study tackled several challenges such as imperfect channel estimation, power control, and non-orthogonality of pilot sequences. The optimization problems were effectively solved by combining the use of block coordinate descent (BCD) techniques with sequential linear algorithms, resulting in a consistent and high-QoS over the whole network coverage area. In [59], the authors addressed the impact of low-capacity fronthaul connections on system performance, as well as the impact of low-resolution ADC and digital-to-analog converter (DAC) on

UL channel estimation and DL data transmission. To mitigate channel estimation errors, the authors optimized the minimum MSE-achieving codebook associated with the fronthaul compression. Additionally, they proposed an alternative optimization method that alternated between two sub-problems: power allocation and codebook design. This approach aimed to address the max–min fairness problem for the MR and ZF precoding schemes.

4.3.2 System Model and Design

The main objective of this section is to present a model and a design overview for a scalable CF system, that is, with a reduced number of fronthaul connections between the APs and CPU. In the following, it is assumed that each UE transmits a single data stream to the APs and then the resulting signals from the APs are sent to the CPU for combination via limited capacity fronthaul connections. The association of the UE-APs is performed by the CPU, in which the switches $\mathbf{T}_m$ for mth AP control this association procedure, as

$$\hat{\mathbf{s}}_{t,m,k} = \mathbf{T}_m \, \hat{\mathbf{s}}_{hyb,m,k}, \tag{4.6}$$

where $\mathbf{T}_m = \text{diag}(t_{m,1}, \ldots, t_{m,u}, \ldots, t_{m,U})$ represents a diagonal matrix with the diagonal entries set to zero or one. When the entry is zero, it indicates that the associated user is not served by the corresponding AP, while when equal to one, the user is served by that AP. Therefore, the number of nonzero elements in matrix $\mathbf{T}_m$ determines the number of signals sent to the fronthaul by the mth AP. Finally, the CPU data combiner receives these signals and combines them to obtain the final data estimate for the users. This can be expressed as

$$\hat{\mathbf{s}}_{UC,k} = \sum_{m=1}^{M} \mathbf{V}_m \hat{\mathbf{s}}_{t,m,k} = \sum_{m=1}^{M} \mathbf{V}_m \mathbf{T}_m \hat{\mathbf{s}}_{hyb,m,k}, \tag{4.7}$$

where $\mathbf{V}_m$ denotes the CPU combiner weight matrix relative to the mth AP.

In order to maximize the system performance under limited fronthaul capacity, one option is to jointly design the user selectors and CPU data combiner, presented in [51], using a similar to (4.5) MSE optimization problem but with different constraints, formally

$$\begin{aligned}
\underset{\{\mathbf{T}_m, \mathbf{V}_m\}_m}{\text{minimize}} \quad & N_c^{-1} \sum_{k=1}^{N_c} \mathbb{E}\left[\|\hat{\mathbf{s}}_{UC,k} - \mathbf{s}_k\|^2 \right] \\
\text{subject to} \quad & \sum_{m=1}^{M} \mathbf{T}_m \mathbf{V}_m = \mathbf{I}, \ \forall m \in [M] \\
& \sum_{m=1}^{M} \mathbf{T}_m = N_u \mathbf{I}, \ \forall m \in [M] \\
& t_{m,u} \in \{0, 1\}, \quad \forall m \in [M], u \in [U],
\end{aligned} \tag{4.8}$$

where the first constraint enforces zero bias at the CPU, while the second and the third ones determine the number of connections serving a specific UE. The active connections are denoted by $t_{m,u} = 1$ and the inactive ones by $t_{m,u} = 0$.

To find a solution to (4.8), one approach is to employ the data selector (switch)-based method proposed in [51]. This method determines the optimal configuration for the switches $t_{m,u}$ for mth AP and uth UE by sorting and selecting the N_u APs with the highest SINR for each user.

4.3.3 Numerical Results

To emphasize the performance of UC CF mMIMO system, Figure 4.8 compares the BER performance of the work proposed in [51] that solves (4.8) and [45], considering a scenario with $M = 16$, $U = N_{rx}^{\text{RF}} = 2$, and $N_t = N_r = 16$ antennas, with different number of connections per user. The solid line curves present the performance of the switches-based method [51], while the dashed line curves indicate the LSF-based approach [45]. Herein, the curve of 16 APs–UE corresponds to the case of conventional CF mMIMO system, in which each user is served by all APs. The numerical results showed a clear tradeoff between performance and fronthaul requirements. The switches method proposed in [51] achieves very close BER performance compared to the conventional CF system when employing a small subset of APs per UE (4 APs/UE). This leads to reduced fronthaul requirements. However, it can be observed that the BER performance is getting worse as the number of fronthaul connections decreases. The gap widens notably when each

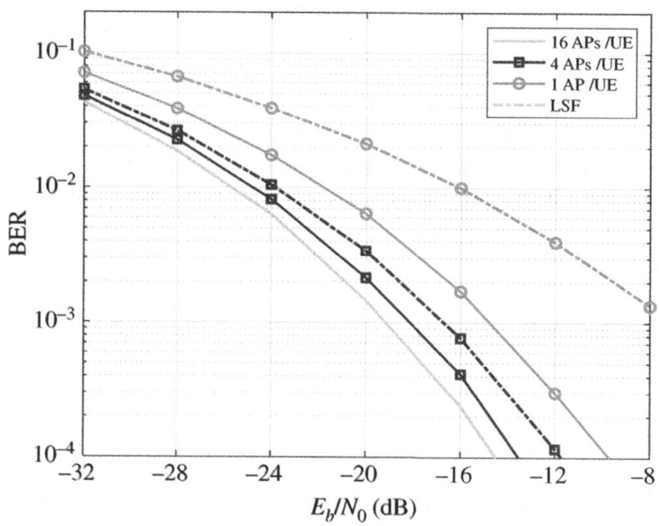

Figure 4.8 BER performance comparison between the UE–APs association method proposed in [51] and the LSF-based approach presented in [45].

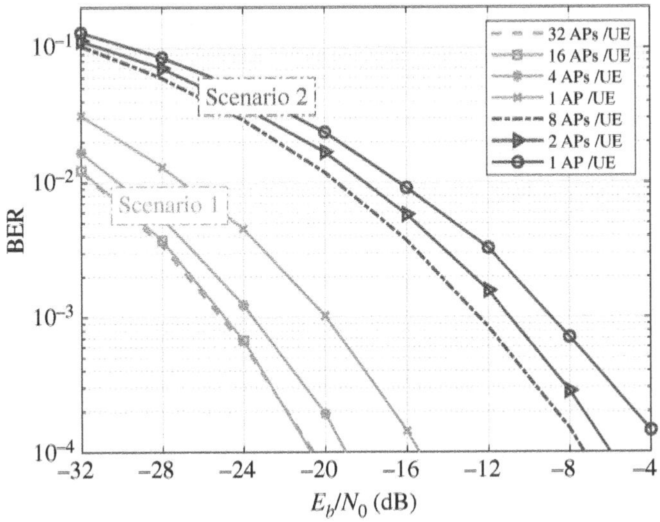

Figure 4.9 BER performance comparison between two scenarios based on the proposed UE–APs association method presented in [51]. Source: Adapted from [51].

UE is served by just one AP, particularly in the case of LSF approach. This goes back to the fact that in [51] the CPU combiner and UE–APs association are jointly performed, whereas in the LSF [45], these two aspects are decoupled. Consequently, the switches-based method of [51] is more resource efficient, requiring fewer fronthaul connections to achieve a specific performance target.

Figure 4.9 presents a comparison of the BER performance between two different scenarios, with all parameters identical to those in Figure 4.8, except for the number of APs. Specifically, we consider 32 and 8 APs for scenario 1 and scenario 2, respectively. The figure clearly shows that as the number of deployed APs serving a particular user increase, the gap between the BER curves is getting higher. However, despite the exponential increase in the deployed APs, the number of connections required does not increase proportionally. For instance, in scenario 1, achieving comparable performance to the full connection setup (32 APs/UE) only requires around 4 APs. This observation highlights the scalability of the switch-based method [51], where a high number of deployed APs does not always imply high fronthaul requirements.

4.4 Radio Stripes Cell-Free Systems

In practical implementation of CF mMIMO systems, the internal connection and cabling between APs can indeed pose significant challenges. However, integrating

RS with such systems can be considered an interesting way to provide a suitable and cost-efficient solution [40]. Therefore, this section starts with the main limitations of CF systems, followed by introducing RS technology as a prospective approach within CF systems.

4.4.1 Cell-Free Systems Limitations

Although CF mMIMO systems offer several advantages, it has some limitations and challenges besides the scalability, such as complex coordination of a large number of APs, reliance on high-quality backhaul connections, potential high energy consumption, interference management issues due to serving multiple users simultaneously, and deployment costs. Despite these limitations, ongoing research and technological advancements aim to address these challenges and make CF systems more practical and efficient for various use cases. Therefore, this section addresses the integration of CF systems with RS technology as a potential cost-effective solution.

4.4.2 Integrating Radio Stripes with Cell-Free Systems

The deployment of RSs, as depicted in Figure 4.10, is appropriate for indoor/pedestrian scenarios such as stadiums, train stations, and shopping centers [60]. As known, conventional CF mMIMO networks necessitate separate cables to connect the APs with the CPU (star topology). This requirement results in high implementation costs and unscalable system. In contrast, in RS systems, the antennas and the antenna processing units (APUs) are located in a serial/sequential manner within a single fronthaul cable [6]. This cable performs synchronization, data transfer, and power supply, all facilitated via a shared bus. Herein, each RS is connected to one or more CPUs. The APs themselves have antenna elements and circuit-mounted chips, including power amplifiers, phase shifters, filters, modulators, and ADC/DAC converters. These APs can cooperate coherently, making the RS effectively operates as a multiple-antenna AP.

The transmit and receive processing of antennas are performed locally [61]. On the transmitter side, each APU receives multiple input data stream from the previous APU via the shared bus. Within each antenna, these streams are appropriately scaled using a pre-calculated precoding vector. Then, the combined signal is transmitted over the channel to the receiver. On the receiver side, the received signal is multiplied by the combining vector that was previously calculated. This multiplication generates multiple data streams. The processed streams are subsequently combined with the data streams received from the shared bus and retransmitted on the shared bus to the next APU in the system. To be more specific, each APU sums its received data streams with the input streams from the previous APU,

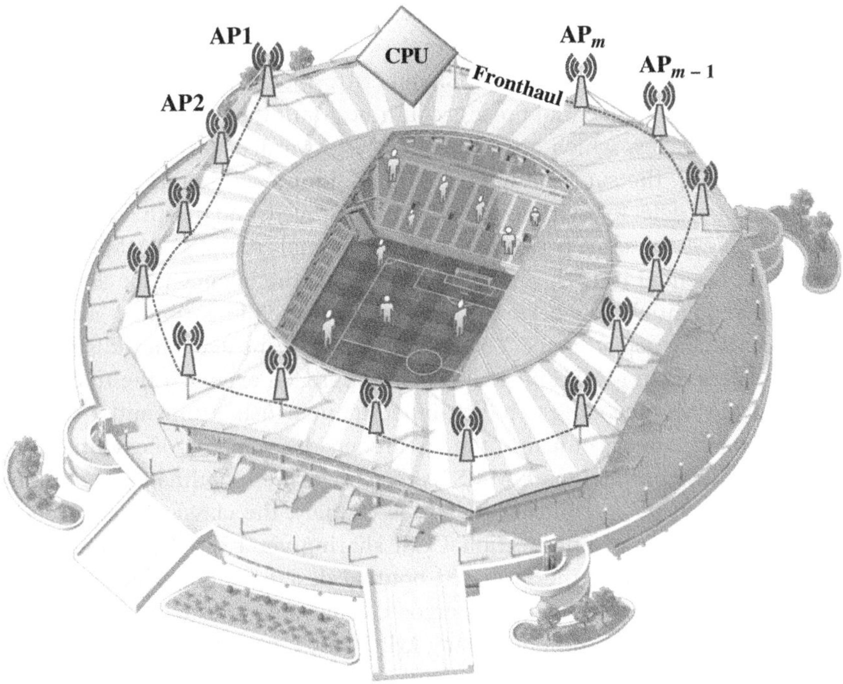

Figure 4.10 Radio stripe deployment in stadium.

which consists of combined signals originating from APUs 1 to $m - 1$, intended for one or more UEs. The purpose of this combining process is to merge the individual streams into a cumulative signal. This cumulative signal is then passed on to the next APU in sequential order. The process of combining signals is accomplished through a simple addition operation performed on each data stream.

The placement of the multiple antennas on the RS relies on the carrier frequency [60]. At higher frequencies, the antennas can be positioned close together due to their compact size, while at lower frequencies, they may need to be spread throughout the RS to guarantee optimal performance. As the system assumes a significant number of antennas, it becomes feasible to maintain a low transmit power for each individual antenna [61]. This configuration offers several advantages, including minimal heat dissipation, which simplifies and reduces the cost of cooling systems, smaller volume and weight, as well as lower overall cost. Moreover, the highly distributed functionality ensures that even if a few RSs fail, the network will be largely unaffected. Therefore, the robust and resilient nature of the RS system leads to reduced maintenance costs [40].

Research focused on RS architecture with CF mMIMO systems is still at an early stage [62–67]. Shaik et al. [62] proposed a sequential UL detection method, where each AP independently computes its own local channel estimates and uses normalized linear MMSE combining to make soft estimates of the desired signals. These soft estimates, along with the CSI and error statistics, are then transmitted to the next AP. This sequential process continues until the last AP computes the final signal estimates, which are subsequently forwarded to the CPU for final decoding. In [63], the same authors of [62] additionally developed an optimal linear sequential detector with a trade-off between the SE and fronthaul signaling, aiming to achieve the same performance of the optimal centralized MMSE implementation, presented in [27]. Then, they addressed the latency issue associated with long sequential networks by introducing a semi-distributed approach without sacrificing performance. The work of [64] investigated a new distributed approach called team MMSE precoding. This approach extended the traditional centralized MMSE precoding to a distributed manner using transmitter-specific CSI. Based on the theory of teams, the researchers derived a set of essential conditions to achieve optimal team MMSE precoding. These optimality conditions were examined in two significant scenarios, namely, the classical case of local CSI and the scenario of unidirectional CSI sharing via a serial fronthaul. The results demonstrated that the proposed optimal design outperformed heuristic methods in terms of achieving higher ergodic achievable rates while adhering to a sum-power constraint. Although using MRC on CF mMIMO systems that are integrated with RS can reduce the fronthaul loads and produce cost-effective APs, the performance is quite bad. To overcome these limitations, the authors of [65] proposed a solution by designing an SVD-based algorithm called quasi-LMMSE, which was derived from standard linear MMSE using only the MRC fronthaul signals, aiming to achieve higher SE while maintaining low fronthaul loading, minimizing AP complexity and latency. In [66], the research was concentrated on designing a joint optimization framework by iterating between power control and AP scheduling under fronthaul bandwidth constraint. The proposed systems aimed to achieve a maximum power to each user while effectively manage the limited shared fronthaul bandwidth. Another approach was considered in [67], where a random AP switch ON/OFF strategy was suggested to reduce power consumption and improve the achievable EE.

4.5 Conclusions and Future Work

This chapter on CF mMIMO technology presented a comprehensive analysis of its capabilities and applications for 6G wireless networks. It highlighted the advantages of CF mMIMO over traditional cellular networks, including enhanced network capacity, coverage, and service quality. Then, it described the system architecture of CF mMIMO, including the recent literature review, as well as the system model and the design overview, which provided a noticeable performance in minimizing intercell interference. The UC approach of CF mMIMO systems has been presented, along with recent research developments to address scalability issues. The chapter has also explored the integration of CF mMIMO systems with other wireless technologies, namely, RSs for improved performance, reduced costs, and simplified deployment.

Looking toward the future of wireless communication networks, CF mMIMO stands out as a highly promising technology for 6G. With its potential to deliver remarkable performance, CF mMIMO offers an appealing solution by ensuring seamless connectivity for all users across the network's coverage area. Therefore, the future direction for CF mMIMO systems can extend beyond RS integration and include investigating how they might function along with other technologies for 6G networks. This may involve combining CF mMIMO with reconfigurable intelligent surface (RIS), large intelligent surface (LIS), unmanned aerial vehicle (UAV), and artificial intelligence (AI) techniques. This is a promising approach that may unlock a diverse range of innovative wireless network applications and services in the upcoming generations of wireless communications systems.

Acknowledgment

This work has received funding from the FCT – Fundação para a Ciencia e a Tecnologia under the PhD research Studentships (SFRH/BD/05674/2020) and REVOLUTION project 2022.08005.PTDC.

References

1 A. J. Paulraj, D. A. Gore, R. U. Nabar and H. Bolcskei, "An overview of MIMO communications-a key to gigabit wireless," *Proceedings of the IEEE*, vol. 92, no. 2, pp. 198–218, Feb. 2004.

2 Wang, C.-X., Haider, F., Gao, X. et al. (2014). Cellular architecture and key technologies for 5G wireless communication networks. *IEEE Communications Magazine* 52 (2): 122–130.

3 Larsson, E.G., Edfors, O., Tufvesson, F., and Marzetta, T.L. (2014). Massive MIMO for next generation wireless systems. *IEEE Communications Magazine* 52 (2): 186–195.

4 K. A. Alnajjar, P. J. Smith and G. K. Woodward, "Co-located and distributed antenna systems: deployment options for massive multiple-input–multiple-output," *IET Microwaves, Antennas and Propagation*, vol. 9, no. 13, pp. 1418–1424, Oct. 2015.

5 Dai, L. (2011). A comparative study on uplink sum capacity with co-located and distributed antennas. *IEEE Journal on Selected Areas in Communications* 29 (6): 1200–1213.

6 Elhoushy, S., Ibrahim, M., and Hamouda, W. (2022). Cell-free massive MIMO: a survey. *IEEE Communications Surveys and Tutorials* 24 (1): 492–523. 1st Quart.

7 Jungnickel, V. et al. (2014). The role of small cells, coordinated multipoint, and massive MIMO in 5G. *IEEE Communications Magazine* 52 (5): 44–51.

8 Nam, W., Bai, D., Lee, J., and Kang, I. (2014). Advanced interference management for 5G cellular networks. *IEEE Communications Magazine* 52 (5): 52–60.

9 Zhang, J., Chen, R., Andrews, J.G. et al. (2009). Networked MIMO with clustered linear precoding. *IEEE Transactions on Wireless Communications* 8 (4): 1910–1921.

10 Irmer, R., Droste, H., Marsch, P. et al. (2011). Coordinated multipoint: concepts, performance, and field trial results. *IEEE Communications Magazine* 49 (2): 102–111.

11 Castanheira, D. and Gameiro, A. (2010). Distributed antenna system capacity scaling [coordinated and distributed MIMO]. *IEEE Wireless Communications* 17 (3): 68–75.

12 You, X.-H., Wang, D.-M., Sheng, B. et al. (2010). Cooperative distributed antenna systems for mobile communications. *IEEE Wireless Communications* 17 (3): 35–43.

13 He, H., Yu, X., Zhang, J. et al. (2021). Cell-free massive MIMO for 6G wireless communication networks. *Journal of Communications and Information Networks* 6 (4): 321–335.

14 Ngo, H.Q., Ashikhmin, A., Yang, H. et al. (2017). Cell-free massive MIMO versus small cells. *IEEE Transactions on Wireless Communications* 16 (3): 1834–1850.

15 Zhang, J., Chen, S., Lin, Y. et al. (2019). Cell-free massive MIMO: a new next generation paradigm. *IEEE Access* 7: 99878–99888.

16 Giordani, M., Polese, M., Mezzavilla, M. et al. (2020). Toward 6G networks: use cases and technologies. *IEEE Communications Magazine* 58 (3): 55–61.

17 Rappaport, T.S. et al. (2013). Millimeter wave mobile communications for 5G cellular: it will work! *IEEE Access* 1: 335–349.

18 Salameh, A.I. and El Tarhuni, M. (2022). From 5G to 6G—challenges, technologies, and applications. *Future Internet* 14 (4): 117.

19 Swindlehurts, A., Ayanoglu, E., Heydari, P., and Capolino, F. (2014). Millimeter-wave massive MIMO: the next wireless revolution? *IEEE Communications Magazine* 52 (9): 52–62.

20 Ahmed, I., Khammari, H., Shahid, A. et al. (2018). A survey on hybrid beamforming techniques in 5G: architecture and system model perspectives. *IEEE Communications Surveys and Tutorials* 20 (4): 3060–3097. 4th Quart.

21 Boccardi, F., Heath, R.W., Lozano, A. et al. (2014). Five disruptive technology directions for 5G. *IEEE Communications Magazine* 52 (2): 74–80.

22 Bashar, M., Cumanan, K., Burr, A.G. et al. (2019). On the uplink max–min SINR of cell-free massive MIMO systems. *IEEE Transactions on Wireless Communications* 18 (4): 2021–2036.

23 Zhang, J., Zhang, J., Ng, D.W.K. et al. (2021). Improving sum-rate of cell-free massive MIMO with expanded compute-and-forward. *IEEE Transactions on Signal Processing* 70: 202–215.

24 Song, H., Goldstein, T., You, X. et al. (2022). Joint channel estimation and data detection in cell-free massive MU-MIMO systems. *IEEE Transactions on Wireless Communications* 21 (6): 4068–4084.

25 Mai, T.C., Ngo, H.Q., and Tran, L.-N. (2022). Energy efficiency maximization in large-scale cell-free massive MIMO: a projected gradient approach. *IEEE Transactions on Wireless Communications* 21 (8): 6357–6371.

26 Nayebi, E., Ashikhmin, A., Marzetta, T.L. et al. (2017). Precoding and power optimization in cell-free massive MIMO systems. *IEEE Transactions on Wireless Communications* 16 (7): 4445–4459.

27 Björnson, E. and Sanguinetti, L. (2019). Making cell-free massive MIMO competitive with MMSE processing and centralized implementation. *IEEE Transactions on Wireless Communications* 19 (1): 77–90.

28 Interdonato, G., Ngo, H.Q., and Larsson, E.G. (2021). Enhanced normalized conjugate beamforming for cell-free massive MIMO. *IEEE Transactions on Communications* 69 (5): 2863–2877.

29 Ataeeshojai, M., Elliott, R.C., Krzymień, W.A. et al. (2022). Iterative matrix inversion methods for precoding in cell-free massive MIMO systems. *IEEE Transactions on Vehicular Technology* 71 (11): 11972–11987.

30 Femenias, G. and Riera-Palou, F. (2019). Cell-free millimeter-wave massive MIMO systems with limited fronthaul capacity. *IEEE Access* 7: 44596–44612.

31 Nguyen, N.T., Lee, K., and Dai, H. (2022). Hybrid beamforming and adaptive RF chain activation for uplink cell-free millimeter-wave massive MIMO systems. *IEEE Transactions on Vehicular Technology* 71 (8): 8739–8755.

32 Hong, S.E. (2021). On the effect of shadowing correlation and pilot assignment on hybrid precoding performance for cell-free mmWave massive MIMO UDN system. *ICT Express* 7 (1): 60–70.

33 Femenias, G. and Riera-Palou, F. (2022). Wideband cell-free mmWave massive MIMO-OFDM: beam squint-aware channel covariance-based hybrid beamforming. *IEEE Transactions on Wireless Communications* 21 (7): 4695–4710.

34 Kassam, J., Castanheira, D., Silva, A. et al. (2022). Distributed hybrid equalization for cooperative millimeter-wave cell-free massive MIMO. *IEEE Transactions on Communications* 70 (8): 5300–5316.

35 Venkatesan, S., Huang, H., Lozano, A., and Valenzuela, R. (2009). A WiMAX-based implementation of network MIMO for indoor wireless systems. *EURASIP Journal on Advances in Signal Processing* 2009 (1): 963547.

36 Kobayashi, M., Debbah, M., and Belfiore, J. (2009). Outage efficient strategies in network MIMO with partial CSIT. *Proc. IEEE ISIT'09*, pp. 249–253.

37 Björnson, E., Zakhour, R., Gesbert, D., and Ottersten, B. (2010). Cooperative multicell precoding: rate region characterization and distributed strategies with instantaneous and statistical CSI. *IEEE Transactions on Signal Processing* 58 (8): 4298–4310.

38 Boldi, M., Tölli, A., Olsson, M. et al. (2011). Coordinated multipoint (CoMP) systems. In: *Mobile and Wireless Communications for IMT-Advanced and Beyond* (ed. A. Osseiran, J. Monserrat, and W. Mohr), 121–155. Wiley.

39 Zhang, Z., Wang, N., Zhang, J., and Mu, X. (2018). Dynamic user-centric clustering for uplink cooperation in multi-cell wireless networks. *IEEE Access* 6: 8526–8538.

40 Kassam, J., Castanheira, D., Silva, A. et al. (2023). A review on cell-free massive MIMO systems. *Electronics* 12 (4): 1001.

41 Björnson, E. and Sanguinetti, L. (2020). Scalable cell-free massive MIMO systems. *IEEE Transactions on Communications* 68 (7): 4247–4261.

42 Dao, H.T. and Kim, S. (2020). Effective channel gain-based access point selection in cell-free massive MIMO systems. *IEEE Access* 8: 108127–108132.

43 Alonzo, M., Baracca, P., Khosravirad, S.R., and Buzzi, S. (2021). Cell-free and user-centric massive MIMO architectures for reliable communications in indoor factory environments. *IEEE Open Journal of the Communications Society* 2: 1390–1404.

44 Gao, X., Li, Y., Cheng, W. et al. (2023). Secure optimal precoding for user-centric cell-free massive MIMO system. *IEEE Wireless Communications Letters* 12 (1): 31–35.

45 Ngo, H.Q., Tran, L.-N., Duong, T.Q. et al. (2018). On the total energy efficiency of cell-free massive MIMO. *IEEE Transactions on Green Communications and Networking* 2 (1): 25–39.

46 Zhou, X., Xia, H., and Wu, S. (2022). A successful transmission probability based AP clustering scheme in user-centric cell-free networks. *China Communications* 19 (12): 118–128.

47 Agheli, P., Emadi, M.J., and Beyranvand, H. (2023). Cognitive RF–FSO fronthaul assignment in cell-free and user-centric mMIMO networks. *IEEE Transactions on Mobile Computing* 22 (5): 2537–2550.

48 Alonzo, M., Buzzi, S., Zappone, A., and D'Elia, C. (2019). Energy-efficient power control in cell-free and user-centric massive MIMO at millimeter wave. *IEEE Transactions on Green Communications and Networking* 3 (3): 651–663.

49 Guo, J.-C., Yu, Q.-Y., Sun, W.-B., and Meng, W.-X. (2021). Robust efficient hybrid pre-coding scheme for mmWave cell-free and user-centric massive MIMO communications. *IEEE Transactions on Wireless Communications* 20 (12): 8006–8022.

50 Demirhan, U. and Alkhateeb, A. (2022). Enabling cell-free massive MIMO systems with wireless millimeter wave fronthaul. *IEEE Transactions on Wireless Communications* 21 (11): 9482–9496.

51 Kassam, J., Castanheira, D., Silva, A. et al. (2023). Joint decoding and UE-APs association for scalable cell-free systems. *IEEE Transactions on Communications* 71 (12): 7303–7315.

52 Ammar, H.A., Adve, R., Shahbazpanahi, S. et al. (2022). User-centric cell-free massive MIMO networks: a survey of opportunities, challenges and solutions. *IEEE Communications Surveys and Tutorials* 24 (1): 611–652. 1st Quart.

53 Masoumi, H. and Emadi, M.J. (2020). Performance analysis of cell-free massive MIMO system with limited fronthaul capacity and hardware impairments. *IEEE Transactions on Wireless Communications* 19 (2): 1038–1053.

54 Femenias, G. and Riera-Palou, F. (2020). Fronthaul-constrained cell-free massive MIMO with low resolution ADCs. *IEEE Access* 8: 116195–116215.

55 Bashar, M., Xiao, P., Tafazolli, R. et al. (2021). Limited-fronthaul cell-free massive MIMO with local MMSE receiver under Rician fading and phase shifts. *IEEE Wireless Communications Letters* 10 (9): 1934–1938.

56 Bashar, M., Ngo, H.Q., Cumanan, K. et al. (2021). Uplink spectral and energy efficiency of cell-free massive MIMO with optimal uniform quantization. *IEEE Transactions on Communications* 69 (1): 223–245.

57 Xu, K., Huang, M., Zhang, D. et al. (2022). User rate optimization for fronthaul-constrained cell-free massive MIMO-OFDM systems: a quadratic transform-based approach. *IEEE Open Journal of the Communications Society* 3: 1244–1251.

58 Ibrahim, M., Elhoushy, S., and Hamouda, W. (2022). Uplink performance of mmWave-fronthaul cell-free massive MIMO systems. *IEEE Transactions on Vehicular Technology* 71 (2): 1536–1548.

59 Kim, I.-s., Bennis, M., and Choi, J. (2022). Cell-free mmWave massive MIMO systems with low-capacity fronthaul links and low-resolution ADC/DACs. *IEEE Transactions on Vehicular Technology* 71 (10): 10512–10526.

60 López, O.L.A., Kumar, D., Souza, R.D. et al. (2022). Massive MIMO with radio stripes for indoor wireless energy transfer. *IEEE Transactions on Wireless Communications* 21 (9): 7088–7104.

61 Interdonato, G., Bjornson, E., Ngo, H.Q. et al. (2019). Ubiquitous cell-free massive MIMO communications. *EURASIP Journal on Wireless Communications and Networking* 2019 (1): 1–13.

62 Shaik, Z.H., Björnson, E. and Larsson, E.G. (2020). Cell-free massive MIMO with radio stripes and sequential uplink processing. *Proc. IEEE Int. Conf. Commun. (ICC Workshops)*, June 2020, pp. 1–6.

63 Shaik, Z.H., Björnson, E., and Larsson, E.G. (2021). MMSE-optimal sequential processing for cell-free massive MIMO with radio stripes. *IEEE Transactions on Communications* 69 (11): 7775–7789.

64 Miretti, L., Björnson, E., and Gesbert, D. (2022). Team MMSE precoding with applications to cell-free massive MIMO. *IEEE Transactions on Wireless Communications* 21 (8): 6242–6255.

65 Yuan, Z., Ma, Y., and Yu, G. (2022). LMMSE processing for cell-free massive MIMO with radio stripes and MRC fronthaul. *Proc. IEEE Wirel. Commun. Netw. Conf. (WCNC)*, April 2022, pp. 1336–1340.

66 Guenach, M., Gorji, A.A., and Bourdoux, A. (2021). Joint power control and access point scheduling in fronthaul-constrained uplink cell-free massive MIMO systems. *IEEE Transactions on Communications* 69 (4): 2709–2722.

67 Mishra, A.K. and Vijayakumar, P. (2021). AP switch ON/OFF strategy in green cell-free massive MIMO based radio stripe network. *Int. Conf. Comput. Commun. Signal Process. (ICCCSP)*, May 2021, pp. 125–130.

5

Localization in Massive MIMO Networks: From Far-Field to Near-Field

Parisa Ramezani[1], Özlem Tuğfe Demir[2], and Emil Björnson[1]

[1] *Department of Computer Science, KTH Royal Institute of Technology, Stockholm, Sweden*
[2] *Department of Electrical and Electronics Engineering, TOBB University of Economics and Technology, Ankara, Türkiye*

5.1 Introduction

Antenna arrays with massive antenna numbers are commonplace at the base stations in 5G networks. These arrays are deployed to enable massive multiple-input multiple-output (MIMO) communications between the multi-antenna base station and a large number of spatially multiplexed user devices [1]. However, these arrays might be used for additional services in the future [2], and this chapter focuses on localization services.

Source localization is an important problem in the field of array signal processing and has a wide range of applications in sonar, radar, seismic exploration, and wireless communications [3–8]. As such, it has drawn considerable attention over the past decades and various algorithms have been developed to address this problem [7–11]. Subspace-based approaches are among the most popular and widely used classical techniques for source localization when using a single antenna array [6]. These approaches utilize the properties of the signal and noise subspaces of the received signals to estimate the locations of the sources. In particular, they are built on the fact that the signal subspace is generally of lower dimension than the total measurement space which helps in separating the signal from the noise. Subspace-based methods often provide high spatial resolution, meaning that they can accurately estimate the location of closely spaced sources. However, their performance is dependent on various factors such as the array geometry, signal-to-noise ratio (SNR) and number of antennas (sensors). Two classical subspace-based localization methods are MUltiple SIgnal Classification (MUSIC) [12, 13] and Estimation of Signal Parameters by Rotational Invariance

Massive MIMO for Future Wireless Communication Systems: Technology and Applications,
First Edition. Edited by Webert Montlouis and Agbotiname Lucky Imoize.

Techniques (ESPRIT) [13, 14], which, respectively, exploit the noise and signal subspace of the received signals for localizing the sources.

When considering large antenna arrays, the sources can be classified into two categories based on their distance from the array: far-field sources and near-field sources. Far-field sources are characterized by planar wavefronts, where the curvature of the wave arriving at the antenna array is negligible and the localization only involves estimating the direction of arrival (DoA) of the source. When the source is located in the radiative near-field region of the antenna array, the spherical wavefront of the received signal can no longer be neglected and the locations of the sources are specified by both the DoA and range [15].

5.1.1 Key Contributions

This chapter provides an overview of source localization considering both far-field and near-field scenarios. We first present the popular MUSIC and ESPRIT algorithms for cases where the sources are situated in the far-field of the antenna array. We then describe the generalized ESPRIT approach, which relaxes the strict shift-invariance requirement of the conventional ESPRIT algorithm and is applicable to a general class of subarray geometries. Afterward, we describe the application of MUSIC and ESPRIT algorithms to near-field localization, in which both the DoA and range must be estimated for localizing the sources. The two-dimensional (2D) MUSIC is first put forth, which simultaneously estimates the DoAs and ranges via a 2D grid search over all possible angles and ranges. This grid search can lead to high computational costs, especially when the grid resolution is fine. To alleviate this issue, some algorithms have been developed to enable the separate estimation of DoAs and ranges. We present two methods based on MUSIC and ESPRIT algorithms that exploit the symmetric geometry of the antenna array to decouple the DoA and range estimation problems. These methods first estimate the DoAs of all sources via a one-dimensional (1D) search over possible angles and then perform multiple 1D searches to obtain the range associated with each estimated DoA. In this way, the location of sources can be estimated through multiple 1D searches, thereby reducing the complexity of the 2D search required by the conventional MUSIC algorithm.

5.1.2 Chapter Organization

In Section 5.2, we introduce MUSIC and ESPRIT algorithms for far-field localization and also present the generalized ESPRIT method that extends the standard ESPRIT algorithm to a more general class of array geometries. Section 5.3 discusses near-field localization and clarifies how the standard 1D MUSIC can be extended to 2D MUSIC for locating near-field sources. This section also introduces symmetry-based localization where the symmetric structure of the array is exploited to decouple DoA and range estimation problems in the near-field. Finally, Section 5.4 provides concluding remarks.

5.1.3 Notations

Scalars are denoted by italic letters, and vectors and matrices are denoted by bold-face lower-case and upper-case letters, respectively. $\mathbb{C}$ represents the set of complex numbers and $\mathcal{N}_{\mathbb{C}}(0, \sigma^2)$ indicates a circularly symmetric complex Gaussian distribution with variance σ^2. $(\cdot)^*$ indicates conjugation, and $(\cdot)^{\mathrm{T}}$, $(\cdot)^{\mathrm{H}}$, and $(\cdot)^{-1}$ represent transpose, conjugate transpose, and inverse of a matrix, respectively. $\mathrm{diag}(\mathbf{x})$ represents a diagonal matrix having $\mathbf{x}$ on its main diagonal, and $\det(\cdot)$ denotes the determinant of a matrix.

5.2 Far-Field DoA Estimation

In this section, we review the conventional MUSIC and ESPRIT algorithms for locating far-field sources and numerically evaluate their performance. We further introduce the generalized ESPRIT algorithm that extends the concept of standard ESPRIT by allowing for arbitrary displacements between the corresponding antennas of the subarrays.

5.2.1 System Model

Assume K uncorrelated narrowband signals with unknown DoAs, $\theta_1, \theta_2, \ldots, \theta_K$, impinge on a uniform linear array (ULA) with M antennas. The sources have a line-of-sight connection to the receiving array and they are assumed to be located in the far-field. The antenna separation is denoted by d. We assume $K < M$, which will later be required by the MUSIC algorithm to construct a noise subspace. At time slot t, the received signal by the antenna array is obtained as

$$\mathbf{y}(t) = \sum_{k=1}^{K} e^{j\psi_k} \mathbf{a}(\theta_k) s_k(t) + \mathbf{n}(t), \tag{5.1}$$

where $\mathbf{y}(t) = [y_1(t), \ldots, y_M(t)]^{\mathrm{T}}$ is the received signal at the antenna array, and $s_k(t)$ is the random signal from the kth source, which is unknown at the receiver. The $\mathbf{n}(t) = [n_1(t), \ldots, n_M(t)]^{\mathrm{T}} \sim \mathcal{N}_{\mathbb{C}}(\mathbf{0}, \sigma^2 \mathbf{I}_M)$ is the additive independent complex Gaussian noise and the term $e^{j\psi_k}$ in (5.1) represents the phase-shift on the first antenna of the array. The $\mathbf{a}(\theta_k)$ is the array response vector for the kth source and expressed as

$$\mathbf{a}(\theta_k) = \left[1, \dots, e^{-j\frac{2\pi}{\lambda}\left(\overline{r}_k^m(\theta_k)-r_k\right)}, \dots, e^{-j\frac{2\pi}{\lambda}\left(\overline{r}_k^M(\theta_k)-r_k\right)}\right]^{\mathrm{T}}, \tag{5.2}$$

where r_k is the distance between the kth source and the first antenna, and $\overline{r}_k^m(\theta_k)$ denotes the distance between the source and the mth antenna ($m \neq 1$). The λ represents the wavelength of the transmitted signals. Consider the antenna array shown in Figure 5.1. According to the figure, $\overline{r}_k^m(\theta_k)$ is obtained as

$$\begin{aligned}
\overline{r}_k^m(\theta_k) &= \sqrt{r_k^2 + (m-1)^2 d^2 - 2r_k(m-1)d\sin(\theta_k)} \\
&= r_k\sqrt{1 + \frac{(m-1)^2 d^2}{r_k^2} - 2\frac{(m-1)d}{r_k}\sin(\theta_k)} \\
&\overset{(a)}{\approx} r_k\left(1 + \frac{(m-1)^2 d^2}{2r_k^2} - \frac{(m-1)d}{r_k}\sin(\theta_k)\right) \\
&\overset{(b)}{\approx} r_k - (m-1)d\sin(\theta_k),
\end{aligned} \tag{5.3}$$

where the first-order Taylor approximation $\sqrt{1+x} \approx 1 + \frac{x}{2}$ for $|x| \ll 1$ is used in (a), and (b) holds because $d^2 \ll r_k^2$ in the far field. Substituting (5.3) into (5.2), we arrive at

$$\mathbf{a}(\theta_k) = \left[1, \dots, e^{j(m-1)\gamma_k}, \dots, e^{j(M-1)\gamma_k}\right]^{\mathrm{T}}, \tag{5.4}$$

where $\gamma_k = 2\pi\frac{d}{\lambda}\sin(\theta_k)$.

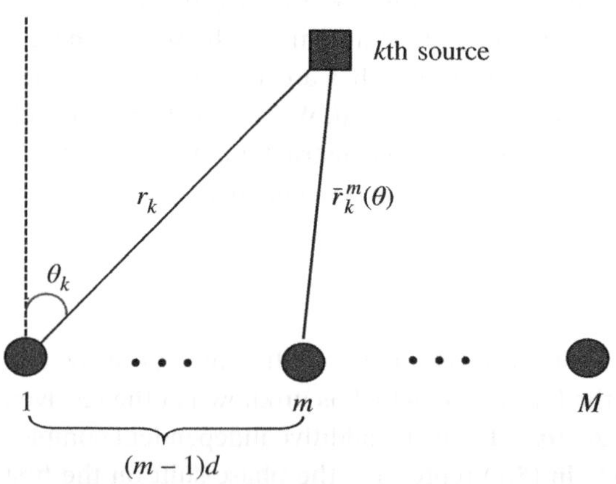

Figure 5.1 ULA with M antennas.

Collecting all array response vectors in matrix $\mathbf{A} = \left[\mathbf{a}(\theta_1), \ldots, \mathbf{a}(\theta_K)\right] \in \mathbb{C}^{M\times K}$ and all source signals in vector $\mathbf{s} = [s_1(t), \ldots, s_K(t)]^{\mathrm{T}} \in \mathbb{C}^K$, Eq. (5.1) can be rewritten as

$$\mathbf{y}(t) = \mathbf{A}\,\mathrm{diag}\left(e^{j\psi_1}, \ldots, e^{j\psi_K}\right)\mathbf{s}(t) + \mathbf{n}(t). \tag{5.5}$$

The covariance matrix of the received signal is given by

$$\mathbf{R} = \mathbb{E}\left\{\mathbf{y}(t)\mathbf{y}^{\mathrm{H}}(t)\right\} = \mathbf{ASA}^{\mathrm{H}} + \sigma^2\mathbf{I}_M, \tag{5.6}$$

where $\mathbf{S} = \mathbb{E}\left\{\mathbf{s}(t)\mathbf{s}^{\mathrm{H}}(t)\right\}$ is the source covariance matrix. The eigenvalue decomposition of the positive semi-definite Hermitian matrix $\mathbf{ASA}^{\mathrm{H}}$ in (5.6) can be written as

$$\mathbf{ASA}^{\mathrm{H}} = \mathbf{U}\dot{\mathbf{\Sigma}}\mathbf{U}^{\mathrm{H}}, \tag{5.7}$$

where the diagonal entries of $\dot{\mathbf{\Sigma}}$ contain the eigenvalues, which are real-valued and positive, arranged in the descending order. The columns of $\mathbf{U}$ correspond to the unit-length eigenvectors. By augmenting a scaled identity matrix, as shown in (5.6), to the product of $\mathbf{U}\dot{\mathbf{\Sigma}}\mathbf{U}^{\mathrm{H}}$, the eigenvectors remain unaltered, while the eigenvalues are increased by σ^2. Hence, the eigenvalue decomposition of $\mathbf{R}$ yields

$$\mathbf{R} = \mathbf{ASA}^{\mathrm{H}} + \sigma^2\mathbf{I}_M = \mathbf{U}\underbrace{\left(\dot{\mathbf{\Sigma}} + \sigma^2\mathbf{I}_M\right)}_{=\mathbf{\Sigma}}\mathbf{U}^{\mathrm{H}} = \mathbf{U}_s\mathbf{\Sigma}_s\mathbf{U}_s^{\mathrm{H}} + \mathbf{U}_n\mathbf{\Sigma}_n\mathbf{U}_n^{\mathrm{H}}, \tag{5.8}$$

where $\mathbf{\Sigma}_s \in \mathbb{C}^{q\times q}$ and $\mathbf{\Sigma}_n \in \mathbb{C}^{(M-q)\times(M-q)}$ are diagonal matrices having the q largest and $(M-q)$ smallest eigenvalues of $\mathbf{R}$ on their diagonal, respectively. Here, q is the rank of the matrix $\mathbf{ASA}^{\mathrm{H}}$, and, thus, the q largest eigenvalues on the diagonal of $\mathbf{\Sigma}_s$ are strictly greater than σ^2. On the other hand, the $M-q$ eigenvalues on the diagonal of $\mathbf{\Sigma}_n$ are exactly equal to σ^2. Matrices $\mathbf{U}_s \in \mathbb{C}^{M\times q}$ and $\mathbf{U}_n \in \mathbb{C}^{M\times(M-q)}$ contain the corresponding eigenvectors. Specifically, the columns of $\mathbf{U}_n$ span the noise subspace of $\mathbf{R}$ which is orthogonal to its signal subspace. This property will be later used for estimating the directions of the sources.

The rank q of $\mathbf{ASA}^{\mathrm{H}}$ is at most K and it achieves the maximum rank K when the rank of $\mathbf{A}$ and $\mathbf{S}$ are both K. If the sources are not fully correlated (coherent), then $\mathbf{S}$ is not rank deficient, and it has a rank of K. When $M > K$, as assumed before, and the condition in the following remark holds, the rank of $\mathbf{A}$ is K.

Remark 5.1 The rank of $\mathbf{A}$ is K for a ULA with $d \leq \lambda/2$ if the K DoAs result in distinctly different values of $\sin(\theta_k)$.

When the columns of the array response matrix $\mathbf{A}$ are linearly independent, meaning that the requirement stated in Remark 5.1 is satisfied, the array is referred to as *unambiguous*. This property allows for the estimation of angles that are unique [16].

In practice, the theoretical signal covariance matrix is not available and it is estimated as the sample average covariance matrix that, for L samples, is obtained as

$$\hat{\mathbf{R}} = \frac{1}{L}\sum_{t=1}^{L}\mathbf{y}(t)\mathbf{y}^{\mathrm{H}}(t). \tag{5.9}$$

5.2.2 MUSIC

Due to the orthogonality between signal and noise subspaces, we have

$$\mathbf{ASA}^{\mathrm{H}}\mathbf{U}_n = \mathbf{0}. \tag{5.10}$$

From linear algebra, we know that the rank of $\mathbf{AS} \in \mathbb{C}^{M\times K}$ is equal to the rank of $\mathbf{A}$ when the source signal covariance matrix $\mathbf{S}$ is non-singular under the assumption $K < M$. Moreover, if the condition in Remark 5.1 holds, then $\mathbf{AS}$ has the maximum rank of K. Then, Eq. (5.10) implies

$$\begin{aligned}\mathbf{A}^{\mathrm{H}}\mathbf{U}_n = \mathbf{0} \Rightarrow \quad & \mathbf{a}^{\mathrm{H}}(\theta_k)\mathbf{U}_n = \mathbf{0}, \quad k = 1, \dots, K \\ \Rightarrow \quad & \mathbf{a}^{\mathrm{H}}(\theta_k)\mathbf{U}_n\mathbf{U}_n^{\mathrm{H}}\mathbf{a}(\theta_k) = 0, \quad k = 1, \dots, K.\end{aligned} \tag{5.11}$$

If we have knowledge of $\mathbf{U}_n$ and the array is unambiguous, we may determine the DoA angles of the sources by identifying K array response vectors that are linearly independent and satisfy the requirement stated in (5.11). The MUSIC algorithm is based on this premise, but it specifically addresses the scenario where the estimation of $\mathbf{U}_n$ is derived from the sample average covariance matrix $\hat{\mathbf{R}}$ in (5.9).

After computing the sample average covariance matrix, we compute its eigenvalue decomposition and obtain the noise subspace matrix $\hat{\mathbf{U}}_n \in \mathbb{C}^{M\times(M-K)}$ whose columns are the unit-length eigenvectors corresponding to the $M - K$ smallest eigenvalues of $\hat{\mathbf{R}}$. Using the fact that $\mathbf{a}^{\mathrm{H}}(\theta)\mathbf{U}_n\mathbf{U}_n^{\mathrm{H}}\mathbf{a}(\theta) = 0$ when considering the DoA of a source, we define the MUSIC spectrum as

$$P_{\mathrm{MUSIC}}^{\mathrm{far}}(\theta) = \frac{1}{\mathbf{a}^{\mathrm{H}}(\theta)\hat{\mathbf{U}}_n\hat{\mathbf{U}}_n^{\mathrm{H}}\mathbf{a}(\theta)} \tag{5.12}$$

for angles $\theta \in \left[-\frac{\pi}{2}, \frac{\pi}{2}\right]$ (assuming the signals impinge on the array by the front side). When the angle θ is in close proximity to a source, the denominator approaches zero, resulting in a peak in the spectrum. If $\hat{\mathbf{U}}_n$ were exactly equal to $\mathbf{U}_n$ (i.e., $\hat{\mathbf{R}} = \mathbf{R}$), then the MUSIC spectrum would be infinite in true DoAs. As we only possess the estimate $\hat{\mathbf{U}}_n$, the peak values and locations are mere approximations. The DoA estimations are determined by identifying the K highest peaks in the MUSIC spectrum, given that the value of K is already known. The number of sources can be determined using the Akaike information or the

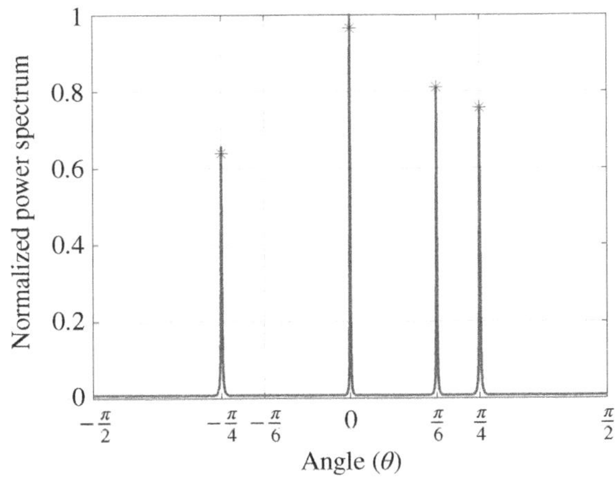

Figure 5.2 The normalized power spectrum for the MUSIC algorithm when the number of antennas is $M = 50$. The source DoAs are given $-\pi/4$, 0, $\pi/6$, and $\pi/4$, respectively, which are shown by stars.

minimum description length criterion [17]. Furthermore, the MUSIC algorithm may detect the number of sources by determining the number of eigenvalues of the estimated covariance matrix, $\hat{\mathbf{R}}$, that significantly exceed σ^2. Utilizing this quantity as the estimation for K, we proceed by determining the K highest peaks of the MUSIC spectrum.

In Figure 5.2, we plot the MUSIC spectrum for a scenario with $\lambda/2$-spacing antenna array with $M = 50$ antennas. There are $K = 4$ sources with independent data streams, that are zero-mean complex Gaussian distributed with unit variance, and equal SNR of 0 dB per antenna and per sample. The DoAs of the sources are given by $-\pi/4$, 0, $\pi/6$, and $\pi/4$, respectively. The number of samples is $L = 100$. As demonstrated in the figure, the correct DoA angles can be almost identified from the four highest peaks of the MUSIC spectrum.

To gain a more comprehensive understanding of the limitations of the MUSIC method in accurately calculating the DoA angles, we will now position the four sources at the specific angles of $-\pi/100$, 0, $\pi/200$, and $\pi/100$ accordingly. Figure 5.3a displays three peaks; moreover, their angles do not correspond to the correct angles. Due to the near proximity of the sources in the angular domain, the resolution offered by a set of $M = 50$ antennas is inadequate. This is because one of the two nearby angles falls inside the beamwidth of a beam pointing at the other DoA. In this case, the two peaks merge. To alleviate this issue, the beamwidth can be made narrower by increasing the number of antennas. In Figure 5.3b, we increase the number of antennas to $M = 100$ while maintaining the same configuration. This time, it is seen that the correct angles can be identified thanks to the increased spatial resolution by doubling the antennas.

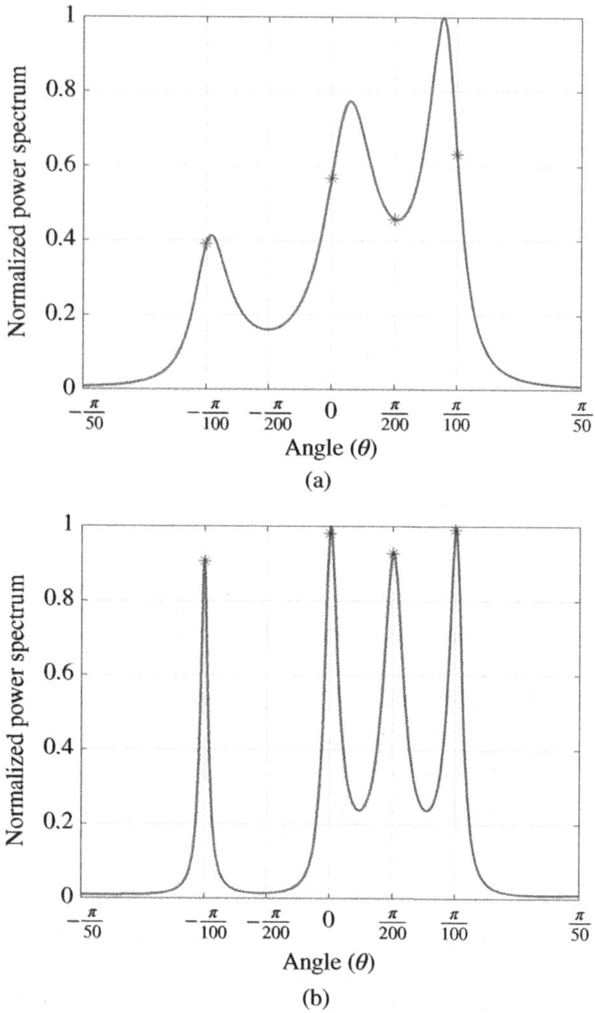

Figure 5.3 The normalized power spectra of the MUSIC algorithm when the four source angles are very close. The DoAs are $-\pi/100$, 0, $\pi/200$, and $\pi/100$, respectively, which are shown by stars. (a) $M = 50$ antennas. (b) $M = 100$ antennas.

5.2.3 ESPRIT

The ESPRIT algorithm differs from the MUSIC algorithm in utilizing the signal subspace instead of the noise subspace to estimate the DoA angles. In order to explain how the angles are determined using the signal subspace estimate obtained from the sample average covariance matrix in (5.9), we will provide an overview of the theoretical framework of the ESPRIT method. We assume $K \leq M$ (unlike MUSIC algorithm, ESPRIT can work when $K = M$) and that $\mathbf{ASA}^{\mathrm{H}}$ has the maximum rank of K, that is, the dimension of the signal subspace is K.

We first multiply Eq. (5.8) from the right by $\mathbf{U}_s$ and obtain

$$\mathbf{R}\mathbf{U}_s = \mathbf{A}\mathbf{S}\mathbf{A}^{\mathrm{H}}\mathbf{U}_s + \sigma^2\mathbf{U}_s = \mathbf{U}_s\boldsymbol{\Sigma}_s\underbrace{\mathbf{U}_s^{\mathrm{H}}\mathbf{U}_s}_{=\mathbf{I}_K} + \mathbf{U}_n\boldsymbol{\Sigma}_n\underbrace{\mathbf{U}_n^{\mathrm{H}}\mathbf{U}_s}_{=\mathbf{0}} = \mathbf{U}_s\boldsymbol{\Sigma}_s$$

$$\Rightarrow \quad \mathbf{A}\mathbf{S}\mathbf{A}^{\mathrm{H}}\mathbf{U}_s = \mathbf{U}_s\underbrace{\left(\boldsymbol{\Sigma}_s - \sigma^2\mathbf{I}_K\right)}_{=\overline{\boldsymbol{\Sigma}}_s}, \tag{5.13}$$

where the diagonal entries of $\overline{\boldsymbol{\Sigma}}_s$ are strictly greater than zero since the signal space eigenvalues appearing on the diagonal of $\boldsymbol{\Sigma}_s$ are strictly greater than σ^2. Hence, $\overline{\boldsymbol{\Sigma}}_s$ is non-singular. Multiplying both sides of (5.13) from the right by $\overline{\boldsymbol{\Sigma}}_s^{-1}$, we obtain

$$\mathbf{U}_s = \mathbf{A}\underbrace{\mathbf{S}\mathbf{A}^{\mathrm{H}}\mathbf{U}_s\overline{\boldsymbol{\Sigma}}_s^{-1}}_{=\mathbf{C}} = \mathbf{A}\mathbf{C}. \tag{5.14}$$

We will now partition ULA into two overlapping subarrays, one containing the antennas 1 to $M-1$ and the other having antennas 2 to M. Let $\mathbf{A}_1$ and $\mathbf{A}_2$ be their corresponding array response matrices such that

$$\mathbf{A} = \begin{bmatrix} \mathbf{A}_1 \\ \text{Last row} \end{bmatrix} = \begin{bmatrix} \text{First row} \\ \mathbf{A}_2 \end{bmatrix}. \tag{5.15}$$

Using the symmetric structure of the ULA, the array response matrix of the second subarray can be represented as

$$\mathbf{A}_2 = \mathbf{A}_1\mathbf{D}, \tag{5.16}$$

where

$$\mathbf{D} = \begin{bmatrix} e^{j\frac{2\pi d}{\lambda}\sin(\theta_1)} & & 0 \\ & \ddots & \\ 0 & & e^{j\frac{2\pi d}{\lambda}\sin(\theta_K)} \end{bmatrix}. \tag{5.17}$$

Similar to (5.15), we can split $\mathbf{U}_s = \mathbf{A}\mathbf{C}$ from (5.14) as

$$\mathbf{U}_s = \begin{bmatrix} \mathbf{U}_{s1} \\ \text{Last row} \end{bmatrix} = \begin{bmatrix} \text{First row} \\ \mathbf{U}_{s2} \end{bmatrix}, \tag{5.18}$$

$$\mathbf{A}\mathbf{C} = \begin{bmatrix} \mathbf{A}_1\mathbf{C} \\ \text{Last row} \end{bmatrix} = \begin{bmatrix} \text{First row} \\ \mathbf{A}_2\mathbf{C} \end{bmatrix}, \tag{5.19}$$

which implies that

$$\mathbf{U}_{s1} = \mathbf{A}_1\mathbf{C}, \tag{5.20}$$

$$\mathbf{U}_{s2} = \mathbf{A}_2\mathbf{C}. \tag{5.21}$$

Substituting $\mathbf{A}_2 = \mathbf{A}_1\mathbf{D}$ into (5.21), we obtain

$$\mathbf{U}_{s2} = \mathbf{A}_1\mathbf{D}\mathbf{C}. \tag{5.22}$$

Since the ranks of $\mathbf{S}$, $\mathbf{A}$, $\mathbf{U}_s$, and $\overline{\boldsymbol{\Sigma}}_s$ are all K, the rank of $K \times K$ matrix $\mathbf{C}$ is also K. Hence, $\mathbf{C}$ is invertible, and we have the relation $\mathbf{A}_1 = \mathbf{U}_{s1}\mathbf{C}^{-1}$ from (5.20). Inserting this into the above expression, we end up with

$$\mathbf{U}_{s2} = \mathbf{U}_{s1}\underbrace{\mathbf{C}^{-1}\mathbf{D}\mathbf{C}}_{\boldsymbol{\Phi}}, \tag{5.23}$$

where the matrix $\boldsymbol{\Phi}$ has the same eigenvalues as $\mathbf{D}$, which are the diagonal entries given in (5.17). This is due to the fact that $\boldsymbol{\Phi}$ is obtained through a *similarity transformation* from $\mathbf{D}$ using the matrix $\mathbf{C}$. Multiplying both sides of the above equation from the left by the left pseudo-inverse of $\mathbf{U}_{s1}$, that is, $\left(\mathbf{U}_{s1}^{\mathrm{H}}\mathbf{U}_{s1}\right)^{-1}\mathbf{U}_{s1}^{\mathrm{H}}$, we arrive at

$$\boldsymbol{\Phi} = \left(\mathbf{U}_{s1}^{\mathrm{H}}\mathbf{U}_{s1}\right)^{-1}\mathbf{U}_{s1}^{\mathrm{H}}\mathbf{U}_{s2} \tag{5.24}$$

whose eigenvalues give the functions $e^{j\frac{2\pi d}{\lambda}\sin(\theta_k)}$ of the DoA angles.

Now, from the eigenvalue decomposition of the sample average covariance matrix $\hat{\mathbf{R}}$ in (5.9), we can obtain $\hat{\mathbf{U}}_{s1} \in \mathbb{C}^{(M-1)\times K}$ and $\hat{\mathbf{U}}_{s2} \in \mathbb{C}^{(M-1)\times K}$ and form the matrix

$$\hat{\boldsymbol{\Phi}} = \left(\hat{\mathbf{U}}_{s1}^{\mathrm{H}}\hat{\mathbf{U}}_{s1}\right)^{-1}\hat{\mathbf{U}}_{s1}^{\mathrm{H}}\hat{\mathbf{U}}_{s2}. \tag{5.25}$$

Assuming that $\mu_1, \dots, \mu_K$ are the eigenvalues of $\hat{\boldsymbol{\Phi}}$ and using $e^{j\frac{2\pi d}{\lambda}\sin(\theta_k)}$, the DoA estimates are obtained as

$$\hat{\theta}_k = \arcsin\left(\angle\mu_k\frac{\lambda}{2\pi d}\right), \quad k = 1, \dots, K, \tag{5.26}$$

where we have implicitly assumed $\lambda/d \leq 2 \Rightarrow d \geq \lambda/2$. For unambiguous DoA estimate, we thus need to set $d = \lambda/2$.

In Tables 5.1 and 5.2, we compare the DoA estimates obtained with the MUSIC and ESPRIT algorithms for the setups in Figures 5.2 and 5.3b, respectively. It should be noted that the performance of the MUSIC algorithm depends on the size of the angular spectrum grid, which is selected as 100,000 points in this simulation. In contrast, ESPRIT is a grid-free algorithm. We observe that

Table 5.1 Comparison of MUSIC and ESPRIT when there are $M = 50$ antennas and the DoAs are $-\pi/4$, 0, $\pi/6$, and $\pi/4$.

True DoA	−0.7854	0	0.5236	0.7854
MUSIC	−0.7859	−0.0004	0.5236	0.7856
ESPRIT	−0.7873	−0.0002	0.5242	0.7851

Table 5.2 Comparison of MUSIC and ESPRIT when there are $M = 100$ antennas and the DoAs are $-\pi/100$, 0, $\pi/200$, and $\pi/100$.

True DoA	−0.0314	0	0.0157	0.0314
MUSIC	−0.0315	0.0003	0.0159	0.0316
ESPRIT	−0.0310	0.0003	0.0150	0.0320

the performance of the MUSIC algorithm is slightly better compared to that of ESPRIT at the cost of increased complexity.

Figure 5.4 shows the steps of DoA estimation using the MUSIC and ESPRIT algorithms.

5.2.3.1 Generalized ESPRIT

The ESPRIT algorithm can be extended to a more general scenario, where the distance between two identically indexed antennas in the subarrays does not need to be the same for all indices [18]. To clarify this, we consider an array with $M = 2N$ antennas that is divided into two subarrays of N antennas. It is assumed that $K \leq N$. The antennas of each subarray can be arbitrarily chosen; therefore, the distance between the nth antenna in the first subarray and the corresponding antenna in the second subarray may be different for different values of n. The array response matrix $\mathbf{A}$ can be expressed as

$$\mathbf{A} = \begin{bmatrix} \mathbf{A}_1 \\ \mathbf{A}_2 \end{bmatrix}, \tag{5.27}$$

where $\mathbf{A}_1$ and $\mathbf{A}_2$ are the array response matrices of the two subarrays. In particular, assuming that $\mathbf{a}_1(\theta)$ is the array response vector of the first subarray for a signal coming from the angle θ, the array response matrices are given by

$$\mathbf{A}_1 = [\mathbf{a}_1(\theta_1), \dots, \mathbf{a}_1(\theta_K)] \in \mathbb{C}^{N\times K}, \tag{5.28}$$

$$\mathbf{A}_2 = [\mathbf{\Lambda}(\theta_1)\mathbf{a}_1(\theta_1), \dots, \mathbf{\Lambda}(\theta_K)\mathbf{a}_1(\theta_K)] \in \mathbb{C}^{N\times K}, \tag{5.29}$$

where

$$\mathbf{\Lambda}(\theta_k) = \begin{bmatrix} e^{j\frac{2\pi}{\lambda}l_1 \sin(\theta_k)} & & 0 \\ & \ddots & \\ 0 & & e^{j\frac{2\pi}{\lambda}l_N \sin(\theta_k)} \end{bmatrix} \tag{5.30}$$

and l_n denotes the distance between the nth antennas of the second and first subarrays. Similar to (5.27), $\mathbf{U}_s = \mathbf{AC}$ can be written as

$$\mathbf{U}_s = \begin{bmatrix} \mathbf{U}_{s1} \\ \mathbf{U}_{s2} \end{bmatrix} = \begin{bmatrix} \mathbf{A}_1\mathbf{C} \\ \mathbf{A}_2\mathbf{C} \end{bmatrix}. \tag{5.31}$$

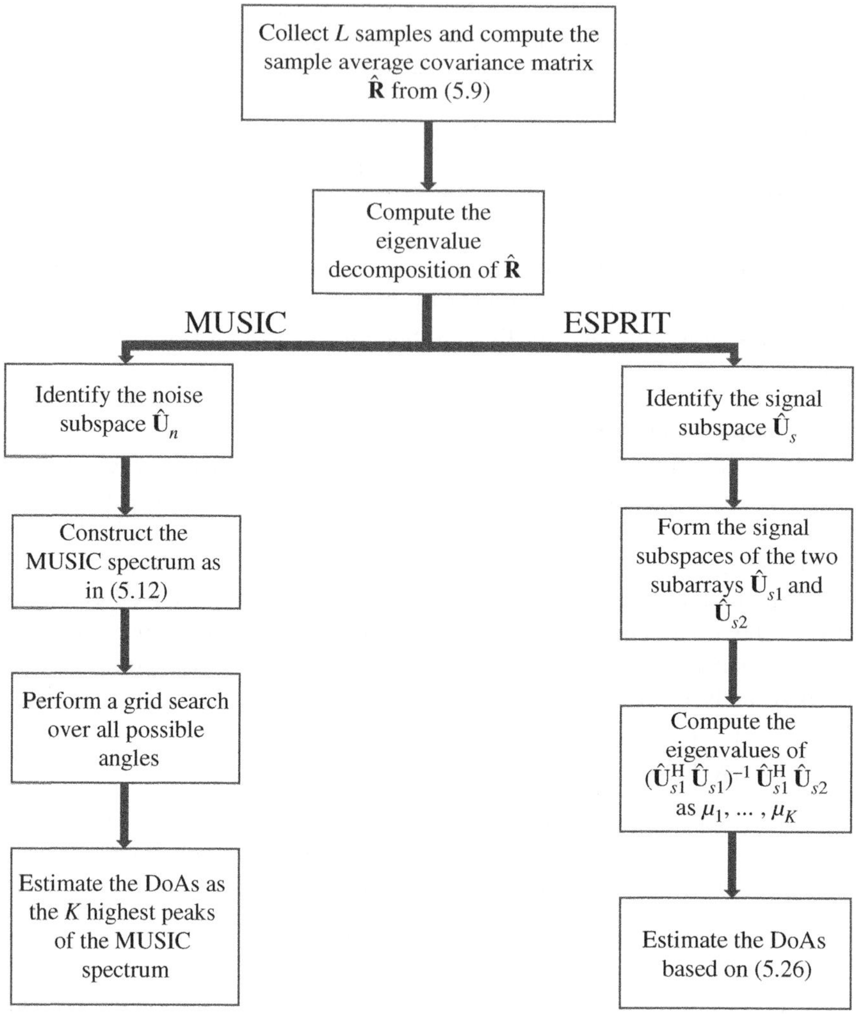

Figure 5.4 Flowchart of the MUSIC and ESPRIT algorithms for DoA estimation.

Introducing the diagonal matrix

$$\boldsymbol{\Psi}(\theta) = \begin{bmatrix} e^{j\frac{2\pi}{\lambda}l_1 \sin(\theta)} & & 0 \\ & \ddots & \\ 0 & & e^{j\frac{2\pi}{\lambda}l_N \sin(\theta)} \end{bmatrix}, \tag{5.32}$$

we form the matrix $\mathbf{F}(\theta) \in \mathbb{C}^{N\times K}$ as

$$\mathbf{F}(\theta) = \mathbf{U}_{s2} - \boldsymbol{\Psi}(\theta)\mathbf{U}_{s1} = \mathbf{QC}, \tag{5.33}$$

where

$$\mathbf{Q} = \left[\left(\boldsymbol{\Lambda}(\theta_1) - \boldsymbol{\Psi}(\theta)\right) \mathbf{a}_1(\theta_1), \dots, \left(\boldsymbol{\Lambda}(\theta_K) - \boldsymbol{\Psi}(\theta)\right) \mathbf{a}_1(\theta_K)\right]. \tag{5.34}$$

When $\theta = \theta_k$, the kth column of $\mathbf{Q}$ becomes zero and $\mathbf{F}(\theta)$ drops in rank. Therefore, matrix $\mathbf{W}^{\mathrm{H}}\mathbf{F}(\theta_k)$ is singular, where $\mathbf{W} \in \mathbb{C}^{N\times K}$ is an arbitrary full-rank matrix. The DoAs can then be estimated by finding the K highest peaks of the following spectrum:

$$f(\theta) = \frac{1}{\det\left(\mathbf{W}^{\mathrm{H}}\mathbf{F}(\theta)\right)}. \tag{5.35}$$

We will use the idea of generalized ESPRIT later for finding the DoAs of near-field sources.

5.3 Near-Field DoA and Range Estimation

This section considers the case where the sources are located in the radiative near-field region of the antenna array. The 2D MUSIC algorithm is first presented for simultaneously estimating the DoAs and ranges of the sources via a 2D search over the grid of possible angles and ranges. Then, two symmetry-based localization algorithms based on MUSIC and generalized ESPRIT are discussed that decompose DoA and range estimation problems and reduce the search complexity by turning a 2D search into multiple 1D searches.

5.3.1 System Model

When the sources lie in the radiative near-field region, the signal wavefront is spherical and the range of the sources must be estimated along with their DoAs for accurate localization. Therefore, far-field DoA estimation methods based on the planar wavefront assumption cannot be used directly for source localization in the radiative near-field region. Denoting by $r_1, r_2, \dots, r_K$ the range of the sources, which are located in the far-field of a single antenna but in the radiative near-field of the whole array, we have

$$r_k \in \left[d_{\mathrm{F}}, d_{\mathrm{FA}}\right], \quad k = 1, 2, \dots, K, \tag{5.36}$$

where d_{F} is the Fraunhofer distance of a single antenna given by

$$d_{\mathrm{F}} = \frac{2D^2}{\lambda}, \tag{5.37}$$

with D being the antenna aperture length. Likewise, the Fraunhofer array distance for an antenna array can be defined as [19]

$$d_{\mathrm{FA}} = \frac{2W^2}{\lambda}, \tag{5.38}$$

where W is the array aperture length. For an aperture ULA of M antennas, the Fraunhofer array distance can be approximated as [20]

$$d_{\text{FA}} \approx \frac{M^2}{2} d_{\text{F}}. \tag{5.39}$$

Similar to the far-field scenario, the received signal at the antenna array can be expressed as

$$\mathbf{y}(t) = \mathbf{A}\text{diag}\left(e^{j\psi_1}, \dots, e^{j\psi_K}\right)\mathbf{s}(t) + \mathbf{n}(t), \tag{5.40}$$

where the array response matrix in the near-field is given by

$$\mathbf{A} = \left[\mathbf{a}(\theta_1, r_1), \dots, \mathbf{a}(\theta_K, r_K)\right] \in \mathbb{C}^{M\times K}, \tag{5.41}$$

and the near-field array response vector for the signal coming from the kth source is

$$\mathbf{a}(\theta_k, r_k) = \left[1, \dots, e^{-j\frac{2\pi}{\lambda}\left(\bar{r}_k^m(\theta_k, r_k) - r_k\right)}, \dots, e^{-j\frac{2\pi}{\lambda}\left(\bar{r}_k^M(\theta_k, r_k) - r_k\right)}\right]^{\text{T}}, \tag{5.42}$$

where the distance between the kth source and mth antenna, $\bar{r}_k^m(\theta_k, r_k)$ is a function of both the DoA and range of the source. Specifically,

$$\begin{aligned}
\bar{r}_k^m(\theta_k, r_k) &= \sqrt{r_k^2 + (m-1)^2 d^2 - 2r_k(m-1)d\sin(\theta_k)} \\
&\overset{(a)}{\approx} r_k\left(1 + \frac{(m-1)^2 d^2}{2r_k^2} - \frac{(m-1)d}{r_k}\sin(\theta_k) - \frac{(m-1)^4 d^4}{8r_k^4}\right. \\
&\quad \left. + \frac{(m-1)^3 d^3}{2r_k^3}\sin(\theta_k) - \frac{(m-1)^2 d^2}{2r_k^2}\sin^2(\theta_k)\right) \\
&\overset{(b)}{\approx} r_k\left(1 - \frac{(m-1)d}{r_k}\sin(\theta_k) + \frac{(m-1)^2 d^2}{2r_k^2}\left(1 - \sin^2(\theta_k)\right)\right),
\end{aligned} \tag{5.43}$$

where the second-order Taylor approximation $\sqrt{1+x} \approx 1 + \frac{x}{2} - \frac{x^2}{8}$, for $|x| \ll 1$ is used in (a) and (b) is obtained by assuming $d^3 \ll r_k^3$ and $d^4 \ll r_k^4$. Therefore, the near-field array response vector can be expressed as

$$\mathbf{a}(\theta_k, r_k) = \left[1, \dots, e^{j\left((m-1)\gamma_k + (m-1)^2\phi_k\right)}, \dots, e^{j\left((M-1)\gamma_k + (M-1)^2\phi_k\right)}\right]^{\text{T}}, \tag{5.44}$$

where $\phi_k = -\pi\frac{d^2}{\lambda r_k}\cos^2(\theta_k)$.

Remark 5.2 According to (5.44), the array is said to be unambiguous if $d \leq \lambda/2$. In such a case, the entries of the array response vector will be distinct for different values of θ_k.

Recall from Section 5.2 that the covariance matrix of the received signal is obtained as

$$\mathbf{R} = \mathbb{E}\left\{\mathbf{y}(t)\mathbf{y}^{\text{H}}(t)\right\} = \mathbf{A}\mathbf{S}\mathbf{A}^{\text{H}} + \sigma^2\mathbf{I}_M = \mathbf{U}_s\mathbf{\Sigma}_s\mathbf{U}_s^{\text{H}} + \mathbf{U}_n\mathbf{\Sigma}_n\mathbf{U}_n^{\text{H}}, \tag{5.45}$$

and the sample average covariance matrix using L samples is constructed as

$$\hat{\mathbf{R}} = \frac{1}{L}\sum_{t=1}^{L}\mathbf{y}(t)\mathbf{y}^{\mathrm{H}}(t). \tag{5.46}$$

5.3.2 Two-Dimensional MUSIC

2D MUSIC extends 1D MUSIC by performing the grid search over two dimensions, that is, DoA and range. Similar to the 1D MUSIC method described in Section 5.2.2, 2D MUSIC exploits the orthogonality between the signal and noise subspaces for estimating the location of the sources. According to (5.11), we have

$$\begin{aligned}\mathbf{A}^{\mathrm{H}}\mathbf{U}_n = \mathbf{0} \Rightarrow \quad & \mathbf{a}^{\mathrm{H}}(\theta_k, r_k)\mathbf{U}_n = \mathbf{0}, \quad k = 1, \dots, K \\ \Rightarrow \quad & \mathbf{a}^{\mathrm{H}}(\theta_k, r_k)\mathbf{U}_n\mathbf{U}_n^{\mathrm{H}}\mathbf{a}(\theta_k, r_k) = 0, \quad k = 1, \dots, K.\end{aligned} \tag{5.47}$$

Using the estimate $\hat{\mathbf{U}}_n$ that contains the eigenvectors corresponding to the $M-K$ smallest eigenvalues of $\hat{\mathbf{R}}$, the angles and ranges of the sources can be obtained by identifying the K highest peaks of the spectrum

$$P_{\mathrm{MUSIC}}^{\mathrm{near}}(\theta, r) = \frac{1}{\mathbf{a}^{\mathrm{H}}(\theta, r)\hat{\mathbf{U}}_n\hat{\mathbf{U}}_n^{\mathrm{H}}\mathbf{a}(\theta, r)}, \tag{5.48}$$

which can be done via a 2D grid search over angles $\theta \in \left[-\frac{\pi}{2}, \frac{\pi}{2}\right]$ and ranges $r \in [d_{\mathrm{F}}, d_{\mathrm{FA}}]$.

Figure 5.5 evaluates the performance of the 2D MUSIC when there are $K = 4$ near-field sources to be localized. Two cases for number of antennas at the array is considered: $M = 10$ and $M = 50$. The spacing between the antennas is assumed to be $d = \lambda/2$, the number of samples is set as $L = 100$, and the SNR is 0dB per antenna and per sample. The angle–range pairs of the sources are given by $(-\pi/4, 10\,d_{\mathrm{F}})$, $(0, 20\,d_{\mathrm{F}})$, $(\pi/6, 30\,d_{\mathrm{F}})$, and $(\pi/4, 40\,d_{\mathrm{F}})$. The figures demonstrate the power spectrum in (5.48) over different angles and ranges, when the peak value of the spectrum is normalized to one. The stars show the locations of the sources. In case of $M = 10$ (Figure 5.5a), the peaks of the spectrum do not exactly match with the location of the sources. When we increase the number of antennas to $M = 50$, the algorithm can estimate the locations more accurately due to the higher spatial resolution in the power spectrum, as depicted in Figure 5.5b.

5.3.3 Symmetry-Based Localization

The MUSIC algorithm incurs high complexity for localization of near-field sources due to performing a 2D grid search over all possible angles and ranges. Decomposing the localization problem into separate DoA and range estimation problems can remarkably reduce the search complexity. However, the DoAs and

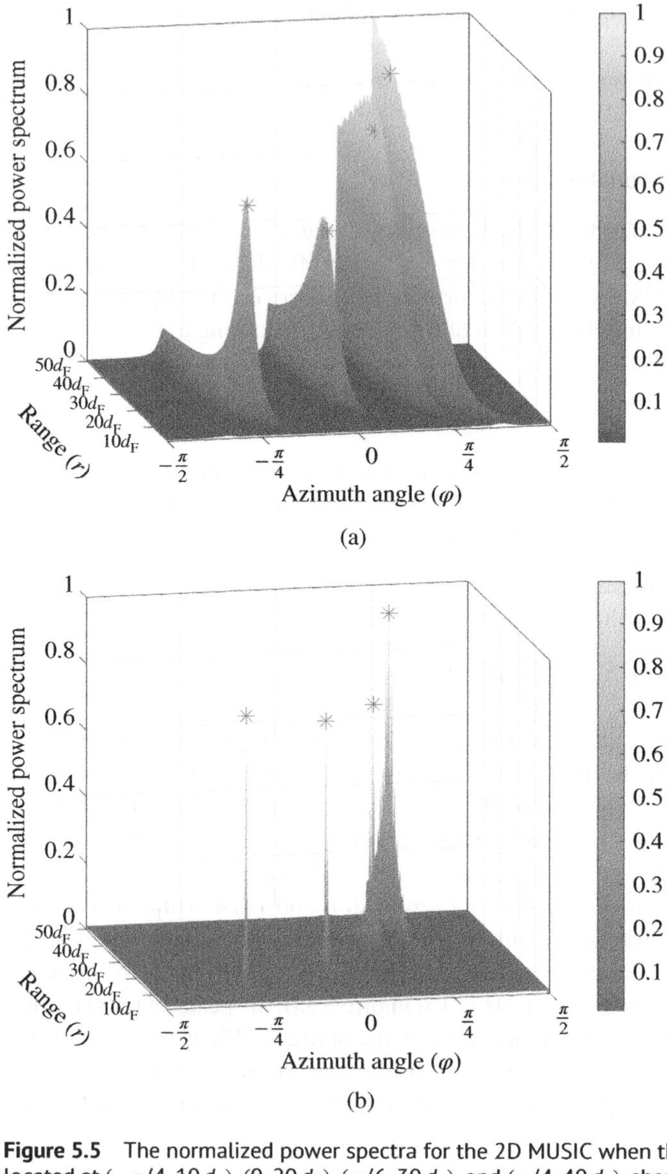

(a)

(b)

Figure 5.5 The normalized power spectra for the 2D MUSIC when there are four sources located at $(-\pi/4, 10\,d_F)$, $(0, 20\,d_F)$, $(\pi/6, 30\,d_F)$, and $(\pi/4, 40\,d_F)$, shown by stars on the figures. (a) $M = 10$. (b) $M = 50$.

ranges of the sources are coupled in the expression of the array response vector in (5.44), which means that the two problems cannot be directly separated. To deal with this issue, some methods have been proposed that, utilizing the symmetry property of the array, decouple the DoA and range estimation problems and solve the localization problem through multiple 1D grid searches [21–23]. The strategy is to first find the DoAs of the K sources by forming a spectrum function that can be utilized for identifying the DoAs independently of the ranges. The ranges are then obtained by performing the standard MUSIC algorithm once for each of the estimated DoAs, repeated K times. Here, we will cover two methods, which decouple the DoA and range estimation problems by resorting to the symmetric structure of the antenna array.

Consider a ULA consisting of $M = 2N + 1$ antennas, where the antennas are indexed from $-N$ to N and the antenna at the center of the array is considered as the reference antenna. With this setup, the array response vector in (5.44) is modified as

$$\mathbf{a}(\theta_k, r_k) = \left[e^{j(-N\gamma_k+N^2\phi_k)}, \dots, 1, \dots, e^{j(N\gamma_k+N^2\phi_k)}\right]^{\mathrm{T}}. \tag{5.49}$$

The received signal at this symmetric antenna array can be expressed as

$$\mathbf{y}(t) = \begin{bmatrix} y_{-N}(t) \\ \vdots \\ y_N(t) \end{bmatrix} = \begin{bmatrix} e^{j(-N\gamma_1+N^2\phi_1)} & \dots & e^{j(-N\gamma_K+N^2\phi_K)} \\ \vdots & \ddots & \vdots \\ e^{j(N\gamma_1+N^2\phi_1)} & \dots & e^{j(N\gamma_K+N^2\phi_K)} \end{bmatrix} \begin{bmatrix} e^{j\psi_1} s_1(t) \\ \vdots \\ e^{j\psi_K} s_K(t) \end{bmatrix} + \begin{bmatrix} n_{-N}(t) \\ \vdots \\ n_N(t) \end{bmatrix}. \tag{5.50}$$

5.3.3.1 Modified MUSIC for Near-Field Localization

After computing the covariance matrix of the received signal $\mathbf{R} = \mathbb{E}\left\{\mathbf{y}(t)\mathbf{y}^{\mathrm{H}}(t)\right\} = \mathbf{A}\mathbf{S}\mathbf{A}^{\mathrm{H}} + \sigma^2\mathbf{I}_M$, its anti-diagonal entries are given by

$$\begin{aligned} \mathbf{R}[n, 2N+2-n] &= \mathbb{E}\left\{\left(\sum_{k=1}^{K} e^{j((n-N-1)\gamma_k+(n-N-1)^2\phi_k)} e^{j\psi_k} s_k(t)\right)\right. \\ &\quad \left.\times\left(\sum_{k=1}^{K} e^{-j((N-n+1)\gamma_k+(N-n+1)^2\phi_k)} e^{-j\psi_k} s_k^*(t)\right)\right\} \\ &\quad + \mathbb{E}\left\{n_{n-N-1}(t) n_{N-n+1}^*(t)\right\} \\ &= \sum_{k=1}^{K} p_k e^{-j2(N-n+1)\gamma_k} + \sigma^2\delta_{n,2N+2-n}, \quad n = 1, \dots, 2N+1, \end{aligned} \tag{5.51}$$

where $\mathbf{R}[i,j]$ denotes the entry in the ith row and the jth column of $\mathbf{R}$, $\delta_{i,j}$ is the Kronecker delta function which returns 1 when $i = j$ and 0 otherwise. We have assumed that the source signals are independent and each has a zero-mean. The variance of $s_k(t)$ is denoted by p_k. We now construct a vector $\overline{\mathbf{y}} \in \mathbb{C}^{2N+1}$ where its nth entry is $\mathbf{R}[n, 2N+2-n]$ without the noise variance, that is,

$$\overline{\mathbf{y}} = \left[\sum_{k=1}^{K} p_k\, e^{-j2N\gamma_k}, \sum_{k=1}^{K} p_k e^{-j2(N-1)\gamma_k}, \ldots, \sum_{k=1}^{K} p_k\, e^{j2(N-1)\gamma_k}, \sum_{k=1}^{K} p_k\, e^{j2N\gamma_k}\right]^{\mathrm{T}}. \quad (5.52)$$

The entries of $\overline{\mathbf{y}}$ only contain the angle information, which signifies that DoA estimation can be performed independently of the range estimation. To this end, $\overline{\mathbf{y}}$ is split into J overlapping subvectors, each containing $2N+2-J$ entries. Specifically, the ith subvector is formed as

$$\overline{\mathbf{y}}_i = \left[\sum_{k=1}^{K} p_k\, e^{-j2(N-i+1)\gamma_k}, \ldots, \sum_{k=1}^{K} p_k\, e^{-j2(J-i-N)\gamma_k}\right], \quad (5.53)$$

which can be decomposed as

$$\overline{\mathbf{y}}_i = \mathbf{B}\mathbf{p}_i, \quad (5.54)$$

where

$$\mathbf{B} = [\mathbf{b}(\gamma_1), \ldots, \mathbf{b}(\gamma_K)] \in \mathbb{C}^{(2N+2-J)\times K}, \quad (5.55)$$

$$\mathbf{p}_i = \left[p_1\, e^{j2i\gamma_1}, \ldots, p_K\, e^{j2i\gamma_K}\right]^{\mathrm{T}} \in \mathbb{C}^K, \quad (5.56)$$

with

$$\mathbf{b}(\gamma_k) = \left[e^{-j2\gamma_k(N+1)}, \ldots, e^{-j2\gamma_k(J-N)}\right]^{\mathrm{T}}. \quad (5.57)$$

Remark 5.3 To avoid DoA ambiguity in (5.57), the antenna spacing must satisfy $d \leq \lambda/4$.

Averaging over the outer products of the J subvectors with themselves, we obtain

$$\overline{\mathbf{R}} = \frac{1}{J}\sum_{i=1}^{J} \overline{\mathbf{y}}_i \overline{\mathbf{y}}_i^{\mathrm{H}} = \frac{1}{J}\mathbf{B}\underbrace{\left(\sum_{i=1}^{J} \mathbf{p}_i \mathbf{p}_i^{\mathrm{H}}\right)}_{\triangleq \mathbf{R}_p}\mathbf{B}^{\mathrm{H}} = \frac{1}{J}\mathbf{B}\mathbf{R}_p\mathbf{B}^{\mathrm{H}}, \quad (5.58)$$

and the eigenvalue decomposition of $\overline{\mathbf{R}}$ yields

$$\overline{\mathbf{R}} = \overline{\mathbf{U}}_s \overline{\mathbf{\Sigma}}_s \overline{\mathbf{U}}_s^{\mathrm{H}} + \overline{\mathbf{U}}_n \overline{\mathbf{\Sigma}}_n \overline{\mathbf{U}}_n^{\mathrm{H}}. \quad (5.59)$$

Employing the MUSIC algorithm, the K DoAs can be estimated by finding the K peaks of the 1D spectrum

$$f(\theta) = \frac{1}{\mathbf{b}^{\mathrm{H}}(\gamma)\overline{\mathbf{U}}_n \overline{\mathbf{U}}_n^{\mathrm{H}} \mathbf{b}(\gamma)}. \quad (5.60)$$

The above MUSIC-based algorithm was first developed in [22]. We refer to this algorithm as *modified MUSIC* in the rest of the chapter.

Remark 5.4 The modified MUSIC algorithm described above only works when $\text{rank}(\mathbf{B}) = K$ and $\text{rank}(\mathbf{R}_p) = K$. The former is satisfied when $K < 2N + 2 - J$. Furthermore, matrix $\mathbf{R}_p$ is sum of J rank-one matrices; thus, it becomes a full-rank matrix when $J > K$. These two conditions imply that $K < N + 1$. Therefore, the number of resolvable sources is at most $K = N = \frac{M-1}{2}$ with the modified MUSIC algorithm.

After estimating the DoAs, the standard MUSIC algorithm is applied for finding the corresponding ranges. In particular, for each DoA estimate $\hat{\theta}_k$, the range r_k is estimated as

$$\hat{r}_k = \arg\max_{r} \frac{1}{\mathbf{a}^{\mathrm{H}}(\hat{\theta}_k, r)\mathbf{U}_n\mathbf{U}_n^{\mathrm{H}}\mathbf{a}(\hat{\theta}_k, r)}, \quad k = 1, \dots, K. \tag{5.61}$$

Figure 5.6 shows the result of DoA and range estimation using the modified MUSIC algorithm. The number of antennas is assumed to be $M = 51$ with $d = \lambda/4$ inter-antenna spacing, and the number of overlapping subvectors is set as $J = 20$. $L = 100$ time samples are used for the estimation. There are $K = 4$ sources with their angle–range pairs given by $(-\pi/3, 60\,d_{\mathrm{F}})$, $(-\pi/6, 80\,d_{\mathrm{F}})$, $(\pi/6, 20\,d_{\mathrm{F}})$, and $(\pi/3, 40\,d_{\mathrm{F}})$. The SNR is 0 dB per antenna and per sample. Figure 5.6a illustrates the power spectrum in (5.60) whose peak value is normalized to 1. It is observed that the DoAs have been accurately identified. Based on the identified DoAs, the estimator in (5.61) finds the corresponding ranges of the sources. The results, depicted in Figure 5.6b, show the effectiveness of the MUSIC in estimating the range of near-field sources.

The number of sources to be localized is limited to $K \leq N = \frac{M-1}{2}$ with the modified MUSIC algorithm, thus incurring an array aperture loss. Next, we describe another DoA estimation approach based on the generalized ESPRIT algorithm presented in Section 5.2.3.1, which can alleviate this issue.

5.3.3.2 Generalized ESPRIT for Near-Field Localization

Building on the symmetric structure of the array response vector in (5.49), the antenna array can be divided into two subarrays each having $J < M$ elements, where the first subarray contains the first J antennas in ascending order and the second subarray consists of the last J antennas in descending order. Based on (5.50), the signals received at the two subarrays are given by

$$\begin{aligned}\mathbf{y}_1(t) &= \left[y_{-N}(t), y_{-N+1}(t), \dots, y_{-N+(J-1)}(t)\right]^{\mathrm{T}} \\ &= \mathbf{A}_1 \operatorname{diag}\left(e^{j\psi_1}, \dots, e^{j\psi_K}\right)\mathbf{s}(t) + \mathbf{n}_1(t),\end{aligned} \tag{5.62}$$

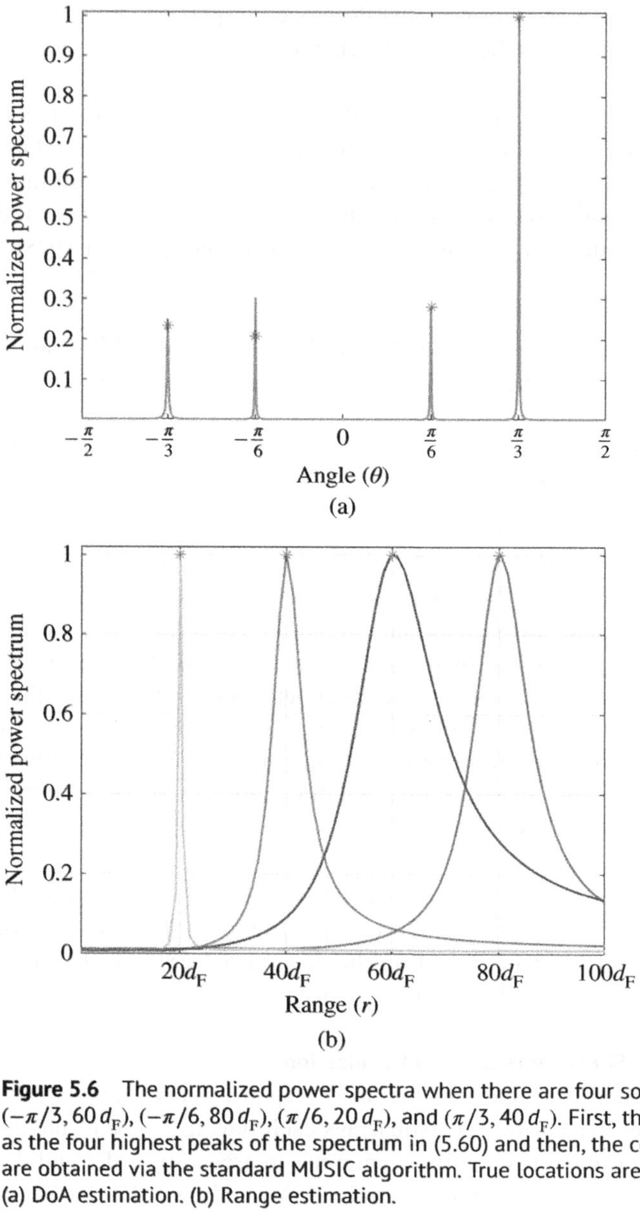

Figure 5.6 The normalized power spectra when there are four sources located at $(-\pi/3, 60\,d_F)$, $(-\pi/6, 80\,d_F)$, $(\pi/6, 20\,d_F)$, and $(\pi/3, 40\,d_F)$. First, the DoAs are estimated as the four highest peaks of the spectrum in (5.60) and then, the corresponding ranges are obtained via the standard MUSIC algorithm. True locations are marked by stars. (a) DoA estimation. (b) Range estimation.

$$\mathbf{y}_2(t) = \left[y_N(t), y_{N-1}(t), \dots, y_{N-(J-1)}(t)\right]^{\mathrm{T}}$$
$$= \mathbf{A}_2 \mathrm{diag}\left(e^{j\psi_1}, \dots, e^{j\psi_K}\right)\mathbf{s}(t) + \mathbf{n}_2(t), \tag{5.63}$$

where

$$\mathbf{A}_1 = \left[\mathbf{a}_1(\theta_1, r_1), \dots, \mathbf{a}_1(\theta_K, r_K)\right] \in \mathbb{C}^{J\times K}, \tag{5.64}$$

$$\mathbf{A}_2 = \left[\mathbf{a}_2(\theta_1, r_1), \dots, \mathbf{a}_2(\theta_K, r_K)\right] \in \mathbb{C}^{J\times K}, \tag{5.65}$$

and the array response vectors of the subarrays are given by

$$\mathbf{a}_1(\theta_k, r_k) = \left[e^{j(-N\gamma_k+N^2\phi_k)}, \dots, e^{j((-N+J-1)\gamma_k+(-N+J-1)^2\phi_k)}\right]^{\mathrm{T}}, \tag{5.66}$$

$$\mathbf{a}_2(\theta_k, r_k) = \left[e^{j(N\gamma_k+N^2\phi_k)}, \dots, e^{j((N-J+1)\gamma_k+(N-J+1)^2\phi_k)}\right]^{\mathrm{T}}. \tag{5.67}$$

The array response matrix $\mathbf{A}$ can thus be divided as

$$\mathbf{A} = \begin{bmatrix} \mathbf{A}_1 \\ \text{Last } (M-J) \text{ rows} \end{bmatrix} = \begin{bmatrix} \text{First } (M-J) \text{ rows} \\ \mathbf{E}\mathbf{A}_2 \end{bmatrix}, \tag{5.68}$$

where $\mathbf{E}$ is the exchange matrix having ones on its anti-diagonal and zeros elsewhere, and $\mathbf{E}^2 = \mathbf{I}_J$. The array response vector of the second subarray can be represented as

$$\mathbf{A}_2 = \left[\mathbf{D}(\theta_1)\mathbf{a}_1(\theta_1, r_1), \dots, \mathbf{D}(\theta_K)\mathbf{a}_1(\theta_K, r_K)\right], \tag{5.69}$$

where

$$\mathbf{D}(\theta_k) = \begin{bmatrix} e^{j\frac{4\pi d}{\lambda}\sin(\theta_k)N} & & 0 \\ & \ddots & \\ 0 & & e^{j\frac{4\pi d}{\lambda}\sin(\theta_k)(N-J+1)} \end{bmatrix}. \tag{5.70}$$

Recall from (5.14) that $\mathbf{U}_s = \mathbf{AC}$. Similar to (5.68), we can split $\mathbf{U}_s$ and $\mathbf{AC}$ as

$$\mathbf{U}_s = \begin{bmatrix} \mathbf{U}_{s1} \\ \text{Last } (M-J) \text{ rows} \end{bmatrix} = \begin{bmatrix} \text{First } (M-J) \text{ rows} \\ \mathbf{U}_{s2} \end{bmatrix}, \tag{5.71}$$

$$\mathbf{AC} = \begin{bmatrix} \mathbf{A}_1\mathbf{C} \\ \text{Last } (M-J) \text{ rows} \end{bmatrix} = \begin{bmatrix} \text{First } (M-J) \text{ rows} \\ \mathbf{E}\mathbf{A}_2\mathbf{C} \end{bmatrix}, \tag{5.72}$$

which implies that

$$\mathbf{U}_{s1} = \mathbf{A}_1\mathbf{C}, \tag{5.73}$$

$$\mathbf{E}\mathbf{U}_{s2} = \mathbf{A}_2\mathbf{C}. \tag{5.74}$$

Following the generalized ESPRIT method in Section 5.2.3.1, the diagonal matrix $\boldsymbol{\Psi}$ is introduced as

$$\boldsymbol{\Psi}(\theta) = \begin{bmatrix} e^{j\frac{4\pi d}{\lambda}\sin(\theta)N} & & 0 \\ & \ddots & \\ 0 & & e^{j\frac{4\pi d}{\lambda}\sin(\theta)(N-J+1)} \end{bmatrix} \tag{5.75}$$

and matrix $\mathbf{F}(\theta)$ is formed as

$$\mathbf{F}(\theta) = \mathbf{E}\mathbf{U}_{s2} - \boldsymbol{\Psi}(\theta)\mathbf{U}_{s1} = \mathbf{Q}\mathbf{C}, \tag{5.76}$$

with

$$\mathbf{Q} = \left[\left(\mathbf{D}(\theta_1) - \boldsymbol{\Psi}(\theta)\right)\mathbf{a}_1(\theta_1, r_1), \dots, \left(\mathbf{D}(\theta_K) - \boldsymbol{\Psi}(\theta)\right)\mathbf{a}_1(\theta_K, r_K)\right]. \tag{5.77}$$

The matrix $\mathbf{F}(\theta_k)$ is rank deficient, which means that for an arbitrary full column rank matrix $\mathbf{W} \in \mathbb{C}^{J\times K}$, the matrix $\mathbf{W}^{\mathrm{H}}\mathbf{F}(\theta_k)$ is singular and its determinant is zero. The DoAs of the sources can therefore be found as the K highest peaks of the spectrum

$$f(\theta) = \frac{1}{\det\left(\mathbf{W}^{\mathrm{H}}\mathbf{F}(\theta)\right)}. \tag{5.78}$$

This generalized ESPRIT-based approach for finding the DoAs of near-field sources was first developed in [21].

Remark 5.5 Using the generalized ESPRIT method, $K \leq J$ sources can be resolved, and because $J < M$, the maximum number of resolvable sources with this method is $K = M - 1$, which is twice the number of resolvable sources compared with the modified MUSIC algorithm.

After estimating the DoAs of the sources as $\hat{\theta}_1, \dots, \hat{\theta}_K$, the ranges can be found from (5.61).

Figure 5.7 shows the normalized spectra for DoA and range, where generalized ESPRIT is utilized for DoA estimation and standard MUSIC is used for range estimation. The number of antennas is $M = 51$ with $d = \lambda/4$ spacing and the number of samples is set to be $L = 100$. Furthermore, the antenna array is divided into two subarrays, each having $J = 50$ antennas. There are $K = 5$ sources with the angle–range pairs of $(-\pi/3, 20\,d_{\mathrm{F}})$, $(-\pi/6, 40\,d_{\mathrm{F}})$, $(0, 60\,d_{\mathrm{F}})$, $(\pi/6, 80\,d_{\mathrm{F}})$, and $(\pi/3, 100\,d_{\mathrm{F}})$. We assume that the number of sources, K, has already been specified [17]. The SNR is set as 0dB per antenna and per sample. We set $\mathbf{W} = \mathbf{F}(\theta)$ since $\mathbf{F}(\theta)$ is full column rank when $\theta \neq \theta_k$, $k = 1, \dots, K$. From Figure 5.7a, it can be observed that the generalized ESPRIT algorithm effectively resolves the DoAs of the five sources. While there are also some peaks at other angles, their power is much smaller than the power of the peaks at the true DoAs. Since the DoAs are accurately estimated in the first step, the MUSIC algorithm can find the ranges of near-field sources with a good accuracy, as observed in Figure 5.7b.

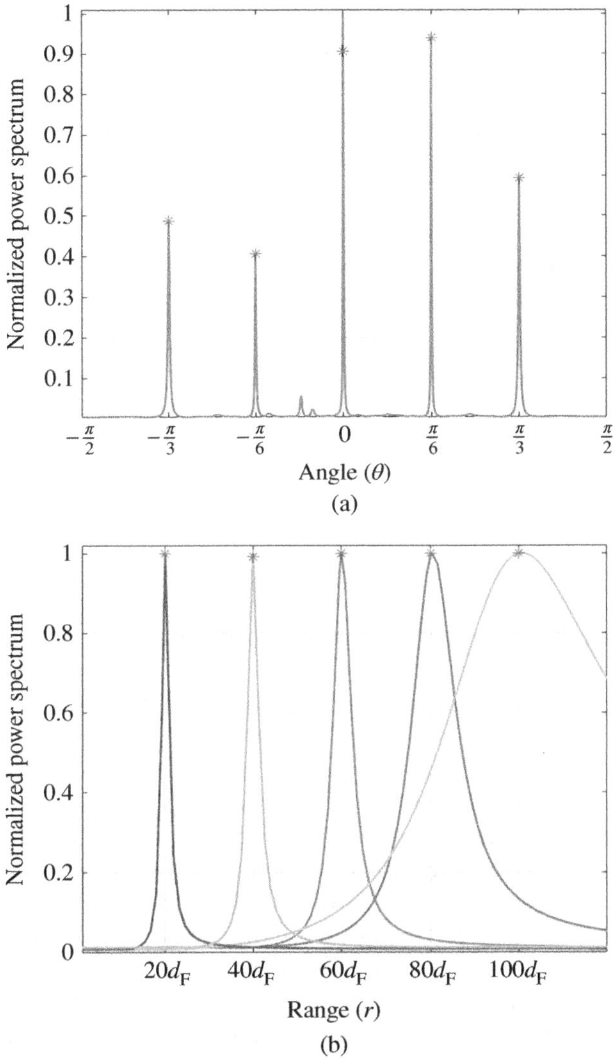

Figure 5.7 The normalized power spectra when there are five sources located at $(-\pi/3, 20\,d_F)$, $(-\pi/6, 40\,d_F)$, $(0, 60\,d_F)$, $(\pi/6, 80\,d_F)$, and $(\pi/3, 100\,d_F)$. First, the DoAs are estimated as the five highest peaks of the spectrum in (5.78) and then, the corresponding ranges are obtained via the standard MUSIC algorithm. True locations are marked by stars. (a) DoA estimation. (b) Range estimation.

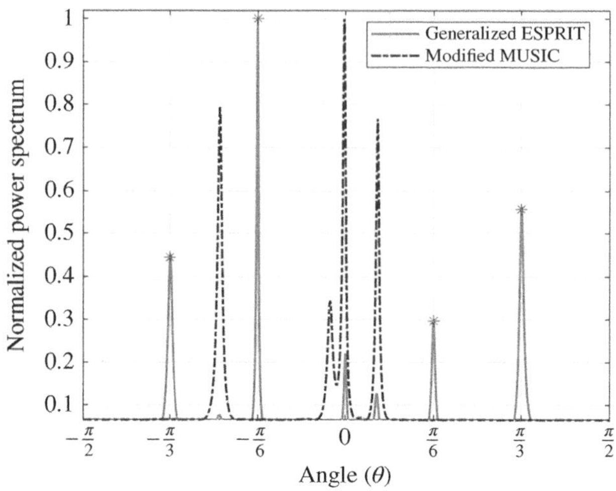

Figure 5.8 The normalized power spectra for the modified MUSIC and generalized ESPRIT algorithms when the correlation coefficient between source signals is 0.9. The setup is the same as in Figure 5.6. True angles are marked by stars.

We have so far assumed that the signals arriving from different sources at the ULA are perfectly uncorrelated. However, in practice, the existence of multipath propagation due to reflection, refraction, and scattering may result in coherent signals that are strongly correlated. We now consider the correlation between source signals and compare the performance of the modified MUSIC and generalized ESPRIT algorithms in the presence of correlation. We consider the same scenario as the one in Figure 5.6 and the correlation coefficient between each pair of sources is assumed to be 0.9. Figure 5.8 shows the normalized power spectra for the modified MUSIC and generalized ESPRIT algorithms. We can observe that the modified MUSIC algorithm fails to estimate the DoAs of the sources and produces four peaks that are far from the true DoAs. On the other hand, the peaks generated by the generalized ESPRIT algorithm are close to the true DoAs, which suggests that this algorithm can better handle source correlations compared to the modified MUSIC algorithm.

5.4 Conclusions

As massive MIMO systems expand to accommodate more users through spatial multiplexing, the increased size of antenna arrays is also beneficial for localization services. This is because having more antennas enhances spatial resolution and allows for more accurate localization of closely spaced sources. In this chapter, we covered two important subspace-based localization algorithms, namely, MUSIC and ESPRIT, for localizing far-field sources, and introduced their variants that are employed for near-field localization. These algorithms can be directly implemented on massive MIMO base stations, and the source signals might be uplink data transmitted by user devices. We observed that both MUSIC and ESPRIT provide high-resolution DoA estimation, with MUSIC slightly outperforming ESPRIT at the expense of increased complexity, which stems from evaluating the spatial spectrum at various grid points corresponding to possible DoAs. Unlike ESPRIT, the MUSIC algorithm can be directly applied to near-field localization where the conventional 1D grid of possible DoAs is extended to a 2D grid involving both DoA and range values and a 2D grid search is subsequently performed for localizing the sources. We further introduced two methods based on MUSIC and ESPRIT algorithms that, by decoupling DoA and range estimation problems, reduce the complexity of the 2D MUSIC algorithm. The presented localization algorithms find applications in various fields where determining the location of signal sources is crucial. These fields include radar systems, wireless communications, sonar systems, astronomy, seismic monitoring, etc.

As a promising future research direction, one could explore the application of the presented algorithms to the scenario where the antenna array is a uniform planar array and the near-field localization problem involves the estimation of azimuth and elevation DoAs as well as the range of the sources. In such cases, a new search dimension is added to the localization problem, which further increases the computational complexity and necessitates the development of efficient algorithms. Inspired by the symmetry-based near-field localization algorithms, it is interesting to investigate new ways for decomposing the localization problem into separate subproblems of azimuth, elevation, and range estimation problems, without compromising the estimation accuracy. Furthermore, as the same signals can be used for both localization and communication purposes, the newly emerged topic of joint localization and communication is worth further investigation.

In addition to the methods discussed in this chapter, other widely employed localization techniques are available in the literature. Table 5.3 presents some of these techniques along with relevant references.

Table 5.3 Other widely used localization techniques.

Technique	References
Time difference of arrival	Wei et al. [24], Wang and Ho [25], and Sun et al. [26]
Linear prediction	Grosicki et al. [27] and Zuo et al. [28]
Maximum likelihood	Chen et al. [29] and Cheng et al. [30]
Sparse signal reconstruction	Malioutov et al. [31] and Wang et al. [32]

References

1 Björnson, E., Hoydis, J., and Sanguinetti, L. (2017). Massive MIMO networks: spectral, energy, and hardware efficiency. *Foundations and Trends® in Signal Processing* 11 (3–4): 154–655.

2 Chen, H., Keskin, M.F., Sakhnini, A. et al. (2023). 6G localization and sensing in the near field: fundamentals, opportunities, and challenges, *arXiv preprint arXiv:2308.15799*.

3 Wax, M. and Kailath, T. (1983). Optimum localization of multiple sources by passive arrays. *IEEE Transactions on Acoustics, Speech, and Signal Processing* 31 (5): 1210–1217. https://doi.org/10.1109/TASSP.1983.1164183.

4 Ziskind, I. and Wax, M. (1988). Maximum likelihood localization of multiple sources by alternating projection. *IEEE Transactions on Acoustics, Speech, and Signal Processing* 36 (10): 1553–1560. https://doi.org/10.1109/29.7543.

5 Liu, W., Haardt, M., Greco, M.S. et al. (2023). Twenty-five years of sensor array and multichannel signal processing: a review of progress to date and potential research directions. *IEEE Signal Processing Magazine* 40 (4): 80–91. https://doi.org/10.1109/MSP.2023.3258060.

6 Pesavento, M., Trinh-Hoang, M., and Viberg, M. (2023). Three more decades in array signal processing research: an optimization and structure exploitation perspective. *IEEE Signal Processing Magazine* 40 (4): 92–106.

7 Zheng, Z., Fu, M., Wang, W.-Q., and So, H.C. (2021). Symmetric displaced coprime array configurations for mixed near-and far-field source localization. *IEEE Transactions on Antennas and Propagation* 69 (1): 465–477. https://doi.org/10.1109/TAP.2020.3005203.

8 Zhang, J., Li, Y., Qi, J., and Li, Y. (2023). Symmetric extended nested array for passive localization of mixture near-and far-field sources. *IEEE Transactions on Circuits and Systems II: Express Briefs* 70 (3): 1244–1248. https://doi.org/10.1109/TCSII.2022.3223940.

9 Liang, J. and Liu, D. (2009). Passive localization of mixed near-field and far-field sources using two-stage MUSIC algorithm. *IEEE Transactions on Signal Processing* 58 (1): 108–120.

10 Wang, B., Zhao, Y., and Liu, J. (2013). Mixed-order MUSIC algorithm for localization of far-field and near-field sources. *IEEE Signal Processing Letters* 20 (4): 311–314. https://doi.org/10.1109/LSP.2013.2245503.

11 Friedlander, B. (2019). Localization of signals in the near-field of an antenna array. *IEEE Transactions on Signal Processing* 67 (15): 3885–3893. https://doi.org/10.1109/TSP.2019.2923164.

12 Schmidt, R. (1986). Multiple emitter location and signal parameter estimation. *IEEE Transactions on Antennas and Propagation* 34 (3): 276–280. https://doi.org/10.1109/TAP.1986.1143830.

13 Stoica, P. and Moses, R.L. (2005). *Spectral Analysis of Signals*, vol. 452. Upper Saddle River, NJ: Pearson Prentice Hall.

14 Roy, R., Paulraj, A., and Kailath, T. (1986). ESPRIT–A subspace rotation approach to estimation of parameters of cisoids in noise. *IEEE Transactions on Acoustics, Speech, and Signal Processing* 34 (5): 1340–1342. https://doi.org/10.1109/TASSP.1986.1164935.

15 Rockah, Y. and Schultheiss, P. (1987). Array shape calibration using sources in unknown locations–Part II: Near-field sources and estimator implementation. *IEEE Transactions on Acoustics, Speech, and Signal Processing* 35 (6): 724–735. https://doi.org/10.1109/TASSP.1987.1165222.

16 Krim, H. and Viberg, M. (1996). Two decades of array signal processing research: the parametric approach. *IEEE Signal Processing Magazine* 13 (4): 67–94.

17 Wax, M. and Kailath, T. (1985). Detection of signals by information theoretic criteria. *IEEE Transactions on Acoustics, Speech, and Signal Processing* 33 (2): 387–392. https://doi.org/10.1109/TASSP.1985.1164557.

18 Gao, F. and Gershman, A.B. (2005). A generalized ESPRIT approach to direction-of-arrival estimation. *IEEE Signal Processing Letters* 12 (3): 254–257. https://doi.org/10.1109/LSP.2004.842276.

19 Ramezani, P. and Björnson, E. (2023). Near-field beamforming and multiplexing using extremely large aperture arrays. In *Fundamentals of 6G Communications and Networking* (ed. X. Lin, J. Zhang, Y. Liu, and J. Kim), 317–349. Springer.

20 Ramezani, P., Kosasih, A., Irshad, A., and Björnson, E. (2023). Exploiting the depth and angular domains for massive near-field spatial multiplexing. *IEEE BITS the Information Theory Magazine* 3 (1): 14–26. https://doi.org/10.1109/MBITS.2023.3322670.

21 Zhi, W. and Chia, M.Y.-W. (2007). Near-field source localization via symmetric subarrays. *IEEE Signal Processing Letters* 14 (6): 409–412. https://doi.org/10.1109/LSP.2006.888390.

22 He, J., Swamy, M.N.S., and Ahmad, M.O. (2012). Efficient application of MUSIC algorithm under the coexistence of far-field and near-field sources. *IEEE Transactions on Signal Processing* 60 (4): 2066–2070. https://doi.org/10.1109/TSP.2011.2180902.

23 Liu, G. and Sun, X. (2013). Efficient method of passive localization for mixed far-field and near-field sources. *IEEE Antennas and Wireless Propagation Letters* 12: 902–905. https://doi.org/10.1109/LAWP.2013.2273451.

24 Wei, H.-W., Peng, R., Wan, Q. et al. (2010). Multidimensional scaling analysis for passive moving target localization with TDOA and FDOA measurements. *IEEE Transactions on Signal Processing* 58 (3): 1677–1688. https://doi.org/10.1109/TSP.2009.2037666.

25 Wang, Y. and Ho, K.C. (2017). TDOA positioning irrespective of source range. *IEEE Transactions on Signal Processing* 65 (6): 1447–1460. https://doi.org/10.1109/TSP.2016.2630030.

26 Sun, Y., Ho, K.C., and Wan, Q. (2019). Solution and analysis of TDOA localization of a near or distant source in closed form. *IEEE Transactions on Signal Processing* 67 (2): 320–335. https://doi.org/10.1109/TSP.2018.2879622.

27 Grosicki, E., Abed-Meraim, K., and Hua, Y. (2005). A weighted linear prediction method for near-field source localization. *IEEE Transactions on Signal Processing* 53 (10): 3651–3660. https://doi.org/10.1109/TSP.2005.855100.

28 Zuo, W., Xin, J., Liu, W. et al. (2019). Localization of near-field sources based on linear prediction and oblique projection operator. *IEEE Transactions on Signal Processing* 67 (2): 415–430. https://doi.org/10.1109/TSP.2018.2883034.

29 Chen, J.C., Hudson, R.E., and Yao, K. (2002). Maximum-likelihood source localization and unknown sensor location estimation for wideband signals in the near-field. *IEEE Transactions on Signal Processing* 50 (8): 1843–1854. https://doi.org/10.1109/TSP.2002.800420.

30 Cheng, C., Liu, S., Wu, H., and Zhang, Y. (2022). An efficient maximum-likelihood-like algorithm for near-field coherent source localization. *IEEE Transactions on Antennas and Propagation* 70 (7): 6111–6116. https://doi.org/10.1109/TAP.2022.3161269.

31 Malioutov, D., Cetin, M., and Willsky, A.S. (2005). A sparse signal reconstruction perspective for source localization with sensor arrays. *IEEE Transactions on Signal Processing* 53 (8): 3010–3022. https://doi.org/10.1109/TSP.2005.850882.

32 Wang, B., Liu, J., and Sun, X. (2012). Mixed sources localization based on sparse signal reconstruction. *IEEE Signal Processing Letters* 19 (8): 487–490. https://doi.org/10.1109/LSP.2012.2204248.

6

Energy-Efficient Uplink Transmission in RIS-Aided M-MIMO IoT Systems*

David William Marques Guerra[1], José Carlos Marinello[2], Ekram Hossain[3], and Taufik Abrão[1]

[1] *Department of Electrical Engineering, State University of Londrina, Londrina, Paraná, Brazil*
[2] *Electrical Engineering Department, UTFPR, Cornélio Procópio, Paraná, Brazil*
[3] *Department of Electrical and Computer Engineering, University of Manitoba, Winnipeg, MB, Canada*

6.1 Introduction

In this introductory section, we discuss the relevance of spectral efficiency (SE) and energy efficiency (EE) optimization in massive multiple-input multiple-output (M-MIMO) Internet of Things (IOT) systems aided by the reconfigurable intelligent surface (RIS)-enabling structures. The use of IoT applications in different fields is intensifying, like in scenarios such as smart cities, smart grids, industry 4.0, and autonomous vehicles. To allow IoT devices to connect, interact, and exchange data anywhere and anytime, they have to be interconnected wirelessly. Hence, wireless access technology providing reliable communications is essential for IoT applications [1]. Besides, while some IoT applications are characterized by a huge amount of simple devices demanding sporadic and low-speed connections, like in monitoring and tracking applications, other scenarios are characterized by the demand for high-speed and reliable coverage, like in driverless vehicles, factory automation, remote surgery, and others.

In parallel to this, it becomes each time more important the need for sustainable development and better usage of energy in many fields, including telecommunications. Climate change has achieved alarming levels, imposing the necessity

*This work was supported by the National Council for Scientific and Technological Development (CNPq) of Brazil, Grants: 405301/2021-9; 401065/2019-7; 310681/2019-7, by CAPES, Grant: FC001, by Londrina State University (UEL), Brazil and by UTFPR, Brazil. The work of E. Hossain was supported by a Discovery Grant from the Natural Sciences and Engineering Research Council of Canada (NSERC).

Massive MIMO for Future Wireless Communication Systems: Technology and Applications,
First Edition. Edited by Webert Montlouis and Agbotiname Lucky Imoize.

of restricting global warming at the limit of 1.5° above the preindustrial marks, as well as reducing the gas emissions by half until 2030 [2]. In this context, it is worth developing the next-generation telecommunication standard, the so-called beyond fifth generation (B5G) or sixth generation (6G), as a highly energy-efficient system able to fulfill the expected requirements, as peak throughputs of 1Tbps and latencies in the order of μs [3].

RIS has been recognized as a promising candidate technology to be integrated into the sixth generation 6G standard [4, 5]. Such surfaces are composed of reflecting elements whose reflection coefficients can be individually adjusted, aiming to improve the propagation features of the environment. The most appealing RIS application scenario is the passive operation, that is, when the reflecting RIS elements cannot amplify the reflected signal, alleviating the need to be equipped with power-hungry radio frequency (RF) chains [6]. In [7], authors have shown that implementing RISs in conjunction with M-MIMO technology can circumvent several problems, such as channel rank deficiency in propagation environments with few scatterers. Besides, the passive implementation enables such benefits at the cost of very little energy consumption, being thus an excellent approach under the perspective of EE [4, 5]. A remarkable challenge in the operation of RISs is how to suitably find the phase shifts of the reflection elements while enabling an effective operation according to certain performance metrics. An obstacle toward this goal is the reduced amount of available information in such contexts. For example, under the passive operation regime, it is very likely that the RISs do not have estimates for the isolated channels, that is, between their reflecting elements and the base station (BS) and between them and the users [5]. This is justified since usually RF chains would be required at the RIS to acquire such isolated estimates, increasing the energy consumption and the overhead associated with the training stage, which is proportional to the number of reflecting elements at the RIS [6, 8], which is usually very high.

Given the above scenario, we focus in this chapter on employing an RIS to reduce the uplink (UL) transmit power of an energetically efficient M-MIMO IOT network. For this purpose, we propose and investigate several schemes optimizing the devices' transmit powers, the RIS phase shifts, and the combiner matrix at the BS. Besides, we give particular attention to the manifold optimization schemes since the unit-modulus constraint of the RIS reflection optimization makes applying such a framework very convenient.

6.1.1 SE and EE General Formulation in M-MIMO

In this subsection, we introduce two critical performance metrics crucial for analyzing efficient communication systems. We initiated the discussion with the concept of SE and expanded into EE.

To assess the system performance of a communication system, it is crucial to evaluate the capacity for transmitting bits per second per Hertz (SE) – a fundamental metric that measures the efficiency of information transmission through a channel. The SE reflects the average number of bits transmitted per channel use across varying fading conditions, serving as a typical metric for evaluating communication system capabilities. Since for the RIS-assisted systems, the channel is a generalization of the conventional M-MIMO systems, the SE can be equally represented as a function of the signal-to-interference-plus-noise ratio (SINR) as the following.

$$\mathrm{SE} = \log_2(1 + \mathrm{SINR}). \qquad \left[\frac{\mathrm{bit}}{\mathrm{s} \cdot \mathrm{Hz}}\right] \tag{6.1}$$

The metric designed to evaluate the sustainable capacity of a system is referred to as EE. This metric essentially quantifies the number of bits that can be reliably transmitted per unit of energy, expressed as the following:

$$\mathrm{EE} = \frac{\text{Throughput [bit/s]}}{\text{System Power Consumption [W]}} = \frac{B \cdot \log_2(1 + \mathrm{SINR})}{P_\mathrm{T}}, \qquad \left[\frac{\mathrm{bit}}{\mathrm{Joule}}\right] \tag{6.2}$$

where B is the system bandwidth and P_T is the total power required for the communication system to operate. Notice that the power consumption for the RIS-assisted M-MIMO systems requires more detailed consideration than traditional M-MIMO systems since although the RIS technology achieves low-power consumption, its consumption is not entirely negligible and should be thoroughly examined.

6.1.2 RIS and Metasurfaces

An RIS is a programmable surface structure that can be used to control the reflection of electromagnetic (EM) waves by changing the electric and magnetic properties of the surface. RIS is a planar *metasurface* with many low-cost reflecting elements. RISs have been envisioned to reduce energy consumption and improve the SE of wireless networks by artificially reconfiguring the wireless propagation environment of EM waves.

Metasurfaces represent a novel class of artificial materials known as metamaterials, with a microstructure designed to display special EM properties not seen in nature and previously thought to be physically impossible [9]. Atoms make up the periodic structure of any material found in nature. An artificial periodic structure is made and grafted onto a host material to form metamaterials. The constituent elements of this manufactured periodic structure are larger than the atoms of the host material. This size difference effectively overrides the host material's natural reaction to the incoming EM radiation. As a result, the penetration, reflection loss, and reflection angle are unique to the artificial structure rather than the host material's atomic structure.

RIS-based transmission is supported by a large number of small, low-cost, and passive elements (metamaterial) on the RIS panel that only reflect the incident signal with an adjustable reflection amplitude and phase shift without requiring a dedicated energy source for radio frequency (RF) processing, decoding, or encoding, being completely different from existing active relays.

RIS is an enabling technology for 5G-advanced and 6G systems composed of voltage-controlled metamaterial elements that allow manipulation of the wireless channel state response; as a result, the primary characteristics of RIS are wireless system coverage enhancement, channel rank enhancement, interference suppression, and Doppler spread reduction. There are many possible applications, but TeraHertz signal enhancement is the most promising. Implementation issues are paramount, where the most crucial issue is creating workable protocols for real-time reconfigurability. Models and methods to deal with EM signal propagation in the presence of RIS should include large-scale path loss near-field and far-field approximation and a hybrid ray tracing method.

6.1.3 Manifolds Optimization in RIS-Aided M-MIMO

RMs are mathematical objects that generalize the notion of Euclidean space to more complex and curved geometries. They are widely used in optimization [10]. An RM is a space that locally looks like Euclidean space, but a Riemannian metric tensor defines the distances and angles between points. This tensor assigns a positive definite inner product to each tangent space of the manifold, which allows us to explain concepts such as length, angle, and curvature. A couple of key definitions and concepts in Riemannian geometry, including *Riemannian gradient* $\nabla_{\mathcal{M}} f(p)$ and *retraction operation* $\text{Retr}_p(v)$, are provided in Section 6.4.

Manifold optimization is concerned with the following optimization problem:

$$\min_{x\in\mathcal{M}} f(x), \tag{6.3}$$

where $\mathcal{M}$ is an RM and the objective function (OF) f is a real-valued function on $\mathcal{M}$, which can be non-smooth.

If additional constraints other than the manifold constraint are applied, one can add in the OF an indicator function of the feasible set of such additional constraints. Hence, the optimization in (6.3) covers a general formulation for manifold optimization. The existence of the manifold constraint is one of the main difficulties in algorithmic design and theoretical analysis.

One of the main challenges in manifold optimization usually is the non-convexity of the manifold constraints. By utilizing the geometry of the manifold, a large class of constrained optimization problems can be viewed as unconstrained optimization problems on the manifold [11].

RISs are typically employed to improve the channel gain between the BS and the user equipments (UEs), which obstacles of the environment can obstruct.

The most appealing operation scenario for the RISs is the passive configuration, in which the device is not equipped with active RF elements for EE purposes. Therefore, the RIS cannot amplify the incoming signal nor even perform further elaborated processing on it. Besides, since the signal departing from the RIS is not originated by it, but indeed coming from a transmitter in a different position, the channel intermediated by the RIS suffers from the so-called double attenuation effect, or multiplicative path-loss (PL), since the signal attenuates in both transmitter-RIS and RIS-receiver links. Aiming to mitigate this effect, it is convenient to have RISs with the largest aperture possible, as well as to make the reflection coefficients magnitude as high as possible, which is one under the passive operation. Therefore, most of the RIS literature constrains the reflection coefficients to be of unit modulus.

One of the most common manifolds is the complex circle one, in which all the elements of the optimization variable should have a unit modulus. This is the case for the problems involving the RIS phase-shift optimization under passive operation mode. Therefore, one can see that the manifold optimization framework is well-suited for the RIS problems in general, which justifies the particular attention we devote to such techniques in this chapter.

6.1.4 Contributions of the Chapter

In this chapter, we focus on the UL of a multiple access massive multiple-input multiple-output (M-MIMO) system, where K IoT single-antenna devices transmit simultaneously to a BS equipped with M antennas, representing a typical M-MIMO scenario, Figure 6.1. Strategically positioning an RIS can improve the system reliability, facilitating the communication link between users and the BS. The RIS contains N reconfigurable reflecting elements. The RIS delivers a phase-shifted version of the transmitted signal, maximizing the composite channel gain. Besides that, it can be noticed that the downlink (DL) dual problem can be considered similarly; hence, for the sake of compactness, we do not introduce the DL dual problem in this chapter.

The primary *Contribution* of this chapter is formulating the EE power minimization problem in the UL RIS-aided M-MIMO IoT network and establishing a solution methodology based on the **Riemannian manifold optimization** approach. Specifically, the contributions of this chapter are fourfold, and they can be summarized as follows:

1) Develop an optimization methodology for EE power minimization problem in different RIS-aided M-MIMO system configurations by deploying specific manifold optimization formulations and tools.
2) Conduct comprehensive analyses, encompassing key performance indicators, for the optimization problems related to RIS-aided M-MIMO systems. These

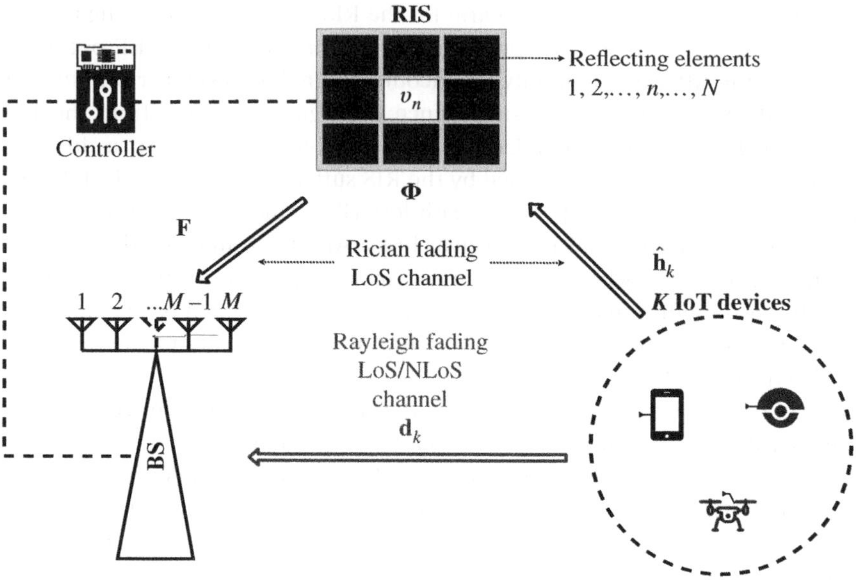

Figure 6.1 Passive RIS-aided MU M-MIMO system model with LoS/NLoS BS-IoT device links.

problems include maximization of system SE and EE, design of precoding/combining techniques, as well as effective power allocation strategies for M-MIMO systems.

3) Investigate channel modeling for RIS-aided M-MIMO wireless communication systems.
4) Thorough analysis of numerical data that support the analytical formulations and provide fresh insights and hypotheses on the progress of these technical subjects.

6.1.5 Organization

The remainder of this chapter is organized as follows. We commence our discourse by formulating a general multipurpose system model for RIS-aided M-MIMO systems in Section 6.3. In the subsequent sections, we aggregate specific considerations and particularities in the system model, presenting the formulation and development of new strategies and algorithms for the EE optimization in RIS-aided M-MIMO system configuration scenarios.

In this chapter, the focus of analysis is the maximization of the EE in M-MIMO systems aided by RIS; we deploy specific optimization tools based on **manifold** optimization procedures and methodologies, confirming the promising results attainable by applying Manifold's optimization tools. Section 6.4 discusses the concepts and definitions of RMs optimization. Section 6.5 discusses an optimization technique for optimizing the RIS phase shifts and the BS combining vectors. This technique is designed to minimize the UL transmit power of the user terminals (UTs) in the context of an IoT network supported by RIS. Initially, we formulate the joint optimization for the RIS phase-shift and BS combining vectors to improve the overall system EE by minimizing the UL transmit power while guaranteeing a minimum Quality of Service (QoS) for the devices. To tackle the non-convexity of the problem, **RM**-based iterative alternating optimization (i-AO) technique was deployed. Numerical results reveal that the Riemannian manifold iterative alternating optimization (RM-AO) algorithm significantly increases the system EE values and resource efficiency (SE-EE tradeoff) compared with the non-RIS multi-user (MU) M-MIMO system. Finally, Section 6.6 brings the main conclusions and findings associated with this research work.

6.2 Related Works

In the following, we scan the literature on the RIS-aided M-MIMO. Particularly, we discuss two optimization problems in multiuser M-MIMO systems: (a) the design of precoding in the BS transmitter (combiner at the user's equipment [UE] receiver) and (b) the SE and/or EE in RIS-aided M-MIMO systems.

A summary of the main related works published recently pointing out the challenges and solutions in RIS-aided M-MIMO systems is depicted in Table 6.2. Several parameters in RIS-aided M-MIMO systems can be optimized, such as power consumption, spectrum deployment, and latency. These parameters generate specific quantitative figures of merit, such as SE, EE, and resource efficiency (RE). Section 6.1.1 discusses the general definitions of the SE and EE.

The authors in [12] present a DL beamforming method for transmit power minimization in RIS-aided MIMO systems under imperfect channel state information (CSI). In [13], for DL transmission in cell-free M-MIMO systems, the authors design a distributed precoding scheme that eliminates the need for CSI exchange via backhaul signaling. The authors in [14] and [15] consider active and passive beamforming subject to users' average transmit power limitations. Besides, the authors of [15] focus on optimizing UL beamforming at the BS and passive beamforming at the RIS, employing a low-complexity max–min strategy. A joint beamforming optimization design for a multicell system aiming to maximize SE and minimize the mean squared error (MSE) of the users' received symbols is developed in [16]. The above works do not consider UT side beamforming for multi-antenna devices. We consider a scenario where mobile users maintain quasi-stationary positions to benefit from user-side precoding. They establish connections to the BS aided by an outdoor or indoor RIS.

A scheme for enhancing wireless networks using RIS in conjunction with joint active and passive beamforming is proposed in [17]. The scheme is applied to a single-cell system with multiple single-antenna users in a DL scenario, where direct and reflected links between access point (AP) and user are considered. An *alternating optimization* (AO) approach combined with semi-definite programming (SDP) formulation is used to minimize the total power at the AP subject to user signal-to-interference-plus-noise ratio (SINR) and phase shift constraints at the RIS. The proposed solution achieves an interesting performance improvement over conventional approaches.

In a DL RIS-aided MU multiple-input single-output (MISO) system, where the BS serves single-antenna users, the authors in [18] present an energy-efficient solution for the transmit power distribution and the phase shifts. To do so, the authors of [18] proposes two computationally feasible solutions that make use of alternating maximization, gradient descent search, and sequential fractional programming. In addition, perfect CSI knowledge is available at the BS, which makes it possible to perform interference cancellation among users by employing ZF precoding. However, in practical scenarios, this assumption is not realistic.

In [19], the authors suggest that by attempting to reduce inter-cell interference, the weighted sum rate (WSR) in RIS-aided DL multicell can be maximized. In order to achieve this aim, the authors of [19] jointly optimize the active precoding matrices at BSs and the phase shifts at the RIS subject to each BSâs power constraint and unit modulus constraint, respectively. In this case, the optimization problem does not consider any QoS restriction, for example, a minimum rate requirement for the users. Hence, this solution tends to maximize the rate of users with better channel conditions.

The EE of RIS-aided communications has been investigated recently in [20]. The EE maximization problem is addressed by considering UL MU single-antenna devices and MISO RIS-aided systems. The joint optimization (JO) involves the transmit power at users, phase shift at the RIS, and combining matrix at BS. The proposed block coordinate descent (BCD) iterative procedure can handle one variable during each iteration while the others are held fixed. Numerical results indicate that the proposed scheme outperforms baselines that optimize only two of the three variables. Differently from [20], in [21], a RIS-aided MU MIMO UL system is considered. The authors assume that there is no multiuser interference (MuI) and then propose an optimization framework for RE maximization, which aims to find a balance between EE and SE by maximizing the bi-objective function $\text{EE}/\text{BW} + \beta\text{SE}/P_{\text{tot}}$ in [bits/Joule/Hz], with $\beta > 0$, while BW is the system bandwidth, and P_{tot} is the overall available power budget. To tackle this problem, an AO-based technique for maximizing the RE is proposed. This technique obtains the covariance matrices and the RIS phase shift matrix. Numerical results demonstrate the effectiveness of this technique when compared to baselines that adopt equal power allocation (EPA) or a fixed phase shift matrix.

More recently, the authors of [22] investigated a cooperative beamforming design for DL multi-RIS-assisted MU MISO systems. An optimization problem is proposed to maximize the sum rate (SR) of users in this multi-reflection link system. Due to the non-convexity and coupling of variables, the problem is first decoupled into three subproblems, and closed-form expressions for decision variables are derived. A low-complexity cooperative beamforming algorithm is proposed, and simulation results corroborate its effectiveness against two baselines: the semi-definite relaxation (SDR)-based and the alternating direction method of multipliers (ADMM)-based algorithms.

In [23], a RIS-aided MU MISO UL system is considered by applying the Riemannian conjugate gradient-based joint optimization (RCG-JO) technique, which jointly optimizes the RIS phase shifts and the BS beamforming vectors to minimize UL transmit power in a RIS-aided IoT network. The technique uses the Riemannian conjugate gradient (RCG) method to exploit the product RM structure of the sets of unit-modulus phase shifts and unit-norm beamforming vectors. The non-convex UL power minimization problem is converted into an unconstrained problem, allowing for an optimal solution over the product of RMs. From that, it is shown that the proposed RCG-JO algorithm is more efficient in reducing the UL transmit power of RIS-assisted IoT networks compared to the SDR-based and deep reinforcement learning (DRL)-based schemes.

The work in [24] deals with the EE problem in RIS-assisted communication systems within the context of green wireless communications. In this study, the authors investigate EE maximization through the JO of power consumption for both the BS and the RIS. Notably, the research addresses a gap in the existing literature by considering the disparate power consumption characteristics of the ON and OFF states of RIS elements, aligning more closely with the real-world behavior of physical components. The authors introduce a novel ON–OFF aware power dissipation model for RIS elements, and subsequently they reformulate the EE optimization problem. The paper proposes low-complexity algorithms grounded in the mathematical structures of the problem. Simulation results underscore the effectiveness of the proposed algorithms across a diverse array of scenarios, contributing valuable insights to the ongoing discourse on optimizing EE in RIS-aided systems. In [25], the enhancement of EE in unmanned aerial vehicle (UAV)-enabled wireless power transfer (WPT) systems incorporating multiple ground sensors is explored. It introduces the use of RIS to minimize the total energy consumption of the UAV while fulfilling the energy requirements of each sensor. The study presents two scenarios: one employing a fly-hover-broadcast (FHB) protocol and the other allowing RF signals during flight. The proposed algorithms, based on successive convex approximation (SCA) and path discretization (PD) protocols, effectively optimize UAV trajectory, hovering time, and RIS reflection coefficients, demonstrating by simulation results a substantial energy savings. Additionally, the authors of [26] focus on the emergence of sixth generation 6g wireless communication following the widespread commercialization of 5G networks. Emphasizing the importance of precise modeling in the terahertz (THz) spectrum, the research employs 3-D ray tracing to analyze realistic indoor THz propagation with human blockers. Introducing a distributed RISs framework (DRF), the study addresses the EE problem formulation in THz wireless communication. The simulation results demonstrate the efficacy of DRF in overcoming mobile human blockage, enhancing THz signal coverage, and improving signal-to-noise ratios (SNRs) and QoS. The work envisions practical deployment, considering hardware constraints and highlighting the potential application of DRF in efficient THz communications for IoT networks.

Differently from [23], in this chapter, we consider several RM-based approaches to solve the EE UL transmission, where we consider heuristic solutions as well as solutions based on gradient descent and trust region (TR) methods. Additionally, a mixed solution employing RM-based techniques to optimize the RIS phase shifts while adopting a ZF combining scheme also proved to be a promising solution for the RIS-aided M-MIMO scenario considered in this chapter.

Our previous related works and contributions on the RE optimization, EE and SE in M-MIMO systems aided or not by RIS include: [27–36]. In this chapter, we propose to deploy the RM optimization combined with i-AO methodology and convex optimization to solve the three coupled variables (UTs power vector, RIS phase shifts, and the BS combining matrix) in the energy-efficient optimization problem considering the context of UL RIS-aided multiuser M-MIMO systems (Table 6.1).

6.3 System Model for RIS-Aided M-MIMO

The analyzed scenario consists of a RIS-aided M-MIMO UL communication system with K IoT devices, as illustrated in Figure 6.1. Herein, we assume M antennas at the BS side while the IoT devices are equipped with *single-antenna*, where $M > K$, and the number of RIS elements is denoted as N. We also consider that there is a direct link between the mobile IoT devices and the BS, as illustrated in Figure 6.1. Let $\hat{\mathbf{H}} = [\hat{\boldsymbol{h}}_1, \dots, \hat{\boldsymbol{h}}_K] \in \mathbb{C}^{N\times K}$ represents channel between the IoT devices and the RIS, $\mathbf{F} \in \mathbb{C}^{M\times N}$ denote the channel between the RIS and the BS and $\mathbf{D} = [\mathbf{d}_1, \dots, \mathbf{d}_K] \in \mathbb{C}^{M\times K}$ describe the direct channel between the mobile devices and the BS.

Table 6.1 Related work addressing the current open issues in EE, SE, and RE M-MIMO.

Setup (year)	Contributions	Problem	Method	References
RIS M-MIMO (2023)	(a) Maximize the EE in RIS-assisted M-MIMO scenario under zero-forcing (ZF) precoding formulating a non-convex optimization problem encompassing the userâs transmit power, the quantity of active base station antennas, and the phase shift angle unity module constraints. (b) Different from the previous works, the formulated EE maximization problem relies upon only the statistical channel state information (CSI) estimates, which can be beneficial in high mobility and short channel coherence time wireless scenarios, reducing the overhead imposed by the optimization procedure of these variables in every channel coherence time.	max-EE under RIS M-MIMO with ZF precoder.	(a) To solve this problem, we apply the Lagrangian dual method with fractional programming (FP) techniques to solve the non-convex problem. Specifically for the RIS phase shift problem, we proposed an innovative FP-based RIS phase shift optimization, which enables us to find a closed-form expression for each element of RIS. (b) To support our findings, the complexity of the proposed antenna, power allocation, and RIS phase shift optimization algorithm has been accurately investigated.	[37]
RIS-aided MU M-MISO (2019)	Authors propose (a) an energy-efficient solution for the transmit power distribution and (b) design of optimal RIS phase-shift elements.	Optimize the downlink (DL) of a RIS-aided multiuser M-MISO.	(a) Two computationally feasible solutions use alternating maximization (AO), gradient descent (GD) search, and sequential fractional programming. (b) Perfect CSI knowledge is assumed at the BS, which makes it possible to perform interference cancellation among users by employing ZF precoding. However, in practical scenarios, this assumption is not realistic.	[18]

(Continued)

Table 6.1 *(Continued)*

Setup (year)	Contributions	Problem	Method	References
Passive **RIS MU M-MIMO**: hybrid precoder, combiner, and low-resolution ADCs (2023)	(a) A novel information-directed branch-and-prune (IDBP) algorithm is used to derive the phase settings of the RIS. (b) Use an information-theoretic measure to decide on the pruning rules in a tree-search algorithm to arrive at the RIS phase-setting solution, which is vastly different compared to the traditional branch-and-bound algorithm that uses bounds of the cost function to define the pruning rules. (c) Theoretical derivation of the near-optimality of the RIS phase-setting solution thus obtained using the Asymptotic Equipartition property, ensuring near-optimal throughput and MSE performance.	SE and EE: hybrid precoder design, combiner, and RIS phase settings optimization.	(a) Derive the expression for the Cramer–Rao lower bound (CRLB) of the mean squared error (MSE) of the received and combined signal at the intended receiver under interference. By appropriate design of the hybrid precoder, combiner, and RIS phase settings, it can be shown that the MSE achieves the CRLB. (b) Prove that minimizing the MSE w.r.t. the phase settings of the RIS is equivalent to maximizing the throughput and energy efficiency of the system.	[38]
RIS IoT Network (2022)	(a) Analysis of the UL channel of a cellular network of IoT devices assisted by a RIS with a limited number of reflecting angle configurations. (b) Develop an expression for the expected battery lifetime (EBL), which explicitly depends on the machine type devices (MTD) transmit power.	UL-EE.	(a) Expressions for the required transmit power of MTDs to attain a target SNR, considering a channel model that accounts for the RIS discretization into sub-wavelength reflecting elements. (b) Numerical simulations for the EE evaluated via the EBL demonstrate the benefits of adopting RISs to enable energy-efficient IoT networks.	[29]

RIS-aided MU MISO UL (2022)	(a) Propose a Riemannian CG-based joint optimization (RCG-JO) technique that jointly optimizes the RIS phase shifts and the BS beamforming vectors. (b) It is shown that the proposed RCG-JO algorithm is more efficient in reducing the UL transmit power of RIS-assisted IoT networks compared to the SDR-based and deep reinforcement learning (DRL)-based schemes.	(a) Jointly RIS phase shifts and beamforming optimization; (b) minimize UL Tx power in a RIS-aided IoT network.	(a) Applying the Riemannian CG-based joint optimization (RCG-JO) technique; (b) RCG method exploits the Riemannian manifold product structure of the sets of unit-modulus phase shifts and unit-norm beamforming vectors; (c) the non-convex UL power minimization problem is converted into an unconstrained problem, allowing for an optimal solution over the product of RMs.	[23]
Multicell M-MIMO (2022)	Low-complexity method for multicellular networks pilot and power allocation under linear combiners.	Pilot and power allocation multicell M-MIMO	2-Step pilot assignment and power allocation using stateless multi- and single-agent*Q*-Learning approaches for suboptimal performance with low complexity.	[31]
XL-MIMO (2020)	Reformulation of SINR performance expressions; evaluation of the total EE XL-MIMO system; proposition of four low-complexity AS procedures to maximize total EE. Schemes are based on long-term fading parameters and remain valid for larger time/frequency intervals. Derivation of approximated closed-form EE expressions; iterative method for finding the optimal number of selected antennas to maximize EE.	Antenna selection (AS) for total EE maximization in XL-MIMO.	Analytical iterative algorithm based on Newton–Raphson (NR) method to determine the optimal number of activated antennas in XL-MIMO, which maximizes the approximated EE expression.	[39]

(Continued)

Table 6.1 *(Continued)*

Setup (year)	Contributions	Problem	Method	References
UL MU M-MIMO; IoT single-antenna devices (2024)	(a) Design of the RIS phase shift and BS beamforming (combining) to minimize the uplink transmit power of the RIS-aided IoT network. (b) Numerical results corroborate that the proposed RM-AO algorithm significantly increases the system EE values for M-MIMO systems, mainly when the ZF combiner is adopted, by ensuring a fixed SE for IoT devices while the uplink transmit power of the devices is minimized.	UL MU M-MIMO; IoT single-antenna devices.	(a) Formulation of the joint optimization for the EE IoT power control for the RIS-aided M-MIMO communication system. (b) System EE UL power minimization problem is treated by applying the **Riemannian manifold**-based alternating optimization (AO) technique due to its non-convexity structure.	This work
RIS-aided M-MIMO (2022)	(a) Phase shift and BS beamforming optimization techniques that can minimize the UL Tx power in a RIS-aided IoT network. (b) The proposed scheme relies on the Riemannian conjugate gradient-based joint optimization (RCG-JO), optimizing the RIS phase shifts and the BS beamforming. (c) RCG-JO technique is promising, reducing to $\approx$ half of the transmit power UL over the conventional scheme without RIS.	Joint RIS phase shift, beamforming, and UEs' power allocation optimization	(a) Exploit the product Riemannian manifold structure of the sets of unit-modulus phase shifts and unit-norm beamforming vectors. (b) The manifold method converts the nonconvex UL power minimization problem into an unconstrained problem and then finds the optimal solution over the product RM.	[23]

Also, we have the phase shift matrix of the RIS as $\mathbf{\Phi} = \text{diag}(\boldsymbol{v}) = \text{diag}\left(e^{j\theta_1}, \dots, e^{j\theta_N}\right)$, where θ_n represents the phase shift of the nth RIS element. Subsequently, the cascaded user-RIS-BS channel can be expressed as $\mathbf{G} = [\mathbf{g}_1, \dots, \mathbf{g}_K] = \mathbf{F}\mathbf{\Phi}\hat{\mathbf{H}} \in \mathbb{C}^{M\times K}$, and the aggregated channel from IoT devices to the BS can be expressed as $\mathbf{H} = \mathbf{G} + \mathbf{D} \in \mathbb{C}^{M\times K}$. Notably, this aggregated channel $\mathbf{H}$ shares the same dimension as the conventional M-MIMO system.

Based on the provided definitions, we will elaborate on the $\mathbf{H}$ channel model in detail. To address the potential blockage of direct links, we opt for the Rayleigh channel model for $\mathbf{D}$, expressed as follows:

$$\mathbf{D} = \overline{\mathbf{D}}\mathbf{\Omega}_d^{1/2}, \tag{6.4}$$

where $\mathbf{\Omega}_d = \text{diag}\left(\gamma_1, \dots, \gamma_K\right)$, with γ_k denoting the distance-dependent PL factor. Each element of the matrix $\overline{\mathbf{D}} \in \mathbb{C}^{N\times K}$ is an independent and identically distributed (i.i.d.) complex Gaussian random variable, having a zero mean and a unit variance.

Considering that the RIS is situated near the users and often installed above ground, we assume line-of-sight (LoS) paths for the IoT devices-RIS channels. We represent this as:

$$\hat{\mathbf{H}} = \left[\sqrt{\alpha_1}\overline{\mathbf{h}}_1, \dots, \sqrt{\alpha_K}\overline{\mathbf{h}}_K\right], \tag{6.5}$$

where α_k represents the path loss, and $\overline{\mathbf{h}}_k$ is specified using the two-dimensional uniform planar array (UPA) model. The spatial steering vector for a $N_h \times N_v$ RIS can be defined as

$$\begin{aligned}\mathbf{a}(\vartheta_k, \varphi_k) = [1,\ & e^{j\frac{2\pi d_s}{\lambda}\sin(\vartheta_k)\sin(\varphi_k)}, \dots, e^{j\frac{2\pi d_s}{\lambda}(N_h-1)\sin(\vartheta_k)\sin(\varphi_k)}]^T \dots \\ & \otimes [1,\ e^{j\frac{2\pi d_s}{\lambda}\cos(\varphi_k)}, \dots, e^{j\frac{2\pi d_s}{\lambda}(N_v-1)\cos(\varphi_k)}]^T,\end{aligned} \tag{6.6}$$

where $\otimes$ represents the Kronecker product, λ denotes the signal wavelength, d_s is the antenna spacing, ϑ_k and φ_k representing the azimuth and elevation angles of arrival (AoA) of the kth user at the RIS. Consequently, $\overline{\mathbf{h}}_k = \mathbf{a}(\vartheta_k, \varphi_k)$, and $\hat{\mathbf{h}}_k = \sqrt{\alpha_k}\overline{\mathbf{h}}_k$.

Since the RIS may be located at a considerable distance from the BS, and both the RIS and the BS have certain heights, the RIS-BS channel is not guaranteed to be purely LoS. Therefore, we adopt the Rician model for the RIS-BS channel, allowing us to study the impact of scatterers in RIS-aided systems. In this case, the Rician model is defined as:

$$\mathbf{F} = \sqrt{\frac{\beta\kappa}{\kappa+1}}\overline{\mathbf{F}} + \sqrt{\frac{\beta}{\kappa+1}}\tilde{\mathbf{F}}. \tag{6.7}$$

Here, β represents the path loss, and κ denotes the Rician factor, varying from 0 to ∞, which characterizes the strength ratio between LoS and non-LoS (NLoS) paths.

The NLoS path $\tilde{\mathbf{F}}$ consists of i.i.d. complex Gaussian random variables with a zero mean and unit variance. While the LoS path, $\overline{\mathbf{F}}$, is expressed as:

$$\overline{\mathbf{F}} = \mathbf{a}(\vartheta_R)\mathbf{a}^H(\vartheta_T, \varphi_T), \tag{6.8}$$

where ϑ_T and φ_T represent the azimuth and elevation angles of departure (AoD) from the RIS, while ϑ_R is the azimuth AoA at the BS. Note that, as the BS is a uniform linear array (ULA), the corresponding steering vector simplifies to $\mathbf{a}(\vartheta) = [1, \ldots, e^{j\frac{2\pi d_s}{\lambda}(N_{\text{BS}}-1)\sin(\vartheta)}]^T$.

From this, the cascade communication link between the IoT devices and the RIS can be expressed as follows:

$$\mathbf{g}_k = \mathbf{F}\boldsymbol{\Phi}\hat{\mathbf{h}}_k. \tag{6.9}$$

Finally, the full channel between the IoT devices and the BS can be expressed as:

$$\mathbf{h}_k = \mathbf{d}_k + \mathbf{F}\boldsymbol{\Phi}\hat{\mathbf{h}}_k = \mathbf{d}_k + \mathbf{g}_k. \tag{6.10}$$

Moreover, the $M \times 1$ received signal vector at the BS is denoted as

$$\mathbf{y} = \mathbf{H}\mathbf{x} + \mathbf{n} = (\mathbf{F}\boldsymbol{\Phi}\hat{\mathbf{H}} + \mathbf{D})\mathbf{x} + \mathbf{n}, \tag{6.11}$$

where $\mathbf{n} \sim \mathcal{CN}(\mathbf{0}, \sigma^2\mathbf{I}_M)$ is the noise vector, the transmitted signal $\mathbf{x}$ is given by

$$\mathbf{x} = \mathbf{P}^{\frac{1}{2}}\mathbf{s}, \tag{6.12}$$

in which $\mathbf{s}$ includes the transmit symbols from the K users, such that $\mathbb{E}[|[\mathbf{s}]_k|^2] = 1, \forall k \in \{1, \ldots, K\}$, $\mathbf{P} = \text{diag}([p_1, p_2, \ldots, p_K])$ is the UL power transmission of the IoT devices.

The BS processes the signal utilizing a combining vector $\mathbf{c}_k \in \mathbb{C}^{M\times 1}$ with respect to the kth IoT equipment for symbol recovery. This results in the following equation denoting the combining operation:

$$\mathbf{c}_k^H\mathbf{y} = \underbrace{\mathbf{c}_k^H\mathbf{h}_k s_k}_{\text{desired signal}} + \underbrace{\sum_{i\in\mathcal{K}\setminus\{k\}}^{K} \mathbf{c}_k^H\mathbf{h}_i s_i}_{\text{interference}} + \underbrace{\mathbf{c}_k^H\mathbf{n}}_{\text{noise}}, \tag{6.13}$$

which represents the summation of the combined signal of interest, interference from the other IoT devices, and the additive white Gaussian noise (AWGN).

6.4 Definitions on Riemannian Manifolds

An RM is a locally looking space like an Euclidean space, but a Riemannian metric tensor defines the distances and angles between points. This tensor assigns a positive definite inner product to each tangent space of the manifold, which allows us to define concepts such as length, angle, and curvature. Basically, the tangent space $\mathcal{T}_p\mathcal{M}$ at a point p of a RM $\mathcal{M}$ is the set of all tangent vectors at point p. Intuitively, $\mathcal{T}_p\mathcal{M}$ is the space of all directions in which one can move away from p on $\mathcal{M}$, where the curve μ of $\mathcal{M}$ is a mapping from $\mathbb{C}$ to $\mathcal{M}$.

A key concept in Riemannian geometry is the *Riemannian gradient* $\nabla_{\mathcal{M}}f(p)$ of a smooth function $f : \mathcal{M} \rightarrow \mathbb{C}$ at a point $p \in \mathcal{M}$. This gradient is a vector in $\mathcal{T}_p\mathcal{M}$ that points in the direction of the steepest descent of f at p, taking into account the curvature of $\mathcal{M}$. The Riemannian gradient can be obtained by projecting the Euclidean gradient of f onto $\mathcal{T}_p\mathcal{M}$. Therefore, the Riemannian gradient can be expressed as:

$$\nabla_{\mathcal{M}}f(p) = \text{Proj}\{\mathcal{T}_p\mathcal{M}(\nabla f(p))\}, \tag{6.14}$$

where $\nabla f(p)$ is the Euclidean gradient of f at p, and $\text{Proj}\{\mathcal{T}_p\mathcal{M}\}$ is the projection operator onto the tangent space $\mathcal{T}_p\mathcal{M}$. The Riemannian gradient is an important tool in optimization on RMs. In particular, it is used in gradient-based algorithms that require the computation of a descent direction at each iteration.

Another key operator in many optimization algorithms on manifolds is the *retraction operation*. The retraction operation is a way of moving from a point on a manifold to a nearby point while staying on the manifold. Mathematically, the retraction is a smooth map that takes a point on the tangent space at a certain point on the manifold and returns a point on the manifold. The retraction map $\text{Retr}_p : \mathcal{T}_p\mathcal{M} \rightarrow \mathcal{M}$ is a smooth map that satisfies the following properties [10]:

1) $\text{Retr}_p(0_p) = p$, where 0_p denotes the zero element in $\mathcal{T}_p\mathcal{M}$.
2) The *directional derivative* of the retraction map at 0_p is the identity map, that is, $D\,\text{Retr}\,p(0_p) = \text{id}_{\mathcal{T}_p\mathcal{M}}$, where $\text{id}_{\mathcal{T}_p\mathcal{M}}$ is the identity mapping on $\mathcal{T}_p\mathcal{M}$.
3) $\text{Retr}_p(v)$ is a point on the manifold for all $v \in \mathcal{T}_p\mathcal{M}$.

For example, in the case of the *complex circle manifold*, a common choice for the retraction map is the exponential map, denoted by Exp_p for a point p on the manifold. The exponential map takes a point v in the tangent space at p and returns a nearby point on the manifold. Therefore, the retraction operator via an exponential map for the *complex circle manifold* is given by:

$$\text{Retr}_p(v) = \text{Exp}_p(v) = p \cdot e^{iv}, \tag{6.15}$$

where p is a complex number with magnitude 1 (i.e., $|p| = 1$), v is a real number, and i is the imaginary unit. Hence, we have the complex multiplication operation to rotate p by an angle v. This operation returns a point on the complex circle manifold with magnitude 1, so it stays on the manifold. It is important to note that there are several ways to construct a retraction map, depending on the manifold.

Next, we provide different RM-based solution methods to solve the EE IoT power control for the RIS-aided M-MIMO communication problem described previously.

6.5 Energy-Efficient Uplink Transmission in RIS M-MIMO IoT Systems

In this section, an UL M-MIMO network of IoT devices assisted by a RIS is considered. This study introduces a technique for optimizing RIS phase shifts and BS combining vectors to minimize the UL transmit power in an RIS-aided IoT network. Since IoT devices are battery limited, it is paramount to develop an energy-efficient UL power control mechanism based on the RIS-aided system. The proposed method, known as RM-based alternating optimization (RM-AO), focuses on separately optimizing RIS phase shifts and BS combining vectors using the **RCG** technique. By leveraging the RM structure of sets involving unit-modulus phase shifts and unit-norm beamforming vectors, the non-convex UL power minimization problem is transformed into an unconstrained problem.

6.5.1 Problem Formulation

The cascaded channel between the IoT device k, the RIS, and the BS in (6.9) can be rewritten as

$$\mathbf{h}_k = \mathbf{d}_k + \mathbf{F}\boldsymbol{\Phi}\hat{\mathbf{h}}_k, \tag{6.16}$$

$$= \mathbf{d}_k + \mathbf{F}\mathrm{diag}(\hat{\mathbf{h}}_k)\boldsymbol{v}, \tag{6.17}$$

$$= \mathbf{d}_k + \mathbf{F}\hat{\mathbf{H}}_k\boldsymbol{v}, \tag{6.18}$$

where $\hat{\mathbf{H}}_k = \mathrm{diag}(\hat{\mathbf{h}}_k)$.

By considering known (or accurately estimated) instantaneous CSI, the BS performs UL power allocation and controls the RIS phase shift based on the acquired channel information. Subsequently, the BS transmits the UL transmit power indicator (p_k) and the RIS phase-shift vector ($\boldsymbol{v}$) to the IoT device and RIS corresponding to the kth device. Ultimately, each IoT device transmits data to the BS through the UL channel.

Let $x_k = \sqrt{p_k}s_k$ denote the data signal of the kth IoT device, where s_k represents the normalized data symbol, and $p_k \geq 0$ is the transmit power of the kth IoT device.

Based on the combining operation given by (6.13), the SE of the kth user results in

$$\mathrm{SE}_k = \log_2\left(1 + \frac{p_k|\mathbf{c}_k^H(\mathbf{d}_k + \mathbf{F}\hat{\mathbf{H}}_k\boldsymbol{v})|^2}{\sum_{j\neq k}^{K} p_j|\mathbf{c}_k^H(\mathbf{d}_j + \mathbf{F}\hat{\mathbf{H}}_j\boldsymbol{v})|^2 + \sigma^2}\right) \tag{6.19}$$

The UL power minimization problem to optimize the RIS phase shift vector $\boldsymbol{v}$, the BS beamforming matrix $\mathbf{C} = [\mathbf{c}_1 \ldots \mathbf{c}_K]$, and the device power vector

$\mathbf{p} = [p_1 \dots p_K]$ is formulated as

$$\underset{\boldsymbol{v},\mathbf{C},\mathbf{p}}{\text{minimize}} \sum_{k=1}^{K} p_k \tag{6.20a}$$

$$\text{s.t.} \ \frac{p_k|\mathbf{c}_k^H(\mathbf{d}_k+\mathbf{F}\hat{\mathbf{H}}_k\boldsymbol{v})|^2}{\sum_{j\neq k}^{K} p_j|\mathbf{c}_k^H(\mathbf{d}_j+\mathbf{F}\hat{\mathbf{H}}_j\boldsymbol{v})|^2+\sigma^2} \geq 2^{\overline{\text{SE}}_k} - 1, \quad \forall k \in \mathcal{K}, \tag{6.20b}$$

$$|v_n| = 1 \quad \forall n \in \{1, \dots, N\}, \tag{6.20c}$$

$$\| \mathbf{c}_k \| = 1 \quad \forall k \in \mathcal{K}, \tag{6.20d}$$

$$0 \leq p_k \leq P_{\max,k}, \quad \forall k \in \mathcal{K}, \tag{6.20e}$$

where $\mathcal{K}$ is the set of IoT devices, $\overline{\text{SE}}_k$ is the QoS requirement, and $P_{\max,k}$ is the maximum transmit power of the kth IoT device. To solve this problem, an AO approach is considered. When the power allocation for the IoT devices is fixed, the RIS phase shift values and the BS combining vectors can be obtained via manifold-based solution by applying a RM algorithm [40] solving the following optimization problem:

$$\text{Find}(\boldsymbol{v}, \mathbf{C}) \tag{6.21a}$$

$$\text{s.t.} \ \frac{p_k A_{k,k}(\boldsymbol{v},\mathbf{c}_k)}{\sum_{j\neq k}^{K} p_j A_{j,k}(\boldsymbol{v},\mathbf{c}_k)+\sigma^2} \geq 2^{\overline{\text{SE}}_k} - 1, \quad \forall k \in \mathcal{K}, \tag{6.21b}$$

$$|v_n| = 1 \quad \forall n \in \{1, \dots, N\}, \tag{6.21c}$$

$$\| \mathbf{c}_k \| = 1 \quad \forall k \in \mathcal{K}, \tag{6.21d}$$

$$0 \leq p_k \leq P_{\max,k}, \quad \forall k \in \mathcal{K}, \tag{6.21e}$$

where $A_{j,k}(\boldsymbol{v}, \mathbf{c}_k) = |\mathbf{c}_k^H(\mathbf{d}_j + \mathbf{F}\hat{\mathbf{H}}_j\boldsymbol{v})|^2 \ \forall j, k \in \mathcal{K}$.

Note that the Lagrangian relaxation can be employed to deal with the complicated rate constraint [23]. Then, the modified objective function is

$$L(\boldsymbol{v}, \mathbf{C}, \boldsymbol{\lambda}) = \sum_{k=1}^{K} \lambda_k \left(-\frac{p_k}{2^{\overline{\text{SE}}_k} - 1} A_{k,k}(\boldsymbol{v}, \mathbf{c}_k) + \sum_{k\neq 1}^{K} p_j A_{j,k}(\boldsymbol{v}, \mathbf{c}_k) + \sigma^2 \right) \tag{6.22}$$

where $\boldsymbol{\lambda} = [\lambda_1, \dots, \lambda_K]^T$ is the Lagrangian multiplier vector obtained by solving the corresponding dual problem. Using the Lagrangian function, the optimization problem is relaxed into

$$\underset{\boldsymbol{v},\mathbf{C},\boldsymbol{\lambda}}{\text{minimize}} \sum_{k=1}^{K} \lambda_k \left(-\frac{p_k}{2^{\overline{\text{SE}}_k} - 1} A_{k,k}(\boldsymbol{v}, \mathbf{c}_k) + \sum_{k\neq 1}^{K} p_j A_{j,k}(\boldsymbol{v}, \mathbf{c}_k) + \sigma^2 \right) \tag{6.23a}$$

$$\text{s.t.} \quad |v_n| = 1 \quad \forall n \in \{1, \dots, N\}, \tag{6.23b}$$

$$\| \mathbf{c}_k \| = 1 \quad \forall k \in \mathcal{K}, \tag{6.23c}$$

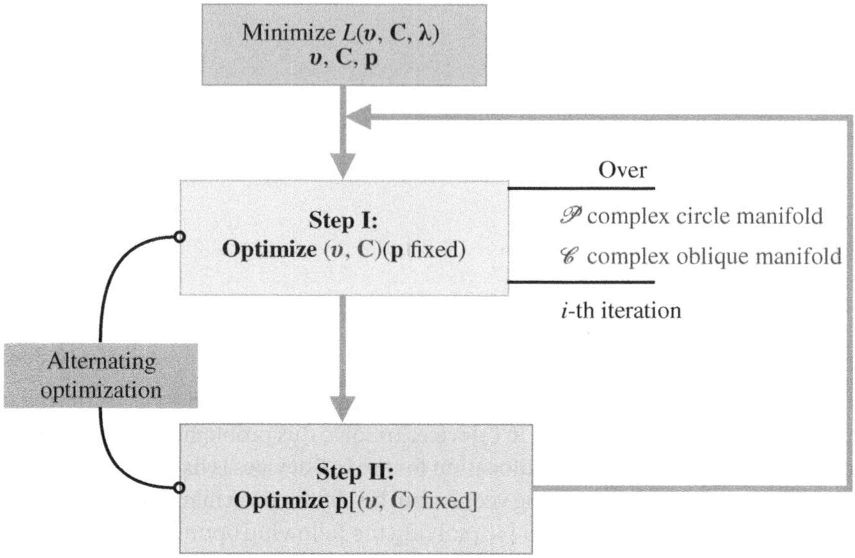

Figure 6.2 Iterative-alternating optimization (i-AO) steps for the proposed algorithm attains convergence based on AO technique regarding variables $(\boldsymbol{v}, \mathbf{C})$ and $\mathbf{p}$.

$$\lambda_k \geq 0 \quad \forall k \in \mathscr{K}, \tag{6.23d}$$

Observe that the constraints in (6.23b) and (6.23c) represent the unit-modulus of the RIS phase shift and the unit-norm constraint of the combining vector for the IoT devices, respectively. Due to the non-convexity of the unit-modulus and unit-norm constraints, (6.23) is still a non-convex problem.

To deal with the issues mentioned above, we can sequentially and alternately solve the joint coupled $[\boldsymbol{v}, \mathbf{C}, \mathbf{p}]$ optimization problem, Figure 6.2. A constraint manifold approach formulates each optimization variable as $\mathscr{P}$ and $\mathscr{C}$, where $\mathscr{P}$ is the *Complex circle manifold*, and $\mathscr{C}$ is the *complex oblique manifold*, given by

$$\mathscr{P} = \{\boldsymbol{v} \in \mathbb{C}^{N\times 1} : |v_n| = 1, \ \forall n \in \{1, \dots, N\}\}, \tag{6.24}$$

$$\mathscr{C} = \{\mathbf{C} \in \mathbb{C}^{M\times K} : \| \mathbf{c}_1 \| = \cdots = \| \mathbf{c}_K \| = 1\}. \tag{6.25}$$

With the RM constraints and its smooth structure, we can address the constraints (6.23b) and (6.23c) of the EE power control problem (6.23). We can do this by applying an AO technique to $\mathbf{p}$ and $(\boldsymbol{v}, \mathbf{C})$ to minimize the modified objective function on the RM. To do this, in the AO approach, we first fix the UL transmit power of the IoT devices ($\mathbf{p}$), and then sequentially optimize the phase shift vector $\boldsymbol{v}$ and the BS combining matrix $\mathbf{C}$ using one of the RM methods. Next, we use the obtained optimal values for $(\boldsymbol{v}, \mathbf{C})$ as fixed input and update the Lagrangian multiplier λ. In the second step, we keep the previous variables fixed to optimize the IoT device's transmit power vector $\mathbf{p}$.

Step I: For a fixed UL transmit power $\mathbf{p}$ and under RM constraints, the optimization problem can be simplified as:

$$\min_{\boldsymbol{v}\in\mathscr{P},\mathbf{C}\in\mathscr{C}} L(\boldsymbol{v}, \mathbf{C}). \tag{6.26}$$

Note that $(\boldsymbol{v}, \mathbf{C})$ in problem (6.26) can be obtained by optimizing $\boldsymbol{v}$ maintaining $\mathbf{C}$ fixed and vice versa. Hence, when the value of $\mathbf{C}$ is provided, $L(\boldsymbol{v})$ transforms into a quadratic function with respect to $\boldsymbol{v}$. Then, the expression for $L(\boldsymbol{v})$ can be written as follows:

$$\underset{\boldsymbol{v}\in\mathscr{P}}{\text{minimize}} \quad L(\boldsymbol{v}) = \boldsymbol{v}^H\mathbf{B}\boldsymbol{v} + \mathbf{b}^H\boldsymbol{v} + \boldsymbol{v}^H\mathbf{b} + b, \tag{6.27}$$

$$\text{where} : \mathbf{B} = \sum_{k=1}^{K}\lambda_k\left(-\frac{p_k}{2^{\overline{\text{SE}}_k}-1}\hat{\mathbf{H}}_k^H\mathbf{F}^H\mathbf{c}_k\mathbf{c}_k^H\mathbf{F}\hat{\mathbf{H}}_k + \sum_{j\neq k}^{K}\hat{\mathbf{H}}_j^H\mathbf{F}^H\mathbf{c}_k\mathbf{c}_k^H\mathbf{F}\hat{\mathbf{H}}_j\right),$$

$$\mathbf{b} = \sum_{k=1}^{K}\lambda_k\left(-\frac{p_k}{2^{\overline{\text{SE}}_k}-1}\hat{\mathbf{H}}_k^H\mathbf{F}^H\mathbf{c}_k\mathbf{c}_k^H\mathbf{d}_k + \sum_{j\neq k}^{K}\hat{\mathbf{H}}_j^H\mathbf{F}^H\mathbf{c}_k\mathbf{c}_k^H\mathbf{d}_j\right), \text{ and}$$

$$b = \sum_{k=1}^{K}\lambda_k\sigma^2.$$

Now, with $\boldsymbol{v}$ fixed, we can optimize problem (6.26) to obtain $\mathbf{C}$. After updating the RIS phase shift vector and the BS combining matrix, the Lagrangian multiplier λ is updated as in [23].

Step II: After attaining convergence for $\boldsymbol{v}$ and $\mathbf{C}$ in (6.26), the optimization problem of variable $\mathbf{p}$ is formulated simply as a linear programming (LP) problem, which can be easily solved using the convex optimization techniques, for example, `cvx` optimizer [41]. If $\boldsymbol{v}$ and $\mathbf{C}$ are given (fixed), the optimization problem becomes a power allocation problem as follows:

$$\min_{\mathbf{p}\in\mathfrak{R}_+} \sum_{k=1}^{K} p_k \tag{6.28a}$$

$$\text{s.t.} \frac{p_k}{2^{\overline{\text{SE}}_k}-1}A_{k,k}(\boldsymbol{v}, \mathbf{c}_k) - \sum_{j\neq k}^{K} p_j A_{j,k}(\boldsymbol{v}, \mathbf{c}_k) \geq \sigma^2,$$
$$\forall k \in \mathscr{K}, \tag{6.28b}$$

$$0 \leq p_k \leq P_{\max,k}, \quad \forall k \in \mathscr{K}, \tag{6.28c}$$

which is a convex optimization problem and then can be solved by the `cvx` solver [42].

We repeat Steps I and II (AO procedure) until the tuple $\{\boldsymbol{\upsilon}, \mathbf{C}, \mathbf{p}\}$ solution converges to fixed values or the difference between consecutive alternate steps achieves a specific threshold ε value.

6.5.2 Proposed Solution and Algorithms Based on Manifolds

In this section, we present the main, representative algorithms based on manifold optimization, namely: (a) manifold particle swarm optimization (PSO)-based method; (b) manifold based on conjugate-gradient (CG); (c) RM TR; and (d) manifold steepest-descent (SD). Hence, in Section 6.5.4, we compare the performance of such algorithms considering typical RIS-aided M-MIMO channel and system scenarios.

6.5.2.1 Manifold PSO-Based Solution Method

The manifold PSO-based algorithm relies on an iterative optimization process that combines the heuristic PSO algorithm with the RM [43]. PSO is a population-based heuristic evolutionary optimization algorithm that uses a set of particles to explore the solution space, searching for a global or a local solution. In the case of manifold PSO-based, the manifold is used to restrict the search of the particles to a specific RM subspace. This is useful in optimization problems where the solutions have a geometric structure, for example, when the problem is subject to constraints or when the optimal solution lies in a specific manifold. By limiting the search of particles to a RM, the PSO-based algorithm can perform a more efficient and precise search in the space of solutions, considering the constraints imposed by the manifold.

The manifold PSO-based algorithm is very similar to the standard PSO algorithm with modifications to account for optimization on RMs. The algorithm's objective is to minimize a function $f(\boldsymbol{\zeta})$ on a RM $\mathcal{M}$, where $\boldsymbol{\zeta}$ is a particle in $\mathcal{M}$. The PSO algorithm updates the velocity and position of each particle using respectively [44]:

$$\mathbf{V} = w\mathbf{V} + c_1\mathbf{R}_1 \odot (\mathbf{M}_{\text{PB}} - \boldsymbol{\Psi}) + c_2\mathbf{R}_2 \odot (\mathbf{M}_{\text{GB}} - \boldsymbol{\Psi}), \tag{6.29}$$

$$\boldsymbol{\Psi} = \boldsymbol{\Psi} + \mathbf{V}, \tag{6.30}$$

where $\odot$ denotes the Hadamard product,[1] N_{dim} represents the dimensional of the problem, w represents the inertia factor, and matrices $\mathbf{R}_1$ and $\mathbf{R}_2$ are composed by elements drawn from the uniform distribution $\mathcal{U}[0, 1]$. Matrices $\boldsymbol{\Psi} \in \mathbb{R}^{N_{\text{dim}} \times N_{\text{pop}}}$ and $\mathbf{V} \in \mathbb{R}^{N_{\text{dim}} \times N_{\text{pop}}}$ store the position and velocity of N_{pop} particles of the swarm in each column, such that $\boldsymbol{\Psi} = [\boldsymbol{\psi}_1 \dots \boldsymbol{\psi}_{N_{\text{pop}}}]$ and $\mathbf{V} = [\mathbf{v}_1 \dots \mathbf{v}_{N_{\text{pop}}}]$. The matrix $\mathbf{M}_{\text{PB}}$ is constructed with the personal best position of each particle, while the best position matrix is given by $\mathbf{M}_{\text{GB}} = [\boldsymbol{\psi}_{\text{GB}} \cdots \boldsymbol{\psi}_{\text{GB}}] \in \mathbb{R}^{N_{\text{dim}} \times N_{\text{pop}}}$, where vector $\boldsymbol{\psi}_{\text{GB}} \in \mathbb{R}^{N\,\text{dim} \times 1}$

---

1 Element-wise multiplication.

denotes the best position in the swarm, that is, the global best position that minimizes the fitness function.

In the initialization of the implemented manifold PSO-based algorithm, the positions of the particles $\boldsymbol{\Psi}$ are generated randomly following a uniform distribution inside the search space. Those positions are set as the personal best position of the particle in the matrix $\mathbf{M}_{\text{PB}}$. The fitness function is evaluated ($\zeta_k = \boldsymbol{\Psi}_k, k = 1, \ldots, N_{\text{iter}}$) and the position of the particle that produces the lowest value (since we are dealing with a minimization problem) is set as the global best position $\boldsymbol{\Psi}_{\text{GB}}$, and the matrix $\mathbf{M}_{\text{GB}}$ is formed [45]. After evaluation of Eqs. (6.29) and (6.30), matrices $\mathbf{M}_{\text{PB}}$ and $\mathbf{M}_{\text{GB}}$ are updated and the process is repeated till the stop criterion is met; herein, the criterion of maximum number of evaluations N_{iter} is deployed. The output of the algorithm is the vector of the best position $\boldsymbol{\psi}_{\text{GB}}$. A pseudocode summarizing the procedure described for the evolutionary heuristic PSO algorithm is presented in Algorithm 6.1.

Algorithm 6.1 Manifold PSO-based scheme

Input: $c_1, c_2, w, N_{\text{pop}}, N_{\text{iter}}$
Generate initial positions $\boldsymbol{\Psi}$;
Fitness function evaluation and initialization of $\mathbf{M}_{\text{PB}}, \mathbf{M}_{\text{GB}}$;
for $k = 1$ until N_{iter} **do**
  Calculate velocity, Eq. (6.29);
  Calculate position, Eq. (6.30);
  Update each particle's position by applying the retraction operation;
  Evaluate fitness function for all particles $\boldsymbol{\psi}_k$;
  Update personal best matrix $\mathbf{M}_{\text{PB}}$;
  Update global best matrix $\mathbf{M}_{\text{GB}}$;
end for
Outputs: $\boldsymbol{\psi}_{\text{GB}}$

6.5.2.2 Manifold CG-Based Algorithm

In order to solve the optimization problem, we also can apply a modified version of the classical *conjugate gradient* (CG) *method* [46] by using the RM concepts, namely RCG. To achieve this goal, we update the conjugate direction $\mathbf{D}$ and the point $\mathbf{X}$ as follows:

$$\mathbf{D}_k = -\nabla_{\mathcal{M}} f(\mathbf{X}_k) + \beta_k \text{Proj}\{\mathcal{T}_{\mathbf{X}_k}\mathcal{M}(\mathbf{D}_{k-1})\}, \tag{6.31}$$

$$\mathbf{X}_{k+1} = \text{Retr}_{\mathbf{X}_k}(\alpha_k \mathbf{D}_k), \tag{6.32}$$

where $\nabla_{\mathcal{M}} f(\mathbf{X}_k)$ denotes the Riemannian gradient of the objective function at $\mathbf{X}_k$, with $\mathbf{X}_k$ being the phase shift matrix $\boldsymbol{\Phi}$ or the beamforming matrix $\mathbf{W}$, $\text{Retr}_p(v)$ is the *retraction operator*, which can be an exponential map in v for the case of the complex circle manifold as defined, for example, in Eq. (6.15), β_k is the

Fletcher–Reeves CG parameter, and α_k is the step size at the kth iteration of the RCG algorithm [46]. In addition, the CG parameter β_k is calculated as:

$$\beta_k = \frac{\|\nabla_{\mathcal{M}} f(\mathbf{X}_k)\|^2}{\|\nabla_{\mathcal{M}} f(\mathbf{X}_{k-1})\|^2}. \tag{6.33}$$

Besides, the step size α_k at the kth iteration is calculated using a line search algorithm, aiming to find a step size α_k that reduces the objective function sufficiently while satisfying the Armijo rule, a sufficient condition for the step size to be accepted, given by [46]:

$$f(\mathrm{Retr}_{\mathbf{X}_k}(\alpha_k \mathbf{D}_k)) \leq f(\mathbf{X}_k) + c_1 \alpha_k \langle \nabla_{\mathcal{M}} f(\mathbf{X}_k), \mathbf{D}_k \rangle, \tag{6.34}$$

where $\langle X, Y \rangle$ is the inner product between X and Y, and c_1 is a constant chosen to be a small positive number, for example, $c_1 = 10^{-4}$.

The line search algorithm starts with an initial step size α_0 (e.g., $\alpha_0 = 1$) and iteratively adjusts the step size until the Armijo rule is satisfied. Then, the *backtracking line search* algorithm is used as follows [47]:

1) Set $\alpha_k = \alpha_0$, and $\mathbf{X}_{\mathrm{new}} = \mathbf{X}_k + \alpha_k \mathbf{D}_k$.
2) If the Armijo rule is satisfied, then return α_k.
3) Otherwise, decrease the step size by a factor $\rho \in (0, 1)$, i.e., $\alpha_k \leftarrow \rho\alpha_k$, and set $\mathbf{X}_{\mathrm{new}} = \mathbf{X}_k + \alpha_k \mathbf{D}_k$.
4) Repeat steps 2 and 3 until the Armijo rule is satisfied or the step size becomes too small.

If the step size becomes too small, the line search algorithm terminates with the current iterate $\mathbf{X}_k$. Otherwise, the line search algorithm returns the step size α_k that satisfies the Armijo rule.

The step-by-step procedure for the manifold CG-based algorithm is provided in Algorithm 6.2.

6.5.2.3 Riemannian Manifold TR-Based Scheme

The Riemannian TR solution is an optimization procedure that iteratively solves a trust-region subproblem to obtain a search direction for updating the current iteration. The TR subproblem involves minimizing a quadratic objective function within a TR defined by a radius parameter. The solver seeks to find a point that improves the objective function within the TR, and the radius parameter is adjusted based on the quality of the updated values at each iteration [10].

The TR starts with an initial point $\mathbf{X}_0$ and a radius parameter $\Delta_0 > 0$. At each iteration k, the algorithm solves the TR subproblem:

$$\underset{\boldsymbol{\eta} \in \mathcal{T}_{\mathbf{X}_k}\mathcal{M}}{\text{minimize}} m_k(\boldsymbol{\eta}) = f(\mathbf{X}_k) + \langle \nabla_{\mathcal{M}} f(\mathbf{X}_k), \boldsymbol{\eta} \rangle + \frac{1}{2} \langle \boldsymbol{\eta}, \mathbf{H}_k \boldsymbol{\eta} \rangle,$$
$$\text{s.t.} ||\boldsymbol{\eta}||^2 \leq \Delta_k^2,$$

Algorithm 6.2 Manifold CG-based scheme

Input: $c_1, \epsilon, \mathcal{M}, N_{\text{iter}}$
$k = 0$;
$\beta_0 = 0$;
Initialize $\mathbf{X}_0$ randomly from the manifold $\mathcal{M}$;
$\mathbf{D}_0 = -\nabla_{\mathcal{M}} f(\mathbf{X}_0)$;
while $k \leq N_{\text{iter}}$ and $||\mathbf{X}_k - \mathbf{X}_{k-1}||^2 \leq \epsilon$ **do**
  $k = k + 1$;
  $\nabla_{\mathcal{M}} f(\mathbf{X}_k) = \text{Proj}\{\mathcal{T}_{\mathbf{X}_k}\mathcal{M}(\nabla f(\mathbf{X}_k))\}$;
  Update β_k according to Eq. (6.33);
  Update $\mathbf{D}_k$ according to Eq. (6.31);
  Find the step size $\alpha_k > 0$ via backtracking line search algorithm;
  Update $\mathbf{X}_k$ according to Eq. (6.32);
end while
Outputs: $\mathbf{X}_k$

where $\mathbf{H}_k$ is the Hessian of the objective function at $\mathbf{X}_k$. Then, the new candidate is obtained as $\mathbf{X}_{k+1} = \text{Retr}_{\mathbf{X}_k}(\boldsymbol{\eta}_k)$, where $\boldsymbol{\eta}_k$ is the solution of the TR subproblem at the kth iteration. The TR radius Δ_k is updated based on the ratio of the actual to the predicted reduction in the objective function as:

$$\rho_k = \frac{f(\mathbf{X}_k) - f(\mathbf{X}_{k+1})}{m_k(\mathbf{0}_{\mathbf{X}_k}) - m_k(\boldsymbol{\eta})}. \tag{6.35}$$

If ρ_k is small, the update is rejected, and the TR radius Δ_k is reduced. If ρ_k is large, the update is accepted, and the TR radius is increased. If ρ_k is in a middle range, the update is accepted without changing the TR radius. Therefore, the TR solver finishes when a stopping criterion is met, such as reaching a maximum number of iterations N_{iter} or achieving a small gradient norm. A pseudocode for the manifold TR-based solution is presented in Algorithm 6.3.

6.5.2.4 Manifold SD-Based Scheme

The SD algorithm for RMs (RSD) uses a first-order gradient method to solve an optimization problem. Given a smooth objective function f on a RM $\mathscr{M}$, the algorithm finds a point $\mathbf{X} \in \mathscr{M}$ that minimizes f. The SD method works by taking steps in the direction of the negative gradient of f at the current point [10]. The SD algorithm begins with an initial guess $\mathbf{X}_0$ on the manifold $\mathscr{M}$. At each iteration, it calculates the gradient of f at the current point $\mathbf{X}_k$ by using the Riemannian gradient operator. Then, it takes a step in the direction of the negative gradient scaled by a step size α_k, that is, $-\alpha_k \nabla_{\mathscr{M}} f(\mathbf{X}_k)$. The step size is selected by a line search algorithm satisfying the Armijo rule, such as a backtracking line search, similar to

Algorithm 6.3 Manifold TR-based scheme

Input: $\overline{\rho} \in [0, \frac{1}{4})$, Δ_0, $\Delta_{\max} > 0$, $\mathcal{M}$, N_{iter}
$k = 0$;
Initialize $\mathbf{X}_0$ randomly from the manifold $\mathcal{M}$;
while $k < N_{\text{iter}}$ **do**
  Obtain $\boldsymbol{\eta}_k$ by solving the subproblem in (6.35) (e.g., applying the CG method);
  Evaluate ρ_k according to Eq. (6.35);
  if $\rho_k > 1/4$ **then**
    $\Delta_{k+1} = \frac{1}{4}\Delta_k$;
  end if
  if $\rho_k > 3/4$ and $||\boldsymbol{\eta}_k|| = \Delta_k$ **then**
    $\Delta_{k+1} = \min(2\Delta_k, \Delta_{\max})$;
  else
    $\Delta_{k+1} = \Delta_k$;
  end if
  if $\rho_k > \overline{\rho}$ **then**
    $\mathbf{X}_{k+1} = \text{Retr}_{\mathbf{X}_k}(\boldsymbol{\eta}_k)$;
  else
    $\mathbf{X}_{k+1} = \mathbf{X}_k$
  end if
  $k = k + 1$;
end while
Outputs: $\mathbf{X}_k$

what occurs in the RCG algorithm. The algorithm then updates the point on the manifold to

$$\mathbf{X}_{k+1} = \text{Retr}_{\mathbf{X}_k}(-\alpha_k \nabla_{\mathcal{M}} f(\mathbf{X}_k)). \tag{6.36}$$

This procedure repeats until convergence or the maximum number of iterations is achieved. A pseudocode for the RSD solution is given in Algorithm 6.4.

6.5.3 Manopt Toolbox

Manifold optimization approach brings the advantage of a better theoretical, geometric, and global knowledge of the problem [43]. Manopt toolbox for optimization on manifolds and matrices is suitable for non-convex problems with establishing constraints, symmetries, or invariance properties in the form of manifolds [47]. Manopt stands out as an exceptional optimization toolbox for MATLAB, tailored specifically for optimization problems on manifolds

Algorithm 6.4 Manifold SD-based

Input: $\mathcal{M}$, N_{iter}
$k = 0$;
Initialize $\mathbf{X}_0$ randomly from the manifold $\mathcal{M}$;
while $k \leq N_{\text{iter}}$ **do**
  $\nabla_{\mathcal{M}} f(\mathbf{X}_k) = \text{Proj}\{\mathcal{T}_{\mathbf{X}_k}\mathcal{M}(\nabla f(\mathbf{X}_k))\}$;
  Find the step size $\alpha_k > 0$ via backtracking line search algorithm;
  $\mathbf{X}_{k+1}$ according to Eq. (6.36);
  $k = k + 1$;
end while
Outputs: $\mathbf{X}_k$

[47]. Ideal for applications in machine learning, computer vision, physics, and communication systems, Manopt is designed to handle optimization challenges where variables are situated on spaces with intrinsic geometries.

This toolbox seamlessly integrates into MATLAB, offering a versatile suite of tools to address optimization problems on a diverse range of manifolds. Users can effortlessly work with familiar spaces like Stiefel and Grassmann manifolds or create custom manifolds tailored to their specific needs. The toolbox is equipped with a collection of optimization algorithms [47], including gradient descent, CG, TR methods, and quasi-Newton methods, all adapted for efficient operation on manifolds.

One of the Manopt's strengths lies in its intuitive problem definition approach. Users can succinctly specify the cost function, its gradient, constraints, and additional parameters, streamlining the optimization process. The toolbox also excels in managing geodesic computations on manifolds, ensuring optimization algorithms respect the underlying geometry.

Operational advantages include the toolbox's ability to leverage manifold-specific optimization algorithms, fostering better convergence and accuracy for non-Euclidean problems. Manopt's high-level abstraction simplifies intricate details, allowing users to concentrate on problem formulation. Additionally, the supportive user community and open-source nature contribute to continuous development.

On the other hand, users might face challenges when dealing with less common manifolds, necessitating the creation of custom definitions. However, the provided documentation and examples help mitigate these difficulties.

6.5.4 Numerical Results and Discussions

The effectiveness of the proposed EE UL power control in RIS-aided M-MIMO systems is corroborated by the numerical results, where the multiuser combining

Table 6.2 Simulation system parameter values, metrics, and optimization variables.

Parameter	Value
κ	∞
PL factor (IoT device-BS link)	3.8
PL factor factor (IoT device-RIS link)	2.2
PL factor (RIS-BS link)	2.0
IoT device-BS distance	[57–65] [m]
RIS-BS distance	65 [m]
IoT device-RIS distance	[7–11] [m]
PL at the reference distance	$C_0 = -30$ [dB]
Reference distance	$D_0 = 1$ [m]
Number of antennas per IoT device	1
Number of antennas at BS	64
Number of RIS elements	$N = 64$
Number of IoT devices	$K = 8$ [3 : 9]
Carrier frequency	$f = 2.5$ [GHz]
Max power per IoT device	$P_{k,\max} = 1$ [W], $\forall k$
Number of maximum iterations	$\mathcal{I}_{\max} = 5$
Monte-Carlo simulations (MCs)	$\mathcal{T} = 1000$ trials
Optimization variables	**Method| tools**
Multiuser combining matrix (BS), $\mathbf{C}$	R. Manifold Optz. | `Manopt` & ZF
RIS phase-shifts vector, $\boldsymbol{v}$	R. Manifold Optz. | `Manopt`
UTs' power transmitting vector, $\mathbf{p}$	Convex Optz. | `cvx`
Metrics	**Range**
Minimum SE requirement ($\overline{\text{SE}}_k$)	$[0.1 : 0.1 : 0.7]; 0.3; \left[\frac{\text{bps}}{\text{Hz}}\right]$
Background noise variance @BS	$\sigma^2 = -80$; $[-100 : 5 : -70]$ [dBm]

($\mathbf{C}$), and the RIS phase-shifts vector ($\boldsymbol{v}$) were optimized under RM methods, while the IoT device power allocation vector ($\mathbf{p}$) is obtained via convex optimization with `cvx` solver. Table 6.2 outlines the main system and channel parameter values used in this section based on [20, 21, 23]. Moreover, the PL is modeled as: $\text{PL} = C_0(\frac{d}{D_0})^{-\alpha}$, with d being the distance, α the PL exponent, and $C_0 = -30$ dB the PL at the reference distance $D_0 = 1$ m.

6.5.4.1 Total UL Power Transmission

In the next, we explore the total UL power transmission metric (P_{UT}), which is given by:

$$P_{\text{UT}} = \sum_{k=1}^{K} p_k, \qquad \text{averaged over } \mathscr{T} = 1000 \text{ MCS trials} \tag{6.37}$$

varying the minimum SE requirement ($\overline{\text{SE}}_k$) and the noise variance at the BS (σ^2).

The i-AO structure adopted in the RM-based algorithms is depicted in Figure 6.2. In the UL total power transmission analysis, Eq. (6.37), we have considered different optimization setups to evaluate the achievable performance of the manifold-based algorithms. Figures 6.3 and 6.4 reveal the performance results for the following schemes:

i) UL M-MIMO system under minimum mean squared error (MMSE) combiner ($\mathbf{C}_{\text{MMSE}}$) but without RIS.

Also, we have adopted different RM-based optimization solutions for both optimization variables $\boldsymbol{\upsilon}$ and $\mathbf{C}$, precisely:

i) PSO manifold,
ii) TR manifold,
iii) steepest descent (SD) manifold, and
iv) CG manifold.
v) CG manifold joint-optimization of [23] (RCJ-JO).

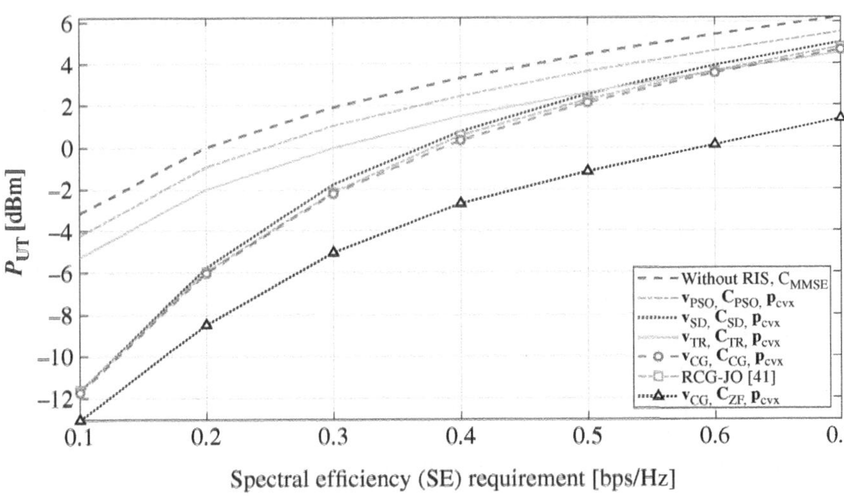

Figure 6.3 Uplink transmit power (P_{UT}) vs minimum spectral efficiency requirement ($\overline{\text{SE}}_k$).

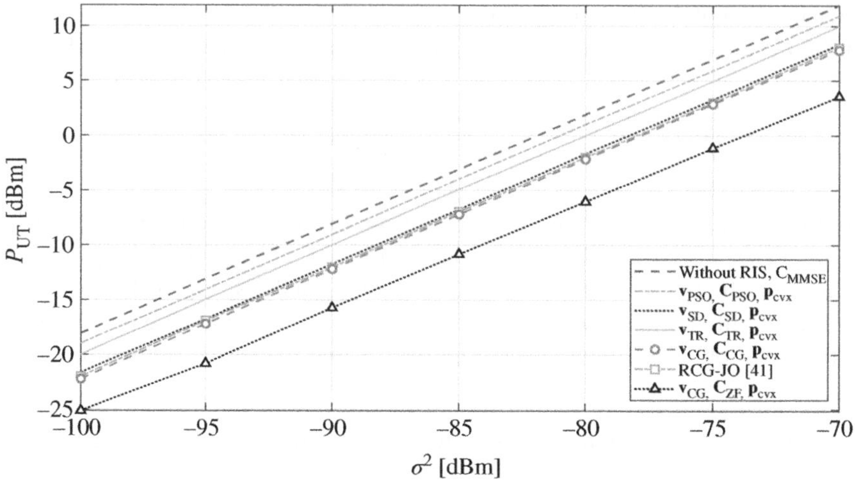

Figure 6.4 Uplink transmit power (P_{UT}) vs noise power (σ^2) with $\overline{\text{SE}}_k = 0.3$ [bits/s/Hz].

Finally, we have considered the following approach:

i) CG manifold for $\boldsymbol{v}$ and ZF combiner ($\mathbf{C}_{ZF}$).

In contrast, cvx solver [42] is deployed to allocate the optimal power vector $\mathbf{p}$ to the IoT devices.

In Figure 6.3, we analyze the required UL transmit power of the IoT devices to achieve the minimum QoS criterion for all of them ($\overline{\text{SE}}_k$), which was varied from 0.1 to 0.7 [bps/Hz]; in this case, we set the noise power as $\sigma^2 = -80$ dBm. Furthermore, in Figure 6.4, we have set $\overline{\text{SE}}_m = 0.3$ [bps/Hz], while noise power at the BS is varied in the range $\sigma^2 \in [-100; -70]$ dBm. While in Figure 6.5, we have examined the dependence of the UL transmit power with the quantity of active IoT devices. To conduct this analysis, we established $\overline{\text{SE}}_m = 0.3$ [bps/Hz], noise power as $\sigma^2 = -80$ [dBm], and systematically adjusted the number of UTs within the range of 3–9.

In both scenarios analyzed in Figures 6.3 and 6.4, it is evident that the system configuration without RIS represents the worst-case UT transmit power scenario, requiring the highest amount of P_{UT}. Furthermore, as depicted in Figure 6.3, even with the heuristic PSO-based solution, a noteworthy enhancement is achieved compared to the M-MIMO without RIS, resulting in a power reduction of $\approx 22\%$ in this comparison. Notably, the manifold CG-based solution with the ZF combining consistently proves to be the choice that achieves the minimum UL power transmission across the entire $\overline{\text{SE}}_k$ range. For instance, when $\overline{\text{SE}}_k = 0.1$ bps/Hz, the CG approach + ZF combiner yields an impressive power saving of 89% compared to the conventional M-MIMO system. In the RIS-aided configuration, the CG solutions (AO and JO approaches) and also the manifold SD scheme for both

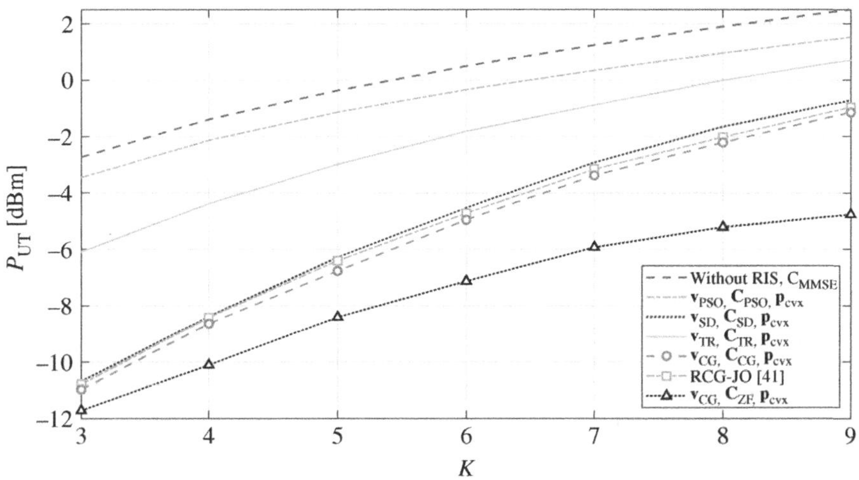

Figure 6.5 Uplink transmit power (P_{UT}) vs number of IoT devices (K) being served, with $\overline{\text{SE}}_k = 0.3$ [bits/s/Hz] and $\sigma^2 = -80]$ dBm.

$\boldsymbol{v}$ and $\mathbf{C}$ optimization exhibits power reductions of 82%, 77%, and 61% against the PSO, and TR, respectively, but the approach *vii)* still present a power saving of around 82% compared with approaches (iv), (v), and (vi). Additionally, it is observed that for $\overline{\text{SE}}_k \geq 0.5$ bps/Hz, the TR, SD, and both CG schemes tend to exhibit similar performance, achieving a power saving of around 35% over the conventional M-MIMO system, that is, without RIS aided. At the same point, solution (vii) shows a transmit power reduction of $\approx$ 52% against the manifold approaches for both $\boldsymbol{v}$ and $\mathbf{C}$ optimization. This can be explained by the fact that with ZF combiner and perfect CSI, there is no MuI; hence, the SINR of the UTs is significantly improved, resulting in power reduction to guarantee the same QoS requirement. This behavior is not observed with the other approaches that are not capable to completely eliminating the MuI after optimizing the combining matrix $\mathbf{C}$.

In Figure 6.4, the manifold CG-based approach with the ZF combining once again emerges as the best solution in terms of UTs' transmission power saving, showcasing superior performance compared to other methods. For instance, when we compare the solution (vii) against solution (i), there is a power saving of around 83%; while comparing approach (vii) in relation to approaches (iv), (v), and (vi), there is still a power saving of around 55% across all over the analyzed range of σ^2.

Finally, Figure 6.5 highlights the observable widening performance gap between the mixed solution, utilizing the CG-based technique with ZF combining, and the RM solutions for both $\boldsymbol{v}$ and $\mathbf{C}$ as the number of devices served by the BS increase. For instance, when $K = 3$, approach (vii) exhibits a power reduction of approximately 15% compared to approaches (iv), (v), and (vi). In this scenario, where the number of served devices is relatively small, the adoption of ZF combining does not yield a substantial gain due to the relatively low level of inter-user interference. Nevertheless, as the number of users increases to 9, the aforementioned power reduction further escalates to about 55%.

6.5.4.2 Complexity-Performance Trade-Off

Optimizing on a manifold ($\mathscr{P}$,$\mathscr{C}$) is not essentially different from optimizing in $\mathbb{R}^n$ [48]. Herein, we compare the computation complexity of running time (seconds) per iteration of the different design approaches. Also, for this analysis, we set $K = 8$, $\overline{\text{SE}}_k = 0.3$ [bits/s/Hz], while $\sigma^2 = -80$ dBm. As a result, the algorithmic complexity is compared via *running time* in Table 6.3. All simulation outcomes were obtained using Matlab (MATLAB R2021a environment) on hardware equipped with an AMD Ryzen 5 5600 CPU @ 3.6 GHz, 32 GB RAM, and an AMD RX 6600 GPU.

From Table 6.3, it is possible to notice that the TR-based optimization method exhibits the highest computational complexity among the presented approaches. The increased running time may indicate the additional computational load introduced by certain features or processes specific to the TR algorithm. Furthermore, it is evident that both CG-based solutions (JO and AO) exhibit nearly identical computational complexity. This similarity arises from the fact that, despite RCG-JO utilizing the product of manifolds to jointly optimize $\boldsymbol{v}$ and $\mathbf{C}$, all the update operations for these variables must be performed independently (involving Riemannian gradients and retraction operations). Consequently, the computational complexity, in terms of running time, remains almost the same for these approaches. Additionally, it is evident that the SD-based technique exhibits slightly lower complexity compared to the CG counterparts, a fact supported

Table 6.3 Computational complexity regarding running time per iteration, in seconds.

Algorithm	Running time [s]
(i) MIMO + $\mathbf{C}_{\text{MMSE}}$ w/o RIS	0.0082
(ii) PSO-RM	0.0498
(iii) TR-RM	0.3096
(iv) SD-RM	0.2066
(v) CG-RM	0.2462
(vi) RCG-JO [23]	0.2341
(vii) CG-RM ($\boldsymbol{v}$) + $\mathbf{C}_{\text{ZF}}$	0.0947

by a direct comparison of Algorithms 6.2 and 6.4. A straightforward analysis of these algorithms reveals that the SD solution involves fewer operations than the CG approach. Finally, we have illustrated that approach (vii), employing the ZF combiner and the CG-RM algorithm, demands less computational time while delivering substantial performance gains. This efficiency is attributed to the fact that in this solution, the ZF combiner is computed only once, and only the RIS phase-shift power allocation needs to be recalculated in each iteration. In contrast, the RM-based approaches necessitate updating the combining matrix in every iteration, contributing to a higher computational load.

6.6 Conclusion and Research Directions

We have introduced the RM technique for optimizing the RIS phase shift and the BS beamforming, aiming to reduce the transmit power in the UL of a RIS-assisted M-MIMO IoT network. The fundamental concept behind the proposed RIS phase shift and BS combining optimization algorithms is to employ RM methods for optimizing such $\boldsymbol{v}$ vector and $\mathbf{C}$ matrix variables alternately and separately. Additionally, the ZF combining method was taken into consideration. Subsequently, the phase shifts of the RIS were optimized based on the instantaneous CSI using the CG algorithm. The simulation results indicate that this mixed approach is promising to minimize the required UL transmission power to meet the minimum QoS requirements for IoT applications.

As we discussed in this chapter, improving the EE of RIS-aided M-MIMO IOT networks is a quite appealing problem, and the use of the manifold optimization framework is very promising for this objective.

Interesting research topics to further develop the investigated scenario include: first, although we have assumed a far-field propagation channel in our system model, extending the analysis to near-field propagation channels is a very relevant topic for future works. This is justified since the use of higher frequencies, such as mmWave and terahertz communication, in conjunction with the use of large aperture RISs in an attempt to circumvent the multiplicative PL effect, increase the Rayleigh distance, which is widely recognized as the threshold under which the near-field propagation is more prevalent. Furthermore, new opportunities arise in the near-field context, for example, the beam-focusing effect, in which not only the direction of the reflected beam can be adjusted but also the distance of maximum channel gain. Developing strategies to improve the EE of RIS-aided M-MIMO IOT networks leveraging the beam-focusing effect for near-field scenarios is thus of significant relevance.

Besides, another important research perspective involves the analysis of multiple RISs. In a typical sixth generation 6g scenario, multiple RISs can be deployed within a cell, and thus, how to coordinate the operation of these devices is another appealing problem. From an EE perspective, it is also important to obtain the optimal number of RISs and the respective number of RIS elements to be deployed since although the throughput performance can be improved when increasing the number of deployed RISs and/or number of RIS elements, the energy consumption also increases for running the controllers of such devices, also for coordination, and for activation (polarization) of each RIS element.

References

1 Chen, X., Ng, D.W.K., Yu, W. et al. (2021). Massive access for 5G and beyond. *IEEE Journal on Selected Areas in Communications* 39 (3): 615–637. https://doi.org/10.1109/JSAC.2020.3019724.

2 Ericsson (2022). Ericsson Mobility Report. Technical report. Stockholm, Sweden: Ericsson. https://www.ericsson.com/4ae28d/assets/local/reports-papers/mobility-report/documents/2022/ericsson-mobility-report-november-2022.pdf (accessed 27 September 2024).

3 Huang, T., Yang, W., Wu, J. et al. (2019). A survey on green 6G network: architecture and technologies. *IEEE Access* 7: 175758–175768. https://doi.org/10.1109/ACCESS.2019.2957648.

4 Liu, Y., Liu, X., Mu, X. et al. (2021). Reconfigurable intelligent surfaces: principles and opportunities. *IEEE Communications Surveys & Tutorials* 23 (3): 1546–1577. https://doi.org/10.1109/COMST.2021.3077737.

5 Wu, Q., Zhang, S., Zheng, B. et al. (2021). Intelligent reflecting surface-aided wireless communications: a tutorial. *IEEE Transactions on Communications* 69 (5): 3313–3351. https://doi.org/10.1109/TCOMM.2021.3051897.

6 Björnson, E., Ozdogan, O., and Larsson, E.G. (2020). Reconfigurable intelligent surfaces: three myths and two critical questions. *IEEE Communications Magazine* 58 (12): 90–96. https://doi.org/10.1109/MCOM.001.2000407.

7 Van Chien, T., Ngo, H.Q., Chatzinotas, S., and Ottersten, B. (2022). Reconfigurable intelligent surface-assisted massive MIMO: favorable propagation, channel hardening, and rank deficiency [lecture notes]. *IEEE Signal Processing Magazine* 39 (3): 97–104. https://doi.org/10.1109/MSP.2021.3128352.

8 Wang, Z., Liu, L., Zhang, S., and Cui, S. (2023). Massive MIMO communication with intelligent reflecting surface. *IEEE Transactions on Wireless Communications* 22 (4): 2566–2582. https://doi.org/10.1109/TWC.2022.3212537.

9 Jevremovic, V. Metamaterials and reconfigurable intelligent surfaces. iBwave, White Paper. http://info.ibwave.com/metamaterials-reconfigurable-intelligent-surfaces/ (accessed 27 September 2024).

10 Absil, P.-A., Mahony, R., and Sepulchre, R. (2007). *Optimization Algorithms on Matrix Manifolds*. Princeton, New Jersey, USA: Princeton University Press. ISBN 0691132984.

11 Hu, J., Liu, X., Wen, Z.-W., and Yuan, Y.-X. (2020). A brief introduction to manifold optimization. *Journal of the Operations Research Society of China* 8 (2): 199–248. https://doi.org/10.1007/s40305-020-00295-9.

12 Zhou, G., Pan, C., Ren, H. et al. (2020). A framework of robust transmission design for IRS-aided MISO communications with imperfect cascaded channels. *IEEE Transactions on Signal Processing* 68: 5092–5106.

13 Gouda, B., Atzeni, I., and Tolli, A. (2020). Distributed precoding design for cell-free massive MIMO systems. *2020 IEEE 21st International Workshop on Signal Processing Advances in Wireless Communications (SPAWC)*, 1–5. https://doi.org/10.1109/SPAWC48557.2020.9154279.

14 Wu, Q. and Zhang, R. (2019). Beamforming optimization for intelligent reflecting surface with discrete phase shifts. *ICASSP 2019-2019 IEEE International Conference on Acoustics, Speech and Signal Processing (ICASSP)*, 7830–7833. IEEE.

15 Xie, H., Xu, J., and Liu, Y.-F. (2020). Max-min fairness in IRS-aided multi-cell MISO systems with joint transmit and reflective beamforming. *IEEE Transactions on Wireless Communications* 20 (2): 1379–1393.

16 Rehman, H.U., Bellili, F., Mezghani, A., and Hossain, E. (2021). Joint active and passive beamforming design for IRS-assisted multi-user MIMO systems: a VAMP-based approach. *IEEE Transactions on Communications* 69 (10): 6734–6749.

17 Wu, Q. and Zhang, R. (2019). Intelligent reflecting surface enhanced wireless network via joint active and passive beamforming. *IEEE Transactions on Wireless Communications* 18 (11): 5394–5409. https://doi.org/10.1109/TWC.2019.2936025.

18 Huang, C., Zappone, A., Alexandropoulos, G.C. et al. (2019). Reconfigurable intelligent surfaces for energy efficiency in wireless communication. *IEEE Transactions on Wireless Communications* 18 (8): 4157–4170. https://doi.org/10.1109/TWC.2019.2922609.

19 Pan, C., Ren, H., Wang, K. et al. (2020). Multicell MIMO communications relying on intelligent reflecting surfaces. *IEEE Transactions on Wireless Communications* 19 (8): 5218–5233. https://doi.org/10.1109/TWC.2020.2990766.

20 Zeng, M., Bedeer, E., Dobre, O.A. et al. (2021). Energy-efficient resource allocation for IRS-assisted multi-antenna uplink systems. *IEEE Wireless Communications Letters* 10 (6): 1261–1265. https://doi.org/10.1109/LWC.2021.3063554.

21 You, L., Xiong, J., Ng, D.W.K. et al. (2021). Energy efficiency and spectral efficiency tradeoff in RIS-aided multiuser MIMO uplink transmission. *IEEE*

Transactions on Signal Processing 69: 1407–1421. https://doi.org/10.1109/TSP.2020.3047474.

22 Ma, X., Fang, Y., Zhang, H. et al. (2022). Cooperative beamforming design for multiple RIS-assisted communication systems. *IEEE Transactions on Wireless Communications* 21 (12): 10949–10963. https://doi.org/10.1109/TWC.2022.3188350.

23 Wu, J., Kim, S., and Shim, B. (2022). Energy-efficient power control and beamforming for reconfigurable intelligent surface-aided uplink IoT networks. *IEEE Transactions on Wireless Communications* 21 (12): 10162–10176. https://doi.org/10.1109/TWC.2022.3182773.

24 Li, Z., Zhang, J., Zhu, J., and Dai, L. (2023). RIS energy efficiency optimization with practical power models. *2023 International Wireless Communications and Mobile Computing (IWCMC)*, 1172–1177. https://doi.org/10.1109/IWCMC58020.2023.10183034.

25 Ren, H., Zhang, Z., Peng, Z. et al. (2023). Energy minimization in RIS-assisted UAV-enabled wireless power transfer systems. *IEEE Internet of Things Journal* 10 (7): 5794–5809. https://doi.org/10.1109/JIOT.2022.3150178.

26 Huo, Y., Dong, X., and Ferdinand, N. (2023). Distributed reconfigurable intelligent surfaces for energy-efficient indoor terahertz wireless communications. *IEEE Internet of Things Journal* 10 (3): 2728–2742. https://doi.org/10.1109/JIOT.2022.3214471.

27 de Souza, J.H.I., Marinello Filho, J.C., Amiri, A., and Abrão, T. (2023). QoS-aware user scheduling in crowded XL-MIMO systems under non-stationary multi-state LoS/NLoS channels. *IEEE Transactions on Vehicular Technology* 72 (6): 7639–7652. https://doi.org/10.1109/TVT.2023.3243488.

28 Marinello Filho, J.C., Brante, G., Souza, R.D., and Abrão, T. (2022). Exploring the non-overlapping visibility regions in XL-MIMO random access and scheduling. *IEEE Transactions on Wireless Communications* 21 (8): 6597–6610. https://doi.org/10.1109/TWC.2022.3151329.

29 de Souza, J.H.I., Marinello Filho, J.C., Abrão, T., and Panazio, C. (2022). Reconfigurable intelligent surfaces to enable energy-efficient IoT networks. In *2022 Symposium on Internet of Things (SIoT)*, 1–4. https://doi.org/10.1109/SIoT56383.2022.10070317.

30 Guerra, D.W.M. and Abrão, T. (2022). Clustered double-scattering channel modeling for XL-MIMO with uniform arrays. *IEEE Access* 10: 20173–20186. https://doi.org/10.1109/ACCESS.2022.3152704.

31 dos Santos, H.L., Marinello, J.C., Panazio, C.M., and Abrão, T. (2022). Machine learning-aided pilot and power allocation in multi-cellular massive MIMO networks. *Physical Communication* 52: 101646. https://doi.org/10.1016/j.phycom.2022.101646.

32 de Souza, J.H.I., Amiri, A., Abrão, T., de Carvalho, E., and Popovski, P. (2021). Quasi-distributed antenna selection for spectral efficiency maximization in subarray switching XL-MIMO systems. *IEEE Transactions on Vehicular Technology* 70 (7): 6713–6725. https://doi.org/10.1109/TVT.2021.3081462.

33 Ubiali, G.A., Marinello, J.C., and Abrão, T. (2021). Energy-efficient flexible and fixed antenna selection methods for XL-MIMO systems. *AEU - International Journal of Electronics and Communications* 130: 153568. https://doi.org/10.1016/j.aeue.2020.153568.

34 Alves, T.A.B. and Abrão, T. (2023). NOMA-based random access in MMTC XL-MIMO. *IEEE Access* 11: 1944–1954. https://doi.org/10.1109/ACCESS.2022.3233942.

35 Alves, T.A.B. and Abrão, T. (2022). Massive MIMO and NOMA bits-per-antenna efficiency under power allocation policies. *Physical Communication* 51: 101588. https://doi.org/10.1016/j.phycom.2021.101588.

36 Taniguchi, L.M., de Souza, J.H.I., Guerra, D.W.M., and Abrão, T. (2021). Resource efficiency and pilot decontamination in XL-MIMO double-scattering correlated channels. *Transactions on Emerging Telecommunications Technologies* 32 (12): e4365. https://doi.org/10.1002/ett.4365.

37 de Souza Junior, W. and Abrao, T. (2023). Energy efficiency maximization for intelligent surfaces aided massive MIMO with zero forcing. *IEEE Transactions on Green Communications and Networking* 8 (2): 802–814. https://doi.org/10.1109/TGCN.2023.3346367.

38 Ahmed, I.Z., Sadjadpour, H.R., and Yousefi, S. (2023). An information-theoretic branch-and-prune algorithm for discrete phase optimization of RIS in massive MIMO. *IEEE Transactions on Vehicular Technology* 72 (6): 7395–7410. https://doi.org/10.1109/TVT.2023.3237682.

39 Marinello, J.C., Abrão, T., A, A., de Carvalho, E., and Popovski, P. (2020). Antenna selection for improving energy efficiency in XL-MIMO systems. *IEEE Transactions on Vehicular Technology* 69 (11): 13305–13318. https://doi.org/10.1109/TVT.2020.3022708.

40 Boumal, N. (2023). *An Introduction to Optimization on Smooth Manifolds.* Cambridge University Press. https://doi.org/10.1017/9781009166164.

41 Shen, X., Diamond, S., Gu, Y., and Boyd, S. (2016). Disciplined convex-concave programming. *2016 IEEE 55th Conference on Decision and Control (CDC),* 1009–1014. https://doi.org/10.1109/CDC.2016.7798400.

42 Grant, M. and Boyd, S. (2014). CVX: Matlab Software for Disciplined Convex Programming, version 2.1. http://cvxr.com/cvx (accessed 27 September 2024).

43 Trendafilov, N. and Gallo, M. (2021). *Multivariate Data Analysis on Matrix Manifolds (with Manopt),* 1e. Cham: Springer Series. ISBN 9783030769734. https://doi.org/10.1007/978-3-030-76974.

44 Guerra, D.W.M., Fukuda, R.M., Kobayashi, R.T., and Abrão, T. (2018). Efficient detectors for MIMO-OFDM systems under spatial correlation antenna arrays. *ETRI Journal* 40 (5): 570–581. https://doi.org/10.4218/etrij.2018-0005.

45 Fukuda, R.M., Guerra, D.W.M., Kobayashi, R.T., and Abrão, T. (2018). DE/PSO-aided hybrid linear detectors for MIMO-OFDM systems under correlated arrays. *Transactions on Emerging Telecommunications Technologies* 29 (12): e3495. https://doi.org/10.1002/ett.3495.

46 Dai, Y.-H. (2011). *Nonlinear Conjugate Gradient Methods*. Wiley. ISBN 9780470400531. https://doi.org/10.1002/9780470400531.eorms0183. https://onlinelibrary.wiley.com/doi/abs/10.1002/9780470400531.eorms0183.

47 Boumal, N., Mishra, B., Absil, P.-A., and Sepulchre, R. (2014). Manopt, a Matlab toolbox for optimization on manifolds. *Journal of Machine Learning Research* 15 (42): 1455–1459. https://www.manopt.org.

48 Boumal, N., Absil, P.-A., and Cartis, C. (2018). Global rates of convergence for nonconvex optimization on manifolds. *IMA Journal of Numerical Analysis* 39 (1): 1–33. https://doi.org/10.1093/imanum/drx080.

7

Energy Efficiency in RIS-Aided Massive MIMO and XL-MIMO Communication Systems

Wilson Souza Junior[1], João Henrique Inacio de Souza[1], José Carlos Marinello[2], and Taufik Abrão[1]

[1] *Department of Electrical Engineering, State University of Londrina, Londrina, Paraná, Brazil*
[2] *Department of Electrical Engineering, UTFPR, Cornélio Procópio, Paraná, Brazil*

7.1 Introduction

Massive multiple-input multiple-output (M-MIMO) is a fundamental technology for contemporary mobile networks, equipping the base station (BS) with hundreds to thousands of antennas. A large amount of antennas enables the transmission and reception to and from many user's equipment (UEs) within the same time–frequency resources through spatial multiplexing. Furthermore, the so-called *channel hardening* phenomenon and the *favorable propagation* condition can be exploited to improve the spectral efficiency (SE) and lessen the inter-cell interference. Nevertheless, increasing the number of antennas at the BS means also increasing the number of network components such as *transceiver chains* and *power amplifiers*, enhancing substantially the hardware cost and, importantly, the network power consumption. Therefore, to reach global scalability with low environmental impact and low operational cost, instead of following the standard capacity maximization approach, the design of the M-MIMO systems must be guided by the principle of energy efficiency (EE) maximization, oriented by the fulfillment of the UEs' quality of service (QoS) requirements. The system optimization guided by such a principle has resulted in efficient beamforming and power allocation (PA) schemes, which, based on channel state information (CSI), adequately assign the transmit power level at each BS antenna so as to reduce the power amplifiers consumption [1]. Furthermore, to reduce the overall power consumption, strategies to shut down transitorily unneeded network components

Massive MIMO for Future Wireless Communication Systems: Technology and Applications, First Edition. Edited by Webert Montlouis and Agbotiname Lucky Imoize.

have been proposed [2, 3], making progress toward the objective of achieving zero power consumption at zero system load. Despite the recent advances, there is still a need for efficient resource allocation methods and system optimization schemes to deploy energy-efficient M-MIMO systems. Aiming to contribute on this subject, this chapter discusses two contributions involving EE, one embracing M-MIMO systems aided by reconfigurable intelligent surface (RIS), and the other addressing extra-large-scale M-MIMO (XL-MIMO) systems, where the number of BS antennas is pushed to the extreme with the adoption of physically large arrays. The discussion herein not only offers a comprehensive description of schemes to deploy energy-efficient M-MIMO systems but also provides a basic mathematical model and a set of optimization tools and analytical frameworks to develop novel schemes, that can be combined with other technologies and applied to other environments.

M-MIMO is a crucial technology for networks of the fifth generation (5G), equipping the BS with hundreds or thousands of antennas. Increasing the antenna separation is paramount to make a real advantage from an array dimension of thousands of antennas. The so-called channel hardening phenomenon and the favorable propagation conditions are utilized in M-MIMO to improve SE performance and lessen inter-cell interference with the help of several antennas. In addition, with the development of wireless communications and greater communications demands, the number of antennas and array aperture are predicted to increase. One potential approach is integrating the antenna array into large structures called XL-MIMO. However, centralized processing architectures face a challenging complexity when employing vast arrays. A promising solution is to divide the antenna array into disjoint units, referred to as subarray (SA), with individual processing units. XL-MIMO is emerging as a promising solution.

XL-MIMO uses a large number of antennas in a small space. Existing works have examined two mainstream implementation methods for XL-MIMO:

i) **Discrete aperture method**: compactly deploys many antennas with spacing significantly less than the half-wavelength. Thousands of antennas can be used to improve system performance. The array aperture is discrete because XL-MIMO uses discrete antenna deployment.
ii) **Continuous aperture (CAP) MIMO**: embed an approximately unlimited number of antennas in a compact space in a dense manner so that metamaterials can make the array aperture appear continuous. Perfect CAP MIMO uses a spatially continuous electromagnetic volume with infinite and infinitesimal antennas.

The first realization strategy has mature analytical modeling and practical implementation [4]. However, compact deployment causes strong mutual coupling between discrete antenna parts. Emerging metamaterials can assist in building the CAP in the second implementation strategy.

Based on these two implementation approaches, several XL-MIMO designs with diverse architectures and terminologies have been researched, including holographic MIMO, large intelligent surfaces (LISs), extremely large-scale antenna arrays (ELAAs), and CAP MIMO. These XL-MIMO solutions seem promising for 6G and future wireless communications.

In this chapter, we deal with M-MIMO channels while modifying the conventional M-MIMO channel model, aiming to take two implications of the extreme array dimensions into account: the spherical wavefronts, which is called near–field propagation, and the concentration of the majority of energy received from a specific user on a small portion of the array, namely, spatial non-stationarity. Considering the latter, it is inefficient from an energy perspective to activate the antennas with lower channel gains to transmit/receive the signal to/from a given user. Thus, antenna selection (AS) methods are paramount in the scenario. We explore AS methods for XL-MIMO: the adjustable flexible antenna selection (FAS), and the fixed subarray selection (FSS). Numerical results demonstrate that, by judiciously selecting the antennas subset used to detect the signal of each user in the uplink (UL), the number of active antennas can be reduced considerably in a typical massive number of BS antenna scenarios, reducing the power consumption without compromising the throughput. Moreover, the EE is hugely increased in such scenarios. Finally, with a more straightforward hardware implementation, the FSS scheme achieved the same performance as FAS.

In this introductory section, we contextualize the relevance of antenna selection (AS) methods, resource efficiency (RE), SE, and EE optimization in XL-MIMO and M-MIMO systems aided or not by the RIS enabling structures.

While M-MIMO technology brings about substantial performance improvements, it can suffer a noteworthy degradation in scenarios where the direct communication links between the BS and UE face obstruction. To tackle this challenge cost–effectively, one viable solution is to establish alternative signal paths. RIS emerges as a promising option, offering a practical and feasible way of enhancing communication systems. Composed of a dielectric substrate embedded with conductive metamaterial elements of sub-wavelength size and spacing, RIS empowers the impedance adjustment, thus facilitating advantageous alterations in amplitude and phase responses to incoming electromagnetic waves.

A novel class of artificial materials known as metamaterials has been developed, with a microstructure designed to display unique electromagnetic properties not seen in nature and previously thought to be physically impossible [5]. Atoms make up the periodic structure of any material found in nature. An artificial periodic structure is made and grafted onto a host material to form metamaterials. The constituent elements of this manufactured periodic structure are larger than the atoms of the host material. This size difference effectively overrides the host material's natural reaction to the incoming electromagnetic radiation. As a result, the penetration, reflection loss, and reflection angle are unique to the artificial structure rather than the host material's atomic structure.

An RIS panel is composed of voltage–controlled metamaterial elements that allow manipulation of the wireless channel state response; as a result, the primary characteristics of RISs are wireless system coverage enhancement, channel rank enhancement, interference suppression, and Doppler spread reduction. There are many possible applications, but TeraHertz signal enhancement is the most promising. Implementation issues are paramount, where the most crucial issue is creating workable protocols for real–time reconfigurability. Models and methods to deal with electromagnetic (EM) signal propagation in the presence of RIS should include large–scale path loss, near–field and far–field approximation, and a hybrid ray tracing method.

7.1.1 Related Works

A summary of the main related works published recently investigating RE (EE and SE) in M-MIMO and/or XL-MIMO systems aided by RIS, as well as highlighting the advances and deployed methods, is depicted in Table 7.1. Our previous related works and contributions on the RE, EE, and SE analysis and optimization in XL-MIMO and M-MIMO systems aided or not by RIS include [2, 6–15].

7.1.2 Contributions of the Chapter

In this chapter, we present two studies regarding the design of energy-efficient M-MIMO systems, aiming to reduce the system power consumption while providing the UEs' QoS requirements; one of the M-MIMO is aided by RIS. In detail, our contributions are threefold:

- Develop optimization methods to improve the EE in RIS-aided M-MIMO and XL-MIMO communication systems by formulating specific optimization problems and using tools from convex optimization and evolutionary heuristic to solve them.
- Conduct comprehensive analyses for two optimization problems related to RIS-aided M-MIMO and XL-MIMO systems, identifying the key performance indicators, investigating the associated technical challenges, and proposing potential solutions with low computational complexity. These problems include RIS phase–shift optimization, PA, optimization of the number of BS antennas, and AS.

Table 7.1 Recent works addressing RE (EE and SE) and related open issues in RIS-aided M-MIMO and XL-MIMO.

Setup (year)	Contributions	Problem	Method	References
Multicell MIMO (2022)	PA and PS in multi cell MIMO system under pilot contamination. It uses a reinforcement learning (RL) approach to implement random access (RA) policies. The paper compares Q-Learning with classical methods and analyzes SE and EE. Simulations confirm the effectiveness and efficiency of the proposed sequentially combined PS-PA method.	Pilot and PA	The new RA for MIMO uses an RL approach to solve NP-Hard pilot assignment (PS) and non-convex PA problems with low-complexity algorithms. Non-asymptotic antenna configurations and asynchronous interference are considered. The proposed algorithm uses a Q-Learning variation, reducing computational complexity and self-adaptivity.	[10]
XL-MIMO (2023)	QoS-aware joint user scheduling and PA technique to enhance the number of served users in crowded downlink (DL) XL-MIMO with min-QoS requirements guarantee.	DL user scheduling and PA	Sequential procedures: (a) Clique search-based scheduling (CBS) for user scheduling; (b) Optimal PA with transmit power budget and minimum achievable rate per user constraints.	[6]
XL-MIMO (2022)	(a) Joint protocol for RA and user scheduling: non-overlapping visibility region (VR) XL-MIMO (NOVR-XL) RA protocol. (b) NOVR-XL RA protocol leverages the different VRs of the UE to improve RA performance. (c) Compared to conventional RA systems, the approach yields notable sum-rate increases.	User scheduling and RA protocol	(a) XL-MIMO array exhibits spatial non-stationarities and VRs: its huge dimension is of the order of the distances to the UEs. (b) The paucity of pilots and the increased density of UE in typical XL-MIMO system applications pose challenges to the design of RA protocols and scheduling algorithms.	[7]

(Continued)

Table 7.1 (Continued)

Setup (year)	Contributions	Problem	Method	References
XL-MIMO (2021)	Low-complexity and scalable joint AS and PA method for SE maximization in XL-MIMO systems implemented with a subarray switching architecture, where each subarray is equipped with fewer radio frequency (RF) chains than antennas.	AS in XL-subarrays for max-SE	Evolutionary algorithms: (a) Centralized genetic algorithm (GA) for joint AS and PA; (b) Distributed GA to split the computational complexity among the subarrays and reduce the interconnection bandwidth between subarrays and the central processing unit. (c) Greedy heuristic for AS and PA based on the selection of the antennas that present the highest values of total received power.	[11]
XL-MIMO (2021)	Proposition of two different AS approaches under realistic XL-MIMO channels to maximize the EE without compromising the SE.	Fixed-AS max-EE XL-MIMO	FAS: assigns the BS subset of antennas to perform receive combining and transmit precoding to each user; FSS: divides the antenna array into several fixed-size subarrays, which can be implemented in a distributed way, as separate hardware entities, while selecting one of the subarrays to communicate with each user.	[12]
XL-MIMO (2020)	Reformulation of SINR performance expressions; evaluation of the total EE XL-MIMO system; proposition of four low-complexity AS procedures to maximize total EE. Schemes are based on long-term fading parameters and remain valid for larger time/frequency intervals. Derivation of approximated closed-form EE expressions; iterative method for finding the optimal number of selected antennas to maximize EE.	AS for total EE maximization in XL-MIMO	Analytical iterative algorithm based on Newton Raphson method to determine the optimal number of activated antennas in XL-MIMO, which maximizes the approximated EE expression.	[2]

Non-orthogonal multiple access (NOMA) XL-MIMO (2023)	(a) Grant-based (GB)-RA XL-MIMO protocol operating in power-domain NOMA and exploring the overlapping VRs (NOVR-XL) combines the advantages of XL-MIMO and NOMA, allowing multiple users to select the same pilot sequence in at least one SA. (b) Numerical optimization procedure for the scale factor $\delta_{\text{novr-xl}}$ aiming to maximize the system sum rate; (c) By exploring the NOMA successive interference cancellation (SIC), the NOVR-XL allows multiple users to select the same pilot sequence: in case of possible false positives, it is resolved to convert them into successful attempts. (d) Aiming to maximize the system sum rate, a look-at-table procedure is proposed for different loading users and scenarios.	GA RA Protocol for crowded massive machine type communications (mMTC) XL-MIMO NOMA	(a) VRs and spatial non-stationarities arise throughout the massive extra large array expansion. The proposed GB-RA XL-MIMO protocol for crowded mMTC combines the NOMA benefits with the strongest user collision resolution in extra-large arrays, namely NOVR-XL. (b) Several colliding users can access the same XL SA using the NOVR-XL method and the same pilot. As the number of SA rises, the proposed NOVR-XL GB-RA protocol can reduce the number of tries to reach the network while boosting the average sum rate.	[13]
XL-MIMO double-scat. correlated channels (2021)	Analyze the effectiveness of the minimum mean squared error (MMSE) combiner in XL-MIMO systems in terms of SE, EE, RE, and the capability of eliminating the intra-cell pilot contamination problem. UE sharing pilot sequences may have overlapped VRs. Such reuse increases intra-cell interference. The MMSE combiner can substantially reduce pilot contamination, justifying its attainable excellent SE and massive connectivity.	RE and pilot decontamination	(a) By comparing the MMSE combiner performance-complexity with the linear zero-forcing (ZF) combiner and the randomized Kaczmarz regularized ZF (rKZF), which consists of the RZF combiner emulated by the rKA approach and a flexible number of iterations, allowing balances the performance and complexity of linear combiners for M-MIMO. (b) Double-scattering correlated channel model, which considers clusters at the BS and UE sides, is deployed to evaluate the spatial correlation between the antennas.	[15]

(Continued)

Table 7.1 (Continued)

Setup (year)	Contributions	Problem	Method	References
RIS MIMO	(a) Maximize the EE in RIS-assisted M-MIMO scenario under ZF precoding formulating a non-convex optimization problem encompassing the user's transmit power, the quantity of active BS antennas, and the phase–shift angle unity module constraints. (b) Different from previous works, the formulated EE maximization problem relies upon only the statistical CSI estimates, which can be beneficial in high mobility and short channel coherence time wireless scenarios, reducing the overhead imposed by the optimization procedure of these variables in every channel coherence time.	max-EE with RIS M-MIMO with ZF.	(a) Apply the Lagrangian dual method with fractional programming (FP) techniques to solve the non-convex problem. Specifically for the RIS phase–shift problem, we proposed an innovative fractional programming (FP)-based RIS phase–shift optimization, which enables us to find a closed-form expression for each element of RIS. (b) To support our findings, the complexity of the proposed antenna, PA, and RIS phase–shift optimization algorithm has been accurately investigated.	[16]
RIS Internet of things (IoT) Ntw (2022)	(a) Analysis of the UL channel of a cellular network of IoT devices assisted by an RIS with a limited number of reflecting angle configurations. (b) Develop an expression for the expected battery lifetime (EBL), which explicitly depends on the machine-type device (MTD) transmit power.	UL-EE	(a) Analytical expressions for the required transmit power of the MTDs to attain a target SNR, considering a channel model that accounts for the RIS discretization into sub-wavelength reflecting elements. (b) Numerical simulations for the EE evaluated via the EBL demonstrate the benefits of adopting RISs to enable energy efficient IoT networks.	[8]

RE in RIS M-MIMO	Explore the EE–SE tradeoff in MIMO UL communications using an RIS with discrete phase shifters.	EE–SE tradeoff in RIS-aided M-MIMO	The transmission strategy design uses partial CSI to reduce signaling overhead and energy consumption. Framework for optimizing UEs' transmit precoding and RIS passive beamforming to maximize RE. UEs' precoding design is simplified using closed-form solutions and an asymptotic deterministic objective expression. RIS phases are designed using an iterative MMSE.	[17]
XL-MIMO Double scattering channels (2022)	Double scattering channel modeling for crowded XL-MIMO system subject to many obstacles, where the BS is equipped with XL ULA and UPA. One scattering cluster emerges stochastically near the BS side, while the second cluster also stochastically appears near the user side.	Ch. Model.	(a) Birth and death of the correlated BS-clusters and UE-clusters modeled via Poisson processes, generating a space time evolutionary process of scatterers/clusters that results in quasi-static and dynamic spatial correlated XL-MIMO channels. (b) Simulations validate the XL-MIMO channel model via the birth and death Poisson point processes and Poisson clustering process, the effect of interference, and correlation.	[9]
XL-MIMO channel features	Survey on the new channel properties and low-cost designs for XL-MIMO systems.	Ch. Model.	(a) Mathematical analysis to explain why the new features appear and illustrate how they affect the system model. (b) Summarize and compare the low-cost designs from various perspectives.	[18]

(Continued)

Table 7.1 (Continued)

Setup (year)	Contributions	Problem	Method	References
M-MIMO (2023)	Proposition of precoding methods to minimize the consumed power in wideband and narrowband M-MIMO systems subject to zero inter-user interference constraint. The introduced precoders account for a realistic power consumption model for the BS power amplifiers, following the square-root behavior observed in class B amplifiers operating with adequate back-off.	Precoding, PA, and energy consumption minimization	Closed-form precoding solutions for the cases of narrowband system under low-load, wideband system under low-load, and asymptotic wideband system under low-load. For the two first cases, a fixed point method is developed to compute the PA among the BS antennas. Differently, for the third case, a method to determine the optimal number of active antennas is proposed.	[1]
M-MIMO (2023)	Joint beamforming and AS to minimize the power consumption, satisfying the UEs' QoS constraints in M-MIMO systems with imperfect CSI available at the BS. Initially, a branch and bound framework is introduced to solve optimally the joint beamforming and AS problem. Then, aiming to improve the computational complexity of the scheme, a classifier based on machine learning is proposed to accelerate the branch and bound procedure.	Precoding, PA, AS, and energy consumption minimization	(a) Optimal joint beamforming and AS via branch and bound. (b) Classifier based on a graph neural network design to accelerate the execution of the branch and bound procedure.	[3]

mmWave M-MIMO (2023)	Proposition of AS strategies for the UL and DL of M-MIMO systems aiming to maximize the system capacity. Especially, the effect of quantization noise due to limited-resolution analog-to-digital and digital-to-analog converters is taken into account.	AS and capacity maximization	(a) Quantization-aware greedy with square maximum-volume (QAG-SMV) scheme for AS in the UL where, initially, the antennas are selected by an SMV-based algorithm without considering the impact of the finite-resolution converters; then, the remaining antennas are selected by the QAG algorithm. (b) Group-selection (GS) scheme for AS in the DL, selecting the antennas based on their contribution to the enhancement of the system capacity.	[19]
M-MIMO (2023)	Development of a scheme for joint AS and user scheduling in M-MIMO systems to maximize the expected long-term SR. Particularly, the scheme relies on partial CSI, since the BS is equipped with fewer transceiver chains than the total number of antennas.	AS, user scheduling, and SE maximization	The optimization problem is modeled as a partially observable Markov decision process. Given that, the myopic policy is derived by quantizing the squared absolute values of the channel coefficients, namely, the Markov chain states, into two levels.	[20]
M-MIMO (2023)	Maximization of the EE in an M-MIMO system via an heuristic scheme for AS and precoding. Specifically, the array of BS antennas has a partially connected hybrid architecture, where each transceiver chain is connected to a limited set of antennas, and each antenna is connected to a dedicated phase shifter. The proposed scheme is evaluated both under the assumptions of perfect and imperfect CSI at the BS. In summary, the scheme achieves performance levels close to the optimal exhaustive search, but with lower computational complexity.	Precoding, AS, and EE maximization	AS via an algorithm based on the cyclic binary particle swarm optimization, while the precoding is calculated by successive interference cancellation.	[21]

- Through a theoretical framework and with the aid of numerical simulations, provide fresh insights and hypotheses to contribute to the development of energy-efficient M-MIMO systems.

7.1.3 Organization

The remainder of this chapter is organized as follows. We commence our discourse by formulating a general multi-purpose system model for RIS-aided M-MIMO systems in Section 7.2. The main figures of merit deployed along this chapter, including SE, EE, and RE in M-MIMO and XL-MIMO systems, are discussed in Section 7.2.4. In the subsequent sections, we aggregate specific considerations and particularities to the system model, presenting the formulation and development of efficient algorithms for two different M-MIMO systems, aided or not by RIS, namely, RIS-aided XL-MIMO and M-MIMO system. In this chapter, the focus of analysis is maximizing the RE (EE, SE, or EE–SE trade-off) in M-MIMO systems by deploying different optimization tools, such as convex, heuristic evolutionary, or manifold optimization procedures and methodologies. The diagram in Figure 7.1 shows the topics explored in this chapter.

The EE maximization problem in the DL of a multiuser M-MIMO RIS-aided communication system under ZF precoding is developed in Section 7.3. Since systematically optimizing the RIS phase-shift in each channel coherence time

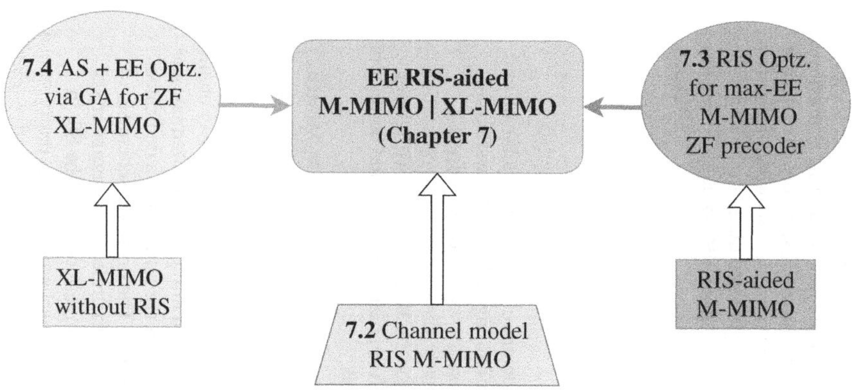

PA: power allocation	PS: pilot assignment	PC: pilot contamination
EE: energy efficiency	SE: spectral efficiency	RE: resource efficiency
M-MIMO: massive MIMO	XL-MIMO	QL: Q-learning
AS: antenna selection	GA: genetic algorithm	Optz: optimization
UE: unit equipment	RIS: reconfigurable intelligent surfaces	

Figure 7.1 Chapter organization: two RE problems, which arise in different M-MIMO and XL-MIMO scenarios aided by RIS or not.

interval is challenging and burdensome, we employ the statistical CSI-based optimization strategy to alleviate this overhead. We can reduce the computational complexity by treating the RIS phase–shift matrix as a constant over multiple channel coherence time intervals while maintaining suitable performance. To circumvent the non-convexity of the EE optimization problem, we explore the **sequential optimization** approach where the PA vector ρ, the number of antennas M, and the RIS phase–shift vector $\boldsymbol{v}$ are separate and sequentially solved iteratively until convergence. With the help of the Lagrangian dual method and fractional programming (FP) techniques, insightful closed-form expressions for all the optimization variables are derived. The proposed approach can reduce the power consumption and the number of active BS antennas, demonstrating the potential of RIS deployment and optimization in M-MIMO systems under simple linear precoding. Section 7.4 examines the DL of a ZF-precoded XL-MIMO system, focusing on load-dependent implementations that can dynamically turn off network components to improve the system EE. To this end, from CSI, the BS deploys a joint AS and PA strategy aiming to maximize the system EE. A GA, a heuristic based on the channel gain, and a local search algorithm are proposed, showing that remarkable EE gains can be attained when the system load is low. Finally, Section 7.5 points out the main conclusions, lessons learned, and findings associated with this extensive work and research developed over the last years.

7.2 General System Model for RIS-Aided M-MIMO

In this section, we describe a general model for M-MIMO systems potentially assisted by an RIS. This model is based on [18, 22, 23]. The proposed RIS-aided M-MIMO channel model is capable of emulating a variety of scenarios, being able to simulate different M-MIMO and XL-MIMO setups at the link level. It can handle both co-located and distributed/cell-free architectures. Moreover, with a channel based on the spherical wavefront propagation assumption, the model reproduces the propagation aspects observed in the near-field scenarios as well as in the far-field regions of the BS array. We start the section by describing the scenario setup and the main assumptions made. Then, we introduce the M-MIMO channel model and an extension for the scenario with communication assisted by an RIS. In the sequence, we provide the models for the UL and DL received baseband signals. Next, we particularize the M-MIMO channel model for XL-MIMO configurations. Based on these models, we finish the section by formulating expressions for SE, EE, and RE for RIS-aided M-MIMO systems. The notation deployed in this chapter is summarized in Table 7.2.

Consider an M-MIMO wireless communication system with $K \in \mathbb{Z}_+$ UEs served by a single BS, equipped with an array of $M \in \mathbb{Z}_+$ antennas such that

Table 7.2 List of symbols for the M-MIMO and XL-MIMO systems modeling.

Symbol	Description
$M \in \mathbb{Z}_+$	Number of BS antennas
$K \in \mathbb{Z}_+$	Number of UEs
$k \in \{1, \ldots, K\}$	UE index
$m \in \{1, \ldots, M\}$	BS antenna index
$\lambda > 0$	Carrier wavelength
$\beta_{km} \in \mathbb{R}$	Large-scale fading coefficient of the channel from UE k to the BS antenna m
$\mathbf{h}_k \in \mathbb{C}^M$	Channel of the UE k
$\overline{\mathbf{h}}_k \in \mathbb{C}^M$	Line-of-sigh (LoS) component of the UE k channel
$\tilde{\mathbf{h}}_k \in \mathbb{C}^M$	Multipath component of the UE k channel
$\kappa_k \in [0, \infty)$	Rician κ-factor of the UE k channel
$\mathbf{R}_k \in \mathbb{R}^{M\times M}$	Spatial correlation matrix of the multi–path component of the UE k channel
$\mathbf{V}_k \in \{0, 1\}^{M\times M}$	Visibility matrix of the UE k channel
$\boldsymbol{\beta}_k \in \mathbb{C}^{M\times M}$	Matrix with the UE k channel gains
$\mathbf{a}_k \in \mathbb{C}^M$	Steering vector of the BS array
$\mathbf{p}_k^{\text{ue}} \in \mathbb{R}^3$	UE k position
$\mathbf{p}_m^{\text{bs}} \in \mathbb{R}^3$	BS antenna m position
$\mathcal{P}^{\text{bs}} \in \wp(\mathbb{R}^3)$	Set with the positions of the BS antennas

$M \gg K$. We assume the block fading model, so the duration of a single UL or DL transmission is less than the channel *coherence time*, and the bandwidth of the transmitted signals is less than the channel *coherence bandwidth*. Therefore, the channel is flat (non-selective) in frequency and remains constant for the whole transmission time (slow), simplifying the channel estimation and equalization. Also, we assume the TDD transmission mode, where UL and DL transmissions are time multiplexed at each channel realization. In this case, due to *channel reciprocity*, the channel estimated from UL pilot signals can be used for both UL and DL equalization if both the UL and DL operations are realized within the channel coherence time.

In TDD mode, the CSI for UL or DL directions can be estimated in a channel coherence time-basis interval, being part of a transmission scheme with instantaneous CSI-based optimization. On the other hand, a transmission scheme based on statistical CSI optimization is deployed when, typically, the estimation of the

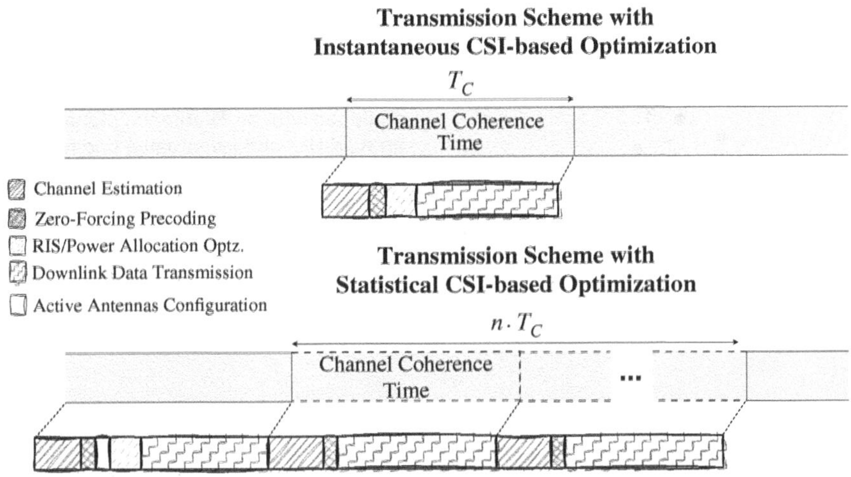

Figure 7.2 Temporal frame for TDD transmission by adopting instantaneous and statistical CSI-based estimation. In the latter, the RIS phase-shift vector, the transmit PA, and the number of BS antennas (M) remain fixed for the UEs over multiple channel coherence times (nT_c); differently, adopting the instantaneous CSI-based approach, the channel estimation, precoding, RIS phase-shift, and transmit power must be updated at each T_c time interval.

second moment of the channel is updated periodically, that is, a sequence of consecutive channel coherence time intervals $n \cdot T_c$ is deployed for estimating the channel covariance matrix. Figure 7.2 illustrates the tasks carried out in a T_c interval-basis for an instantaneous CSI against a $n \cdot T_c$ interval-basis for a statistical CSI. Indeed, Figure 7.2 reveals the reduction of complexity by adopting statistical CSI for executing different tasks such as channel estimation, precoding, PA, RIS element phase-shift, and BS antenna reconfiguration/activation.

7.2.1 M-MIMO Array Location

Adopting the rectangular coordinate system of Figure 7.3, let $\mathbf{p}_k^{\text{ue}} \in \mathbb{R}^3$ denote the vector describing the position of UE k. Similarly, let $\mathbf{p}_m^{\text{bs}} \in \mathbb{R}^3$ denote the vector with the coordinates of the BS antenna-element m. The set of vectors $\mathcal{P}^{\text{bs}} = \{\mathbf{p}_m^{\text{bs}}\}_{m=1}^{M}$ describes the geometry of the BS array, and is set according to the considered M-MIMO architecture. For instance, in collocated setups, $\mathcal{P}^{\text{bs}}$ can describe the localization of a uniform planar array (UPA) with half-wavelength spaced antennas. On the other hand, in cell-free setups, subsets of the total antennas form individual arrays, so $\mathcal{P}^{\text{bs}}$ can describe multiple UPAs geographically spread over the coverage area.

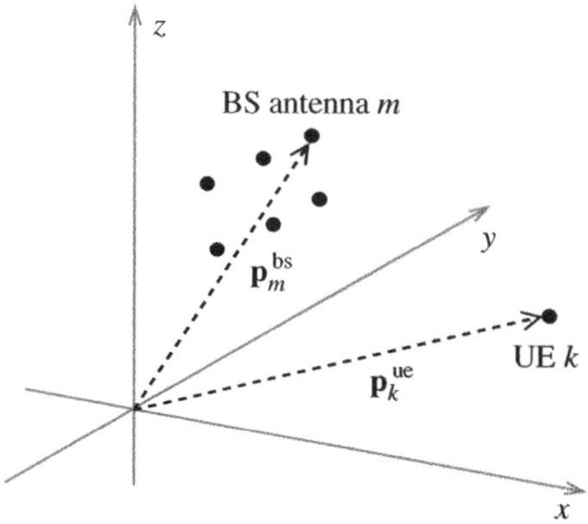

Figure 7.3 Three-dimensional (3-D) geometric representation of the BS, UEs, and RIS localization in the multiuser M-MIMO system modeling. The BS is equipped with an M-antenna planar array; K UEs and (eventually) one full passive RIS panel.

7.2.2 Wireless M-MIMO Channel Model

Now, we introduce a three-dimensional (3-D) correlation-based channel model for the investigated communication system. We start by describing the propagation loss faced by the signals traveling from the source to the destination. Initially, to cover the near-field propagation scenario, we assume the UEs are potentially in the near-field region of the BS array but always in the far-field region of the antennas individually. In this case, the channel power associated with the UE in the near-field region is unevenly distributed across the BS array. Therefore, the propagation loss must be defined considering the distances from each UE to every antenna independently. On this basis, the large-scale coefficient $\beta_{km} \in (0, 1)$ corresponding to the channel gain from the UE k to the BS antenna m is defined such that

$$10 \log_{10}\beta_{km} = 10 \log_{10}\overline{\beta} - 10\gamma\log_{10}\|\mathbf{p}_m^{\text{bs}} - \mathbf{p}_k^{\text{ue}}\|_2 + F_k, \tag{7.1}$$

where $\overline{\beta} \in (0, 1)$ denotes the channel gain at the reference distance, $\gamma \geq 2$ the path loss exponent, $\|\mathbf{p}_m^{\text{bs}} - \mathbf{p}_k^{\text{ue}}\|_2$ is the 3-D distance from the UE to the BS antenna array, and $F_k \sim \mathcal{N}(0, \sigma_{\text{sf}}^2)$ corresponds to the shadow fading term. Notice that, for an UE in the far-field region of the BS array, the large-scale coefficient does not depend on the antenna index since the distance from every antenna element to the UE is almost the same, that is, $\|\mathbf{p}_m^{\text{bs}} - \mathbf{p}_k^{\text{ue}}\|_2 \to c, \forall m \in \{1, \dots, M\}$. Therefore, such a model for the propagation loss generalizes the near- and far-field regimes, depending only on the geometry of the BS array and the distribution of the UEs across the coverage area.

Under the near-field regime, where the array aperture is comparable to the distance from the UE to the BS, it is not only the per-antenna propagation loss that produces the uneven distribution of the channel power across the array. Indeed, some propagation paths are potentially blocked by obstacles, while others are favored by environment scattering. To account for the former aspect, we introduce the concept of VR of an UE.

Definition 7.1 ***(Visibility region)*** The VR of an UE comprises the BS antenna elements that concentrate most of the channel power. Since the BS antennas outside the UE's VR have negligible channel power due to the blocked propagation paths, they cannot be used for reliably communicating with such an UE.

More specifically, we represent the VRs in the channel model by indicating the antennas of the BS array that has nonzero (above a threshold) channel power. Let the diagonal matrix $\mathbf{V}_k \in \{0, 1\}^{M \times M}$ denotes the visibility matrix for the UE k. It contains 1s only at the diagonal entries corresponding to the antennas inside the UE's VR, that is, $[\mathbf{V}_k]_{m,m} = 1$ for the antennas with nonzero channel power, and $[\mathbf{V}_k]_{m,m} = 0$ otherwise.

As a result, the channel response of an UE is a combination of a LoS propagation path and scattered paths due to physical environment scattering. More precisely, the channel of the UE k follows the Rician-fading model

$$\mathbf{h}_k = \mathbf{V}_k \left(\sqrt{\frac{\kappa_k}{\kappa_k + 1}} \bar{\mathbf{h}}_k + \sqrt{\frac{1}{\kappa_k + 1}} \tilde{\mathbf{h}}_k \right), \tag{7.2}$$

where $\bar{\mathbf{h}}_k$ corresponds to the deterministic LoS channel component, $\tilde{\mathbf{h}}_k$ to the Rayleigh-fading multi–path component, and $\kappa_k \geq 0$ denotes the Rician κ-factor for the kth UE. Notice that the Rician κ-factor determines the power ratio between the LoS and multi–path components, so $\kappa > 1$ yields a channel dominated by LoS, while $\kappa < 1$ yields a multi–path-dominated channel.

The Rayleigh-fading multi–path component follows a correlated complex Gaussian distribution such that $\tilde{\mathbf{h}}_k \sim \mathcal{CN}(\mathbf{0}_M, \mathbf{R}_k)$, where $\mathbf{R}_k \in \mathbb{R}^{M \times M}$ is the spatial correlation matrix for the kth UE. Specifically, $\mathbf{R}_k$ is semi–definite positive and its diagonal entries are filled with the channel gains defined in Eq. (7.1), that is, $[\mathbf{R}_k]_{m,m} = \beta_{km}, \forall m \in \{1, \dots, M\}$. In addition to the propagation loss, the spatial correlation matrix represents other long-term propagation effects resulting from aspects such as the distribution of multi–path directions, the geometry of the BS array, and the radiation pattern of the transmit and receive antennas.

The deterministic LoS component follows the spherical wave model

$$\bar{\mathbf{h}}_k = \boldsymbol{\beta}_k^{\frac{1}{2}} \mathbf{a}_k, \tag{7.3}$$

where $\boldsymbol{\beta}_k^{\frac{1}{2}} = \mathrm{diag}([\sqrt{\beta_{k1}} \ \cdots \ \sqrt{\beta_{kM}}]^T)$ denotes the channel gains matrix of the kth UE and $\mathbf{a}_k \in \mathbb{C}^M$ is the steering vector of the BS array for the UE k, defined by

$$\mathbf{a}_k = [e^{-j\frac{2\pi}{\lambda}\|\mathbf{p}_1^{\mathrm{bs}} - \mathbf{p}_k^{\mathrm{ue}}\|_2} \ \dots \ e^{-j\frac{2\pi}{\lambda}\|\mathbf{p}_M^{\mathrm{bs}} - \mathbf{p}_k^{\mathrm{ue}}\|_2}]^T. \tag{7.4}$$

From Eqs. (7.3) and (7.4), one can perceive that the BS antenna elements experience distinct channel powers, while the phase difference among them is non-linear. This is due to the spherical wavefront propagation model, which aims to represent accurately the channel behavior in the near-field propagation regime or with high-aperture arrays. However, it is essential to mention that the spherical wave model is a generalization of the model adopted for the far-field scenario. Therefore, the introduced channel model representation converges to the planar wavefront assumption for UEs located at the far-field region of the BS array.

The planar wavefront propagation is held if $d \geq d_{\text{Rayl}}$, while a spherical wavefront propagation condition is assumed when $d < d_{\text{Rayl}}$, where d is the distance between BS antennas and the UE, and d_{Rayl} is the Rayleigh distance. In massive antenna arrays, the Rayleigh distance is associated with array aperture D as [24, 25]:

$$d_{\text{Rayl}} = q\frac{D^2}{\lambda_c}, \tag{7.5}$$

where q is a dimensionless constant that depends on the phase tolerance along the array extension, and λ_c is the carrier wavelength. Usual q and associated $\Delta\phi$ values are $q = 2$ and $\Delta\phi = 22.5°$, respectively. In M-MIMO and XL-MIMO arrays, the aperture of radiation D can be considered as the length of the BS SA:

$$D = M_b d_r \lambda_a, \tag{7.6}$$

where $M_b = M/S$ antennas per linear SA, and S is the number of uniform SAs in the XL-MIMO panel,[1] d_r is the relative antenna spacing factor, in multiples of λ_a, where λ_a is the wavelength used to define the antenna spacing (*array aperture built wavelength*). Generically, for a M-MIMO/XL-MIMO system, the Rayleigh distance is calculated as [25]:

$$d_{\text{Rayl}} = q\frac{M_b^2 d_r^2 \lambda_a^2}{\lambda_c}. \tag{7.7}$$

For the particular case of uniform arrays, where the BS array[2] is built considering $\lambda_a = \lambda_c$, Eq. (7.7) reduces to:

$$z = qM_b^2 d_r^2 \lambda_c. \tag{7.8}$$

To illustrate both conditions, Figure 7.4 reveals the difference in the distribution of the channel gain and phase of received signal along the BS array for a single UE in the near-field (left) and then in the far-field (right) region of the array. For the case of the UE in the far-field region, the channel power is practically constant. At the same time, the phase–shift is linear, following the widespread planar wavefront model assumed for collocated M-MIMO setups.

1 In M-MIMO scenarios, the array is constituted simply defining $M_b = M$ antennas, that is, $S = 1$.

2 ULA, UPA, USA, UCA, etc. configuration antenna arrays.

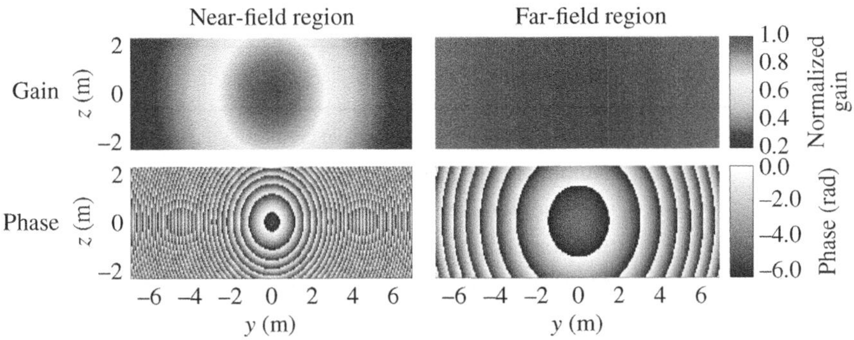

Figure 7.4 Normalized channel gain and channel phase-shift of the deterministic LoS component, considering a single UE in the near-field region of the BS array, and then in the far-field region. The array is an UPA with center at the origin of the coordinate system and placed along the *yz*-plane, with 240 rows and 80 columns of antennas evenly distributed in a rectangular area of 13.8 m by 4.6 m. In the near-field case, the UE is positioned 5 m from the array center, while, in the far-field case, it is placed 30 m from the array center.

7.2.3 RIS-Aided M-MIMO Model

In this section, we focus on extending the M-MIMO system model of Section 7.2.2, by deploying an RIS to improve the wireless channel reliability, especially within the context of M-MIMO and XL-MIMO systems. We envision similar physical geometric channel scenarios described in Sections 7.2.1 and 7.2.2, in which the RIS structure is aggregated to the channel, taking the form of a thin uniform squared planar array (USPA), featuring $N = N_h \times N_v$ total elements, with $N_h = N_v = \sqrt{N}$ elements in the horizontal and vertical direction, with $\sqrt{N} \in \mathbb{Z}_+$ being a positive integer. Furthermore, the BS antenna elements are vertically and horizontally equally spaced with spacing d, respectively. Moreover, in this section, we have assumed a direct LoS link between the RIS and the BS through a dedicated return link serving a programmable controller at the RIS, enabling the BS to manage both phase-shift and amplitude coefficients of each RIS elements. Furthermore, we assume that the configuration time of the RIS is substantially shorter than the *channel coherence time*, Figure 7.2, meaning that under the perspective of instantaneous CSI estimation, at the start of each time slot, the BS re-configures the phase-shift of each RIS element for its properly operation in dynamic channel controllability. Table 7.3 summarizes the values of the main parameters adopted in this second RIS-aided M-MIMO scenario.

7.2.3.1 RIS-Aided M-MIMO Correlated Channel Modeling

For more accurate modeling of RIS-assisted M-MIMO channels, in this subsection, we model the channel correlation effect. The system and channel model presented

Table 7.3 List of symbols used in the RIS-aided M-MIMO model, Section 7.2.3.

Symbol	Description
$N \in \mathbb{Z}_+$	Number of total RIS reflective elements
$\sqrt{N} \in \mathbb{Z}_+$	Number of horizontal and vertical reflective elements
$\mathbf{F} \in \mathbb{C}^{N\times M}$	Channel between the BS and RIS
$\overline{\mathbf{F}} \in \mathbb{C}^{N\times M}$	LoS component of the BS and RIS channel
$\tilde{\mathbf{F}} \in \mathbb{C}^{N\times M}$	Multipath component of the BS and RIS channel
$\mathbf{g}_k \in \mathbb{C}^N$	Channel between the RIS and UE k
$\overline{\mathbf{g}}_k \in \mathbb{C}^N$	LoS component of the RIS and UE k channel
$\tilde{\mathbf{g}}_k \in \mathbb{C}^N$	Multipath component of the RIS and UE k channel
$\hat{\kappa} \in \mathbb{R}^+$	Rician factor for BS-RIS link
$\tilde{\kappa} \in \mathbb{R}^+$	Rician factor for the RIS-UE link
$\kappa \in \mathbb{R}^+$	Rician factor for BS-UE link
$\boldsymbol{v} \in \mathbb{C}^N$	RIS amplitude/phase-shift vector
$\boldsymbol{\Phi} \in \mathbb{C}^{N\times N}$	RIS diagonal amplitude/phase-shift matrix

in this subsection are based on [16]. Hence, we start making the assumption of the far-field regime between the BS and the RIS panels, as well as between the RIS and the UEs, that is, the planar wavefront is assumed in both links. Furthermore, in the context of RIS-assisted communications, we need to account for two additional links that require estimation and significantly impact the system's performance. To be more specific, we establish the channel between the BS and the RIS based on the Rician-fading model

$$\mathbf{F} = \sqrt{\frac{\beta_F \hat{\kappa}}{\hat{\kappa}+1}}\overline{\mathbf{F}} + \sqrt{\frac{\beta_F}{\hat{\kappa}+1}}\tilde{\mathbf{F}}, \tag{7.9}$$

where $\overline{\mathbf{F}} \in \mathbb{C}^{N\times M}$ corresponds to the deterministic LoS channel component, $\tilde{\mathbf{F}} \in \mathbb{C}^{N\times M}$ to the Rayleigh-fading multi-path component, and $\hat{\kappa} \geq 0$ denotes the Rician κ-factor and β_F is the path-loss. Specifically, the deterministic LoS component can be described through the steering vector of a two-dimensional USPA, and expressed as:

$$\mathbf{a}_{\text{ris}}(\mathbf{p}) = \left[e^{-j\frac{2\pi}{\lambda}\|\mathbf{p}_1^{\text{ris}}-\mathbf{p}\|_2} \;\ldots\; e^{-j\frac{2\pi}{\lambda}\|\mathbf{p}_{N_h}^{\text{ris}}-\mathbf{p}\|_2}\right]^T \otimes \left[e^{-j\frac{2\pi}{\lambda}\|\mathbf{p}_1^{\text{ris}}-\mathbf{p}\|_2} \;\ldots\; e^{-j\frac{2\pi}{\lambda}\|\mathbf{p}_{N_v}^{\text{ris}}-\mathbf{p}\|_2}\right]^T, \tag{7.10}$$

where $\mathbf{p}_n^{\text{ris}}$ and $\mathbf{p}$ denote the position of the nth RIS element and a generic point in the 3-D space, respectively. Furthermore, in this subsection, the BS is assumed to be equipped with an uniform linear array (ULA) with a steering vector given as:

$$\mathbf{a}_{\text{bs}}(\mathbf{p}) = \left[e^{-j\frac{2\pi}{\lambda}\|\mathbf{p}_1^{\text{bs}}-\mathbf{p}\|_2} \; \dots \; e^{-j\frac{2\pi}{\lambda}\|\mathbf{p}_M^{\text{bs}}-\mathbf{p}\|_2} \right]^T, \tag{7.11}$$

with $\mathbf{p}_m^{\text{bs}}$ being the position of mth antenna at BS.

Thus, the deterministic LoS component $\tilde{\mathbf{F}}$ can be given as:

$$\overline{\mathbf{F}} = \mathbf{a}_{\text{bs}}(\mathbf{p}_{\text{o}}^{\text{ris}})\,\mathbf{a}_{\text{ris}}(\mathbf{p}_{\text{o}}^{\text{bs}})^H, \tag{7.12}$$

with $\mathbf{p}_{\text{o}}^{\text{ris}}$ and $\mathbf{p}_{\text{o}}^{\text{bs}}$ denoting the position of central point of RIS and BS, respectively, since far-field is assumed. Moreover, the multi-path component follows an uncorrelated Rayleigh distribution such that $\tilde{\mathbf{f}}_m \sim \mathcal{CN}(\mathbf{0}_N, \mathbf{I}_N)$, $\forall m = 1, \dots, M$, with $\tilde{\mathbf{f}}_m$ being the mth column of matrix $\tilde{\mathbf{F}}$. Figure 7.5 illustrates a typical three-links RIS-aided M-MIMO scenario, which includes the direct BS-UE link.

The direct link communication channel between the UEs and the BS can be expressed as Eq. (7.2), where we consider complete VR, that is, $\mathbf{V}_k = \mathbf{I}_M$ [16], leading to

$$\mathbf{h}_k = \sqrt{\frac{\beta_k \kappa_k}{\kappa_k + 1}}\overline{\mathbf{h}}_k + \sqrt{\frac{\beta_k}{\kappa_k + 1}}\tilde{\mathbf{h}}_k. \tag{7.13}$$

Furthermore, we also have the link between the RIS and UE k equipped with a single antenna, which is assumed to be Rician fading distributed, and expressed as $\mathbf{G} = [\mathbf{g}_1, \mathbf{g}_2, \dots, \mathbf{g}_K]$, with:

$$\mathbf{g}_k = \sqrt{\frac{\beta_{g,k} \tilde{\kappa}_k}{\tilde{\kappa}_k + 1}}\overline{\mathbf{g}}_k + \sqrt{\frac{\beta_{g,k}}{\tilde{\kappa}_k + 1}}\tilde{\mathbf{g}}_k, \tag{7.14}$$

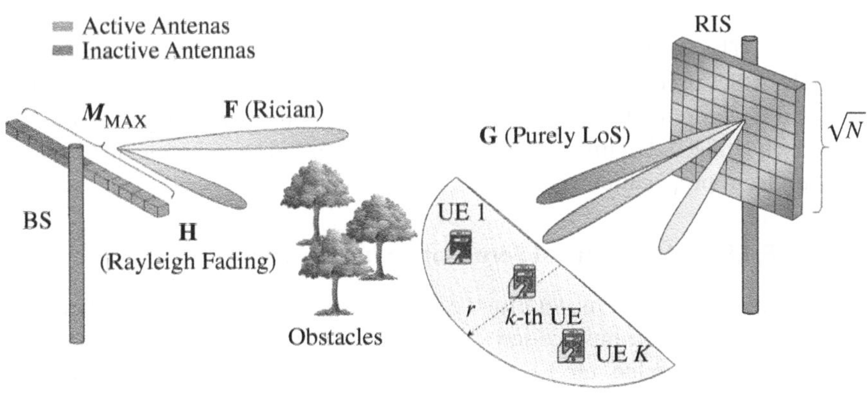

Figure 7.5 RIS-aided M-MIMO multi-user scenarios illustrating the three link types: RIS-UEs, BS-RIS, and BS-UEs. Particularly, this setup is illustrated with $N = 64$ reflective elements at RIS and $M_{\text{max}} = 12$ BS antennas; in this specific configuration, there are $M = 6$ actives elements at BS.

where $\bar{\mathbf{g}}_k \in \mathbb{C}^N$ corresponds to the deterministic LoS channel component between the RIS and UE k, $\tilde{\mathbf{g}}_k \in \mathbb{C}^N$ to the Rayleigh-fading multi–path component between RIS and UE k, and $\tilde{\kappa}_k \geq 0$ denotes the Rician κ-factor for the RIS-UE k link. $\beta_{g,k}$ is the path-loss of kth UE with the RIS. Similar to the BS and RIS channel model, the deterministic LoS component $\overline{\mathbf{G}} \in \mathbb{C}^{M\times K}$ has its columns given by the steering vector in Eq. (7.10)

$$\bar{\mathbf{g}}_k = \mathbf{a}_{\text{ris}}(\mathbf{p}_k), \tag{7.15}$$

with $\mathbf{p}_k$ being the position of UE k. Furthermore, the multi–path component follows an uncorrelated Rayleigh distribution such that $\tilde{\mathbf{g}}_k \sim \mathcal{CN}(\mathbf{0}_N, \mathbf{I}_N)$, $\forall k = 1, \ldots, K$.

The RIS operation is encapsulated in the complex vector $\boldsymbol{v}$:

$$\boldsymbol{v} = \left[\alpha_1 e^{-j\theta_1}, \ldots, \alpha_N e^{-j\theta_N}\right]^H \in \mathbb{C}^N, \qquad \text{where} \quad \alpha_n \in [0,1] \quad \text{and} \quad \theta_n \in [0, 2\pi] \tag{7.16}$$

denoting the passive amplitude and phase–shift reflection imposed by the nth reflecting element unit on the incident signal, respectively. Furthermore, we define the diagonal matrix $\boldsymbol{\Phi} \triangleq \text{diag}(\boldsymbol{v})$. Therefore, the **cascaded channel** between the UE k, the RIS and BS can be written as a function of the amplitude/phase–shift imposed by the RIS and the direct link $\mathbf{H}$, Figure 7.5, as following:

$$\hat{\mathbf{h}}_k = \mathbf{h}_k + \mathbf{F}^H \boldsymbol{\Phi} \mathbf{g}_k. \tag{7.17}$$

From Eq. (7.17), we can see the influence of deploying the RIS on the entire channel linking the BS and user k. Furthermore, it becomes apparent that the RIS is subject to a double effect. Hence, carefully considering its strategic placement is essential to attain significant improvements.

Remark: The far-field hypothesis is usually confirmed on the BS-RIS link since, typically, RIS is used to strengthen the channel because of obstacles on the BS-UE link, and it would not be very practical for it to be located extremely near the BS. Conversely, the RIS-UE link would typically be governed by a channel with spherical wavefront propagation (near-field).

7.2.4 SE, EE, and RE General Formulation in M-MIMO

In this section, we introduce a range of critical performance metrics crucial for analyzing efficient communication systems. We initiated the discussion with the concept of SE, expanded into EE, and finally delved into the definition of RE.

7.2.4.1 Spectral Efficiency

To assess the system performance of a communication system, it is crucial to evaluate the capacity for transmitting bits per second per Hertz (SE) – a fundamental metric that measures the efficiency of information transmission through a channel. The SE reflects the average number of bits transmitted per channel use across varying fading conditions, serving as a common metric for evaluating communication system capabilities. Since for the RIS-assisted systems, the channel is a generalization of the conventional M-MIMO systems, the SE can be equally represented as the following:

$$\text{SE} = \log_2(1 + \text{SINR}) \qquad \left[\frac{\text{bit}}{\text{s} \cdot \text{Hz}}\right]. \tag{7.18}$$

7.2.4.2 Energy Efficiency

The metric designed to evaluate the sustainable capacity of a system is referred to as EE. This metric essentially quantifies the number of bits that can be reliably transmitted per unit of energy and expressed as the following:

$$\text{EE} = \frac{\text{Throughput [bit/s]}}{\text{System Power Consumption [W]}} = \frac{B \cdot \log_2(1 + \text{SINR})}{P_{\text{T}}} \qquad \left[\frac{\text{bit}}{\text{Joule}}\right], \tag{7.19}$$

where B is the system bandwidth and P_{T} is the total power required for the communication system to operate. Notice that the power consumption for RIS-assisted M-MIMO systems requires more detailed consideration than traditional M-MIMO systems since although the RIS technology achieves low–power consumption, its consumption is not entirely negligible and should be thoroughly examined.

7.2.4.3 Resource Efficiency

RE stands for the efficient use of natural resources and the relationship between the resources used and the benefits derived from them. In the telecommunication context, "natural resource" primarily refers to spectrum and energy. Hence, since SE and EE represent conflicting objectives, from the perspective of multi-objective optimization, the RE optimization can be considered as the trade-off between EE and SE. Such an EE–SE tradeoff in multiuser M-MIMO aided by an RIS equipped with discrete phase shifters is not trivial.

Elaborating further, the RE as the EE–SE trade-off optimization problem in the RIS-aided M-MIMO context can be formulated as a bi-objective optimization problem [26], consisting of maximizing simultaneously two conflicting objectives:

$$\begin{aligned}
&\underset{M,\ \boldsymbol{\Phi},\ \rho\in\mathbb{R}^K}{\text{maximize}} \ \text{SE} \ \left[\frac{\text{bit}}{\text{s.Hz}}\right] \quad \text{and} \quad \underset{M,\ \boldsymbol{\Phi},\ \rho\in\mathbb{R}^K}{\text{maximize}} \ \text{EE} \ \left[\frac{\text{bit}}{\text{J}}\right],\\
&\text{s.t.} \quad \text{(c.1)} \quad \gamma_i \geq \gamma_i^*, \quad \forall i,\\
&\qquad\ \ \text{(c.2)} \quad r_i \geq r_{i,\min}^{\text{serv}},\\
&\qquad\ \ \text{(c.3)} \quad p_{\min} \leq p_i \leq p_{\max},\\
&\qquad\ \ \text{(c.4)} \quad M_{\min} < M \leq M_{\max},\\
&\qquad\ \ \text{(c.5)} \quad |[\boldsymbol{\Phi}]_{n,n}| = 1, \quad \forall n = 1, \ldots, N,
\end{aligned} \tag{7.20}$$

where the optimization variables are composed of the transmit power vector $\boldsymbol{\rho}$, the number of active BS antennas M, and the RIS phase-shift elements matrix $\boldsymbol{\Phi}$. The constraint (c.1) refers to the signal-to-interference-plus-noise ratio (SINR) for the ith user; in this conflicting bi-objective optimization, (c.1) indicates a minimum QoS for the ith UE; constraint (c.2) is the attainable ergodic rate (ER) for the ith user; it must be equal or greater than the minimum information rate required to guarantee the QoS of a specific user; formally (c.2) can be satisfied into (c.1); thus, the problem will be simplified; constraint (c.3) indicates that power allocated to the ith UE p_i cannot exceed a maximum transmission power available to the ith user; (c.4) implies that the number of active BS antennas does not exceed the number of antennas installed at the BS and must be higher than a minimum number of antennas, under ZF precoding, $M_{\min} = K$, where K is the number of UEs; finally, constraint (c.5) indicates that the RIS is a passive reflecting structure while making the reflected signal energy as high as possible.

Some strategies are available in the literature to improve RE. One can reduce the required power consumption by reducing the signaling overhead and energy consumption to improve the RE [17]. Also, the authors use partial CSI, particularly the statistical CSI between the RIS and UEs, and the instantaneous CSI between the RIS and the BS. Hence, to investigate the EE–SE tradeoff in the UL of RIS-aided M-MIMO systems, the authors develop a framework for the joint optimization of UEs' transmit precoding and RIS reflective beamforming to maximize the RE performance metric. The design of UE's precoding is simplified with the aid of some closed-form solutions. In [17], to avoid the high complexity in computing the nested integrals involved in the expectations, the authors derive an asymptotic and deterministic objective expression. For the design of the RIS phases, iterative optimization methods, such as the MMSE and majorization minimization, can be deployed.

7.3 Energy Efficiency in RIS-Aided M-MIMO

This section discusses the EE maximization problem in the DL of a multiuser RIS-aided M-MIMO communication system under ZF precoding. Since systematically optimizing the RIS phase-shift in each channel coherence time interval is challenging and burdensome, we employ the statistical CSI-based optimization strategy to alleviate this overhead. We can reduce the computational complexity by treating the RIS phase-shift matrix as a constant over multiple channel coherence time intervals while maintaining suitable performance. To circumvent the non-convexity of EE optimization problem, we explore the **sequential optimization** approach where the PA vector $\boldsymbol{\rho}$, the number of antennas M, and the RIS angle phase-shift $\boldsymbol{v}$ are separate and sequentially solved iteratively until convergence. With the help of the Lagrangian dual method and FP techniques, insightful closed-form expressions for all the optimization variables are derived. The proposed approach can substantially reduce the power consumption and the number of active BS antennas, demonstrating the potential of RIS deployment and optimization in M-MIMO systems under simple linear precoding.

7.3.1 Problem Formulation

We consider an RIS -assisted M-MIMO downlink communication scenario, assuming far-field propagation for BS to RIS and BS to UEs links, and near-field wavefront propagation for RIS to UEs links. This can be justified in practice since a quite appealing application scenario for the RIS is to improve the communication channel of UEs whose direct link to the BS is obstructed; thus, it is convenient to deploy the RIS close to the UEs but in a position with good visibility to the BS. Besides, the direct channel link matrix $\mathbf{H}$ is assumed to be purely non-line-of-sight (NLoS), that is, $\kappa = 0$. We also consider that the RIS is positioned strategically resulting in the $\mathbf{G}$ (RIS to UEs) channel being purely LoS, that is, $\tilde{\kappa}_k = \infty, \forall k = 1, \dots, K$. The BS array is equipped with a maximum of $M_{\max}$ antennas, each with its own RF transceiver, resulting in substantial energy consumption. Moreover, considering the significant configuration overhead introduced by the RIS, the objective is to minimize it to enhance the performance of the communication system. In this context, our focus lies in optimizing three fundamental aspects:

- ***Antenna activation***: We are concerned with determining the optimal number of antennas to be activated, denoted as M^*.
- ***RIS phase-shift*** matrix optimization, where we seek to optimize the phase-shift matrix $\mathbf{\Phi}$ associated with the RIS.
- UEs' ***power allocation*** vector $\boldsymbol{\rho} = [\rho_1, \rho_2, \dots, \rho_K]$.

This holistic approach addresses the efficient utilization of two precious resources: spectrum and energy. Furthermore, here we take a long-term perspective into account, which is driven by the goal of reducing the configuration overhead imposed by the RIS configuration in our communication system; besides, this approach can be adaptive concerning the number of UEs that are served by the BS.

The DL received signal at the UEs considering the BS active antennas, which are activated/deactivated, is given as

$$\mathbf{y} = \hat{\mathbf{H}}^H \mathbf{x} + \mathbf{n}, \tag{7.21}$$

where $\hat{\mathbf{H}} \in \mathbb{C}^{M \times K}$ is the complete communication channel, whose columns are given by (7.17), which is composed of the direct link between the BS and the

UEs $\mathbf{H} \in \mathbb{C}^{M \times K}$ and the composed channel between the BS and UEs intermediated by the RIS, $\mathbf{F}^H \boldsymbol{\Phi} \mathbf{G}$. Besides, $\mathbf{x} \in \mathbf{C}^M$ is the signal transmitted by the BS and $\mathbf{n} \in \mathbb{C}^K$, with $\mathbf{n} \sim \mathcal{CN}\left(\mathbf{0}, \sigma^2 \mathbf{I}_K\right)$ is the additive white Gaussian noise (AWGN) at the receivers. In order to null the interference between the UEs, the BS employs the ZF precoder given as the following

$$\mathbf{W}_{\text{ZF}} = \hat{\mathbf{H}}(\hat{\mathbf{H}}^H \hat{\mathbf{H}})^{-1}. \tag{7.22}$$

Therefore, the transmitted signal $\mathbf{x}$ can be given as

$$\mathbf{x} = \mathbf{W}_{\text{ZF}} \mathbf{P}^{\frac{1}{2}} \mathbf{s}, \tag{7.23}$$

where $\mathbf{s}$ is the transmitted data symbol such that $\mathbb{E}[|[\mathbf{s}]_k|^2] = 1, \forall k \in \{1, \dots, K\}$, $\mathbf{P} = \text{diag}(\boldsymbol{\rho}) = \text{diag}([\rho_1, \rho_2, \dots, \rho_K])$ is the power allocated to the UEs, such as $\text{tr}(\mathbf{P}) = P_{\max}$, with $P_{\max}$ being the maximum available power (BS power budget).

The instantaneous SE for the kth UE with the ZF precoder in the RIS-aided M-MIMO system can be written as

$$R_k = \log_2 \left(1 + \frac{\rho_k}{\sigma^2 \left[\left(\hat{\mathbf{H}}^H \hat{\mathbf{H}} \right)^{-1} \right]_{k,k}} \right) \quad \text{[bits]}. \tag{7.24}$$

As our focus lies in long-term perspective optimization, it becomes valuable to acquire insights into the statistical behavior of the SE of the studied wireless communication system. In [27], a rigorous *lower bound* for the ER employing ZF precoding in the context of RIS-aided M-MIMO systems is derived, which can be computed for the kth UE as

$$\tilde{R}_k = \log_2 \left(1 + \frac{\rho_k (M-K) \boldsymbol{v}^H \mathbf{B} \boldsymbol{v}}{\sigma^2 (\hat{\kappa} + 1) \boldsymbol{v}^H \mathbf{A} \boldsymbol{v}} \right) \leq \mathbb{E}\left[R_k\right], \tag{7.25}$$

where $\mathbf{A}$ and $\mathbf{B}$ are constants related to the RIS deployment (RIS-array gain) given as

$$\mathbf{B} = \frac{1}{N} \mathbf{I}_K + \beta_F \hat{\kappa} \text{diag}(\mathbf{a}_{\text{ris}}^H) \mathbf{G} \boldsymbol{\Lambda}^{-1} \mathbf{G}^H \, \text{diag}(\mathbf{a}_{\text{ris}}), \tag{7.26}$$

$$\mathbf{A}_k = \left[\boldsymbol{\Lambda}^{-1}\right]_{k,k} \mathbf{B} - \hat{\kappa} \beta_F \mathbf{q}_k \mathbf{q}_k^H, \tag{7.27}$$

$$\mathbf{q}_k^H = \left[\boldsymbol{\Lambda}^{-1} \mathbf{G}^H \text{diag}(\mathbf{a}_{\text{ris}})\right]_{(k,:)}, \tag{7.28}$$

$$\boldsymbol{\Lambda} = \beta_F \mathbf{G}^H \mathbf{G} + (\hat{\kappa} + 1) \boldsymbol{\beta}, \tag{7.29}$$

with $\boldsymbol{\beta} = \text{diag}([\sqrt{\beta_1}, \sqrt{\beta_2}, \dots, \sqrt{\beta_K}]^T)$ being the path-loss factors of BS-UEs links. It should be noticed that since we consider far-field communication for such links, all antennas of BS array experience the same path-loss, that is, $\beta_{km} = \beta_k$, $\forall m \in \{1, \dots, M\}$. Furthermore, when the RIS is not deployed ($\beta_F = 0$), Eq. (7.25) reduces to the traditional ER equation for M-MIMO with ZF precoding, where **G** is the RIS-UEs channel.

The EE maximization problem can be formulated as the following

$$\underset{M,\rho,\boldsymbol{v}}{\text{maximize}} \quad \eta_{\text{EE}} = \frac{\sum_{k=1}^{K} \log_2\left(1 + \frac{\rho_k (M-K)\boldsymbol{v}^H \mathbf{B} \boldsymbol{v}}{\sigma^2(\hat{\kappa}+1)\boldsymbol{v}^H \mathbf{A}_k \boldsymbol{v}}\right)}{\varrho \sum_{k=1}^{K} \rho_k + P_{\text{FIX}} + M P_{\text{BS}} + N P_{\text{RIS}}}, \tag{7.30}$$

$$\text{subject to} \quad \sum_{k=1}^{K} \rho_k \leq P_{\max}, \tag{7.30a}$$

$$\log_2\left(1 + \frac{\rho_k (M-K)\boldsymbol{v}^H \mathbf{B} \boldsymbol{v}}{\sigma^2(\hat{\kappa}+1)\boldsymbol{v}^H \mathbf{A} \boldsymbol{v}}\right) \geq R_{\min}, \quad \forall k = 1, \dots, K, \tag{7.30b}$$

$$K < M \leq M_{\max}, \tag{7.30c}$$

$$|[\boldsymbol{v}]_n| = 1, \quad \forall n = 1, \dots, N. \tag{7.30d}$$

Constraint (7.30a) is the maximum RF transmit power constraint at BS; constraint (7.30b) guarantees the minimum achievable downlink rate for all UEs, where $R_{\min}$ is a predefined parameter which depends on QoS requirements; constraint (7.30c) guarantees the applicability of the ZF in the ER lower bound equation, and constraint (7.30d) assumes that $\alpha_n = 1$, $\forall n$, in Eq. (7.16), since it is desirable to make the reflected signal energy as high as possible, but without any signal amplification due to the passive operation of each RIS reflecting element.

The denominator in (7.30) is the total power consumed by the system, being composed of four terms:

a) $\varrho \cdot \rho_k$ is the power consumed by the kth RF power amplifier, with ϱ being its inefficiency.
b) P_{FIX} is the fixed consumed power at the BS.
c) $M \cdot P_{\text{BS}}$ is the power consumed by all active antennas at the BS.
d) $N \cdot P_{\text{RIS}}$ is the power consumed by the RIS for switching and changing the impedance of reflecting elements on RIS in order to redirect the incident wave.

The problem in (7.30) is quasi-convex regarding variable $\boldsymbol{\rho}$ since it is the ratio between a convex function (sum of ER) and an affine function in $\boldsymbol{\rho}$. Furthermore, regarding to $\boldsymbol{v}$, it is a non-convex problem with a non-convex constraint (7.30d).

7.3.2 Solution and Algorithms

In order to solve the optimization problem (7.30), an alternating optimization (AO) method is proposed. Basically, the solution is given in two major steps:

Step 1: RIS phase–shift optimization; and
Step 2: AS and PA jointly optimization.

Basically, for Step 1, FP techniques and Lagrangian dual transform (LDT) are utilized. Moreover, two approaches are compared: (a) closed-form solution and (b) sequential fractional programming (SFP) solution. Furthermore, for Step 2, we leverage the Lagrangian dual method to obtain a low-complexity solution. In general, for all the optimization variables discussed in this section, we will provide a closed-form solution.

7.3.2.1 Optimization Techniques

In order to comprehensively elucidate the proposed optimization strategy for tackling the EE RIS-aided M-MIMO optimization problem, as described in (7.30), it is imperative to delve into several key techniques that act in our complete solution. These techniques play an instrumental role in our goal to address the problem entirely, and a firm grasp of their principles is essential for a careful understanding of the proposed solution. In the following, we provide a detailed review of these foundational techniques, including LDT, FP sum-of-ratios (Dinkelbach's method), and SFP, among others, which serve as the building blocks of our complete proposed solution.

Lagrangian Dual Transform In addressing the challenge posed by the sum-of-logarithms problem, specific techniques prove to be valuable in easing the management of the optimization problem, thereby enabling the possibility of finding a closed-form, low-complexity solution. Given that communication system-related problems frequently involve logarithmic function summation, finding any form of solution can be exceedingly challenging as the presence of logarithmic functions hinders the analytical tractability and, consequently, optimization. As a result, an optimization technique introduced by Shen and Yu [28], namely LDT leverages the Lagrangian duality to externalize the argument of logarithm functions to the outside, eliminating the dependence of the variable of interest with the logarithmic functions, resulting in easier management and analytical tractability of the problem. To further understand the LDT, let us define the following target maximization problem:

$$\underset{\mathbf{x}}{\text{maximize}} \quad \sum_{\ell=1}^{L} \log\left(1 + \frac{f_\ell(\mathbf{x})}{g_\ell(\mathbf{x})}\right), \tag{7.31}$$

$$\text{subject to} \quad h_i(\mathbf{x}) \leq 0, \quad i \in \{1, \ldots, I\}, \tag{7.31a}$$

where $f_\ell(\mathbf{x}) : \mathbb{C}^N \rightarrow \mathbb{R}_+$ is a non-negative function, and $g_\ell(\mathbf{x}) : \mathbb{C}^N \rightarrow \mathbb{R}_{++}$ is a positive function, $\forall \ell \in \{1, \dots, L\}$. It is important to observe that the ratio $f_\ell(\mathbf{x})/g_\ell(\mathbf{x})$ can have a physical interpretation as SINR in our context. The problem (7.31) is assumed to be a non-convex problem, with a non-convex constraint (7.31a). According to [29], the problem (7.31) can be transformed in the following equivalent problem:

$$\underset{\mathbf{x}, \boldsymbol{\alpha}}{\text{maximize}} \quad \sum_{\ell=1}^{L} \log(1+\alpha_\ell) - \sum_{\ell=1}^{L} \alpha_\ell + \sum_{\ell=1}^{L} \frac{(1+\alpha_\ell) f_\ell(\mathbf{x})}{f_\ell(\mathbf{x}) + g_\ell(\mathbf{x})}, \tag{7.32}$$

$$\text{subject to} \quad h_i(\mathbf{x}) \le 0, \qquad \forall i \in \{1, \dots, I\}, \tag{7.32a}$$

$$\alpha_\ell \ge 0, \qquad \forall \ell \in \{1, \dots, L\}, \tag{7.32b}$$

where $\boldsymbol{\alpha} = [\alpha_1, \alpha_2, \dots, \alpha_L]^T$ is an auxiliary variable introduced for each ratio term $f_\ell(\mathbf{x})/g_\ell(\mathbf{x})$. It's important to note that the two problems, (7.31) and (7.32), are equivalent. In other words, the solution $\mathbf{x}$ to (7.31) is identical to the solution to (7.32), and their respective optimal objective values are also equal [28, III-B]. Furthermore, it is worth highlighting that, according to the KKT conditions we can obtain the following:

$$\frac{\partial \left(\sum_{\ell=1}^{L} \log(1+\alpha_\ell) - \sum_{\ell=1}^{L} \alpha_\ell + \sum_{\ell=1}^{L} \frac{(1+\alpha_\ell) f_\ell(\mathbf{x})}{f_\ell(\mathbf{x}) + g_\ell(\mathbf{x})} \right)}{\partial \alpha_\ell} = 0, \tag{7.33}$$

$$\alpha_\ell^\star = \frac{f_\ell(\mathbf{x})}{g_\ell(\mathbf{x})}. \tag{7.34}$$

This equivalence allows for a seamless transition between the two problem formulations while preserving the optimally.

Fractional Problem Sum-of-Ratios FP is a typical class of optimization problems generally viewed in the most diverse areas of application. Basically, they are divided into two classes, being it single-ratio problems and multiple-ratio problems. For single-ratio problems, many transforms deal consistently with this type of problem; the most classical is named *Dinkelbach's* Transform [29]. Let us define the following fractional optimization problem:

$$\underset{\mathbf{x}}{\text{maximize}} \quad \frac{f(\mathbf{x})}{g(\mathbf{x})}, \tag{7.35}$$

$$h_i(\mathbf{x}) \le 0, \qquad \forall i \in \{1, \dots, I\}, \tag{7.35a}$$

where $f(\mathbf{x}) : \mathbb{C}^N \rightarrow \mathbb{R}_+$ is a non-negative function and $g(\mathbf{x}) : \mathbb{C}^N \rightarrow \mathbb{R}_{++}$ is a positive function. The conventional approach for dealing with this FP is decoupling the numerator and denominator and treating them jointly. The *Dinkelbach's* transform reformulates the single-ratio problem (7.35) as the following:

$$\underset{\mathbf{x},\alpha}{\text{maximize}} \quad f(\mathbf{x}) - \alpha g(\mathbf{x}), \tag{7.36}$$

$$h_i(\mathbf{x}) \leq 0, \qquad \forall i \in \{1, \dots, I\}, \tag{7.36a}$$

where α is an auxiliary variable that is iteratively updated by:

$$\alpha^{(t+1)} = \frac{f(\mathbf{x}^{(t)})}{g(\mathbf{x}^{(t)})}, \tag{7.37}$$

where t is the iteration index.

Although the classic *Dinkelbach's* transform works well for single-ratio problems, they cannot be easily extended to the multiple-ratio problem since for single-ratio problems, the exact optimal solution is the same for the original FP and the transformed problem, but no the value of the objective function of both problems. Aiming to solve the multi-ratio problem, several different optimization techniques have been proposed, such as *Quadratic Transform* [28], in which for each ratio term, an auxiliary variable is considered. Let us define the following multi-ratio problem:

$$\underset{\mathbf{x}}{\text{maximize}} \quad \sum_{\ell=1}^{L} \frac{f_\ell(\mathbf{x})}{g_\ell(\mathbf{x})}, \tag{7.38}$$

$$\text{subject to} \quad h_i(\mathbf{x}) \leq 0, \qquad i \in \{1, \dots, I\}, \tag{7.38a}$$

where $f_\ell(\mathbf{x}) : \mathbb{C}^N \rightarrow \mathbb{R}_+$ is a non-negative function and $g_\ell(\mathbf{x}) : \mathbb{C}^N \rightarrow \mathbb{R}_{++}, \forall \ell \in \{1, \dots, L\}$ is a strictly positive function. In [30], the authors proposed an iterative algorithm to solve problem (7.38), where the equivalent problem can be given as:

$$\underset{\mathbf{x},\mathbf{u},\alpha}{\text{maximize}} \quad \sum_{\ell=1}^{L} u_\ell \left(f_\ell(\mathbf{x}) - \alpha_\ell g_\ell(\mathbf{x}) \right), \tag{7.39}$$

$$\text{subject to} \quad h_i(\mathbf{x}) \leq 0, \qquad i \in \{1, \dots, I\}, \tag{7.39a}$$

where $\mathbf{u}$ and $\boldsymbol{\alpha}$ are auxiliary variables which are iteratively updated as:

$$u_\ell^{(t+1)} = \frac{1}{g(\mathbf{x}^{(t)})}, \quad \forall \ell = 1, \dots, L, \tag{7.40}$$

$$\alpha_\ell^{(t+1)} = \frac{f(\mathbf{x}^{(t)})}{g(\mathbf{x}^{(t)})}, \quad \forall \ell = 1, \dots, L. \tag{7.41}$$

Sequential Fractional Programming An attractive and robust method for obtaining computationally efficient solutions for challenging optimization problems is referred to as SFP. The fundamental concept underlying SFP is to determine the KKT points of the original problems by solving a series of approximate (simpler) sub problems under the condition that these sub problems meet certain assumptions. To elaborate further, consider the following optimization problem:

$$\underset{\mathbf{x}}{\text{minimize}} \quad f(\mathbf{x}), \tag{7.42}$$

$$\text{subject to} \quad h_i(\mathbf{x}) \leq 0, \quad \forall i \in \{1, \dots, I\}. \tag{7.42a}$$

Suppose we have a sequence $\mathbf{x}^{(t)}$ of feasible points for (7.42), and let us consider the following alternative objective function denoted as $\hat{f}\left(\mathbf{x}, \mathbf{x}^{(t)}\right)$. If the following conditions are satisfied for each feasible point $\mathbf{x}^{(t)}$:

1) $\hat{f}\left(\mathbf{x}|\mathbf{x}^{(t)}\right) \geq f(\mathbf{x})$ for all feasible $\mathbf{x}$;
2) $\hat{f}\left(\mathbf{x}^{(t)}|\mathbf{x}^{(t)}\right) = f\left(\mathbf{x}^{(t)}\right)$;
3) $\nabla_{\mathbf{x}} \hat{f}\left(\mathbf{x}^{(t)}|\mathbf{x}^{(t)}\right) = \nabla_{\mathbf{x}} f(\mathbf{x}^{(t)})$.

Thus, the optimal sequence $f(\mathbf{x}^{\star})^{(t)}$ exhibits a monotonically decreasing trend and converges [29]. Evidently, the effectiveness of the SFP approach relies on the ability to identify appropriate functions $f(\mathbf{x}|\mathbf{x}^{(t)})$ that satisfy the aforementioned three conditions. Furthermore, these functions must be less complex than (7.42) to ensure the practicality of implementing this method.

Specifically, for the sake of interest, let us consider the following quadratic programming problem:

$$\underset{\mathbf{x}}{\text{minimize}} \quad \mathbf{x}^H \mathbf{A} \mathbf{x}, \tag{7.43}$$

$$\text{subject to} \quad \left|[\mathbf{x}]_n\right| = 1, \quad \forall n \in \{1, \dots, N\}, \tag{7.43a}$$

where $\mathbf{x} \in \mathbb{C}^N$ and $\mathbf{A}$ is a quadratic positive semi-definite matrix of order N, that is, $\mathbf{A} \succcurlyeq 0, \mathbf{A} \in \mathbb{C}^{N \times N}$. Within these conditions, (7.43) is a convex problem with a non-convex constraint (7.43a). By considering the following alternative function:

$$\hat{f}\left(\mathbf{x}|\mathbf{x}^{(t)}\right) = \mathbf{x}^H \mathbf{M} \mathbf{x} - 2\text{Re}(\mathbf{x}^H (\mathbf{M} - \mathbf{A}) \mathbf{x}^{(t)}) + (\mathbf{x}^{(t)})^H \left(\mathbf{M} - \mathbf{A}\right) \mathbf{x}^{(t)}, \tag{7.44}$$

with $\mathbf{M} \triangleq \lambda_{\max}(\mathbf{A}) \mathbf{I}_N$, [31] shown that the problem (7.43) can be equivalently solved as:

$$\underset{\mathbf{x}}{\text{maximize}} \quad 2\text{Re}(\mathbf{x}^H (\lambda_{\max}(\mathbf{A}) \mathbf{I}_K - \mathbf{A}) \mathbf{x}^{(t)}), \tag{7.45}$$

$$\text{subject to} \quad \left|[\mathbf{x}]_n\right| = 1, \quad \forall n \in \{1, \dots, N\}, \tag{7.45a}$$

where $\lambda_{\max}(\mathbf{A})$ is the maximum eigenvalue of matrix $\mathbf{A}$. Evidently, the optimal solution for (7.45) is given when the phases are canceled:

$$\mathbf{x}^{\star} = e^{j\angle(\lambda_{\max}(\mathbf{A})\mathbf{I}_K - \mathbf{A})\mathbf{x}^{(t)}}. \tag{7.46}$$

7.3.2.2 RIS Phase-Shift Optimization Analytical Solution

In this section, we delve into the proposed solution for optimizing the RIS phase-shift. First, it's important to note from (7.30) that when M and ρ are fixed, the maximization of EE concerning $\boldsymbol{v}$ is equivalent to the maximization of the sum rate (SR). In other words, the optimal value of phase-shift ($\boldsymbol{v}^{\star}$) for EE remains the same as that for the numerator of (7.30). Consequently, optimizing (7.30) with respect to $\boldsymbol{v}$ is equivalent to the following optimization problem:

$$\underset{\boldsymbol{v}}{\text{maximize}} \quad \sum_{k=1}^{K} \log_2 \left(1 + \frac{\rho_k (M-K) \boldsymbol{v}^H \mathbf{B} \boldsymbol{v}}{\sigma^2 (\hat{\kappa}+1) \boldsymbol{v}^H \mathbf{A}_k \boldsymbol{v}} \right), \tag{7.47}$$

$$\text{subject to} \quad |[\boldsymbol{v}]_n| = 1, \quad \forall n \in \{1, \dots, N\}. \tag{7.47a}$$

Observe that the objective function of problem (7.47) can be represented as a sum of logarithms. Consequently, we can readily apply the LDT to externalize the argument. This process facilitates the independence of the variable of interest from the logarithmic function. By applying LDT to (7.47), as demonstrated in (7.32), we derive the following equivalent problem:

$$\underset{\boldsymbol{v},\boldsymbol{\alpha}}{\text{maximize}} \quad \sum_{k=1}^{K} \log_2(1+\alpha_k) - \sum_{k=1}^{K} \alpha_k + \sum_{k=1}^{K} \frac{(1+\alpha_k)\rho_k (M-K) \boldsymbol{v}^H \mathbf{B} \boldsymbol{v}}{\boldsymbol{v}^H \left(\sigma^2 (\hat{\kappa}+1) \mathbf{A}_k + \rho_k (M-K) \mathbf{B} \right) \boldsymbol{v}}, \tag{7.48}$$

$$\text{subject to} \quad \alpha_k \geq 0, \quad \forall k \in \{1, \dots, K\}, \tag{7.48a}$$

$$\left| [\boldsymbol{v}]_n \right| = 1, \quad \forall n \in \{1, \dots, N\}. \tag{7.48b}$$

Accordingly to (7.34), $\boldsymbol{\alpha}$ should be updated as:

$$\alpha_k = \frac{\rho_k (M-K) \boldsymbol{v}^H \mathbf{B} \boldsymbol{v}}{\sigma^2 (\hat{\kappa}+1) \boldsymbol{v}^H \mathbf{A}_k \boldsymbol{v}}, \quad \forall k \in \{1, \dots, K\}. \tag{7.49}$$

With given $\boldsymbol{\alpha}$, one may focus on the following problem:

$$\underset{\boldsymbol{v}}{\text{maximize}} \quad \sum_{k=1}^{K} \frac{(1+\alpha_k)\rho_k (M-K) \boldsymbol{v}^H \mathbf{B} \boldsymbol{v}}{\boldsymbol{v}^H \left(\sigma^2 (\hat{\kappa}+1) \mathbf{A}_k + \rho_k (M-K) \mathbf{B} \right) \boldsymbol{v}}, \tag{7.50}$$

$$\text{subject to} \quad |[\boldsymbol{v}]_n| = 1, \quad \forall n \in \{1, \dots, N\}. \tag{7.50a}$$

Notice that (7.50) is similar to the multiple-ratio problem as in (7.38). Therefore, by utilizing (7.39), we can re-formulate the problem (7.50) in the following way:

$$\underset{\boldsymbol{v},\boldsymbol{\lambda},\boldsymbol{u}}{\text{maximize}} \quad \boldsymbol{v}^H \mathbf{C}(\boldsymbol{\lambda}, \boldsymbol{u}) \boldsymbol{v}, \tag{7.51}$$

$$\text{subject to} \quad \lambda_k \geq 0, \quad \forall k \in \{1, \dots, K\}, \tag{7.51a}$$

$$u_k \geq 0, \quad \forall k \in \{1, \dots, K\}, \tag{7.51b}$$

$$|[\boldsymbol{v}]_n| = 1, \quad \forall n \in \{1, \dots, N\}, \tag{7.51c}$$

in which the matrix **C** is dependent of two auxiliary variables λ and **u** computed as:

$$\mathbf{C}(\lambda, \boldsymbol{u}) = \sum_{k=1}^{K} u_k \left((1+\gamma_k)\rho_k(M-K)\mathbf{B} - \lambda_k \left(\sigma^2(\hat{\kappa}+1)\mathbf{A}_k + \rho_k(M-K)\mathbf{B} \right) \right). \tag{7.52}$$

Notice that the closed-form solution for $[\lambda]_k$ and $[\mathbf{u}]_k$ can be given directly as:

$$u_k^\star = \frac{1}{\boldsymbol{v}^H \left(\sigma^2(\hat{\kappa}+1)\mathbf{A}_k + \rho_k(M-K)\mathbf{B} \right) \boldsymbol{v}}, \tag{7.53}$$

$$\lambda_k^\star = \frac{(1+\alpha_k)\rho_k(M-K)\boldsymbol{v}^H\mathbf{B}\boldsymbol{v}}{\boldsymbol{v}^H \left(\sigma^2(\hat{\kappa}+1)\mathbf{A}_k + \rho_k(M-K)\mathbf{B} \right) \boldsymbol{v}}. \tag{7.54}$$

Theorem 7.1 The optimal solution for Problem (7.51) can be computed individually for each element of $\boldsymbol{v}$ in low-complexity closed-form expression as:

$$\theta_n^\star = \angle \left[\mathrm{Tr}\left(\left([\mathbf{C}]_{1:(n-1),n} \right)^H [\boldsymbol{v}]_{1:(n-1)} \right) + \mathrm{Tr}\left([\boldsymbol{v}]_{(n+1):N} [\mathbf{C}]_{n,(n+1):N} \right) \right], \tag{7.55}$$

$$[\boldsymbol{v}^\star]_n = e^{j\theta_n^\star}, \qquad \forall n \in \{1, \dots, N\}. \tag{7.56}$$

Proof: Initially, we should notice that Problem (7.51), for $\boldsymbol{v}$ optimization, can be equivalently transformed into

$$\max_{\boldsymbol{v}} \ \boldsymbol{v}^H\mathbf{C}\boldsymbol{v} = \max_{\boldsymbol{v}} \ \mathrm{Tr}\left(\mathbf{C}\boldsymbol{V}\right), \tag{7.57}$$

with $\boldsymbol{V} = \boldsymbol{v}\boldsymbol{v}^H$, where the following constraints rank$(\boldsymbol{V}) = 1$, and $[\boldsymbol{V}]_{n,n} = 1, \forall n \in \{1, \dots, N\}$, must be attained. Proceeding, conveniently, we can rewritten $\boldsymbol{v}$ as the following manner

$$\boldsymbol{v} = [\boldsymbol{v}_1^H, e^{-j\theta_n}, \boldsymbol{v}_2^H]^H, \tag{7.58}$$

where $\boldsymbol{v}_1 \in \mathbb{C}^{n-1}$ and $\boldsymbol{v}_2 \in \mathbb{C}^{N-n}$ are, respectively, defined as $\boldsymbol{v}_1 \triangleq [\boldsymbol{v}]_{1:(n-1)}$ and $\boldsymbol{v}_2 \triangleq [\boldsymbol{v}]_{(n+1):N}$. This alternative notation enables us to look for the nth element of $\boldsymbol{v}$, $e^{j\theta_n}$. Therefore, by utilizing Eq. (7.58), one can rewrite $\boldsymbol{V}$ as the following:

$$\boldsymbol{V} = \boldsymbol{v}\boldsymbol{v}^H = \begin{bmatrix} \boldsymbol{v}_1\boldsymbol{v}_1^H & \boldsymbol{v}_1 e^{-j\theta_n} & \boldsymbol{v}_1\boldsymbol{v}_2^H \\ e^{j\theta_n}\boldsymbol{v}_1^H & 1 & e^{j\theta_n}\boldsymbol{v}_2^H \\ \boldsymbol{v}_2\boldsymbol{v}_1^H & \boldsymbol{v}_2 e^{-j\theta_n} & \boldsymbol{v}_2\boldsymbol{v}_2^H \end{bmatrix}. \tag{7.59}$$

By defining the following variables:

$$\begin{cases} \mathbf{C}_1 = [\mathbf{C}]_{1:(n-1),1:(n-1)} & \in \mathbb{C}^{(n-1)\times(n-1)} \\ \mathbf{c}_2 = [\mathbf{C}]_{1:(n-1),n} & \in \mathbb{C}^{(n-1)\times 1} \\ \mathbf{C}_3 = [\mathbf{C}]_{1:(n-1),(n+1):N} & \in \mathbb{C}^{(n-1)\times(N-n)} \\ C_4 = [\mathbf{C}]_{n,n} & \in \mathbb{C} \\ \mathbf{c}_5 = [\mathbf{C}]_{n,(n+1):N} & \in \mathbb{C}^{1\times(N-n)} \\ \mathbf{C}_6 = [\mathbf{C}]_{(n+1):N,(n+1):N} & \in \mathbb{C}^{(N-n)\times(N-n)}, \end{cases} \tag{7.60}$$

the matrix $\mathbf{C}$ can be rewritten in terms of these variables as:

$$\mathbf{C} = \begin{bmatrix} \mathbf{C}_1 & \mathbf{c}_2 & \mathbf{C}_3 \\ \mathbf{c}_2^H & C_4 & \mathbf{c}_5 \\ \mathbf{C}_3^H & \mathbf{c}_5^H & \mathbf{C}_6 \end{bmatrix}. \tag{7.61}$$

Therefore, after some basic algebraic manipulations, we obtain:

$$\begin{aligned} \mathrm{Tr}(\mathbf{C}\boldsymbol{V}) = {} & e^{-j\theta_n}(\mathrm{Tr}(\mathbf{c}_2^H\boldsymbol{v}_1) + \mathrm{Tr}(\mathbf{c}_5\boldsymbol{v}_2)) + e^{j\theta_n}(\mathrm{Tr}(\mathbf{c}_2\boldsymbol{v}_1^H) + \mathrm{Tr}(\mathbf{c}_5^H\boldsymbol{v}_2^H)) + C_4 \\ & + \mathrm{Tr}\left(\mathbf{C}_3\boldsymbol{v}_2\boldsymbol{v}_1^H + \mathbf{C}_1\boldsymbol{v}_1\boldsymbol{v}_1^H\right) + \mathrm{Tr}\left(\mathbf{C}_3^H\boldsymbol{v}_1\boldsymbol{v}_2^H + \mathbf{C}_6\boldsymbol{v}_2\boldsymbol{v}_2^H\right). \end{aligned} \tag{7.62}$$

Interestingly, Eq. (7.62) leads us to see that maximize the problem (7.51) with respect to the nth element of $\boldsymbol{v}$ (θ_n) is equivalent to:

$$\max_{\theta_n} \ \mathrm{Tr}(\mathbf{C}\boldsymbol{V}) = \max_{\theta_n} \ 2\mathbb{R}\{e^{-j\theta_n}(\mathrm{Tr}(\mathbf{c}_2^H\boldsymbol{v}_1) + \mathrm{Tr}(\boldsymbol{v}_2\mathbf{c}_5))\}, \quad \forall n \in \{1,\dots,N\}, \tag{7.63}$$

whose optimal solution is straightforwardly given in closed-form by $\theta_n^\star = \angle\left(\mathrm{Tr}(\mathbf{c}_2^H\boldsymbol{v}_1) + \mathrm{Tr}(\boldsymbol{v}_2\mathbf{c}_5)\right)$. Substituting the values of $\mathbf{c}_2$ and $\mathbf{c}_5$ given by Eq. (7.60), and $\boldsymbol{v}_1$ and $\boldsymbol{v}_2$, the proof is complete. ∎

To facilitate a better understanding of the proposed optimization method outlined in the preceding subsections, in the following, we illustrate its construction using pseudo-codes, given by Algorithm 7.1. Notice that in Algorithm 7.1, we also include the SFP solution for problem (7.51) as well explained in 7.3.2.1.

7.3.2.3 RIS Phase–Shift Algorithm

We have named this algorithm the FP–phase–shift optimization. Fundamentally, it comprises two essential phases, known as the *outer loop* and the *inner loop*. In essence, the *outer loop* is tasked with optimizing the LDT auxiliary variable $\boldsymbol{\alpha}$. Conversely, the *inner loop* is responsible for optimizing the auxiliary variables introduced by the FP technique, in addition to the fundamental variable, the RIS phase–shift $\boldsymbol{v}$. Additionally, we have proposed optimizing $\boldsymbol{v}$ with two distinct solutions, namely, the analytical solution and the SFP solution. These processes are illustrated in Algorithm 7.1.

Algorithm 7.1 FP-based RIS phase-shift optimization

Input: M, ρ, σ^2, K_1, ϵ, $\mathbf{B}$, $\mathbf{A}_k$, $\forall k \in \{1, \dots, K\}$
Set feasible value for $\boldsymbol{v}$;
repeat
  Outer Loop: LDT (Lagrangian Dual Transform) approach
  Step 1: Update $\boldsymbol{\alpha}$ by (7.49);
  repeat
    Inner Loop: FP (Fractional Programming) approach
    Step 2: Compute $\mathbf{u}$ by (7.53);
    Step 3: Update $\boldsymbol{\lambda}$ by (7.55);
    Step 4: Compute $\mathbf{C}$ by (7.52);
    Set $\ell = 1$;
    repeat
      Step 5:
      1) Analytical Solution:
        $[\boldsymbol{v}]_n^{(\ell)} = e^{j\theta_n^\star}$ based on (7.55), $n = 1, \dots, N$;
      2) SFP (Sequential Fractional Programming) Solution:
        Set $\lambda_{\min}$ as the minimum eigenvalue of $\mathbf{C}$;
        $\boldsymbol{v}^{(\ell)} = e^{j\angle(\lambda_{\min}\mathbf{I}_N - \mathbf{C})\boldsymbol{v}^{(\ell-1)}}$;
    until $||\boldsymbol{v}^{(\ell)} - \boldsymbol{v}^{(\ell-1)}|| < \epsilon$
  until the objective function in (7.51) converge
until the objective function in (7.48) converge
Output: $\boldsymbol{v}^\star = [e^{-j\theta_1^\star}, \dots, e^{-j\theta_N^\star}]^H$

7.3.2.4 Number of Antennas and Power Allocation Jointly Optimization

In this subsection, we discuss the proposed solution for the number of active antenna and power allocation jointly optimization. With given $\boldsymbol{v}$ solution available, our focus now is to optimize M and ρ in a simple way. Utilizing *Dinkelbach's* transform, we have the following:

$$\underset{M,\rho,\eta_{\mathrm{EE}}}{\text{maximize}} \quad \sum_{k=1}^{K} \log_2\left(1 + \frac{\rho_k (M-K)\boldsymbol{v}^H \mathbf{B} \boldsymbol{v}}{\sigma^2(\hat{\kappa}+1)\boldsymbol{v}^H \mathbf{A}_k \boldsymbol{v}}\right) - \eta_{\mathrm{EE}}\left(\varrho \sum_{k=1}^{K} \rho_k + P_{\mathrm{FIX}} + MP_{\mathrm{BS}} + NP_{\mathrm{RIS}}\right), \tag{7.64}$$

$$\text{subject to} \quad \sum_{k=1}^{K} \rho_k \le P_{\max}, \tag{7.64a}$$

$$\sum_{k=1}^{K}\log_2\left(1+\frac{\rho_k(M-K)\boldsymbol{v}^H\mathbf{B}\boldsymbol{v}}{\sigma^2(\hat{\kappa}+1)\boldsymbol{v}^H\mathbf{A}_k\boldsymbol{v}}\right)\geq R_{\min},\quad \forall k=1,\dots,K, \tag{7.64b}$$

where η_{EE} can be directly computed as:

$$\eta_{\text{EE}}^{\star}=\frac{\displaystyle\sum_{k=1}^{K}\log_2\left(1+\frac{\rho_k(M-K)\boldsymbol{v}^H\mathbf{B}\boldsymbol{v}}{\sigma^2(\hat{\kappa}+1)\boldsymbol{v}^H\mathbf{A}_k\boldsymbol{v}}\right)}{\displaystyle\varrho\sum_{k=1}^{K}\rho_k+P_{\text{FIX}}+MP_{\text{BS}}+NP_{\text{RIS}}}. \tag{7.65}$$

Since problem (7.55) is a convex problem with convex constraints, we have that the Lagrangian dual method can be applied. In order to obtain a closed-form solution, utilizing (7.25), we recall ρ_k as the following:

$$\rho_k=\frac{(2^{\tilde{R}_k}-1)\sigma^2(\hat{\kappa}+1)\boldsymbol{v}\mathbf{A}_k\boldsymbol{v}}{(M-K)\boldsymbol{v}\mathbf{B}\boldsymbol{v}}, \tag{7.66}$$

where $\tilde{R}_k$ is the *lower bound* for the ER in the context of ZF precoding RIS-aided M-MIMO, as defined in (7.25). Therefore, substituting (7.56) into (7.55) and after some analytical manipulations, we can model the Lagrangian dual problem as a function of M and $\tilde{R}_k$ in the following way:

$$\begin{aligned}\mathcal{L}\left(M,\tilde{\mathbf{R}},\boldsymbol{\mu},\vartheta\right)\\ =\sum_{k=1}^{K}\tilde{R}_k-\eta_{\text{EE}}\left(\varrho\sum_{k=1}^{K}\frac{(2^{\tilde{R}_k}-1)\sigma^2(\hat{\kappa}+1)\boldsymbol{v}^H\mathbf{A}_k\boldsymbol{v}}{(M-K)\boldsymbol{v}^H\mathbf{B}\boldsymbol{v}}+P_{\text{FIX}}+MP_{\text{BS}}+NP_{\text{RIS}}\right)\\ +\sum_{k=1}^{K}\mu_k\left(\tilde{R}_k-R_{\min}\right)+\vartheta\left(P_{\text{TX}}-\sum_{k=1}^{K}\frac{(2^{\tilde{R}_k}-1)\sigma^2(\hat{\kappa}+1)\boldsymbol{v}^H\mathbf{A}_k\boldsymbol{v}}{(M-K)\boldsymbol{v}^H\mathbf{B}\boldsymbol{v}}\right),\end{aligned} \tag{7.67}$$

in which the vector $\boldsymbol{\mu}$ and the scalar ϑ are non-negative Lagrange multipliers. Therefore, the Lagrangian dual problem can be written as:

$$\underset{\boldsymbol{\mu},\vartheta}{\text{minimize}}\quad\underset{M,\tilde{\mathbf{R}}}{\text{maximize}}\quad\mathcal{L}\left(M,\tilde{\mathbf{R}},\boldsymbol{\mu},\vartheta\right), \tag{7.68}$$

$$\text{subject to}\quad \boldsymbol{\mu}\geq\mathbf{0},\quad \vartheta\geq 0. \tag{7.68a}$$

It is important to note that the alternative Lagrangian formulation for $\tilde{R}_k$ instead of ρ_k allows us to find expedite closed-form solutions. Consequently, one can determine the optimal number of transmit antennas $M^{\star}$ and the optimal power allocation $\boldsymbol{\rho}^{\star}$ by satisfying the KKT conditions, which can be directly computed as follows:

$$\frac{\partial\mathcal{L}(M,\tilde{\mathbf{R}},\boldsymbol{\mu},\vartheta)}{\partial M}=0,\qquad\frac{\partial\mathcal{L}(M,\tilde{\mathbf{R}},\boldsymbol{\mu},\vartheta)}{\partial\tilde{R}_k}=0. \tag{7.69}$$

After some mathematical manipulations from the computed derivatives via Eq. (7.69), the following second-order equation can be obtained:

$$- (M-K)^2 \eta_{\rm EE} P_{\rm BS} \log(2) + (M-K) \sum_{k=1}^{K} (\mu_k + 1) - (\eta_{\rm EE} \varrho + \vartheta) \sigma^2 (K_1 + 1) \log(2) \sum_{k=1}^{K} \frac{\boldsymbol{v}^H \mathbf{A}_k \boldsymbol{v}}{\boldsymbol{v}^H \mathbf{B} \boldsymbol{v}} = 0, \tag{7.70}$$

whose solution can be trivially computed as:

$$M^\star = \left[\frac{\sum_{k=1}^{K} (\mu_k + 1) + \sqrt{\left(\sum_{k=1}^{K} (\mu_k + 1) \right)^2 - 4 \eta_{\rm EE} P_{\rm BS} \log(2)^2 \sigma^2 (\hat{\kappa} + 1) \times (\eta_{\rm EE} \varrho + \vartheta) \sum_{k=1}^{K} \frac{\boldsymbol{v} \mathbf{A}_k \boldsymbol{v}}{\boldsymbol{v} \mathbf{B} \boldsymbol{v}}}}{2 \eta_{\rm EE} P_{\rm BS} \log(2)} + K \right]_{K+1}^{M_{\max}}, \tag{7.71}$$

$$\rho_k^\star = \left[\frac{1 + \mu_k}{\log(2)(\eta_{\rm EE} \varrho + \vartheta)} - \frac{\sigma^2(\hat{\kappa} + 1)}{M - K} \frac{\boldsymbol{v}^H \mathbf{A}_k \boldsymbol{v}}{\boldsymbol{v}^H \mathbf{B} \boldsymbol{v}} \right]^+ . \tag{7.72}$$

Lagrange multipliers updating. Although the analytical expressions for the optimal number of transmit antenna, $M^\star$, and power allocation, $\rho_k^\star$, have been derived, these expressions are related to the Lagrangian multipliers. Herein, we apply the widely-utilized sub-gradient method to update the Lagrangian multipliers; it means that μ_k and ϑ should decrease if the gradients are positives, that is, $\nabla_{\mu_k} \mathcal{L} > 0$ and $\nabla_{\vartheta} \mathcal{L} > 0$, and vice versa. The values of μ_k and ϑ at the ℓth iteration will be updated according to

$$\mu_k^{(\ell)} = \left[\mu_k^{(\ell-1)} - \tfrac{(R_k - R_{\min})}{10} \right]^+ \quad \text{and} \quad \vartheta^{(\ell)} = \left[\vartheta^{(\ell-1)} - \tfrac{\sqrt{\ell}}{10} \left(P_{\rm TX} - \sum_{k=1}^{K} \rho_k \right) \right]^+ . \tag{7.73}$$

7.3.2.5 Number of Active Antennas and Power Allocation Algorithm

Similar to optimizing the RIS phase-shift, to attain a deeper comprehension of the proposed algorithm presented in the previous subsection, we demonstrate its behavior using pseudo-code. Essentially, it consists of only one phase. The fundamental concept consists of iteratively updating the optimization variables

M and ρ until the auxiliary variable η_{EE} converges. This process is illustrated in Algorithm 7.2.

Algorithm 7.2 Proposed solution for M and ρ optimization

Input: $\mathbf{v}, \sigma^2, K_1, \epsilon, \mathbf{B}, \mathbf{A}_k, \forall k \in \{1, \dots, K\}$
Set feasible values for M and $\boldsymbol{\rho}$;
Set $\ell = 1$;
Set $\eta_{\text{EE}}^{(\ell)} = 0$;
repeat
  Step 1: Compute $\eta_{\text{EE}}^{(\ell+1)}$ by (7.65);
  Step 2: Compute $M^{(\ell+1)}$ by (7.71);
  Step 3: Compute $\boldsymbol{\rho}^{(\ell+1)}$ by (7.72);
  Step 4: Update the Lagrangian Multipliers by (7.73);
until $|\eta_{\text{EE}}^{(\ell)} - \eta_{\text{EE}}^{(\ell+1)}| \leq \epsilon$
Outputs: $M^\star$ and $\boldsymbol{\rho}^\star$

7.3.3 Numerical Results and Discussions

In this section, we aim to investigate the performance of the multi-user RIS-aided M-MIMO system. We consider a system configuration setup where the users' localization is fixed concerning their positions, angle of arrival (AoA), and angle of departure (AoD) for azimuth/elevation. Besides, for simplicity of analysis, in this numerical results section, we have considered: (a) far-field planar wave-front propagation conditions for the three BS-RIS, RIS-UE, and BS-UE links; and (b) uncorrelated channels. For one single system configuration setup, the instantaneous-CSI optimization-based approach is averaged over $\mathcal{T} = 500$ realizations of direct link $\mathbf{H}$ and $\mathbf{F}_{\text{NLoS}}$. Moreover, we assume the users are uniformly distributed in a circular area with the center in (100; 50) m and a radius of $r = 15\,m$, while the RIS is located at (100; 0) m and the BS is located at (0; 0) m. In each setup, the AoDs were uniformly distributed as: $\phi_{\text{AoD}}^{r,k} \sim \mathcal{U}[0, \frac{\pi}{2}]$, $\varphi_{\text{AoD}}^{r,k} \sim \mathcal{U}[-\frac{\pi}{3}, \frac{\pi}{3}]$, and $\varphi_{\text{AoD}}^{bs} \sim \mathcal{U}[-\frac{\pi}{2}, \frac{\pi}{2}]$; besides, the AoAs $\phi_{\text{AoA}}^{r} \sim \mathcal{U}[0, \frac{\pi}{2}]$ and $\varphi_{\text{AoA}}^{r} \sim \mathcal{U}[-\frac{\pi}{2}, \frac{\pi}{2}]$. All presented results for statistical CSI-optimization have been averaged over $\mathcal{S} = 50$ different setups. The remainder of the parameters are cited in Table 7.4.

Aiming to highlight the effectiveness of the proposed method, we simulate four different variable optimization strategies denoted as

1) $\boldsymbol{\rho}$ **optimization**: only the allocated power for the users, vector $\boldsymbol{\rho}$, is optimized;
2) $\boldsymbol{\rho}$ and $\mathbf{v}$ **optimization**: the power of the users $\boldsymbol{\rho}$, and the RIS phase-shift $\mathbf{v}$ are optimized;

Table 7.4 List of simulation parameters.

Parameter	Description
$P_{TX} \in \{20; 50\}$ dBm	Maximum power budget at BS
$\sigma^2 = -95$ dBm	Noise variance
$K = 10$	Numbers of UEs
$M_{max} = 256$	Maximum number of antennas at BS
$R_{min} = 1$ bps/Hz	Target rate
$N = 100$	Number of reflecting elements
$\varrho = 1.2$	Power inefficiency
$P_{FIX} = 9$ dBW	Fixed power consumption
$P_{BS} = 1$ W	Power consumed by a BS transceiver chain
$P_{RIS} = 10$ dBm	Power of RIS element configuration
$\beta_F = -25\log(d_F)$	RIS-BS path-loss model
$\beta_{g,k} = -10.6 - 20\log(d_{g,k})$	RIS-UEs path-loss model
$\beta_k = -35.6 - 40\log(d_k)$	BS-UEs path-loss model
$\hat{\kappa} = 3.5$	Rician coefficient
$\mathcal{T} = 500$ realizations	MCs
$S = 50$ setups	Setups

3) ρ and M **optimization**: the power of the users, ρ, and the number of antennas, M, are optimized;
4) ρ, $\boldsymbol{v}$, and M **optimization**: the power of the users, ρ, the number of antennas M, and the RIS phase-shift $\boldsymbol{v}$ are optimized sequentially.

It should be noted that when M, $\boldsymbol{v}$, or ρ are not optimized, they are assigned random feasible values. In addition, for comparison purposes, we also consider the **random all** allocation approach, where all variables (ρ, $\boldsymbol{v}$, and M) are assigned random feasible values. Specifically, power vector ρ is assigned values that satisfy constraint (7.30a), M obeys the constraint (7.30c), and $\boldsymbol{v}$ adheres to constraint (7.30d). In the forthcoming sections, the numerical results will be presented as follows: we explore the achievable EE of the RIS-aided M-MIMO system alongside its corresponding average number of active antennas, power consumption, and attained rates. This structured approach aims to provide a clear and organized understanding of the obtained outcomes we are interested in.

EE × Tx power budget: Figure 7.6 illustrates the achievable EE performance as a function of the transmit power budget at BS. We evaluated the performance for the four variable optimization strategies: (1) only ρ; (2) ρ and $\boldsymbol{v}$; (3) ρ and M; and

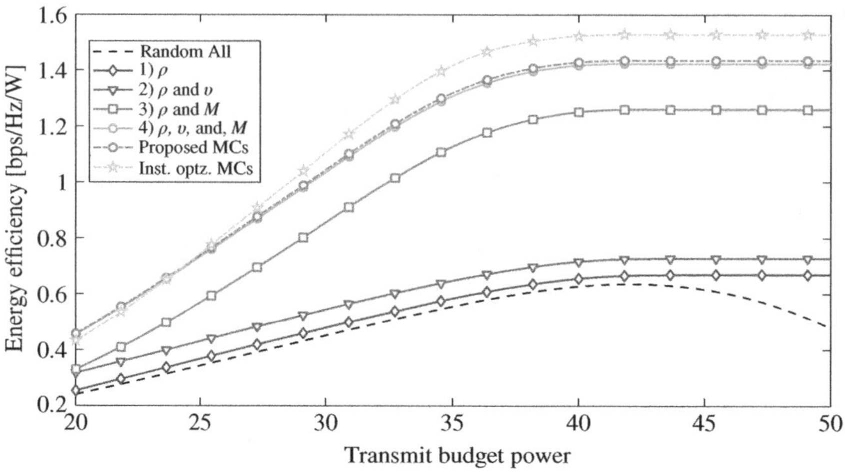

Figure 7.6 Average EE against the transmit power budget [dBm] at the BS (P_{TX}). Performance evaluation of the proposed algorithm for four different approaches: (1) only ρ; (2) ρ and $\boldsymbol{v}$; (3) ρ and M; and (4) ρ, $\boldsymbol{v}$, and M optimization, random ρ, $\boldsymbol{v}$, and M and the instantaneous optimization strategy with the optimization of $\boldsymbol{p}$ and $\boldsymbol{v}$ while considering M obtained by statistical CSI method.

(4) ρ, $\boldsymbol{v}$, and M optimization. As benchmarks, we adopted the "Random all" (ρ, $\boldsymbol{v}$, and M) strategy. Moreover, we adopted the instantaneous CSI-based optimization strategy for gap-performance comparison purposes, in which the optimization of ρ and $\boldsymbol{v}$ are considered. Specifically, the optimization of $\boldsymbol{v}$ has been implemented based on Algorithm 1 proposed in [32], which is designed explicitly for RIS-aided multi-user M-MIMO scenarios with ZF processing, the same adopted herein, the ρ optimization is solved by the CVX-solver for MATLAB, while M is adopted as the solution of statistical-CSI optimization. We also plotted the average instantaneous EE obtained from the proposed statistical CSI-based optimization approach, evaluated from the MCs, namely, "Proposed MCs" aiming to highlight the efficacy of the proposed method, indicated by the circle–dashed line in Figure 7.6. This curve differs from the circle–continuous line curve since, in the latter, the EE is computed analytically using the lower bound provided by Eq. (7.25). It should be noted that 500 MC simulations are carried out for each of the 50 setups. Additionally, we observe the tightness of the lower bound derived in [27].

From Figure 7.6, one can notice the consistency of the EE computed utilizing the ER lower bound, Eq. (7.25), which undoubtedly demonstrates that the ER $R_k > \tilde{R}_k$, $\forall k$. This is evident as the obtained EE MCS curve is slightly higher than the EE evaluated from the ER equation, demonstrating the tightness of the derivation proposed in [27] and confirming the potential of the statistical CSI-based optimization for enhancing EE. Regarding the different strategies for statistical CSI-based optimization, it is evident that the proposed algorithm for strategy (4) can outperform all other strategies with statistical CSI optimization in terms of EE enhancement, achieving a significant improvement by increasing the EE from ≈ 0.61 bps/Hz/W with the random solution (in the maximum value) to ≈ 1.44 bps/Hz/W. This translates to a remarkable EE gain of approximately 136%. Since it optimizes all variables $\boldsymbol{\rho}$, M, and $\boldsymbol{v}$ sequentially and iteratively, such a result was expected. Besides, given the high value assumed by M_{max} as in practical RIS-aided M-MIMO scenarios, there is a significant probability of obtaining a high value of M in the random realizations for both strategies (1) and (2) as depicted in Figure 7.7. Thus, when M is high, the energy consumption significantly increases, implying poor EE. Accordingly, optimizing M becomes of paramount importance.

Also, one can infer from Figure 7.6 that the impact of optimizing the RIS phase-shift angles is not negligible on the attainable EE, even considering the statistical CSI optimization approach. It is noteworthy that the achieved EE improvement in scenarios in which the number of active BS antennas M assumes low values (M optimized) is considerably greater compared to those scenarios where M assumes high values. This can be readily verified by examining the obtained gain in the EE curve of strategy (1) compared to the curve of strategy (2), as well as in the curve of strategy (3) regarding the curve of strategy (4).

Number of active antennas (M) $\times$ Tx power budget: Figure 7.7 shows the number of average active antennas M for the optimization strategies (1), (2), (3), and (4), as well as for the Random all scheme. It is noteworthy that strategies Random all, (1), and (2) exhibit a significantly higher average number of active

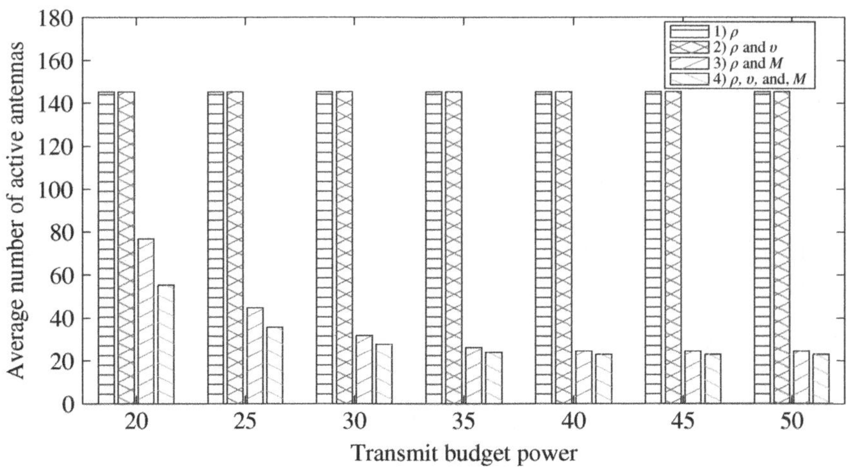

Figure 7.7 Average number of antennas against Tx power budget in dBm for four optimization strategies.

antennas compared to the average number of active antennas in strategies (3) and (4). It occurs due to the random assignment for M in each setup S, and since strategies Random all, (1), and (2) do not optimize M, there is a high probability for M to be higher than that in strategies (3) and (4). This finding helps explain why strategies Random all, (1), and (2) attain poor EE as illustrated in Figure 7.6. Moreover, optimizing the RIS phase-shift in strategy (4) remarkably reduces the average number of active BS antennas, mainly in low-power regimes, for example, by comparing strategy (3) with strategy (4), at $P_{TX} = 20$ dBm, the average number of antennas is reduced from 153 to 79. These results highlight the potential of RIS deployment and optimization in such scenarios.

Remark: Based on Figure 7.6, one can see the potential to achieve higher EE through the statistical CSI optimization approach since this methodology enables us to reduce the overhead associated with the phase-shift optimization during each channel coherence time while maintaining the $\boldsymbol{\upsilon}$ fixed over multiples channel coherence time, with a minimal performance loss. As a result, the EE marginally decreases from 1.53 to 1.44 bps/Hz/W, representing a loss of $\approx 5.88\%$, in high-power regimes.

Percentage of Tx power budget deployed × Tx power budget available: Figure 7.8 depicts the average percentage of the transmit power budget utilized for transmission against the transmit power budget. In the low-budget power regime, all strategies deplete the available power budget; however, as the transmit power budget increases beyond 40 dBm, the algorithm does not allocate to the

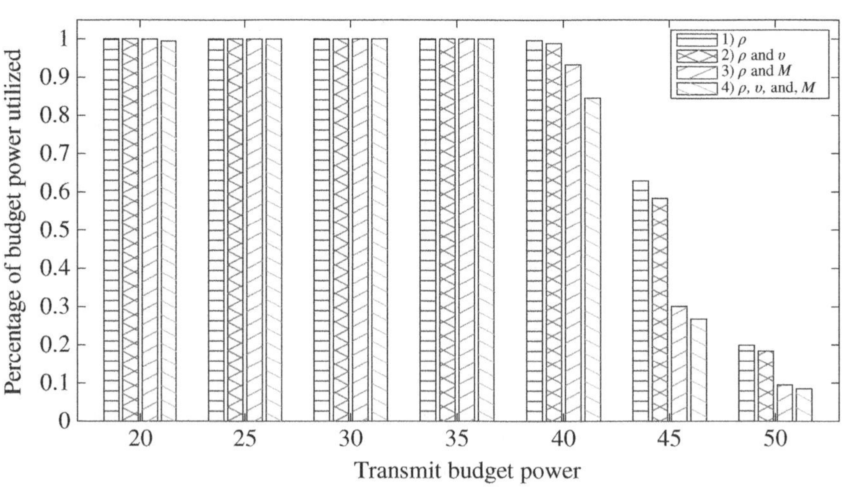

Figure 7.8 Percentage of the transmit power budget dBm utilized against the transmit power budget at BS.

UEs all the power budget available. This behavior is expected since, aiming to achieve the maximum EE, the available power budget should not be fully utilized.

Additionally, this strategy is directly accountable for the constant EE observed in high-power budget regimes, as presented in Figure 7.6. Furthermore, it becomes apparent that optimizing the number of antennas M is of utmost importance to save a significant amount of power, reducing the power consumption substantially, from 100% (as in the Random all Strategy case) to 84%, 26%, and 9% in the best strategy, for $P_{\mathrm{TX}} = 40$, 45, and 50, dBm respectively. Consequently, the proposed strategies (1), (2), (3), and (4) also imply high progressive EE gains, as presented in Figure 7.6. Moreover, one can see the substantial impact on power consumption reduction caused by optimizing the RIS phase-shift. This power reduction corroborates the potential of RIS deployment and configuration, resulting in a remarkable EE gain that cannot be obtained in non-RIS scenarios.

Figure 7.9 reveals the complementary cumulative distribution function (CCDF) of the UE' ER $\tilde{R}_k$ over $\mathcal{S} = 50$ different realizations setups, represented by the solid lines. Besides, dashed lines indicate the computed instantaneous UE' rate, R_k, over $\mathcal{T} = 500$ MCS for direct link $\mathbf{H}$ and $\mathbf{F}_{\mathrm{NLoS}}$, in the range of P_{TX} from 20 to 50 dBm. Notice that the proposed statistical CSI-based optimization method necessarily obeys the constraint (7.30b) for all strategies solutions. This fulfillment is primordial to guarantee appropriate QoS for all UEs. Moreover, one can see that strategies "Random all," (1), and (2) are capable of providing higher rates than strategies (3)

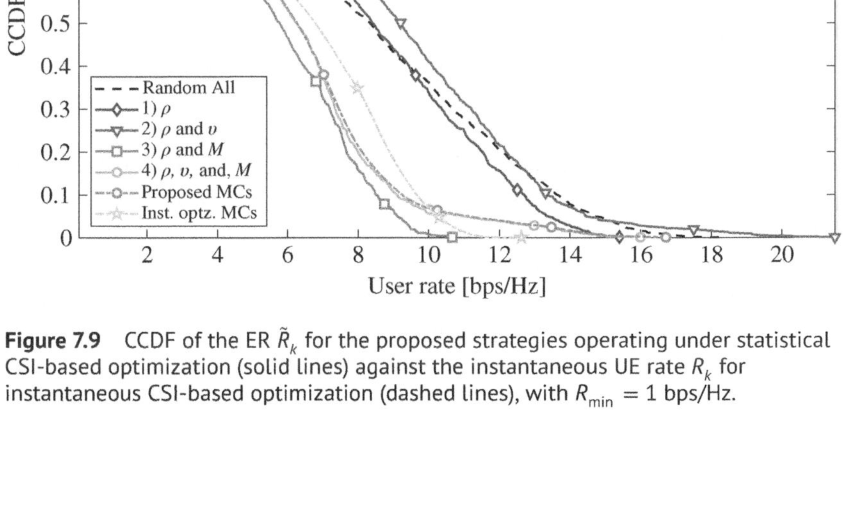

Figure 7.9 CCDF of the ER $\tilde{R}_k$ for the proposed strategies operating under statistical CSI-based optimization (solid lines) against the instantaneous UE rate R_k for instantaneous CSI-based optimization (dashed lines), with $R_{\mathrm{min}} = 1$ bps/Hz.

and (4). This is due to the consumed power in strategies (1) and (2) being higher than that in strategies (3) and (4), as shown in Figure 7.8. However, one can see that although strategy (4) has a much lower number of active antennas and lower-power consumption than strategy (3), strategy (4) still has a feasible probability of attaining higher rates than strategy (3). This is due to the RIS phase–shift optimization, which can enhance some UEs by maximizing the ER. We can also notice from Figure 7.9 that the instantaneous UE rate R_k obtained through statistical CSI-based optimization can achieve lower rates than $R_{\min}$, implying a violation of (7.30b), in contrast to the data rates obtained through the instantaneous CSI-based optimization (dashed curve with star symbol), which are always higher or equal than $R_{\min}$.

7.4 Energy Efficiency in Demand-Adaptive XL-MIMO Systems

To reduce the operational cost and the impact on the environment, mobile networks must implement strategies to adaptively turn on/off system modules according to the network demand and the conditions of the wireless channel. For instance, an XL-MIMO system deployed to serve a densely populated area should be capable of deactivating part of its massive array of antennas to improve the EE in moments of low system load. In this specific case, AS constitutes a pertinent allocation strategy because the power-hungry RF circuits comprising each antenna of the huge XL-MIMO array are significant contributors to the power consumption at the BS. The main challenge is deciding the number of antennas and which specific antennas must be activated to balance the communication performance experienced by the UEs and the energy savings. Given that, this subject has been discussed largely in the recent literature, being efficiently addressed by the contributions of [2, 11]. In short, these works propose methods and algorithms for AS in XL-MIMO to improve the system SE and EE, assuming different constraints involving the network architecture and addressing several communication scenarios. Predominantly, they use heuristic and evolutionary algorithms to design the allocation strategies, since AS configures an intractable decision problem in XL-MIMO due to the massive number of antennas.

In this section, we analyze how AS can be adopted in XL-MIMO systems to improve the EE, with emphasis on network deployments where the number of UEs served by the BS oscillates widely. The analysis is based on the concept of a *demand-adaptive XL-MIMO system*, which, equipped with AS and other resource allocation strategies, scales the BS capability according to the network demand, balancing the communication performance and EE. Specifically, we propose methods for joint AS and PA aiming to maximize the EE in the DL of an XL-MIMO system. The methods include a GA, a low-complexity heuristic based on the channel power attenuation related to each BS antenna, and a local search algorithm. These three methods are numerically assessed and compared in terms of EE and SR, showing that the gains in EE are consistently accompanied by high data rates.

7.4.1 System Model

We consider the XL-MIMO system depicted in Figure 7.10, serving $K \in \mathbb{Z}_+$ single-antenna UEs with a BS equipped with $M \in \mathbb{Z}_+$ antennas, where $M \gg K$, and operating in a narrowband channel with bandwidth $B > 0$. The UL and DL are multiplexed following the TDD protocol, in frames with duration equal to the channel coherence time $T_c > 0$. Our analysis focuses on the DL, where the BS uses a precoding strategy to spatially multiplex the data intended for all K UEs based on their CSI. Essential for attaining reliable transmission, the CSI can be acquired from UL training pilots sent by the UEs at the frame start [23]. However, to simplify the analysis, we assume that the BS has perfect CSI knowledge. Throughout this section, we use $m \in \{1, \dots, M\}$ to denote the indices of the BS antennas and $k \in \{1, \dots, K\}$ to denote the indices of the UEs.

In the BS array, each antenna has a dedicated *transceiver chain*, comprising all the circuits necessary for signal transmission and reception, for example, filters, amplifiers, converters, and mixers. Since around 1 W is required to run

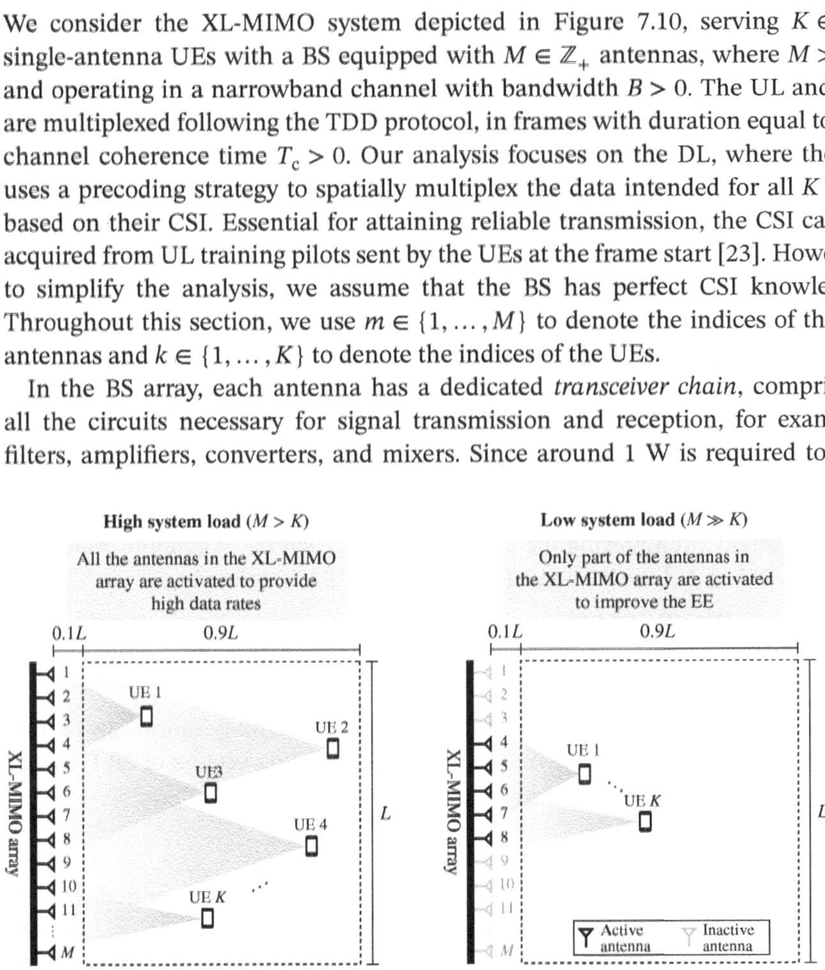

Figure 7.10 Diagram of the studied demand-adaptive XL-MIMO system. Under high system load (left), all the antennas in the XL-MIMO array are activated to provide high data rates in environments with high UE density. In moments of low user load (right), activating only a small subset of antennas is sufficient to provide the required data rate to the few UEs, so the other antennas can be deactivated to improve the system EE.

each transceiver chain, the power consumption of an XL-MIMO array, which is equipped with hundreds to thousands of antennas, represents a significant proportion of the power consumed by the BS. Given this, to save power, the BS turns on only part of the antennas of its array, having less than M active transceiver chains. To determine the active antennas, the BS employs an AS strategy aiming to optimize the system EE. The AS can help to improve the EE in several situations. For instance, antennas with deep faded channels can be turned off, as the low powers received by them contribute marginally to the signal-to-noise ratio (SNR) improvement. Moreover, under low network load, a small subset of antennas can be activated, saving energy and still providing the degrees of freedom required to serve all the UEs. In this case, the challenge is to find the set of active antennas that provides a good trade-off between communication performance, say, SE, and power consumption.

Let $\mathbf{D} \in \{0,1\}^{M \times M}$ be a diagonal matrix with $[\mathbf{D}]_{m,m} = 1$ if the antenna m is active and 0 otherwise. Hence, the set of active antennas is defined by $\mathbf{D}$, while $\text{tr}(\mathbf{D}) \leq M$ is the total active antennas. Since only the active antennas can be used by the BS to transmit and receive signals, the DL received signal at the UEs is

$$\mathbf{y} = \mathbf{H}^H \mathbf{D} \mathbf{x} + \mathbf{n}, \tag{7.74}$$

where $\mathbf{H} \in \mathbb{C}^{M \times K}$ is the matrix with the channels of all the BS antennas, $\mathbf{x} \in \mathbb{C}^M$ is the signal transmitted by the BS, and $\mathbf{n} \in \mathbb{C}^K$ is the AWGN at the receivers such that $\mathbf{n} \sim \mathcal{CN}(\mathbf{0}, \sigma_{\text{ue}}^2 \mathbf{I})$.

To null the interference among the served UEs, the BS employs the ZF precoder, spatially multiplexing the data intended to each UE. Therefore, assuming $\text{rank}(\mathbf{H}^H \mathbf{D} \mathbf{H}) \geq K$,[3] the ZF precoding vector for UE k is defined as:

$$\mathbf{w}_k = \frac{\mathbf{D}\mathbf{H}[(\mathbf{H}^H \mathbf{D} \mathbf{H})^{-1}]_{:,k}}{\sqrt{[(\mathbf{H}^H \mathbf{D} \mathbf{H})^{-1}]_{k,k}}}, \tag{7.75}$$

where the denominator ensures the vector has a unitary norm. By letting $\mathbf{W} = [\mathbf{w}_1 \ \cdots \ \mathbf{w}_K]$ denote the matrix with the precoding vectors of all UEs, the transmitted signal is computed by

$$\mathbf{x} = \mathbf{W} \mathbf{P}^{\frac{1}{2}} \mathbf{s}, \tag{7.76}$$

where $\mathbf{P} \in \mathbb{R}^{K \times K}$ is the diagonal matrix with the BS transmitted powers $[\mathbf{P}]_{k,k} > 0, \forall k$, and $\mathbf{s} \in \mathbb{C}^K$ is the transmitted data such that $\mathbb{E}\{|[\mathbf{s}]_k|^2\} = 1, \forall k$ and $\mathbb{E}\{[\mathbf{s}]_k^* [\mathbf{s}]_{k'}\} = 0, \forall k \neq k'$. Notice that, due to the presence of $\mathbf{D}$ in the precoding vector formulation in (7.75), the entries of $\mathbf{x}$ corresponding to the inactive

3 This condition guarantees that the channel resulted from the active antennas provides enough degrees of freedom to spatially multiplex all K UEs. It can be satisfied by ensuring that $\mathbf{H}^H \mathbf{D} \mathbf{H}$ has linearly-independent columns, which can only be achieved by activating at least K antennas.

antennas are null, since they transmit no signals. Also, the BS transmits with the power budget $P_{\text{max}} > 0$, so the DL powers are allocated meeting the constraint

$$\text{tr}(\mathbf{P}) \leq P_{\text{max}}. \tag{7.77}$$

From the received and transmitted signals in (7.74) and (7.76), respectively, the SINR of the UE k is calculated by

$$\text{SINR}_k = \frac{[\mathbf{P}]_{k,k}}{[(\mathbf{H}^H\mathbf{D}\mathbf{H})^{-1}]_{k,k}\sigma_{\text{ue}}^2}. \tag{7.78}$$

Therefore, for allocated powers that satisfy (7.77), the system SR is calculated by

$$\text{SR}(\mathbf{D},\mathbf{P}) = \sum_{k=1}^{K} B\log_2\left(1 + \frac{[\mathbf{P}]_{k,k}}{[(\mathbf{H}^H\mathbf{D}\mathbf{H})^{-1}]_{k,k}\sigma_{\text{ue}}^2}\right) \text{ [bit/s]}. \tag{7.79}$$

Notice that the SR essentially depends on the allocated power for each UE in $\mathbf{P}$, the channel fading realization given by $\mathbf{H}$, and especially on the active antennas indicated by $\mathbf{D}$. This showcases the impact of activating different sets of antennas on the system performance, requiring proper AS strategies to achieve the target requirements.

7.4.1.1 Power Consumption Model

Now, we describe the power consumption model for the XL-MIMO system. Focusing on the DL, the following terms contribute to the system's power consumption. $P_{\text{tx}}^{\text{d}} > 0$ denotes the power for the BS to irradiate the data signal, while $P_{\text{tx}}^{\text{tr}} > 0$ for the UEs to irradiate the UL training pilots. $P_{\text{ce}}(\mathbf{D}) > 0$ denotes the power for channel estimation at the BS. $P_{\text{c/d}}(\mathbf{D},\mathbf{P}) > 0$ is the power for data coding and decoding at the BS and the UEs, respectively. $P_{\text{bh}}(\mathbf{D},\mathbf{P}) > 0$ is the load-dependent power for backhaul operation, while $P_{\text{pr}}(\mathbf{D}) > 0$ is the power for linear processing, both dissipated at the BS. $P_{\text{tc}}(\mathbf{D}) > 0$ is the power consumed by the transceiver chains at the BS and UEs, while $P_{\text{fix}} > 0$ is the power consumed by the infrastructure and baseband processors at the site. Therefore, the system power consumption can be expressed by

$$\begin{aligned} P(\mathbf{D},\mathbf{P}) &= P_{\text{tx}}^{\text{d}} + P_{\text{tx}}^{\text{tr}} + P_{\text{ce}}(\mathbf{D}) + P_{\text{c/d}}(\mathbf{D},\mathbf{P}) + P_{\text{bh}}(\mathbf{D},\mathbf{P}) \\ &\quad + P_{\text{pr}}(\mathbf{D}) + P_{\text{tc}}(\mathbf{D}) + P_{\text{fix}} \text{ [W]}. \end{aligned} \tag{7.80}$$

$P_{\text{ce}}(\mathbf{D})$ and $P_{\text{pr}}(\mathbf{D})$ depend on the computational complexity of the algorithms adopted for channel estimation and computation of the transmitted signal in (7.76). Such a complexity is proportional to the number of active antennas $\text{tr}(\mathbf{D})$, and is characterized in [2] and [11]. The remaining terms comprising the system power consumption are specified in the sequence.

Starting by the powers for signal irradiation, let $\xi_{\text{bs}}, \xi_{\text{ue}} \geq 1$ denote the inverse of the power amplifier efficiency at the BS and the UEs, respectively. Also, let

$P_{tr} > 0$ denote the transmit power of a single UE pilot. The effective RF power emitted by the transmitters is less than the input power at the power amplifier due to its efficiency. Hence, the power spent by the BS to irradiate the data signal is $P_{tx}^{d} = \xi_{bs} P_{max}$. Similarly, the total power spent by the UEs to transmit their respective UL training pilots is $P_{tx}^{tr} = K\xi_{ue} P_{tr}$.

The powers for coding, decoding, and backhaul operation depend on the system SR. Let $P_{cod} > 0$ denote the coding power density, $P_{dec} > 0$ denote the decoding power density, and $P_{bhd} > 0$ denote the backhaul traffic power density, all measured in W/(bit/s). The power for coding and decoding is computed by

$$P_{c/d}(\mathbf{D}, \mathbf{P}) = \mathrm{SR}(\mathbf{D}, \mathbf{P})(P_{cod} + P_{dec}). \tag{7.81}$$

Meanwhile, the load-dependent power for backhaul operation is computed by

$$P_{bh}(\mathbf{D}, \mathbf{P}) = \mathrm{SR}(\mathbf{D}, \mathbf{P}) P_{bhd}. \tag{7.82}$$

The power consumed by the transceiver chains depends essentially on the number of antennas at the BS and the number of single-antenna UEs. Let $P_{bs} > 0$ denote the power consumed by a single transceiver chain at the BS, while $P_{ue} > 0$ represents the same, but at an UE. Also, let $P_{syn} > 0$ denote the power spent by the local oscillator at the BS. Hence, the power consumed by the transceiver chains is computed by

$$P_{tc}(\mathbf{D}) = P_{syn} + \mathrm{tr}(\mathbf{D}) P_{bs} + K P_{ue}. \tag{7.83}$$

7.4.1.2 Energy Efficiency

Looking at the derived metrics, the system SR is used to evaluate the system performance from the communication perspective, while the system power consumption from the energy and cost perspectives. Now, we derive the EE metric, which can be used to evaluate the trade-off between the communication performance and the energy expenditure the system achieves. From the SR in (7.79) and the power consumption model in (7.80), the system EE is

$$\mathrm{EE}(\mathbf{D}, \mathbf{P}) = \frac{\mathrm{SR}(\mathbf{D}, \mathbf{P})}{P(\mathbf{D}, \mathbf{P})} \ [\text{bit/J}]. \tag{7.84}$$

In summary, the EE provides the number of bits that can be transmitted in the DL for each Joule of energy the system spends. With this metric, we can compare the trade-offs offered by different resource allocation methods.

7.4.2 Problem Formulation

Our objective herein is to develop a method to balance the DL communication performance and the power consumed by the system, leveraging energy resources. Optimizing the system EE in (7.84) is a promising path to reach the balance between these two aspects. On this basis, we propose the following optimization problem for maximizing the system EE by jointly selecting the active antennas at the XL-MIMO array and allocating the transmit power intended for each UE:

$$\underset{\mathbf{D},\mathbf{P}}{\text{maximize}} \quad \text{EE}(\mathbf{D}, \mathbf{P}), \tag{7.85}$$

$$\text{subject to} \quad \text{rank}(\mathbf{H}^H \mathbf{D} \mathbf{H}) \geq K, \tag{7.85a}$$

$$\text{tr}(\mathbf{P}) \leq P_{\max}, \tag{7.85b}$$

$$[\mathbf{D}]_{m,m} \in \{0, 1\}, \ \forall m, \quad [\mathbf{D}]_{m,m'} = 0, \ \forall m \neq m', \tag{7.85c}$$

$$[\mathbf{P}]_{k,k} > 0, \ \forall k, \quad [\mathbf{P}]_{k,k'} = 0, \ \forall k \neq k'. \tag{7.85d}$$

In such a problem, the constraint (7.85a) is a necessary condition for the channels of the active antennas to provide sufficient degrees of freedom to serve all UEs. Meanwhile, (7.85b) is the DL power budget. Moreover, constraints (7.85c) preserve the structure of the matrix with the active antenna indicators, $\mathbf{D}$. In contrast, constraints (7.85d) do the same for the matrix with the allocated powers, $\mathbf{P}$.

There are technical challenges that preclude the computation of an optimal solution for the problem (7.85) if it exists. One of them is the fact that the optimization variables $\mathbf{D}$ and $\mathbf{P}$ are interdependent, as it can be observed in the SR expression, which is present both at the EE numerator and denominator. This non-trivial relationship between the optimization variables and the objective function makes it difficult to manipulate the problem and find a tractable equivalent formulation. Moreover, the optimization variable $\mathbf{D}$ has binary-constrained diagonal entries, which makes (7.85) to be an instance of *integer programming*, a class of NP-complete problems [33]. Hence, a promising path to address (7.85) is to approximate a suboptimal solution by adopting heuristic methods. We follow this approach by firstly decoupling the optimization variables of (7.85) to formulate two different subproblems, one for PA and the other for AS. Then, we propose a framework to approximate a solution for the original problem by sequentially solving the derived subproblems.

To formulate the first subproblem, let $\tilde{\mathbf{P}}$ be the diagonal matrix with the allocated transmit powers already decided by the BS. Therefore, the AS subproblem is defined by

$$\underset{\mathbf{D}}{\text{maximize}} \quad \text{EE}(\mathbf{D}, \tilde{\mathbf{P}}), \tag{7.86}$$
$$\text{subject to} \quad \text{(7.85a) and (7.85c).}$$

Notice that such a problem is still NP-complete since the optimization variable $\mathbf{D}$ is binary-constrained. For this reason, we use algorithms to approximate a suboptimal solution by searching the set of feasible solutions. Analyzing the constraints of the subproblem (7.86), we obtain the set of feasible solutions given by

$$\mathcal{D} = \left\{\mathbf{D} \in \{0,1\}^{M\times M} \mid (7.85\text{a}) \text{ and } (7.85\text{c})\right\}. \tag{7.87}$$

Since M is large in XL-MIMO arrays and $|\mathcal{D}|$ scales exponentially with M, there is no sufficient computational resources at the BS to check all solutions in $\mathcal{D}$ to find one that maximizes the objective function of (7.86). Hence, to approximate a solution with feasible computational complexity, we propose adopting methods such as the GA and local search. These algorithms try to improve the value of the objective function at each iteration by searching sub-spaces of the original feasible space of the problem, created based on solutions observed in the previous iterations. The detailed procedures for these algorithms are provided in the next subsection.

Regarding the second subproblem, let $\tilde{\mathbf{D}}$ denote the matrix with the AS scheme adopted by the BS. The allocated powers in $\mathbf{P}$ affect three terms in the EE expression, the SR at the numerator, in addition to $P_{\text{c/d}}(\mathbf{D},\mathbf{P})$ and $P_{\text{bh}}(\mathbf{D},\mathbf{P})$ at the denominator. Since $\mathbf{P}$ has more impact on the numerator, we formulate the PA subproblem to maximize the numerator of the EE in (7.84), that is,

$$\begin{aligned} \underset{\mathbf{P}}{\text{maximize}} \quad & \text{SR}(\tilde{\mathbf{D}},\mathbf{P}), \qquad (7.88)\\ \text{subject to} \quad & (7.85\text{b}), \\ & [\mathbf{P}]_{k,k} \geq 0,\ \forall k, \quad [\mathbf{P}]_{k,k'} = 0,\ \forall k \neq k'. \qquad (7.88\text{a}) \end{aligned}$$

Importantly, the constraints (7.88a) are relaxations of the original constraints (7.85d), admitting $[\mathbf{P}]_{k,k} = 0$ for any k. When an UE has allocated power equal to zero, we consider that the BS decided not to transmit DL data for it. We assume this hypothesis to get a simple closed-form solution for the PA subproblem. Specifically, problem (7.88) is convex and can be solved optimally by the *water filling* algorithm. The algorithm computes the optimal solution by finding the value of $\mu \geq 0$ that satisfies [6]

$$\sum_{k=1}^{K} \max\left(0, \mu - [(\mathbf{H}^H\tilde{\mathbf{D}}\mathbf{H})^{-1}]_{k,k}\sigma_{\text{ue}}^2\right) - P_{\max} = 0. \tag{7.89}$$

Then, the allocated powers are computed by

$$[\mathbf{P}^\star]_{k,k} = \max\left(0, \mu - [(\mathbf{H}^H\tilde{\mathbf{D}}\mathbf{H})^{-1}]_{k,k}\sigma_{\text{ue}}^2\right), \forall k. \tag{7.90}$$

7.4.3 Proposed Solution and Algorithm

7.4.3.1 Genetic Algorithm

The GA is an optimization method inspired by the concepts of natural selection and genetics from biology [34]. In such a method, individuals described by encoded vectors represent candidate solutions for the optimization problem at hand. Iteratively, these individuals are combined and compared through a cost function, deploying an intelligent search procedure to find a near-optimal solution for the optimization problem.

To approximate a solution for (7.86) using the GA, we need to define a cost function. Used to evaluate the individuals, the values of the cost function are essential to rank the candidate solutions and identify the best one. Based on the objective function of (7.86) and knowing that the standard GA is designed for minimization problems, the adopted cost function is given by

$$f(\mathbf{D}) = -\mathrm{EE}(\mathbf{D}, \mathbf{P}^{\star}), \tag{7.91}$$

where $\mathbf{P}^{\star}$ is as in (7.90), but with $\tilde{\mathbf{D}} = \mathbf{D}$. Hence, in the GA, the output solution is the one that, among all the others generated during the algorithm iterations, yields the smallest value for $f(\mathbf{D})$.

In the GA used to solve problem (7.86), the individuals are represented by binary strings of length M, containing the diagonal of the antenna indicators matrix $\mathbf{D}$. Moreover, the algorithm comprises the three standard steps: *selection*, *crossover*, and *mutation*. The standard GA is devised to address unconstrained and real-valued optimization problems, requiring additional mechanisms to tackle constrained ones such as (7.86). To that end, we adopt the GA of [35], developed to solve constrained mixed integer optimization problems. In this algorithm, *Laplace crossover* is used, generating the offspring based on a random variable that follows a Laplace distribution with location parameter $a_{\mathrm{ga}} \in \mathbb{R}$ and scale parameter $b_{\mathrm{ga}} > 0$. Specifically, the parents submitted to the crossover step are selected via *tournament selection*, with tournaments played by $k_{\mathrm{ga}} \in \mathbb{Z}_+$ individuals. Furthermore, *power mutation* is adopted, generating the muted offspring from an exponentially distributed random variable with parameter $p_{\mathrm{ga}} \in (0, 1)$, called the index of mutation. The detailed expressions for the crossover and mutation steps can be found in [35, Section 2]. Each GA iteration evaluates a total of $N_{\mathrm{ga}} \in \mathbb{Z}_+$ individuals. From the pool of individuals, $N_{\mathrm{ga}}^{\mathrm{e}} \in \mathbb{Z}_+$ are the elite offspring, $N_{\mathrm{ga}}^{\mathrm{c}} \in \mathbb{Z}_+$ are generated by crossover, and $N_{\mathrm{ga}}^{\mathrm{m}} \in \mathbb{Z}_+$ are generated by mutation such that $N_{\mathrm{ga}} = N_{\mathrm{ga}}^{\mathrm{e}} + N_{\mathrm{ga}}^{\mathrm{c}} + N_{\mathrm{ga}}^{\mathrm{m}}$.

Aiming to ensure that the individuals at each iteration correspond to feasible solutions, after crossover and mutation steps, all the optimization variables with non-binary values are assigned either with 0 or 1 with the same probability. Moreover, all individuals that don't satisfy (7.85a), that is, present $\mathrm{rank}(\mathbf{H}^H\mathbf{D}\mathbf{H}) < K$, are replaced by random individuals that meet $\mathrm{rank}(\mathbf{H}^H\mathbf{D}\mathbf{H}) \geq K$. The GA execution is limited to $T_{\mathrm{ga}}^{\mathrm{max}} \in \mathbb{Z}_+$ iterations. Nevertheless, if the minimum cost value observed in the pool of individuals does not improve after $T_{\mathrm{ga}}^{\mathrm{stall}} \in \mathbb{Z}_+$ iterations, the algorithm is interrupted prematurely.

7.4.3.2 Heuristic Based on the Channel Gain

Measuring the power attenuation caused by the wireless channel is a simple way to verify the quality of a radio link. Given that, we propose a low-complexity method to select the active antennas based on their channel gain, activating the antennas that most contribute to the power of the DL signal received by the UEs. To this end, we compute the overall channel gain exerted by the BS antenna m,

$$\varpi_m = \|[\mathbf{H}]_{m,:}\|_2^2. \tag{7.92}$$

The value of ϖ_m can be interpreted as the total power received by the UEs if the BS antenna m transmits solely with unitary power. Therefore, the antennas that have higher ϖ_m are the ones with the best channel condition, considering their contribution to the signals of all UEs jointly. Let $M_a \in \mathbb{Z}_+$ denote the number of antennas to be activated. Hence, the algorithm selects the antennas corresponding to the largest M_a values from the set $\{\varpi_1, \varpi_2, \dots, \varpi_M\}$.

7.4.3.3 Local Search

Local search is a simple optimization strategy that relies solely on randomization to find good-quality solutions to the optimization problem, commonly resulting in low-complexity algorithms [36]. Despite the strategy is tied to the structure of the problem at hand, the local search can be summarized by a three-step procedure. Initially, a base solution for the optimization problem is obtained by some heuristic, and the value of the objective function for this solution is calculated. Then, certain local modifications are randomly applied to the base solution, and the value of the objective function for it is calculated as well. If the modifications improve the objective function, the base solution is replaced by the modified one. Otherwise, the base solution is maintained. Applying modifications and preserving the improved solutions repeats until some convergence criterion is satisfied or if the algorithm reaches the limit number of iterations.

The proposed local search algorithm for AS is built based on the introduced three-step procedure. As in the GA, the solution is represented by a binary string containing the diagonal of the matrix $\mathbf{D}$, while the solution quality is measured by (7.91). Moreover, the initial solution is computed by the heuristic based on the channel gain. Following the approach of [2], the local modifications are parameterized by the Hamming distance $d_{\text{Ham}} \in \mathbb{Z}_+$. Therefore, the modified solutions are calculated by randomly flipping d_{Ham} entries of the base solution. It is worth mentioning that the modified solution is discarded if it doesn't satisfy the constraint (7.85a). The local search is interrupted when the algorithm reaches the predefined number of iterations.

7.4.4 Numerical Results and Discussions

To analyze how the proposed joint AS and PA methods can improve the EE of XL-MIMO systems, we present numerical simulations. The adopted scenario for these simulations is similar to the one in [2], where the communication cell is a square of side $L > 0$, and the BS array is a ULA placed along one of the cell sides. Specifically, the M BS antennas are evenly spaced over the distance L. For simplicity, we assume that the BS array and the UEs lie in the xy-plane of the coordinate system. Moreover, the UEs' x and y coordinates are random for each realization, considering that the UEs are at the minimum distance of $0.1L$ from the BS array and that their positions are uniformly distributed over the cell area. All the simulation parameters, including the ones related to the power consumption, are given in Table 7.5. Among the proposed methods, we evaluate the performance of AS via GA, via local search, and the heuristic based on the channel gain. Regarding the GA, the population-related parameters are $N_{\text{ga}} = 100$, $N_{\text{ga}}^{\text{e}} = 5$, $N_{\text{ga}}^{\text{c}} = 50$, and $N_{\text{ga}}^{\text{m}} = 45$; the parameters related to crossover and selection are $a_{\text{ga}} = 0$, $b_{\text{ga}} = 0.35$, and $k_{\text{ga}} = 4$; the mutation-related parameter is $p_{\text{ga}} = 4$; and the parameters defining the stop criteria are $T_{\text{ga}}^{\text{max}} = 1000$ and $T_{\text{ga}}^{\text{stall}} = 50$. Particularly for the local search, we set $d_{\text{Ham}} = 10$ to generate the modified solutions, and a limit of 100 iterations. Furthermore, as benchmarks, we evaluate the performance achieved by random AS and activating the full array to serve the UEs, which represents the standard XL-MIMO system operation, without AS.

Table 7.5 List of simulation parameters.

Parameter	Description
$L = 30$ m	Length of the cell side
$B = 20$ MHz	Transmission bandwidth
$\sigma_{\text{ue}}^2 = -96$ dBm	Noise power
$\overline{\beta} = 10^{-7}$	Channel gain at the reference distance
$\gamma = 3$	Path loss exponent
$\xi_{\text{bs}} = 2.56$	Inverse of the BS power amplifier efficiency
$\xi_{\text{ue}} = 2.00$	Inverse of the UE power amplifier efficiency
$P_{\text{tr}} = 20$ mW	Transmit power of a single UE pilot
$P_{\text{cod}} = 0.1$ W/Gbps	Coding power density
$P_{\text{dec}} = 0.8$ W/Gbps	Decoding power density
$P_{\text{bhd}} = 0.25$ W/Gbps	Backhaul traffic power density
$P_{\text{bs}} = 1$ W	Power consumed by a BS transceiver chain
$P_{\text{ue}} = 0.1$ W	Power consumed by an UE transceiver chain
$P_{\text{syn}} = 2$ W	Power consumed by the BS oscillator
$P_{\text{fix}} = 18$ W	Power consumed by the site infrastructure and baseband processors

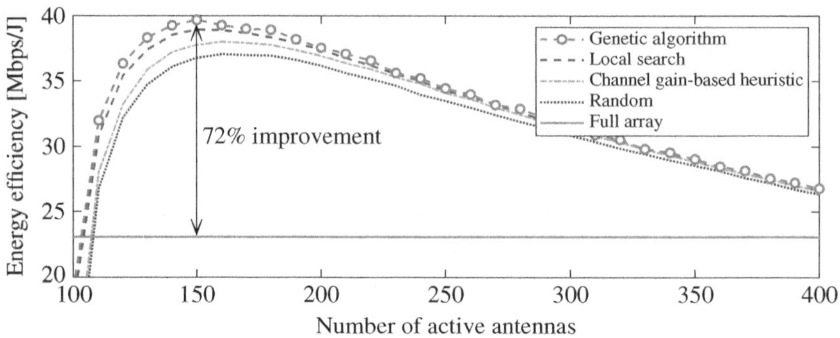

Figure 7.11 EE as a function of the number of active antennas achieved by the proposed joint AS and PA methods. In this result, the number of UEs is $K = 100$, the number of BS antennas is $M = 500$, and the transmit power budget is $P_{max} = 24$ dBm.

Initially, Figure 7.11 depicts the EE obtained by the proposed joint AS and PA methods as a function of the number of active antennas. For each point in the curves, the AS methods are employed to select, from the total antennas of the BS array, the exact number of antennas to be activated. The result shows that the proposed AS methods improve significantly the EE of the XL-MIMO system. Specifically, the AS via GA can improve the EE by up to 72% compared to the one obtained by activating the entire array. Moreover, comparing the four AS methods, the peak EE value achieved by the GA outperforms the local search, the heuristic based on the channel gain, and the random selection by 2%, 4%, and 7%, respectively. In summary, this result corroborates the benefit of AS to leverage the energy resources when the system load is low, reducing the power consumed by the massive XL-MIMO arrays.

In the sequel, Figure 7.12 illustrates the SR achieved by the proposed joint AS and PA methods as a function of the number of active antennas, considering 100 UEs. Observing the point where the GA reaches the EE peak, when 150 antennas are activated, the SR decreases by 39% from the SR achieved with the entire array. Therefore, to exploit the EE improvement offered by AS, we must give up on the potential communication performance of the XL-MIMO array. However, it is worth remembering that, in the studied case, the purpose of AS is to adapt the communication system to save energy in the regime of low system load, that is, when the system does not need to explore the full potential of XL-MIMO.

Now, considering only 10 UEs, Figure 7.13 depicts the EE and SR achieved by the joint AS and PA methods as functions of the number of active antennas. Noticeably, the AS via GA increases the EE by 583% compared to the full array activation, showing that the benefits of AS are enhanced as the system load, calculated by the ratio K/M, decreases. Moreover, the peak EE achieved by GA is

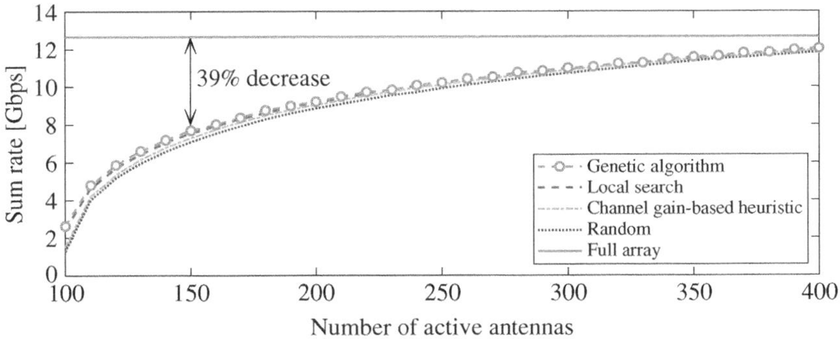

Figure 7.12 SR as a function of the number of active antennas achieved by the proposed joint AS and PA methods. In this result, the number of UEs is $K = 100$, the number of BS antennas is $M = 500$, and the transmit power budget is $P_{max} = 24$ dBm.

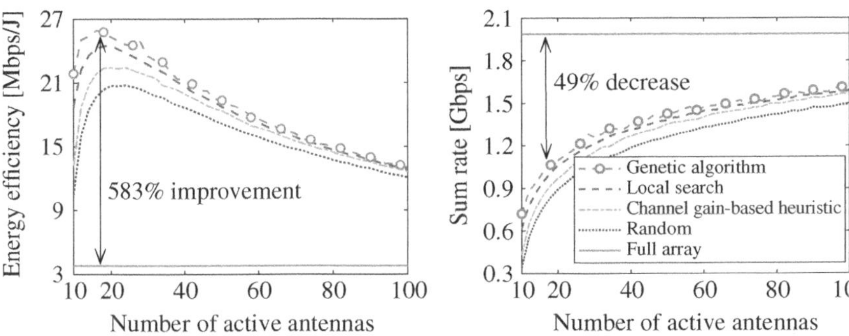

Figure 7.13 EE (left) and SR (right) as a function of the number of active antennas achieved by the proposed joint AS and PA methods. In this result, the number of UEs is $K = 10$, the number of BS antennas is $M = 500$, and the transmit power budget is $P_{max} = 24$ dBm.

6%, 15%, and 25% higher than the one achieved by the local search, the heuristic based on the channel gain, and the random selection, respectively. Furthermore, comparing the EE results from Figures 7.11 and 7.13, we observe that the proposed methods reach their peak values at a different number of active antennas, implying there is a dependence of it with the number of UEs. Specifically, with 100 UEs, the GA reaches its peak with 150 active antennas, while the peak for the same method with 10 UEs occurs with 16 antennas. Therefore, to fully exploit the EE improvement offered by AS, the number of active antennas must be carefully defined according to the number of UEs in the system.

The results of Figures 7.11 and 7.12 show that the EE improvement yielded by AS comes with the price of a reduced SR. To compensate for this SR loss, we propose that, when higher UE data rates are required, the transmit power budget

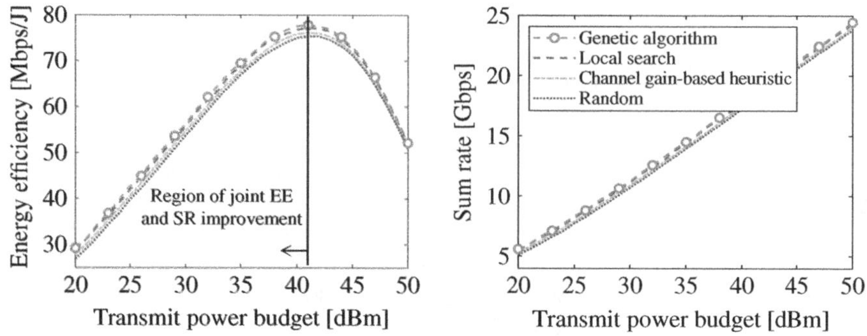

Figure 7.14 EE (left) and SR (right) as a function of the transmit power budget achieved by the proposed joint AS and PA methods. In this result, the number of UEs is $K = 100$, the number of BS antennas is $M = 500$, while 150 antennas are activated.

can be increased without compromising the EE. Specifically, this can be verified in Figure 7.14, which illustrates the EE and SR as functions of the transmit power budget. Notice that both SR and EE practically double by increasing the transmit power budget from 20 to 30 dBm. This happens because, unlike the number of active antennas, the transmit power budget contributes marginally to the system's power consumption. In this case, from the EE perspective, increasing the transmit power budget is more efficient than activating a higher number of antennas.

7.5 Conclusions and Perspective

In this chapter, we have developed two EE problem formulations and solutions by considering different M-MIMO system configurations. The first optimization problem deals with the optimization of three parameters of an RIS-aided M-MIMO system, namely the number of BS antennas M, transmit power vector $\boldsymbol{\rho}$, and RIS phase-shift $\boldsymbol{\Phi}$, aiming to maximize the system EE with low complexity. The second optimization, developed using evolutionary optimization and a low-complexity heuristic, deals with joint AS and PA, aiming to maximize the EE of an XL-MIMO system. A summary of the key quantitative results obtained by each contribution is provided in Table 7.6.

Section 7.3 addressed the EE maximization problem in a downlink RIS-aided M-MIMO communication system operating under generalized Rician BS-RIS channels and ZF precoding. We proposed a statistical CSI-based optimization strategy that treats the RIS phase-shift and PA as a constant over multiple channel coherence time intervals, reducing the overhead of the variables optimization in every channel coherence time. Based on an ER lower-bound closed-form expression, the EE optimization problem with power and QoS constraints is

Table 7.6 Summary of the contributions – key quantitative results.

Contribution	Key results
Section 7.3: Joint RIS phase–shift design, PA, and optimization of the number of active BS antennas to maximize the EE in RIS-aided M-MIMO systems	• EE improvement of up to 136% • The RIS phase–shift optimization reduces the average number of active BS antennas by 48% with a transmit power of 20 dBm • The proposed optimization based on statistical CSI decreases the EE only by 5.88% compared to the case with instantaneous CSI approach
Section 7.4: Joint AS and PA to maximize the EE in XL-MIMO systems	• EE improvement of up to 72% under the system load of 20% • EE improvement of up to 583% under the system load of 2% • GA peak EE up to 25% higher than the benchmarks

formulated. To circumvent the challenging coupled optimization variables and non-convex constraints, the proposed alternating-sequential optimization approach, based on the FP techniques and dual Lagrangian method, optimizes the PA vector ($\boldsymbol{\rho}$), the number of BS antennas (M), and the RIS phase–shift ($\boldsymbol{v}$) in an iterative manner based on analytical closed-form expressions, until convergence. Our approach offers an elegant methodology to gradually improve the system EE performance by optimizing the system variables. Our numerical results validate the effectiveness of the proposed algorithm, providing a performance loss of $\approx 5.88\%$, compared to the instantaneous CSI-based optimization method, therefore achieving promising EE gains with lower overhead.

A limitation of the proposed approach in Section 7.3 is its exclusive applicability with ZF precoding, assuming complete LoS channels between the RIS and the UEs. Achieving this condition in some practical, real-world, scenarios may be sometimes unattainable. Moreover, in scenarios involving extensive RIS arrays, typical of beyond 5-G communications, the energy consumption of the RIS may become significant. Consequently, investigating RIS-element selection mechanisms for optimized EE operation will be a subject of future research.

Section 7.4 proposed schemes for AS and PA to maximize the EE in XL-MIMO systems. The schemes are intended for the particular scenario where massive arrays of antennas originally deployed to operate in areas with high UE density need to adapt their power consumption under low system load. To this end, from CSI, the three proposed methods use different approaches to approximate the set of active antennas that maximizes the EE, including GA, a heuristic method based on the channel gain, and local search. Remarkably, numerical results have demonstrated that the proposed AS methods can improve the system

EE by up to 72% and 583% compared to the full array activation for the system loads of 20% and 2%, respectively. Therefore, AS surges as a promising solution to deploy adaptive XL-MIMO systems for environments with widely oscillating demand, providing consistent communication performance under high system load against high EE under low system load.

The main limitation of the proposed schemes in Section 7.4 is their dependence on instantaneous CSI from all the BS antennas. Indeed, in XL-MIMO systems, the task of acquiring such CSI at every time coherence interval might be costly due to the high number of antennas. Approaches to circumvent this limitation include the design of AS procedures based on partial CSI obtained for a smaller subset of antennas, on past CSI measurements, or on statistical CSI, which changes over a long-term time horizon.

7.5.1 Research Perspectives

We have investigated in this chapter the EE problem in M-MIMO and XL-MIMO systems, aided or not by an RIS. In this context, some points remain as important open problems that lack further research investigation but have good opportunities for promising solutions. For example, we have first investigated here the EE in RIS-aided M-MIMO systems, optimizing the number of activated antennas, the RIS phase-shift, and the BS transmit power. Although we have considered a single RIS, an important research perspective is extending the system model for a multi-RIS scenario, for example, determining how the multiple RIS can cooperatively determine their phase-shifts, as well as how many RISs would be optimal from an EE perspective depending on the system parameters, like cell size, users' density, and others. Besides, obtaining isolated channel estimates for the BS- RIS and/or RIS-UEs links can be burdensome in practice due to the possible absence of active RF elements at the RIS and a large number of reflecting elements. Therefore, devising approaches to optimize the system EE based only on the cascaded channel, or the statistical CSI, is a relevant research direction.

Then, we have investigated in this chapter the AS and PA for EE optimization in XL-MIMO systems, proposing the application of three different schemes: the GA, a heuristic based on the channel gains, and the local search algorithm. In such context, promising research directions involve the application of modern machine learning algorithms, like deep learning and deep RL, and recently proposed generative approaches, like generative diffusion models [37]. Besides, since obstacles in the environment may obstruct the direct links between the XL-MIMO array and the UEs, the integration of RIS to the scenario is a relevant research direction. In this case, how to jointly optimize the set of selected antennas at the XL-MIMO array and the phase-shifts at the RIS is a challenging but promising and opportune research problem.

References

1 Peschiera, E. and Rottenberg, F. (2023). Energy-saving precoder design for narrowband and wideband massive MIMO. *IEEE Transactions on Green Communications and Networking* 7 (4): 1793–1806. https://doi.org/10.1109/TGCN.2023.3293615.

2 Marinello, J.C., Abrão, T., Amiri, A. et al. (2020). Antenna selection for improving energy efficiency in XL-MIMO systems. *IEEE Transactions on Vehicular Technology* 69 (11): 13305–13318. https://doi.org/10.1109/TVT.2020.3022708.

3 Shrestha, S., Fu, X., and Hong, M. (2023). Optimal solutions for joint beamforming and antenna selection: from branch and bound to graph neural imitation learning. *IEEE Transactions on Signal Processing* 71: 831–846. https://doi.org/10.1109/TSP.2023.3244096.

4 Wang, Z., Zhang, J., Du, H. et al. (2023). A tutorial on extremely large-scale MIMO for 6G: fundamentals, signal processing, and applications. *IEEE Communications Surveys & Tutorials* 26 (3): 1560–1605.

5 Jevremovic, V. (2024). Metamaterials and reconfigurable intelligent surfaces. iBwave, White Paper. https://doi.org/http://info.ibwave.com/metamaterials-reconfigurable-intelligent-surfaces/ (accessed 13 November 2024).

6 de Souza, J.H.I., Marinello Filho, J.C., Amiri, A., and Abrão, T. (2023). QoS-aware user scheduling in crowded XL-MIMO systems under non-stationary multi-state LoS/NLoS channels. *IEEE Transactions on Vehicular Technology* 72 (6): 7639–7652. https://doi.org/10.1109/TVT.2023.3243488.

7 Marinello Filho, J.C., Brante, G., Souza, R.D., and Abrão, T. (2022). Exploring the non-overlapping visibility regions in XL-MIMO random access and scheduling. *IEEE Transactions on Wireless Communications* 21 (8): 6597–6610. https://doi.org/10.1109/TWC.2022.3151329.

8 de Souza, J.H.I., Marinello Filho, J.C., Abrão, T., and Panazio, C. (2022). Reconfigurable intelligent surfaces to enable energy-efficient IoT networks. *2022 Symposium on Internet of Things (SIoT)*, 1–4. https://doi.org/10.1109/SIoT56383.2022.10070317.

9 Guerra, D.W.M. and Abrão, T. (2022). Clustered double-scattering channel modeling for XL-MIMO with uniform arrays. *IEEE Access* 10: 20173–20186. https://doi.org/10.1109/ACCESS.2022.3152704.

10 dos Santos, H.L., Marinello, J.C., Panazio, C.M., and Abrão, T. (2022). Machine learning-aided pilot and power allocation in multi-cellular massive MIMO networks. *Physical Communication* 52: 101646. https://doi.org/10.1016/j.phycom.2022.101646.

11 de Souza, J.H.I., Amiri, A., Abrão, T. et al. (2021). Quasi-distributed antenna selection for spectral efficiency maximization in subarray switching XL-MIMO

systems. *IEEE Transactions on Vehicular Technology* 70 (7): 6713–6725. https://doi.org/10.1109/TVT.2021.3081462.

12 Ubiali, G.A., Marinello, J.C., and Abrão, T. (2021). Energy-efficient flexible and fixed antenna selection methods for XL-MIMO systems. *AEU - International Journal of Electronics and Communications* 130: 153568. https://doi.org/10.1016/j.aeue.2020.153568.

13 Bruza Alves, T.A. and Abrão, T. (2023). NOMA-based random access in mMTC XL-MIMO. *IEEE Access* 11: 1944–1954. https://doi.org/10.1109/ACCESS.2022.3233942.

14 Bruza Alves, T.A. and Abrão, T. (2022). Massive MIMO and NOMA bits-per-antenna efficiency under power allocation policies. *Physical Communication* 51: 101588. https://doi.org/10.1016/j.phycom.2021.101588.

15 Taniguchi, L.M., de Souza, J.H.I., Guerra, D.W.M., and Abrão, T. (2021). Resource efficiency and pilot decontamination in XL-MIMO double-scattering correlated channels. *Transactions on Emerging Telecommunications Technologies* 32 (12): e4365. https://doi.org/10.1002/ett.4365.

16 de Souza Junior, W. and Abrao, T. (2023). Energy efficiency maximization for intelligent surfaces aided massive MIMO with zero forcing. *IEEE Transactions on Green Communications and Networking*. https://doi.org/10.1109/TGCN.2023.3346367.

17 You, L., Xiong, J., Ng, D.W.K. et al. (2021). Energy efficiency and spectral efficiency tradeoff in RIS-aided multiuser MIMO uplink transmission. *IEEE Transactions on Signal Processing* 69: 1407–1421. https://doi.org/10.1109/TSP.2020.3047474.

18 Han, Y., Jin, S., Matthaiou, M. et al. (2023). Toward extra large-scale MIMO: new channel properties and low-cost designs. *IEEE Internet of Things Journal* 10 (16): 14569–14594. https://doi.org/10.1109/JIOT.2023.3273328.

19 Wang, S., Zhu, M., Li, Z. et al. (2023). Antenna selections strategies for massive MIMO systems with limited-resolution ADCs/DACs. *IEEE Transactions on Wireless Communications* 22 (11): 8128–8140. https://doi.org/10.1109/TWC.2023.3260002.

20 Sharifi, S. and Shahbazpanahi, S. (2023). A POMDP-based approach to joint antenna selection and user scheduling for multi-user massive MIMO communication. *IEEE Transactions on Communications* 71 (3): 1691–1706. https://doi.org/10.1109/TCOMM.2022.3227304.

21 Abbas, W.B., Khalid, S., Ahmed, Q.Z. et al. (2023). Heuristic antenna selection and precoding for a massive MIMO system. *IEEE Open Journal of the Communications Society* 5: 83–96. https://doi.org/10.1109/OJCOMS.2023.3339402.

22 Demir, O.T., Björnson, E., and Sanguinetti, L. (2021). Foundations of user-centric cell-free massive MIMO. *Foundations and Trends® in Signal Processing* 14 (3–4): 162–472. https://doi.org/10.1561/2000000109.

23 Björnson, E., Hoydis, J., and Sanguinetti, L. (2017). Massive MIMO networks: spectral, energy, and hardware efficiency. *Foundations and Trends® in Signal Processing* 11 (3–4): 154–655. https://doi.org/10.1561/2000000093.

24 Kraus, J.D. (2002). *Antennas for All Applications*. McGraw Hill.

25 Polvani, G. and Abrão, T. (2023). Visibility region and correlation in ELAA-MIMO LoS scenarios under planar and spherical wavefront propagation. *Physical Communication* 60: 102166. https://doi.org/10.1016/j.phycom.2023.102166.

26 Pendeza Martinez, C.A., Abrão, T., and Martinez, A.L.M. (2021). Energy and spectral efficiency trade-off in OCDMA-PON assisted by non-linear programming methods. *Computer Networks* 189: 107920. https://doi.org/10.1016/j.comnet.2021.107920.

27 Zhi, K., Pan, C., Ren, H., and Wang, K. (2022). Ergodic rate analysis of reconfigurable intelligent surface-aided massive MIMO systems with ZF detectors. *IEEE Communications Letters* 26 (2): 264–268. https://doi.org/10.1109/LCOMM.2021.3128904.

28 Shen, K. and Yu, W. (2018). Fractional programming for communication systems—Part II: uplink scheduling via matching. *IEEE Transactions on Signal Processing* 66 (10): 2631–2644. https://doi.org/10.1109/TSP.2018.2812748.

29 Shen, K. and Yu, W. (2018). Fractional programming for communication systems—Part I: power control and beamforming. *IEEE Transactions on Signal Processing* 66 (10): 2616–2630. https://doi.org/10.1109/TSP.2018.2812733.

30 Jong, Y. (2012). An efficient global optimization algorithm for nonlinear sum-of-ratios problem. *Optimization Online* 32: 38–39.

31 Huang, C., Zappone, A., Alexandropoulos, G.C. et al. (2019). Reconfigurable intelligent surfaces for energy efficiency in wireless communication. *IEEE Transactions on Wireless Communications* 18 (8): 4157–4170. https://doi.org/10.1109/TWC.2019.2922609.

32 Miridakis, N.I., Tsiftsis, T.A., and Yao, R. (2023). Zero forcing uplink detection through large-scale RIS: system performance and phase shift design. *IEEE Transactions on Communications* 71 (1): 569–579. https://doi.org/10.1109/TCOMM.2022.3225539.

33 Cormen, T.H., Leiserson, C.E., Rivest, R.L., and Stein, C. (2009). *Introduction to Algorithms*, 3e. Cambridge, MA: The MIT Press. ISBN 0262033844.

34 Goldberg, D.E. (1989). *Genetic Algorithms in Search, Optimization, and Machine Learning*, 1e. Boston, MA: Addison-Wesley Longman Publishing Company. ISBN 0201157675.

35 Deep, K., Singh, K.P., Kansal, M.L., and Mohan, C. (2009). A real coded genetic algorithm for solving integer and mixed integer optimization problems. *Applied Mathematics and Computation* 212 (2): 505–518. https://doi.org/10.1016/j.amc.2009.02.044.

36 Pardalos, P.M., Du, D.-Z., and Graham, R.L. (2013). *Handbook of Combinatorial Optimization*, 2e. New York, NY: Springer. ISBN 978-1-4419-7997-1.

37 Du, H., Zhang, R., Liu, Y. et al. (2023). Beyond deep reinforcement learning: a tutorial on generative diffusion models in network optimization. https://arxiv.org/abs/2308.05384.

8

NOMA-Aided Massive MIMO for Next-Generation Networks

Vinoth Babu Kumaravelu[1], Helen Sheeba J.[1], Dipinkrishnan R.[1], Arthi Murugadass[2], Poongundran Selvaprabhu[1], Rajeshkumar V.[1], Vetriveeran Rajamani[3], Anand Sreekantan Thampy[4], Agbotiname Lucky Imoize[5,6], and Webert Montlouis[7]

[1]*Department of Communication Engineering, School of Electronics Engineering, Vellore Institute of Technology, Vellore, Tamil Nadu, India*
[2]*School of Computer Science and Engineering, Vellore Institute of Technology, Chennai, Tamil Nadu, India*
[3]*Department of Micro and Nano Electronics, School of Electronics Engineering, Vellore Institute of Technology, Vellore, Tamil Nadu, India*
[4]*Centre for Nanotechnology Research, Vellore Institute of Technology, Vellore, Tamil Nadu, India*
[5]*Department of Electrical and Electronics Engineering, Faculty of Engineering, University of Lagos, Lagos, Nigeria*
[6]*Department of Electrical Engineering and Information Technology, Ruhr University, Bochum, Germany*
[7]*Applied Physics Laboratory (APL) and Electrical and Computer Engineering Department, Johns Hopkins University, Baltimore, MD, USA*

8.1 Introduction

The future wireless network expects an explosive growth of connected devices, along with high data demand applications. Researches shows that the technologies need to be capable of handling a traffic capacity of 100 Mbits/s per square meter radius [1]. According to International Telecommunication Union (ITU) report [2], by the end of this decade, the global traffic level will be more than 5000 exabytes per month. The latency requirement is less than a millisecond range, with low jitter rate and it need to support a user moving with a speed, more than 500 km/h with a spectrum extended up to 100 GHz [3]. Artificial intelligence (AI), sensors connected to internet, and global level broadband requirement, all these trend points to an exponential growth in internet connected smart devices called Internet of Everything (IoE) and hence the device density per kilometer will be probably 10^7.

Massive MIMO for Future Wireless Communication Systems: Technology and Applications, First Edition. Edited by Webert Montlouis and Agbotiname Lucky Imoize.

8.1.1 Massive MIMO (mMIMO)

Figure 8.1 depicts the multiuser massive multiple-input multiple-output (mMIMO) system. As the number of antennas increases, so does system capacity, resulting in the maximum data rate for individual users. Improved spatial multiplexing performance results in increased spectral efficiency (SE) and throughput. As the number of antennas rises, the needed transmission power for each antenna can be reduced to a milliwatt level. Network security benefits significantly from the mMIMO. Narrowing the beams decreases attacks, while also making user tracking easier. Using dedicated channels, jammer signals can be cancelled. Reduced interference aids in the refinement of signal processing techniques at the receiver. If the number of antennas is infinite, channel prediction will reach an optimum. Because the number of antennas in mMIMO is greater, it also delivers superior results in predicting expected channel matrix values. In contrast to reflecting intelligence surface (RIS), mMIMO contains an active working mechanism for transmission and reception [4]. Because of the vast number of antenna arrays, the gain utilizing beamforming will be substantial. Complex signal processing modules are required in a mMIMO system with active elements such as radio frequency (RF) components and an RF chain. To reap the maximum benefits of mMIMO, half wavelength spacing between antenna elements is required. mMIMO may be used in conjunction with technologies such as millimeter wave (mmWave) transmission, non-orthogonal multiple access (NOMA), and RIS to achieve high SE in 6G systems. The combination of mMIMO and RIS technology increases energy efficiency (EE), resulting in a green communication paradigm. Beamforming and antenna configuration are concurrent difficulties for mMIMO.

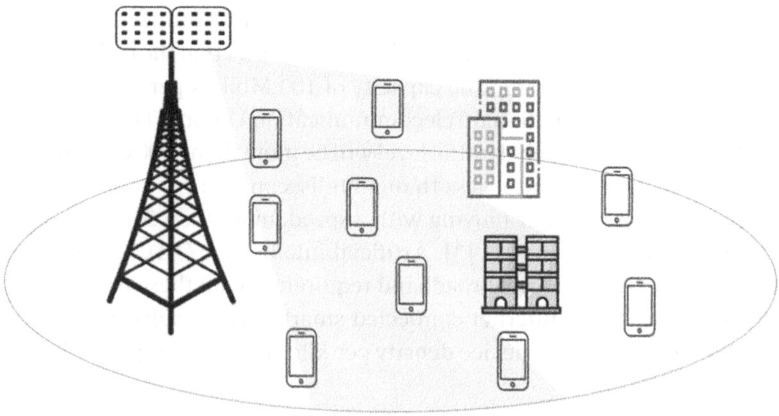

Figure 8.1 Multiuser mMIMO system.

8.1.2 NOMA

Future wireless networks will follow a heterogeneous nature, in which switching between different kinds of network will be encouraged. The massive number of users and smart devices will lead to a demand of spectrally efficient system. The existing orthogonal multiple access (OMA) radio access methods will not be able to support such scenarios. Here NOMA can be considered as apposite technology, which will support the mentioned scenarios with its unhindered nature of spectrum usage. The downlink (DL) NOMA system serving K-users is pictorially represented in Figure 8.2. Minimum outage probability and low latency are also considered as the advantages of NOMA. In NOMA, the allocating power levels for the transmitting signals of different users will be adjusted, according to the corresponding user's channel state information (CSI), which depend on the user position. Hence, the transmission and reception will occur irrespective of time, frequency, and code. Here, the signal will be broadcasted to the group of users and will be the linear combination of different signals with different power levels adjusted by weights and the process is termed as superposition coding (SC). NOMA has a specific decoding procedure called successive interference cancellation (SIC), in which according to the decoding order, each user will decode the data of other user and eliminate it, further decodes its own data. Power allocation (PA) optimization and error propagation in SIC are considered as the major problems related to NOMA system [5].

8.1.3 mMIMO-Aided NOMA

The desire to satisfy the constantly expanding demand for larger data rates, enhanced SE, and better user experiences in future wireless networks serves as

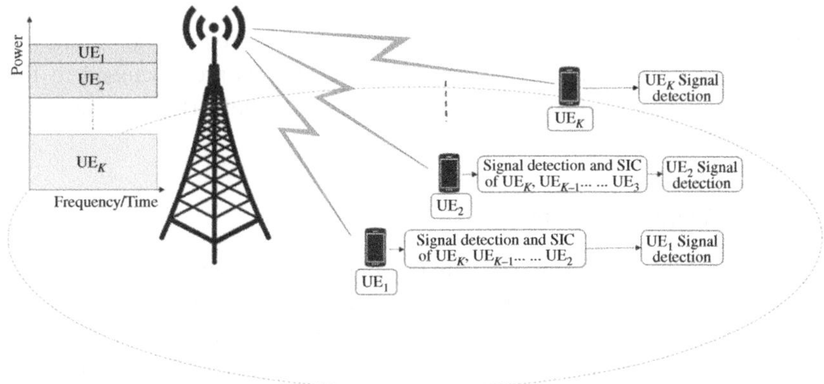

Figure 8.2 Downlink NOMA serving K-users.

the driving force behind investigating the integration of mMIMO with NOMA. mMIMO and NOMA are two cutting-edge technologies that have independently demonstrated appreciable improvements in wireless communication systems. The distinctive characteristics of mMIMO include spatial multiplexing and interference mitigation, while the strengths of NOMA are maximization of SE. Integration of both technologies can offer a better overall performance.

8.1.4 Contributions

The main contributions of this chapter are:

- The system model for NOMA-aided mMIMO is developed and it is validated through Monte-Carlo simulations.
- The sum rate expression for proposed system model is derived and simulations are performed for two-user scenario.
- The proposed system is tested with and without power scaling, as well as with different PA factors, and the impact is observed.

8.1.5 Organization

The organization of this chapter is as follows: Section 8.1 describes motivation and introduction to mMIMO and NOMA. Section 8.2 presents the related works on mMIMO and NOMA variants. System model of NOMA-aided mMIMO is presented in Section 8.3. Section 8.4 depicts the discussions on simulation results. Conclusions and future work are presented in Section 8.5.

8.2 Related Work

The main recommendations, benefits, and challenges related to prior studies on mMIMO, NOMA, and mMIMO-assisted NOMA systems are discussed in this section. Previous studies dealing with limitations such as imperfect SIC (iSIC) are also discussed.

8.2.1 Prior Works on mMIMO

Sadeghi et al. [6] discuss the ergodic sum rate maximization under restricted backhaul capacity in uplink (UL) mMIMO system. It is considered that each antenna in the system transmits data streams to the group of remote radio unit (RRU), which is controlled by a central unit (CU). These CU has restricted backhaul capacity. An exhaustive search mechanism is considered for reaching the maximum sum rate by finding all possible combinations of the number of antennas and quantization levels required. The numerical expression of sum rate is derived under restricted backhaul capacity. The derived expressions are authenticated by simulations and it proves that the suggested system outperforms the centralized case.

A substantial difference is observed by Lu et al. [7] between theoretical and practical mMIMO performance for closely spaced users. The fading of reference symbols received by users is identified as a key issue. The sparsity supplied by mMIMO can help to improve it. This work also focuses on increasing the efficiency of single carrier systems through the utilization of the degree of freedom available. The usage of excess number of pilot symbols leads to the wastage of resources in mMIMO [8]. Throughput and system performance are both increased by reducing these numbers to a minimum, while employing knowledge of sparse channels. The suggested system uses compressive sensing approach to reduce the pilot contaminations. The number of antennas considered for simulation is 128 and the results shows that the system performance is improved with proper CSI.

A precoding method for mMIMO systems is implemented by Feng et al. [9] with the aim of maximizing signal-to-noise ratios (SNRs) to limit the influence of interference. The work considers random interference with bounded values of power, which results in the reduction of outages. The simulation results show that a large reduction in interference is achieved by the suggested precoding system. Björnson et al. [10] suggest a new scheme called dynamic pilot allocation to mitigate the effects of pilot contamination. This scheme dynamically allocates pilots to users based on their channel quality, reducing the interference caused by pilot contamination and further improving the system capacity. The results of their simulations demonstrate that even in scenarios with a high number of antennas, the suggested scheme significantly increases the sum rate and outperforms traditional mMIMO systems. Therefore, the authors believe that capacity limitations in mMIMO systems can be effectively addressed through the combination of precoding and combining techniques like minimum mean squared error (MMSE).

The receiver complexity is seen as a difficult problem for mMIMO systems. In the case of signal detection, the large number of antennas and radio channels creates a complex structure. On the basis of optimal results and complexity, Albreem et al. [11] compare linear and nonlinear methods for symbol detection. High data rate is achieved via the precoding's capabilities for spatial multiplexing. It also eliminates interfering signals and increases EE. Singh et al. [12] suggest filter bank multi-carrier (FBMC) UL mMIMO receiver system and sum rate maximization is analyzed for different types of RF receivers. For the comparison purpose, orthogonal frequency division multiplexing (OFDM)-based system analysis and a study on delay spread for UL is also presented in the paper. An extension of single-cell approach to multicell is performed with the derived form of sum rate expression in UL scenarios. Analysis is also performed for imperfect CSI conditions.

The work mentioned by Albreem et al. [13] compare the complexity of various linear precoding techniques. A brief examination of nonlinear precoding methods is carried out, which elaborates on the performance in terms of computational complexity and accuracy. A solution for interferences caused by multiple users and inter carriers in a mMIMO system is suggested by Sanoopkumar et al. [14]. An equalization method which updates continually in the form of a matrix will be utilized to cancel the interferences. The scheme suits for every carrier assignment method. Dynamic nature of the RF signals is considered for deriving the matrix. The performance of the scheme is validated by comparing the bit error rate (BER) for different SNR values.

The mMIMO is in various phases of development as part of the third-generation partnership project (3GPP). CSI prediction for high mobility systems and beamforming are described by Jin et al. [15] for the most recent stage of mMIMO. The research suggests an ideal beamforming strategy that will improve SE, while reducing the need for reference overhead. The researches are proved practically for a frequency range of 3.5 GHz [15].

Ouyang et al. [16] explore the receive antenna selection schemes in multiuser mMIMO UL system. The low complexity upper bound sum rate expression for multiuser mMIMO is derived assuming large next-generation NodeB (gNB) antennas. The sum rate scales about two times the logarithmic rate with the number of gNB antennas. The expressions are verified using simulations.

A lower limit for the sum rate of mMIMO is determined by Xu et al. [17] using Bussgang decomposition, Gaussian input values and precoding techniques. A diagonal approximation is employed to construct an initial framework for the distortion, by uniform polar quantization, and it is then used to determine the sum rate in closed-form. Numerical findings reveal that, when compared to the Cartesian design, the suggested quantized polar transmitter performs well in terms of sum rate, BER, and outage performance. It is validated through simulations.

Through beamforming, mMIMO mitigates small-scale fading effects and necessary transmission power. The complexity of digital precoding is regarded as a constraint for mMIMO systems. This can be avoided using a hybrid precoding approach that supports beamforming. During precoding, it minimizes the complexity of mMIMO. To reduce the overall number of RF chains, a hybrid precoding solution combining analog and digital precoding is suggested by Nouri et al. [18]. The results are validated experimentally using a testbed of 64 antennas configured for 28 GHz mmWaves.

mMIMO's sparse characteristics result in an increase in the accuracy of CSI prediction. An algorithm is developed by Zhang et al. [19] using sparsity and low rank features for accurate channel estimation. mmWaves are used for transmission, and frequency division duplexing (FDD) is used for duplexing. The methodology helps to reduce the number of pilot symbols by comparing the mean square error (MSE) of several estimation methods for CSI prediction. In mMIMO, the channel is estimated on the basis of beams formed during the transmission of multiple pilot symbols [20]. The evaluation of statistical characteristics of CSI is optimized based on the received signals. MSE is the parameter used to compare the results with existing methods.

8.2.2 Prior Works on NOMA Variants

Do and Nguyen [21] consider DL NOMA system with cooperative communication. In this system, amplify-and-forward (AF) relaying protocol is incorporated with NOMA system. At high SNR, the impact of outage is reduced for minimum data rate requirement of both strong and weak users. Compared to traditional NOMA scheme, the suggested model has better outage performance. The expressions for outage re derived for both strong and weak users. The results are authenticated with the help of Monte-Carlo simulations.

Spectrum sensing in DL scenario of NOMA system is considered by Do et al. [22] and cognitive radio (CR)-NOMA system is presented. The performance of far user (FU) is determined by the resource allocation conditions. Based on the threshold rate requirements, PA is determined which supports the performance gap in the NOMA users. The suggested CR-NOMA-based cooperative communication in DL scenario has better performance in terms of outage and it is verified using simulations.

Yin et al. [23] compare the performance of NOMA-2000 and power-domain NOMA (PD-NOMA) across Rayleigh fading channels before proposing a dynamic user grouping strategy for categorizing users. According to simulations, the PD-NOMA can consistently show lower BER than the NOMA-2000 for various SNR.

In [24], both UL and DL NOMA scenario under Nakagami-m fading environment is considered. With fixed PA, users suffer from poor outage performance. Dao et al. [24] consider an optimal fixed power fraction scheme based on dynamic decoding order (DDO). The simulation results show that the suggested scheme has better outage performance in both UL and DL scenarios when compared with the conventional fixed PA schemes. In NOMA, user interference will dominate, if the distance between the near user (NU) and the FU is too small. Hence, many users would use OMA rather than NOMA. Therefore, in order to encourage massive connection, Mounchili and Hamouda [25] took into account the minimum pairing distance that separates the FU and NU and developed the new user pairing strategy known as minimum distance-NOMA (MD-NOMA) by lowering the pairing distance barrier. Simulation is used to construct and validate the derivation for the pairing distance limit under the assumption of a fixed PA.

A multi-carrier NOMA (MC-NOMA) system is suggested by Al-Obiedollah et al. [26]. Here, the available bandwidth for transmission is split into a number of sub-bands, allowing numerous users in each sub-band to be supplied via PD-NOMA. According to the findings of the simulation, the MC-NOMA with dynamic bandwidth allocation (BA) performs better in terms of attained sum rate than the scheme with equal BA. Thet and Ozdemir [27] presented a user clustering and pairing partial-beam NOMA (PB-NOMA) system for mmWave channels. Due to the widely dispersed nature of users, Thet and Ozdemir [27] presented a fuzzy clustering method using a channel correlation similarity measure.

In [28] to support the sum rate requirements of NOMA users, optimal decoding with user grouping scheme is suggested. Dai et al. [28] consider joint transmission in coordinated multipoint environment. The optimal decoding order is obtained by monotonic optimization. The results show that the optimal decoding order is varied based on the user grouping. The analytical expressions derived for low-complexity algorithm is verified using simulations.

The unique index modulation (IM)-based NOMA DL strategy, known as IM-NOMA, is presented by Almohamad et al. [29]. In this scheme, the gNB utilizes IM idea by which each user is allocated at least one channel, and access method is NOMA. Maximum Likelihood and SIC detection techniques are taken into consideration at the user's end, and their error rates, outage probabilities, and computing complexity are researched. The benefits and the performance of the suggested IM-NOMA over traditional NOMA is supported by simulation results. To increase SE, Zhu et al. [30] implement the idea of IM into NOMA system that is assisted by a simultaneously transmitting and reflecting RIS (STAR-RIS). The Monte-Carlo simulations are used to validate a rough closed-form formula for BER. The suggested IM-based system enhances the transmission rates over traditional STAR-RIS employed NOMA, according to simulation findings.

Zhang et al. [31] consider UL NOMA scenario to achieve maximum sum rate with minimum individual user rate requirement. At first, decoding order is estimated and based on it, the power is optimized. A closed-form solution is derived for the decoding order and optimal power. The suboptimal user grouping is also assigned on the basis of decoding order. The suggested method reaches the near-optimal performance in terms of sum rate. In [32], the decoding order is determined by instantaneous received signal power (IRSP). IRSP-based ordering is dependent on the gNB density and the power splitting coefficients. In [33] unmanned aerial vehicle (UAV) is combined with NOMA to enhance performance. Zhang et al. [33] consider concurrently the localization and power transmitted. Decoding order is considered because of UAV's mobility. Joint localization and power optimization are considered as non-convex subproblems.

By successive optimization, the decoding is determined based on UAV's location. The simulation results prove the effectiveness of suggested scheme.

Based on the power levels of the signal received, Do et al. [34] follow DDO for NOMA UL/DL scenarios. The closed-form outage expressions are derived and validated through simulations. The outage performance of the suggested method is enhanced. Thapar et al. [35] consider decoding order of unsure NOMA users to resolve secrecy problem. All the possible decoding order for every user is determined. By numerical method, the best positive secrecy order is implemented. A suboptimal method is used to overcome the computational complexity. The performance gain is increased by $\approx 37\%$ with suboptimal solution over the existing method and it is verified using simulations.

Do [36] consider cooperative NOMA (CNOMA) scheme under perfect and imperfect hardware noise cancellation conditions. Outage performance of CNOMA with imperfect hardware noise cancellation technique is superior to conventional CNOMA and it is validated through simulations. Cell-edge user (CEU) suffers from poor received signal strength. Under NOMA system, CEUs are given higher power fraction to enhance the received signal strength. Le et al. [37] suggested the system model with multiple antennas at CEU under cooperative communication. To reduce the complexity caused by multiple antennas, transmit antenna selection scheme is incorporated. This results in improved outage performance with reduced computational complexity. The outage expressions are also derived and validated using simulations.

Li et al. [38] consider multiple relaying in CNOMA system using Wyner model for analyzing secrecy performance in direct link and under the presence of eavesdropper. The closed-form solution of outage probability is developed for secrecy rate under two separate schemes namely single-phase and two-phase relay selection. The suggested method is compared with random relay selection. It is proved that the outage performance is higher and there is a significant improvement at high SNR. Nguyen and Do [39] consider Nakagami-m fading channel conditions for DL NOMA system. Cooperative communication between NU and FU follows decode-and-forward (DF) relaying. The enhanced system model shows better performance in terms of outage under perfect channel conditions and optimal PA factors.

In [40], CNOMA protocol while using a durable beamforming architecture for a simultaneous wireless information and power transfer (SWIPT) equipped device is studied. Simulation findings demonstrate that the suggested SWIPT-enabled robust CNOMA system outperforms traditional protocols in terms of system performance for both channel uncertainty models. Results show that the derived algorithm reaches the optimality with minimum iterations.

The BER performance of the DL CNOMA system featuring SWIPT-aided relays is examined by Li et al. [41]. The probability of user grouping error (UGE) of users with relay selection is expressed numerically. The chosen m^{th} relay carries out energy harvesting (EH), amplifies, and retransmits the received signal to the users. For relay selection corresponds to each user's diversity order, the asymptotic UGE expression is constructed. Extensive simulations are used to verify the UGE expressions that are developed.

Nguyen et al. [42] consider EH and best path for cooperation link of CNOMA with multiple relays to support CEU threshold data rate requirement and to enhance outage performance of CNOMA. The optimal fixed PA is considered for direct link. The outage performance of CEU is dependent on power splitting parameter for EH. Numerical results prove that the quality of service (QoS) is enhanced with the relay selection scheme.

Budhiraja et al. [43] implement full duplex (FD)-insisted SWIPT enabled CNOMA under imperfect CSI. The asymptotical closed-form expressions for ergodic capacity at the device-to-device (D2D) transmitters and D2D users are derived. A numerical model for outage is derived and the simulation proves that the suggested method performs well when compared to the existing SWIPT-FD-OMA.

8.2.3 Prior Works on NOMA Variants Considering iSIC

Kara and Kaya [44] analyzed error performance of NOMA-based diamond relaying network (DRN) with iSIC. The exact BER and a tight approximated BER in the closed-form for NOMA-DRN are derived. Huang and Zhu [45] considered DL NOMA system, where the CEU is under heavy shadowing. Therefore, CEU has no direct link from gNB. CEU has to rely on FD relay. Here, the NU performs the iSIC with in-built architecture for power splitting, which energized by simple RF circuits. The designated problem is non-linear and non-convex to determine the optimal global value. Charnes–Cooper's algorithm is used to find the near optimal solution through which the method attains improved EE.

Shi et al. [46] consider iSIC for decoding users in UL scenario by incorporating hierarchical deep reinforcement learning (RL)-based decoding for strong and weak users. Simulation results prove that the suggested algorithm performs well under iSIC and supports the QoS requirements. Do et al. [47] consider efficient usage of spectrum for CR-NOMA system. The recommended model has enhancement under iSIC. The authors proved that the performance of the recommended system has the performance gap which is compensated by power transmitted and level of iSIC. The derived analytical expressions are validated through simulations.

In [48], FD is considered for two known users in CNOMA system. The NU acts as FD relay to CEU with SWIPT-protocol enabled. The system considers iSIC, whereas this imperfection is compensated by optimal power splitting of EH and power fractions. The outage performance and throughput are improved with the suggested system model. The derived numerical expressions are validated through extensive simulations.

8.2.4 Prior Works on mMIMO-Aided NOMA

In [49], the role of NOMA in mMIMO is discussed. NOMA performs better, when the number of gNB antennas and users is almost equal. NOMA gives better results when the user locations are purely random. By considering these results, this paper suggests a hybrid combination of NOMA, beamforming and user grouping. The sparsity property is utilized for channel estimation. With the knowledge of CSI, similar users are grouped together.

NOMA improves SE by integrating with a mMIMO system. The key issue addressed by Hadley and Chatzigeorgiou [50] is the optimization of NOMA power values with the least amount of complexity. The goal of optimization is to maximize SE while limiting transmit power. An asymptotic technique is defined, simulated, and compared to optimal and suboptimal solutions, with the performance analyzed. A gradual resource allocation scheme for mMIMO-NOMA is discussed by Simon et al. [51]. The system model consists of a large number of users using NOMA-based DL radio access. The clustering strategy groups users based on channel correlation and beams to improve system performance. In terms of fairness and rates, the results show a comparison between orthogonal and non-orthogonal techniques, with the NOMA scheme doing better.

mMIMO achieves beamforming with enhanced SE and EE. But the problem is the number of supported devices which is limited by the RF chains. Amirifar and Abouei [52] address this by suggesting beamformed mMIMO-NOMA to support multiuser scenario. In NOMA, PA is the significant factor to achieve maximum sum rate. The PA in suggested scheme is non-convex and nonlinear. The closed-form expression for sum rate maximization using PA optimization is derived. The suggested method is compared with existing MIMO-OMA, beamformed MIMO and fully digitized MIMO (FDM). FDM outperforms the suggested system in terms of SE. However, FDM has least EE. Thus, the suggested system results in nearest sum rate performance of FDM and surpasses existing methods in EE. The performance and closed-form expressions are validated through simulations.

Ahmed et al. [53] considered perfect CSI for user grouping in mMIMO-NOMA system. Based on channel gains, users are clustered with a deep RL approach for beam selection. Finally, optimum PA is allocated to each user. The suggested system with deep RL has $\approx$42% improvement in sum rate and SE over the traditional K-means clustering algorithm. It is validated through simulations. Shili et al. [54] presented CR-based mMIMO-NOMA to obtain maximum sum rate via PA approach to meet each user's minimum data rate requirements. Sum rate expressions are obtained using the optimal PA factors, while accounting for perfect and imperfect CSI. The simulation results show that the maximum sum rate is reached with the optimal PA values in the suggested framework.

A framework for mMIMO-NOMA using cluster-level SIC for multiuser environment is discussed by He et al. [55]. The users are clustered based on distance to the gNB, and NOMA is implemented to suppress inter-cluster interference. A derivation of UL sum rate is highlighted for the suggested system, and numerical expressions and simulation results show that it achieves a higher UL rate than conventional cluster-based MIMO-NOMA systems with user-level SIC.

8.3 System Model

This section derives the mMIMO and mMIMO-NOMA UL system models, received signal-to-interference plus noise ratio (SINR), upper bound on SINR, and achievable rate. Additionally, an analysis is conducted on how power scaling affects SINR and attainable rate.

8.3.1 UL mMIMO System

Assume a mMIMO gNB equipped with M antennas. In the coverage of gNB, there exist K number of single-antenna users. The number of users should be always less than the number of gNB antennas, that is, $K < M$. This scenario is illustrated in Figure 8.3.

The channel vector between gNB and kth user is denoted by [55]

$$\mathbf{h}_k = \sqrt{\beta_k}\mathbf{g}_k \tag{8.1}$$

where,

$$\mathbf{g}_k = \begin{bmatrix} g_{1k} \\ g_{2k} \\ \vdots \\ g_{mk} \\ \vdots \\ g_{Mk} \end{bmatrix}, \quad \mathbf{g}_k \sim \mathcal{CN}(\mathbf{0}_{M\times 1}, \mathbf{I}_M) \tag{8.2}$$

$\mathbf{g}_k$ describes the effect of small-scale Rayleigh fading. The average channel gain of mth element of $\mathbf{h}_k$ is

$$E\left\{\left|h_{mk}\right|^2\right\} = \beta_k \tag{8.3}$$

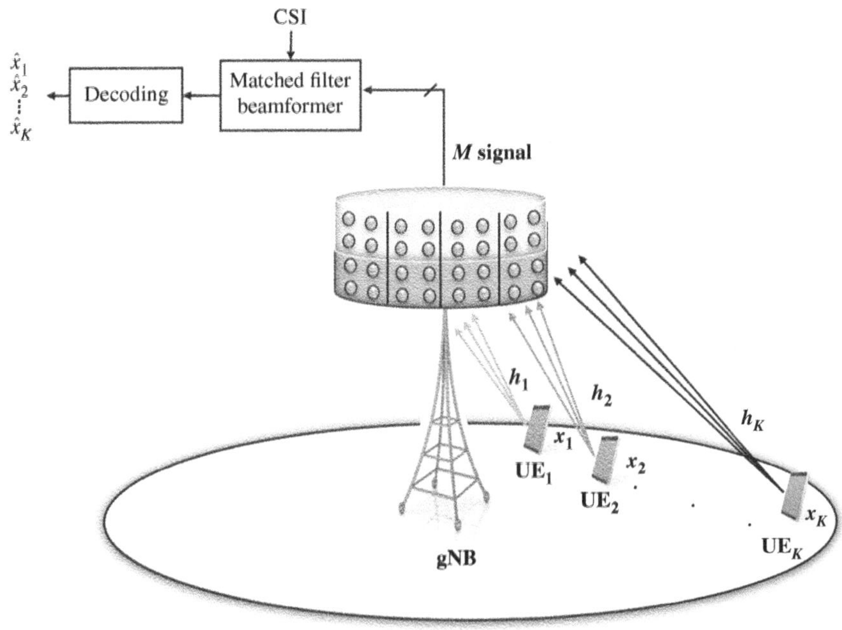

Figure 8.3 UL mMIMO system.

where β_k is the large-scale fading factor, which is given by

$$\beta_k = \left(\frac{d_k}{d_0}\right)^{-\eta} \tag{8.4}$$

where η is the pathloss exponent, d_0 is the reference distance, and d_k is the distance between kth user and gNB.

The signal vector received at the gNB from K single-antenna users is given by [56]

$$\mathbf{y} = \sqrt{P}\sum_{k=1}^{K}\mathbf{h}_k x_k + \mathbf{n} \tag{8.5}$$

where P is the transmit power of user, and x_k is the symbol transmitted by kth user, that is, $E\{x_k x_k{}^*\} = 1,\ k = 1, 2 \ldots K$. $\mathbf{n} \sim \mathcal{CN}(\mathbf{0}_{M\times 1}, \mathbf{I}_M)$ is the additive white Gaussian noise (AWGN) added at the gNB.

If the desired user is jth user, then the received signal can be expressed as

$$\mathbf{y} = \sqrt{P}\mathbf{h}_j x_j + \sqrt{P}\sum_{k=1,k\neq j}^{K}\mathbf{h}_k x_k + \mathbf{n}. \tag{8.6}$$

To detect x_j, a matched filter beamformer is employed at the receiver.

$$r_j = \frac{\mathbf{h}_j^H}{\left\|\mathbf{h}_j\right\|}\mathbf{y}. \tag{8.7}$$

Substituting (8.6) into (8.7)

$$r_j = \frac{\mathbf{h}_j^H}{\left\|\mathbf{h}_j\right\|}\left(\sqrt{P}\mathbf{h}_j x_j + \sqrt{P}\sum_{k=1,k\neq j}^{K}\mathbf{h}_k x_k + \mathbf{n}\right). \tag{8.8}$$

Writing (8.8) in separate terms gives

$$r_j = \sqrt{P}\left\|\mathbf{h}_j\right\| x_j + \sqrt{P}\sum_{k=1,k\neq j}^{K}\frac{\mathbf{h}_j^H}{\left\|\mathbf{h}_j\right\|}\mathbf{h}_k x_k + \frac{\mathbf{h}_j^H}{\left\|\mathbf{h}_j\right\|}\mathbf{n}. \tag{8.9}$$

The SINR for jth user is obtained as

$$\Upsilon_{\mathrm{gNB}}^{x_j} = \frac{P\left\|\mathbf{h}_j\right\|^2}{P\sum_{k=1,k\neq j}^{K}\mathrm{E}\left\{\left|\frac{\mathbf{h}_j^H}{\|\mathbf{h}_j\|}\mathbf{h}_k\right|^2\right\} + E\left\{\left|\frac{\mathbf{h}_j^H}{\|\mathbf{h}_j\|}\mathbf{n}\right|^2\right\}}. \tag{8.10}$$

As per the beamforming gain property of mMIMO,

$$E\left\{\left|\frac{\mathbf{h}_j^H}{\left\|\mathbf{h}_j\right\|}\mathbf{h}_k\right|^2\right\} = \beta_k. \tag{8.11}$$

Since we assumed that the noise samples have unit variance,

$$E\left\{\left|\frac{\mathbf{h}_j^H}{\left\|\mathbf{h}_j\right\|}\mathbf{n}\right|^2\right\} = 1. \tag{8.12}$$

Substituting (8.11) and (8.12) into (8.10)

$$\Upsilon_{\mathrm{gNB}}^{x_j} = \frac{P\left\|\mathbf{h}_j\right\|^2}{P\sum_{k=1,k\neq j}^{K}\beta_k + 1}. \tag{8.13}$$

The above expression is the SINR attained by jth mMIMO user at gNB without power scaling. Let the total transmit power P is reduced by a factor $\frac{1}{M}$, that is,

$$P = \frac{E}{M}. \tag{8.14}$$

Substituting (8.14) into (8.13) gives

$$\Upsilon_{\mathrm{gNB}}^{x_j} = \frac{\mathrm{E}\frac{\|\mathbf{h}_j\|^2}{M}}{\frac{E}{M}\sum_{k=1,k\neq j}^{K}\beta_k + 1}. \tag{8.15}$$

In mMIMO system, the number of gNB antennas is typically larger than 250. When $M \to \infty$, the term $\frac{E}{M}\sum_{k=1,k\neq j}^{K} \beta_k \to 0$. As per the law of large numbers, $\frac{\|\mathbf{h}_j\|^2}{M} \to \beta_j$. This effect is called as channel hardening [56]. The upper bound of SINR attained through matched filtering is given by

$$\Upsilon_{\text{gNB}}^{x_j} \to E\beta_j. \tag{8.16}$$

The scaling mentioned in (8.14) is not preferable when there is CSI uncertainty. It is suggested to use $\frac{E}{\sqrt{M}}$ [56].

The rate achievable by the kth user of mMIMO system is given by [55]

$$R_k = \frac{1}{2}\log_2\left(1 + \Upsilon_{\text{gNB}}^{x_k}\right). \tag{8.17}$$

8.3.2 UL mMIMO-NOMA System

In OMA system, each user should use different frequency for UL transmission. This leads to reduction in SE. When NOMA is employed, all K users can simultaneously transmit at the same time with the same frequency. The users are differentiated by the power levels [57]. The mMIMO-NOMA UL system model is illustrated in Figure 8.4.

The gNB superimposes signals received from K single-antenna users, which is given by [55].

$$\mathbf{y} = \sum_{k=1}^{K} \mathbf{h}_k \sqrt{P_k} x_k + \mathbf{n}, \tag{8.18}$$

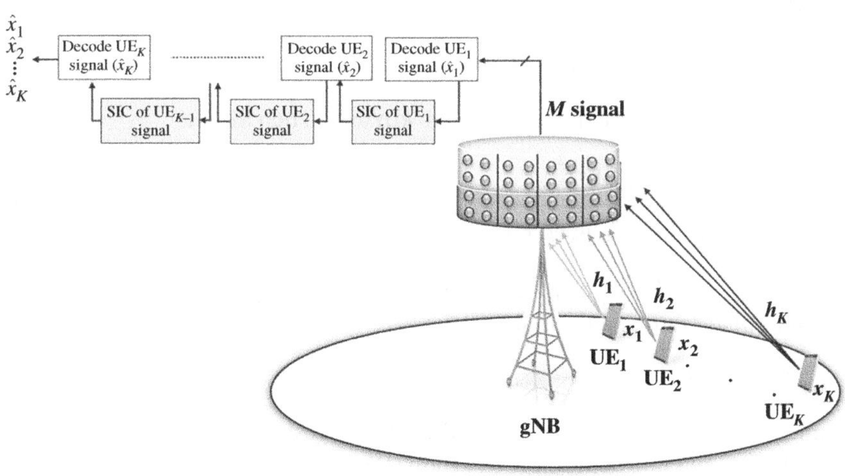

Figure 8.4 UL mMIMO-NOMA system.

where P_k is the power allocated to kth user by the gNB. Based on the CSI, gNB allocates powers to the users. First gNB detects the user allocated with higher power, for example, jth user. As per (8.7), the data corresponding jth user is detected.

Substituting (8.18) into (8.7)

$$r_j = \frac{\mathbf{h}_j^H}{\left\|\mathbf{h}_j\right\|}\left(\sqrt{P_j}\mathbf{h}_j x_j + \sum_{k=1,k\neq j}^{K} \mathbf{h}_k \sqrt{P_k} x_k + \mathbf{n}\right). \tag{8.19}$$

Simplifying (8.19) gives

$$r_j = \sqrt{P_j}\left\|\mathbf{h}_j\right\| x_j + \sum_{k=1,k\neq j}^{K} \sqrt{P_k} x_k \frac{\mathbf{h}_j^H}{\left\|\mathbf{h}_j\right\|}\mathbf{h}_k + \frac{\mathbf{h}_j^H}{\left\|\mathbf{h}_j\right\|}\mathbf{n}. \tag{8.20}$$

The SINR of jth user is obtained as

$$\Upsilon_{\text{gNB}}^{x_j} = \frac{P_j\left\|\mathbf{h}_j\right\|^2}{\sum_{k=1,k\neq j}^{K} P_k \mathrm{E}\left\{\left|\frac{\mathbf{h}_j^H}{\|\mathbf{h}_j\|}\mathbf{h}_k\right|^2\right\} + E\left\{\left|\frac{\mathbf{h}_j^H}{\|\mathbf{h}_j\|}\mathbf{n}\right|^2\right\}}. \tag{8.21}$$

Substituting (8.11) and (8.12) into (8.21) gives

$$\Upsilon_{\text{gNB}}^{x_j} = \frac{P_j\left\|\mathbf{h}_j\right\|^2}{\sum_{k=1,k\neq j}^{K} P_k \beta_k + 1}. \tag{8.22}$$

The above expression is SINR attained by mMIMO-NOMA jth user at gNB without power scaling. Let the power scaling

$$P_k = \alpha_k \frac{E}{M}, \quad \alpha_1 + \alpha_2 + \cdots + \alpha_j + \cdots + \alpha_K = 1 \tag{8.23}$$

where α_k is the power fraction allocated to kth user. Then SINR,

$$\Upsilon_{\text{gNB}}^{x_j} = \frac{\alpha_j \frac{E}{M}\left\|\mathbf{h}_j\right\|^2}{\frac{E}{M}\sum_{k=1,k\neq j}^{K} \alpha_k \beta_k + 1} \tag{8.24}$$

when $M \to \infty$, $\frac{E}{M}\sum_{k=1,k\neq j}^{K} \alpha_k\beta_k \to 0$, and $\frac{\|\mathbf{h}_j\|^2}{M} \to \beta_j$. Substituting these into (8.24), upper bound of SINR can be attained.

$$\Upsilon_{\text{gNB}}^{x_j} = \alpha_j E \beta_j. \tag{8.25}$$

Using CSI-based ordering and SIC, the data corresponding to other users are detected. The data rate achievable by kth mMIMO-NOMA user is given by [55]

$$R_k = \log_2\left(1 + \Upsilon_{\text{gNB}}^{x_k}.\right) \tag{8.26}$$

The sum rate of mMIMO/mMIMO-NOMA system is given by [55]

$$R = \sum_{k=1}^{K} R_k. \tag{8.27}$$

The optimization of power factors is beyond the scope of this chapter. For a simple two-user UL scenario, the SINR of user 1 is given by

$$\Upsilon_{\text{gNB}}^{x_1} = \frac{\alpha_1 \frac{E}{M} \|\mathbf{h}_1\|^2}{\alpha_2 \frac{E}{M} \beta_2 + 1}. \tag{8.28}$$

After direct detection of x_1, SIC is employed and residual signal is obtained.

$$\tilde{\mathbf{y}} \approx \mathbf{h}_2 \sqrt{P_2} x_2 + \mathbf{n}. \tag{8.29}$$

The SNR of detecting second user data at gNB is given by

$$\Upsilon_{\text{gNB}}^{x_2} = \|\mathbf{h}_2\|^2 \quad P_2 = \frac{\alpha_2 \, \|\mathbf{h}_2\|^2 E}{M}. \tag{8.30}$$

The sum rate of two-user scenario is given by

$$R = \log_2 \left(1 + \Upsilon_{\text{gNB}}^{x_1}\right) + \log_2 \left(1 + \Upsilon_{\text{gNB}}^{x_2}\right). \tag{8.31}$$

Substituting (8.28) and (8.30) into (8.31), the sum rate expression can be attained. When the number of gNB antennas is much higher, ideally $M \to \infty$, SINR of users 1 and 2 reaches the upper bounds.

$$\Upsilon_{\text{gNB}}^{x_1} \to \alpha_1 E \beta_1 \tag{8.32}$$

$$\Upsilon_{\text{gNB}}^{x_2} \to \alpha_2 E \beta_2. \tag{8.33}$$

8.4 Simulations and Discussions

Table 8.1 lists the parameters and related values used in simulations. Algorithm 8.1 outlines the procedure used to simulate the performance curves. Simulations are performed for two users. Figure 8.5 compares the UL rate of mMIMO-OMA and mMIMO-NOMA systems for UE_1 by increasing the number of gNB antennas. For mMIMO-NOMA system, power fraction allocated to UE_1 is $\alpha_1 = 0.9$. The power scaling is employed as per (8.14) and (8.23) for mMIMO-OMA and mMIMO-NOMA systems. Based on (8.16) and (8.25), the upper rate bounds of both the systems are also plotted. Actual rate and upper bound rate curves for both systems meet as the number of gNB antennas increases. The increase in gNB antennas results in increase in the data rate of UE_1. When gNB antennas is 400, mMIMO-NOMA system attains the data rate of 1.48 bits/s/Hz, whereas

Table 8.1 Simulation parameters.

Parameters	Values
Number of cell	1
Radius of cell	1000 meters
Number of gNB antennas (M)	10 to 500
Number of users	2
Fixed power per user (E)	10 dB
Power fraction for UE_1 (α_1)	0.9, 0.7
Power fraction for UE_2 (α_2)	0.1, 0.3
Reference distance (d_0)	100 meters
Pathloss exponent (η)	3.8
Mean of log-normal distribution	3
Number of iterations	5000

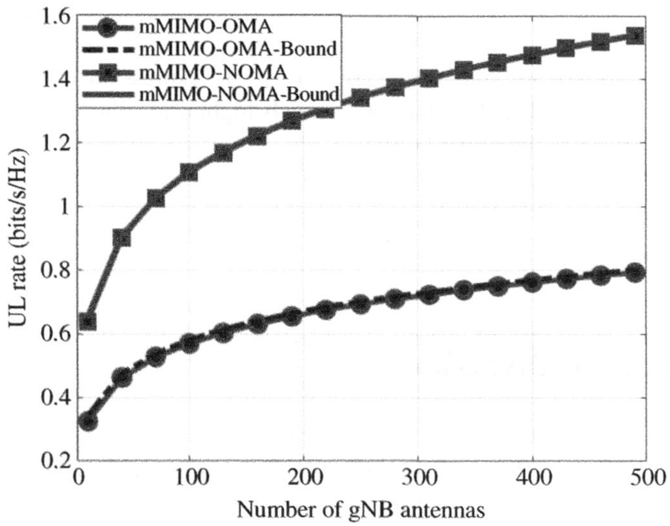

Figure 8.5 UL rate of mMIMO-OMA vs mMIMO-NOMA systems for UE_1 considering $\alpha_1 = 0.9$ and power scaling.

mMIMO-OMA system results 0.76 bits/s/Hz of data rate. The simulation results confirm mMIMO-NOMA outperforms mMIMO-OMA by ≈94.74%.

Figure 8.6 shows the UL rates of mMIMO-OMA and mMIMO-NOMA systems for UE_2 as the number of gNB antennas is increased. For mMIMO-NOMA system,

Algorithm 8.1

Input: M, α_1, α_2, operating frequency.
Output: Data rate of UE_1 and UE_2, Sum rate.
Step 1: Generate random bits and perform constellation mapping.
Step 2: Transmit the symbols from mMIMO-NOMA and mMIMO-OMA setup through uncorrelated channels to the receivers.
Step 3: At the receivers, measure the SNR and SINR.
Step 4: Compute the data rate of UE_1 and UE_2 and sum rate for different M.

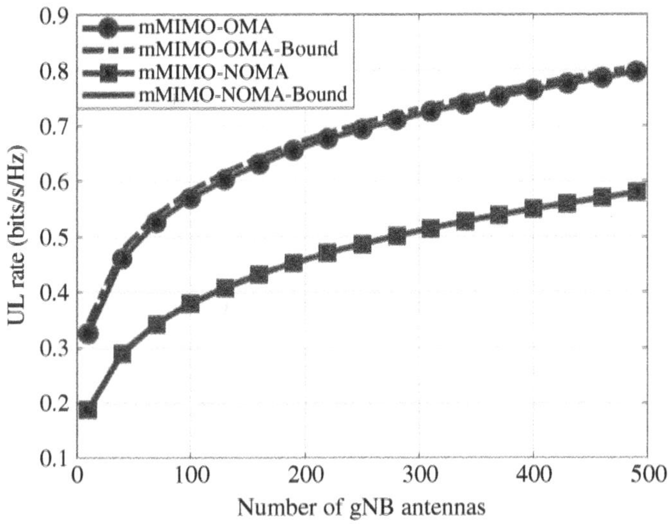

Figure 8.6 UL rate of mMIMO-OMA vs mMIMO-NOMA systems for UE_2 considering $\alpha_2 = 0.1$ and power scaling.

power fraction allocated to UE_2 is $\alpha_2 = 0.1$. The increase in gNB antennas result in increase in the data rate of UE_2. When gNB antennas is 400, mMIMO-OMA system produces the data rate of 0.76 bits/s/Hz, whereas mMIMO-NOMA system results 0.55 bits/s/Hz of data rate. This is because only 0.1 fraction of power is allocated to UE_2 under mMIMO-NOMA.

Figure 8.7 compares the UL sum rate of mMIMO-OMA and mMIMO-NOMA systems by increasing the number of gNB antennas. The increase in gNB antennas result in increase in sum rate. Even though the mMIMO-NOMA rate is lower than the mMIMO-OMA rate for UE_2, the overall sum rate of the mMIMO-NOMA system is higher. When gNB antennas is 400, mMIMO-NOMA system produces the sum rate of 2.03 bits/s/Hz, whereas mMIMO-OMA system results

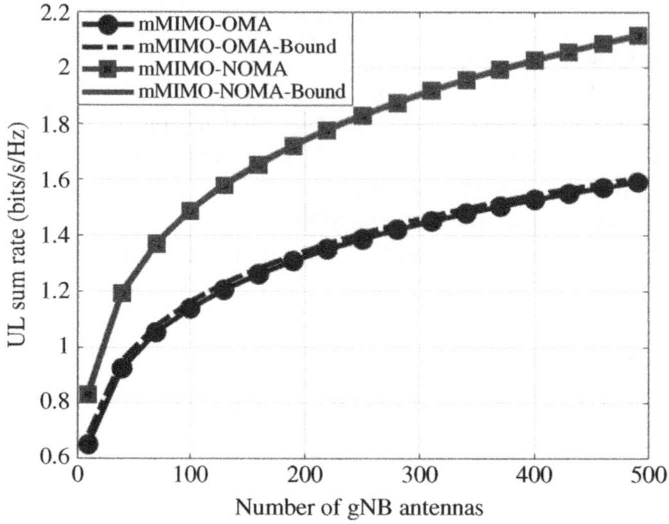

Figure 8.7 UL sum rate of mMIMO-OMA vs mMIMO-NOMA systems considering $\alpha_1 = 0.9$ and $\alpha_2 = 0.1$ and power scaling.

1.52 bits/s/Hz of data rate. The simulation results confirm mMIMO-NOMA outperforms mMIMO-OMA by ≈33.55%.

Figure 8.8 shows the UL rates of mMIMO-OMA and mMIMO-NOMA systems for UE_1 as the number of gNB antennas is increased. For mMIMO-NOMA system, power fraction allocated to UE_1 is $\alpha_1 = 0.7$. For mMIMO-OMA and mMIMO-NOMA systems, power scaling according to (8.14) and (8.23) is used. The upper rate bounds of both systems are plotted using (8.16) and (8.25). As the number of gNB antennas increases, the actual rate and upper bound rate curves for both systems meet. The increase in gNB antennas results in an increase in UE_1 data rate. When gNB antennas is 400, mMIMO-NOMA system attains the data rate of 1.39 bits/s/Hz, whereas mMIMO-OMA system results 0.76 bits/s/Hz of data rate. The simulation results confirm mMIMO-NOMA outperforms mMIMO-OMA by ≈82.89%. When comparing Figures 8.5 and 8.8, it is evident that reducing the PA for UE_1 reduces the rate. The PA has no effect on the mMIMO-OMA user rate.

Figure 8.9 shows the UL rates of mMIMO-OMA and mMIMO-NOMA systems for UE_2 as the number of gNB antennas is increased. For mMIMO-NOMA system, power fraction allocated to UE_2 is $\alpha_2 = 0.3$. The increase in gNB antennas results in increase in the data rate of UE_2. When gNB antennas is 400, mMIMO-OMA system produces the data rate of 0.76 bits/s/Hz, whereas mMIMO-NOMA system results 0.93 bits/s/Hz of data rate. The simulation results confirm mMIMO-NOMA outperforms mMIMO-OMA by ≈22.37%. When comparing Figures 8.6 and 8.9, it

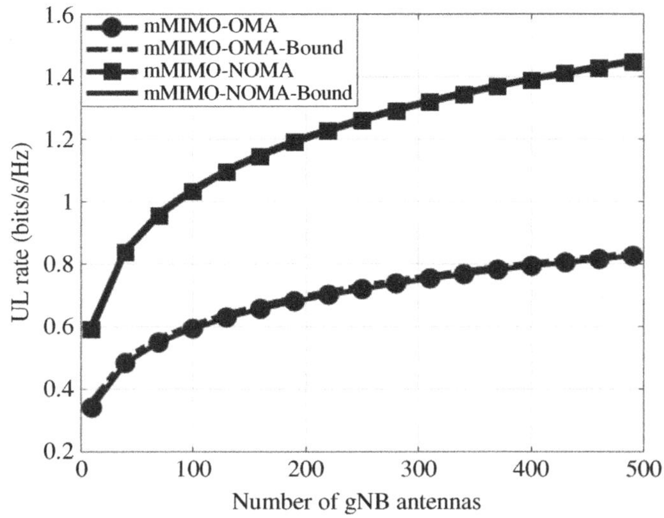

Figure 8.8 UL rate of mMIMO-OMA vs mMIMO-NOMA systems for UE_1 considering $\alpha_1 = 0.7$ and power scaling.

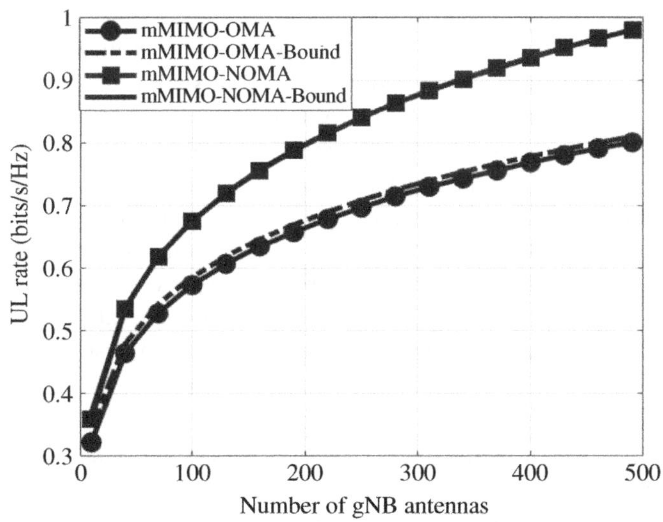

Figure 8.9 UL rate of mMIMO-OMA vs mMIMO-NOMA systems for UE_2 considering $\alpha_2 = 0.3$ and power scaling.

is evident that increasing the PA for UE_2 increases the rate by ≈69.09%. The PA has no effect on the rate of mMIMO-OMA users.

Figure 8.10 compares the UL sum rate of mMIMO-OMA and mMIMO-NOMA systems by increasing the number of gNB antennas, considering $\alpha_1 = 0.7$ and

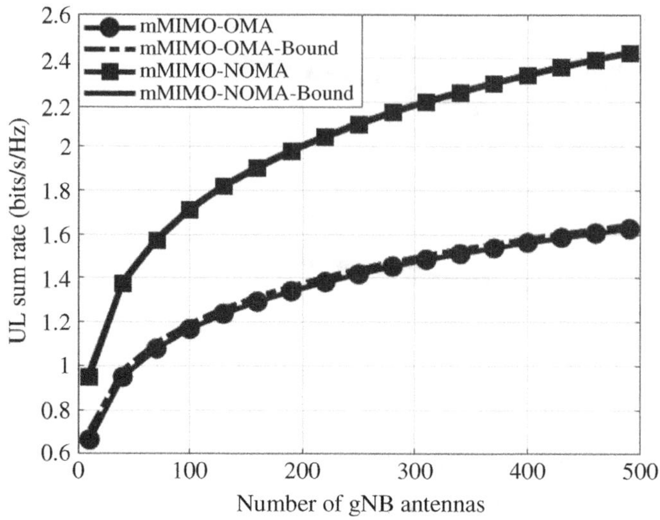

Figure 8.10 UL sum rate of mMIMO-OMA vs mMIMO-NOMA systems considering $\alpha_1 = 0.7$ and $\alpha_2 = 0.3$ and power scaling.

$\alpha_2 = 0.3$ and power scaling. The increase in gNB antennas result in increase in the sum rate. When gNB antennas is 400, mMIMO-NOMA system produces the sum rate of 2.32 bits/s/Hz, whereas mMIMO-OMA system results 1.56 bits/s/Hz of data rate. The simulation results confirm mMIMO-NOMA outperforms mMIMO-OMA by ≈52.63%. By comparing Figures 8.7 and 8.10, it is clear that the change of PA coefficients from $\alpha_1 = 0.9$ and $\alpha_2 = 0.1$ to $\alpha_1 = 0.7$ and $\alpha_2 = 0.3$ increases the sum rate from 2.03 bits/s/Hz to 2.32 bits/s/Hz for 400 antennas. The PA factors are chosen at random. The optimal selection of PA parameters is outside the scope of this chapter. Table 8.2 compares the achievable rate of UE_1 and UE_2 and sum rate under mMIMO-NOMA and mMIMO-OMA systems for the PA factors $\alpha_1 = 0.9$, $\alpha_2 = 0.1$ and $\alpha_1 = 0.7$, $\alpha_2 = 0.3$ at $M = 400$.

Table 8.2 Rate comparison of mMIMO-NOMA and mMIMO-OMA systems for two different PA factors at $M = 400$.

	R_1 ($\alpha_1 = 0.9$)	R_2 ($\alpha_2 = 0.1$)	R	R_1 ($\alpha_1 = 0.7$)	R_2 ($\alpha_2 = 0.3$)	R
mMIMO-NOMA	≈ 1.48	≈0.55	≈2.03	≈1.39	≈0.93	≈2.32
mMIMO-OMA	≈ 0.76	≈0.76	≈1.52	≈0.76	≈0.76	≈1.52
% Increase	≈ 94.74	—	≈33.55	≈82.89	≈22.37	≈52.63

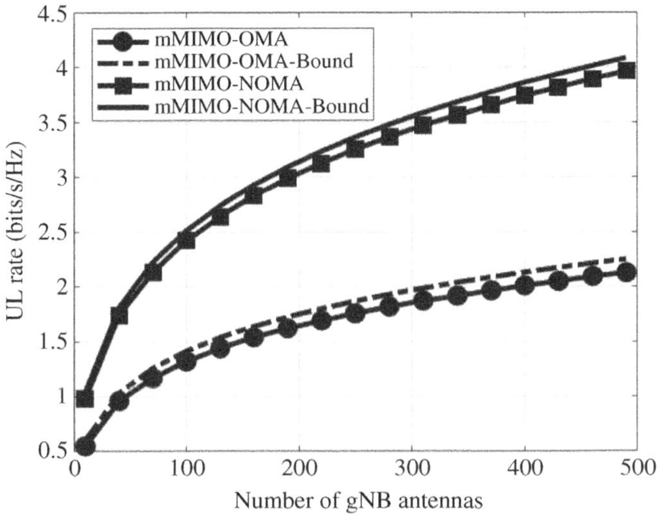

Figure 8.11 UL rate of mMIMO-OMA vs mMIMO-NOMA systems for UE_1 considering $\alpha_1 = 0.7$ and without power scaling.

Figure 8.11 shows the UL rates of mMIMO-OMA and mMIMO-NOMA systems for UE_1 as the number of gNB antennas is increased. For mMIMO-NOMA system, power fraction allocated to UE_1 is $\alpha_1 = 0.7$. For mMIMO-OMA and mMIMO-NOMA systems, no power scaling is used. When gNB antennas is 400, mMIMO-NOMA system attains the data rate of 3.73 bits/s/Hz, whereas mMIMO-OMA system results 2.01 bits/s/Hz of data rate. The simulation results confirm mMIMO-NOMA outperforms mMIMO-OMA by $\approx$85.57%.

Figure 8.12 illustrates the UL rates of mMIMO-OMA and mMIMO-NOMA systems without power scaling for UE_2 as the number of gNB antennas is increased. For mMIMO-NOMA system, power fraction allocated to UE_2 is $\alpha_2 = 0.3$. The increase in gNB antennas result in increase in the data rate of UE_2. When gNB antennas is 400, mMIMO-OMA system produces the data rate of 1.99 bits/s/Hz, whereas mMIMO-NOMA system results 2.95 bits/s/Hz of data rate. The simulation results confirm mMIMO-NOMA outperforms mMIMO-OMA by $\approx$48.24%.

Figure 8.13 compares the UL sum rate of mMIMO-OMA and mMIMO-NOMA systems by increasing the number of gNB antennas, considering $\alpha_1 = 0.7$ and $\alpha_2 = 0.3$ and no power scaling. The increase in gNB antennas result in increase in the sum rate. When gNB antennas is 400, mMIMO-NOMA system produces the sum rate of 6.68 bits/s/Hz, whereas mMIMO-OMA system results 4 bits/s/Hz of data rate. The simulation results confirm mMIMO-NOMA outperforms mMIMO-OMA by $\approx$67%. From Figures 8.11 to 8.13, it is evident that no power

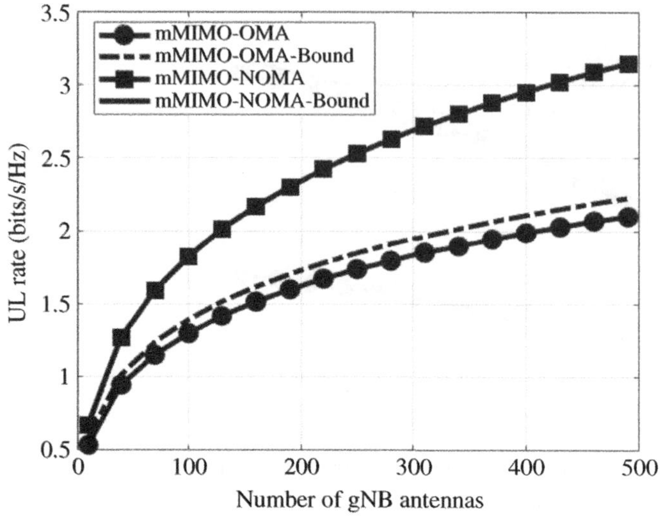

Figure 8.12 UL rate of mMIMO-OMA vs mMIMO-NOMA systems for UE_2 considering $\alpha_2 = 0.3$ and without power scaling.

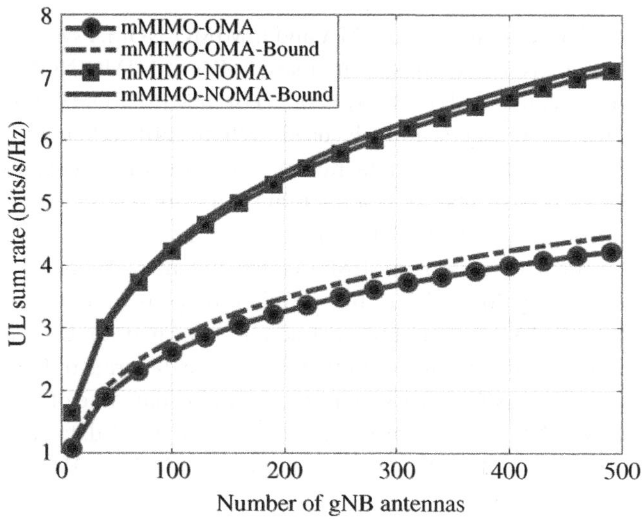

Figure 8.13 UL sum rate of mMIMO-OMA vs mMIMO-NOMA systems considering $\alpha_1 = 0.7$ and $\alpha_2 = 0.3$ and without power scaling.

scaling increases the rate over power scaling. But this leads to reduction of EE of the system. It is also discovered that for all no power scaling scenarios, the actual rates and upper bound rates differ even for higher number of gNB antennas. The attainable individual rates and sum rate of mMIMO-NOMA and mMIMO-OMA users with and without power scaling at $M = 400$ re compared in Table 8.3.

Table 8.3 Rate comparison of mMIMO-NOMA and mMIMO-OMA systems with and without power scaling at $M = 400$

	With power scaling			Without power scaling		
	R_1	R_2	R	R_1	R_2	R
mMIMO-NOMA	≈1.39	≈0.93	≈2.32	≈3.73	≈2.95	≈6.68
mMIMO-OMA	≈0.76	≈0.76	≈1.52	≈2.01	≈1.99	≈4
% Increase	≈82.89	≈22.37	≈52.63	≈85.57	≈48.24	≈67

8.5 Conclusion

The NOMA-assisted mMIMO system is explored in this chapter. The system model for the suggested NOMA-assisted mMIMO UL is formulated with M antennas at gNB and K single-antenna users, where $M \gg K$. The PA factors for the users determine the sum rate of the proposed system. In our work, we varied $\alpha_1 = 0.9$ to $\alpha_1 = 0.7$, which results in ≈14.29% improvement in sum rate. The proposed mMIMO-NOMA outperforms the existing mMIMO-OMA in terms of sum rate by ≈52.63% with power scaling and ≈67% without power scaling for the PA factor of $\alpha_1 = 0.7$ and $\alpha_2 = 0.3$. The proposed system is simulated with power scaling and without power scaling. The system without power scaling performs ≈187.93% better than power scaled system for mMIMO-NOMA. But without power scaling, it cannot support EE requirements of next-generation networks. Our future scope is to extend our work from two-user scenario to multiuser environment. The PA factors utilized are random. An optimization framework for deciding the PA factors can be derived to maximize the sum rate.

References

1 Wikström, G., Persson, P., Parkvall, S. et al. (2022). 6G–connecting a cyber-physical world *White Paper*, 28.

2 IIMT Union (2015). IMT traffic estimates for the years 2020 to 2030. *Report ITU*, 2370.

3 Samsung Research (2020). 6G: The next hyper connected experience for all.

4 Zhang, Z. and Dai, L. (2023). Reconfigurable intelligent surfaces for 6G: nine fundamental issues and one critical problem. *Tsinghua Science and Technology* 28 (5): 929–939.

5 Choi, J. (2014). Non-orthogonal multiple access in downlink coordinated two-point systems. *IEEE Communications Letters* 18 (2): 313–316.

6 Sadeghi, M., Yuen, C., and Chew, Y.H. (2014). Sum rate maximization for uplink distributed massive MIMO systems with limited backhaul capacity. *2014 IEEE Globecom Workshops (GC Wkshps)*, 308–313. IEEE.

7 Lu, L., Li, G.Y., Swindlehurst, A.L. et al. (2014). An overview of massive MIMO: benefits and challenges. *IEEE Journal of Selected Topics in Signal Processing* 8 (5): 742–758.

8 Choi, J.W., Shim, B., and Chang, S.-H. (2015). Downlink pilot reduction for massive MIMO systems via compressed sensing. *IEEE Communications Letters* 19 (11): 1889–1892.

9 Feng, C., Jing, Y., and Jin, S. (2015). Interference and outage probability analysis for massive MIMO downlink with MF precoding. *IEEE Signal Processing Letters* 23 (3): 366–370.

10 Björnson, E., Hoydis, J., and Sanguinetti, L. (2017). Massive MIMO has unlimited capacity. *IEEE Transactions on Wireless Communications* 17 (1): 574–590.

11 Albreem, M.A., Juntti, M., and Shahabuddin, S. (2019). Massive MIMO detection techniques: a survey. *IEEE Communications Surveys & Tutorials* 21 (4): 3109–3132.

12 Singh, P., Mishra, H.B., Jagannatham, A.K. et al. (2020). Uplink sum-rate and power scaling laws for multi-user massive MIMO-FBMC systems. *IEEE Transactions on Communications* 68 (1): 161–176. https://doi.org/10.1109/TCOMM.2019.2950216.

13 Albreem, M.A., Habbash, A.H.A., Abu-Hudrouss, A.M., and Ikki, S.S. (2021). Overview of precoding techniques for massive MIMO. *IEEE Access* (9): 60764–60801.

14 Sanoopkumar, P.S., Muneer, P., and Sameer, S.M. (2022). A joint equalization and decoding technique for multiuser massive MIMO uplink system with transmitter and receiver RF impairments under doubly selective channels. *IEEE Systems Journal* 16 (4): 6142–6152.

15 Jin, H., Liu, K., Zhang, M. et al. (2023). Massive MIMO evolution towards 3GPP release 18. *IEEE Journal on Selected Areas in Communications* 41 (6): 1635–1654.

16 Ouyang, C., Xu, H., Zang, X., and Yang, H. (2023). Sum-rate capacity scaling law in massive MIMO with antenna selection. *2023 IEEE Wireless Communications and Networking Conference (WCNC)*, 1–6. IEEE.

17 Xu, G., Lampu, V., Kosunen, M. et al. (2023). Digital polar transmitters for massive MIMO: sum-rate and power efficiency analysis. *IEEE Transactions on Wireless Communications* 23 (1): 789–805.

18 Nouri, M., Behroozi, H., Bastami, H. et al. (2023). Hybrid precoding based on active learning for mmWave massive MIMO communication systems. *IEEE Transactions on Communications* 71 (5): 3043–3058.

19 Zhang, B., Xu, P., Qiao, B. et al. (2023). Highly accurate millimeter wave channel estimation in massive MIMO system. *IET Communications* 17 (6): 670–680.

20 Lu, A.-A., Chen, Y., and Gao, X. (2023). 2D beam domain statistical CSI estimation for massive MIMO uplink. *IEEE Transactions on Wireless Communications* 23 (1): 749–761.

21 Do, D.-T. and Nguyen, T.-T.T. (2018). Exact outage performance analysis of amplify-and-forward-aware cooperative NOMA. *TELKOMNIKA (Telecommunication Computing Electronics and Control)* 16 (5): 1966–1973.

22 Do, D.-T., Le, C.-B., and Le, A.-T. (2019). Cooperative underlay cognitive radio assisted NOMA: secondary network improvement and outage performance. *TELKOMNIKA (Telecommunication Computing Electronics and Control)* 17 (5): 2147–2154.

23 Yin, Y., Peng, Y., Liu, M. et al. (2019). Dynamic user grouping-based NOMA over Rayleigh fading channels. *IEEE Access* (7): 110964–110971.

24 Dao, V.-L., Hoang, L.-N., Girs, S., and Uhlemann, E. (2020). Outage performance of pairwise NOMA allowing a dynamic decoding order and optimal pairs of power levels. *IEEE Open Journal of the Communications Society* 1: 1886–1906.

25 Mounchili, S. and Hamouda, S. (2020). Pairing distance resolution and power control for massive connectivity improvement in NOMA systems. *IEEE Transactions on Vehicular Technology* 69 (4): 4093–4103.

26 Al-Obiedollah, H., Cumanan, K., Salameh, H.B. et al. (2021). Downlink multi-carrier NOMA with opportunistic bandwidth allocations. *IEEE Wireless Communications Letters* 10 (11): 2426–2429.

27 Thet, N.W.M. and Ozdemir, M.K. (2021). Partial-beam non-orthogonal multiple access (PB-NOMA) with fuzzy clustering. *IEEE Wireless Communications Letters* 10 (6): 1202–1206.

28 Dai, Y., Lyu, L., and Cheng, N. (2023). Impact of decoding order on user pairing in NOMA-CoMP joint transmission. *IEEE Transactions on Vehicular Technology* 72 (7): 9664–9669.

29 Almohamad, A., Hasna, M.O., Althunibat, S., and Qaraqe, K. (2021). A novel downlink IM-NOMA scheme. *IEEE Open Journal of the Communications Society* (2): 235–244.

30 Zhu, J., Gao, P., Chen, G. et al. (2022). Index modulation for STAR-RIS assisted NOMA system. *IEEE Communications Letters* 27 (2): 716–720.

31 Zhang, J., Zhu, L., Xiao, Z. et al. (2019). Optimal and sub-optimal uplink NOMA: joint user grouping, decoding order, and power control. *IEEE Wireless Communications Letters* 9 (2): 254–257.

32 Guo, L., Cong, S., Su, C., and Sun, Z. (2020). Coverage analysis of downlink NOMA-enabled dense networks with ISP-based user ordering. *IEEE Communications Letters* 24 (11): 2440–2444.

33 Zhang, R., Pang, X., Tang, J. et al. (2020). Joint location and transmit power optimization for NOMA-UAV networks via updating decoding order. *IEEE Wireless Communications Letters* 10 (1): 136–140.

34 Do, D.-T., Nguyen, T.-L., Ekin, S. et al. (2020). Joint user grouping and decoding order in uplink/downlink MISO/SIMO-NOMA. *IEEE Access* (8): 143632–143643.

35 Thapar, S., Mishra, D., and Saini, R. (2021). Decoding orders for securing untrusted NOMA. *IEEE Networking Letters* 3 (1): 27–30.

36 Do, D.-T. (2018). Investigation on tradeoff between hardware noise and outage performance in cooperative NOMA. *Journal of Communication* 13 (8): 468–472.

37 Le, C.-B., Dinh-Thuan, D., and Voznak, M. (2019). Wireless-powered cooperative MIMO NOMA networks: design and performance improvement for cell-edge users. *Electronics* 8 (3): 328.

38 Li, H., Chen, Y., Zhu, M. et al. (2020). Secrecy outage probability of relay selection based cooperative NOMA for IoT networks. *IEEE Access* (9): 1655–1665.

39 Nguyen, T.-T. and Do, D.-T. (2021). Performance gap of two users in downlink full-duplex cooperative NOMA. *TELKOMNIKA (Telecommunication Computing Electronics and Control)* 19 (6): 1795–1802.

40 Su, B., Ni, Q., and Yu, W. (2019). Robust transmit beamforming for SWIPT-enabled cooperative NOMA with channel uncertainties. *IEEE Transactions on Communications* 67 (6): 4381–4392.

41 Li, S., Bariah, L., Muhaidat, S. et al. (2020). SWIPT-enabled cooperative NOMA with *m*th best relay selection. *IEEE Open Journal of the Communications Society* (1): 1798–1807.

42 Nguyen, T.-L., Do, D.-T., Mavromoustakis, C.X., and Mastorakis, G. (2021). Opportunistic user selection schemes for energy harvesting-aware cooperative NOMA. *Physical Communication* 44: 101258.

43 Budhiraja, I., Kumar, N., Tyagi, S. et al. (2021). SWIPT-enabled D2D communication underlaying NOMA-based cellular networks in imperfect CSI. *IEEE Transactions on Vehicular Technology* 70 (1): 692–699.

44 Kara, F. and Kaya, H. (2019). Error probability analysis of NOMA-based diamond relaying network. *IEEE Transactions on Vehicular Technology* 69 (2): 2280–2285.

45 Huang, H. and Zhu, M. (2019). Energy efficiency maximization design for full-duplex cooperative NOMA systems with SWIPT. *IEEE Access* (7): 20442–20451.

46 Shi, Z., Xie, X., Lu, H. et al. (2020). Deep reinforcement learning based dynamic user access and decode order selection for uplink NOMA system with imperfect SIC. *IEEE Wireless Communications Letters* 10 (4): 710–714.

47 Do, D.-T., Le, A.-T., and Lee, B.M. (2020). NOMA in cooperative underlay cognitive radio networks under imperfect SIC. *IEEE Access* (8): 86180–86195.

48 Babu, A.V. (2021). Performance analysis of power splitting SWIPT-enabled full duplex cooperative NOMA system with direct link. *IET Communications* 15: 1028–1044.

49 Senel, K., Cheng, H.V., Björnson, E., and Larsson, E.G. (2019). What role can NOMA play in massive MIMO? *IEEE Journal of Selected Topics in Signal Processing* 13 (3): 597–611.

50 Hadley, L. and Chatzigeorgiou, I. (2020). Low complexity optimization of the asymptotic spectral efficiency in massive MIMO NOMA. *IEEE Wireless Communications Letters* 9 (11): 1928–1932.

51 Simon, E.P., Farah, J., Laly, P., and Delbarre, G. (2021). A gradual resource allocation technique for massive MIMO-NOMA. *IEEE Antennas and Wireless Propagation Letters* 21 (3): 476–480.

52 Amirifar, Z. and Abouei, J. (2022). The dynamic power allocation to maximize the achievable sum rate for massive MIMO-NOMA systems. *IET Communications* 16 (17): 2036–2044.

53 Ahmed, I., Shahid, M.K., and Faisal, T. (2022). Deep reinforcement learning based beam selection for hybrid beamforming and user grouping in massive MIMO-NOMA system. *IEEE Access* (10): 89519–89533.

54 Shili, M., Hajjaj, M., and Ammari, M.L. (2022). User clustering and power allocation for massive MIMO with NOMA-inspired cognitive radio. *IEEE Transactions on Vehicular Technology* 71 (7): 7656–7664.

55 He, X., Huang, Z., Wang, H., and Song, R. (2023). Sum rate analysis for massive MIMO-NOMA uplink system with group-level successive interference cancellation. *IEEE Wireless Communications Letters* 12 (7): 1194–1198.

56 Ngo, H.Q., Larsson, E.G., and Marzetta, T.L. (2013). Energy and spectral efficiency of very large multiuser MIMO systems. *IEEE Transactions on Communications* 61 (4): 1436–1449.

57 Kumaravelu, V.B., Imoize, A.L., Soria, F.R.C. et al. (2023). RIS-assisted fixed NOMA: outage probability analysis and transmit power optimization. *Future Internet* 15 (8): 249.

9

Efficient Hybrid Precoding for Millimeter-Wave Massive MIMO-NOMA Systems: A Low-Complexity Approach

Samarendra Nath Sur[1], Arun Kumar Singh[1], Agbotiname Lucky Imoize[2,3], Joseph Bamidele Awotunde[4], Debdatta Kandar[5], and Vinoth Babu Kumaravelu[6]

[1] *Department of Electronics and Communication Engineering, Sikkim Manipal Institute of Technology, Sikkim Manipal University, Sikkim, India*
[2] *Department of Electrical and Electronics Engineering, Faculty of Engineering, University of Lagos, Akoka, Lagos, Nigeria*
[3] *Department of Electrical Engineering and Information Technology, Institute of Digital Communication, Ruhr University, Bochum, Germany*
[4] *Department of Computer Science, Faculty of Information and Communication Sciences, University of Ilorin, Ilorin, Nigeria*
[5] *Department of Information Technology, North Eastern Hill University, Shillong, Meghalaya, India*
[6] *Department of Communication Engineering, School of Electronics Engineering, Vellore Institute of Technology, Vellore-632014 Tamil Nadu, India*

9.1 Introduction

In the realm of burgeoning communication systems for future generations, paramount emphasis is being placed upon the attainment of two pivotal attributes: high-speed transmission and minimal latency [1–5]. Achieving this goal hinges on the seamless integration of millimeter-wave (mmWave) communication and massive multiple-input multiple-output (mMIMO) technology. This integration stands out as a crucial and central factor in fulfilling the objective [6]. This combination of mmWave and mMIMO substantially bolsters the spectral efficiency (SE) and energy efficiency (EE) of the communication setup. Utilization of the mmWave frequency bands enables the accessibility of ample bandwidth. The utilization of extensive antenna arrays at both the transmission and reception ends offers the exciting possibility of achieving higher array gains comparing the conventional MIMO system.

6G mobile networks are set to confront an immense surge in mobile data traffic stemming from the simultaneous usage of both users and machines.

Massive MIMO for Future Wireless Communication Systems: Technology and Applications,
First Edition. Edited by Webert Montlouis and Agbotiname Lucky Imoize.

Alongside this surge, the demand for services necessitating extensive connectivity, unparalleled reliability, exceptionally low latency, and heightened SE [1, 7] is driving the imperative for a revolutionary shift in how resources are allocated orthogonally. In this evolving landscape, the triumph of 6G networks hinges not solely upon conventional communication approaches but also on groundbreaking communication paradigms. These paradigms veer away from the prevailing trends, introducing a fresh perspective. One such paradigm that stands out is non-orthogonal multiple access (NOMA), which distinguishes itself from the conventional orthogonal multiple access (OMA) methods. This ground breaking scheme enables the simultaneous servicing of multiple users (MU) and machines to effectively share the same resources, whether it be time, frequency, or code, which has the potential to significantly improve SE [8]. By embracing NOMA, 6G networks can pave the way for a more efficient and dynamic utilization of resources, ultimately propelling the network's capacity to accommodate the burgeoning demands of a hyper-connected world. Built on the foundation of NOMA, mMIMO-NOMA technology represents a significant development in wireless communication systems. Thus an integration of mMIMO and NOMA enhances the possibility of enhancing the SE of the system even further [9–11]. However there are some challenges related to the mMIMO-NOMA system. Considering a mMIMO-NOMA system, the number of user's equipments (UEs) is larger than the available radio frequency chains (RFCs), resulting in pronounced interference issues both between and within beams, commonly referred to as inter-beam and intra-beam interference [12]. The ramifications of this challenge are substantial, as it directly undermines the intended benefits of mMIMO-NOMA technology. Concurrently, the proliferation of UEs in 6G communication systems introduces a dramatic escalation in energy consumption, accentuating the power allocation predicament [8]. This issue revolves around the efficient distribution of power resources among the surging number of UEs. If not adeptly managed, this heightened power consumption can negate the advancements brought about by mMIMO-NOMA technology, thwarting the expected gains in SE.

Addressing these intricate issues is pivotal for the successful implementation of 6G communication systems. Effective solutions must be devised to mitigate inter-user interference, preserving the envisioned gains in SE, and to develop sophisticated power allocation strategies to counteract the surging energy consumption. Thereby making the system both spectral and energy efficient.

Within the scope of this chapter, one of the possible approaches to address this issues to design advanced precoding and combining techniques. Considering the limitation of the analog and digital beamforming, there exists a compelling need to formulate cost-effective hardware solutions that can effectively leverage the potential multiplexing gains in the mmWave band. Although power allocation technique plays important role but this chapter meant for the design of hybrid precoding (HP) scheme.

The requirement of the low cost, less complex system leads to the development of the HP scheme. Predominantly, researchers have extensively explored three distinct architectures for mmWave-mMIMO systems. These configurations are commonly referred to as the fully digital (FD), the fully connected (FC), and the sub-connected (SC) architecture. In the FD arrangement, each antenna corresponds to a dedicated RFC. Thus for a large antenna system it escalates the system complexity [13]. Addressing this predicament, the concept of HP has been introduced [11, 14, 15] and its performance has been scrutinized over time. Previous research [16, 17] has eloquently demonstrated how HP may provide equivalent performance in mmWave-mMIMO systems while drastically lowering the number of necessary RFCs. Usually, the HP design framework takes both the FC and SC architectures into account. In the FC architecture, fewer RFCs are used, and each RFC is connected to all accessible antennas. The SC architecture, on the other hand, requires connecting each RFC to a portion of the entire antenna pool [18]. The ultimate aim of these architectural choices is to optimize the efficiency and effectiveness of mmWave-mMIMO communication systems. Many academics have taken an interest in SC because of its unique properties that enable it to strike a better balance between performance, cost, and complexity. Hence, in recent times, there has been a notable surge in interest surrounding research on HP algorithms incorporating SC structures.

9.1.1 Contributions

This chapter introduces several contributions, which are summarized as follows:

- The chapter provides an architectural overview on the precoding algorithm for the mmWave-mMIMO system, including the basic mathematical model.
- In the realm of mMIMO systems, the mitigation of computational complexity stands as a pivotal concern. In this work, the researchers have introduced symmetric sequential over-relaxation/complex regularized zero-forcing (SSOR-CRZF) procoder. This novel procoder has been strategically formulated to enhance the performance of mMIMO-NOMA systems, thereby contributing to the advancement of this cutting-edge communication paradigm. And its effectiveness is established through a numeric comparison with the existing precoding schemes.
- The investigation detailed in this work serves to substantiate the incremental performance enhancement yielded by the proposed SSOR-ZF procoder, relative to the established CRZF algorithm along with its associated advantage of diminished computational complexity.

9.1.2 Chapter Organization

The subsequent sections of this chapter are thoughtfully organized to delve into various facets of the proposed research. The structure of the remainder of the paper is outlined as follows: in Section 9.2, an overview of different precoding schemes are discussed with architectural overview. Section 9.3 highlights the related works. Section 9.4 presents the mathematical model for a mMIMO-NOMA system along with the proposed HP algorithm, particularly the SSOR-CRZF algorithm. The numerical results and corresponding analysis are presented in Section 9.5. Section 9.6 highlights the research challenges and future research direction. The conclusion is given in Section 9.7.

9.2 Overview of Basic Precoding Scheme

Analog, digital, and HP are techniques used in wireless communication systems to optimize the transmission of signals from a transmitter to a receiver. These techniques aim to enhance SE and mitigate interference. This section provides a brief overview of each of these precoding schemes along with their mathematical formulations for the precoding matrix.

9.2.1 Analog Beamforming

Point-to-point mmWave systems that make use of large antenna arrays have significantly improved thanks to analog beamforming/ procoding (access point). To broadcast a single data stream, these devices use a single RFC. The key to maximizing antenna array gain and ensuring an effective signal-to-noise ratio (SNR) lies in the implementation of AP. This technique involves manipulating the phases of the original signals, allowing precise control over the signal direction to achieve optimal performance in the mmWave setup. There are several beamforming schemes available, such as beam steering and code book-based beam training. As a part of this chapter, we have highlighted the signal processing aspect of the beam steering approach.

A typical architecture of a mMIMO system with the AP system is presented in Figure 9.1. In case of AP, the steering vector is governed by the array response vectors. The array response vector for the uniform linear array (ULA) with N elements can be presented as [19],

$$\mathbf{a}_{\mathrm{ULA}}(\varphi) = \frac{1}{\sqrt{N}}\left[1, e^{j\frac{2\pi}{\lambda}d_r\ \sin(\varphi)}, \dots, e^{j(N-1))\frac{2\pi}{\lambda}d_r\ \sin(\varphi)}\right] \tag{9.1}$$

where the wavelength of the signal is indicated by λ and d_r is the distance between the elements. As we aware of the fact that array's response is invariant in the elevation domain (ϑ), it is not included in the equation. Similarly for a uniform planar

Figure 9.1 Architecture of analog beamforming for mmWave-mMIMO systems.

array (UPA) with $[N = N_{\text{el}}N_{\text{az}}]$ N_{el} elements in elevation direction and N_{az} elements in azimuth direction can be expressed as [19]

$$\mathbf{a}_{\text{UPA}}(\varphi, \vartheta) = \frac{1}{\sqrt{N}}\left[1, e^{j\frac{2\pi}{\lambda}d_r(n_{\text{az}}\sin(\varphi)\sin(\vartheta)+n_{\text{el}}\cos(\vartheta))}, \ldots, e^{j\frac{2\pi}{\lambda}d_r((n_{\text{az}}-1)\sin(\varphi)\sin(\vartheta)+(n_{\text{el}}-1)\cos(\vartheta))}\right]. \tag{9.2}$$

Considering the mMIMO system diagram as in Figure 9.1, the main objective one designer would be the maximization of SNR through the optimization of the beamforming vector (**F**) at the base station (BS) and combining vector (**W**) at the UE. Let N_T antennas are implemented at the BS with one RFC to broadcast the data toward the UE with N_R antennas. Under this arrangement, the AP optimization problem [20] can be defined as

$$(\mathbf{W}^{\text{opt}}, \mathbf{F}^{\text{opt}}) = \arg\max \left|\mathbf{W}^{H}\mathbf{H}\mathbf{F}\right|^{2} \tag{9.3a}$$

$$s.t. w_k = \sqrt{N_R^{-1}e^{j\varphi_k}},\ \forall k, \tag{9.3b}$$

$$f_m = \sqrt{N_T^{-1}e^{j\varphi_m}},\ \forall m. \tag{9.3c}$$

9.2.2 Digital Precoding

Digital precoding (DP) stands as a well-entrenched technology extensively utilized in low-frequency MIMO systems. The fundamental concept underlying DP involves the manipulation of signal phases and amplitudes, thereby facilitating interference precancellation. This technique can be broadly categorized into two distinct classes: linear and nonlinear precoding. The algorithms associated with these precoding methodologies are succinctly encapsulated in Figure 9.2.

In the context of signal transmission, linear precoding involves creating transmitted signals through a linear combination of the original signals. On the other hand, nonlinear precoding utilizes nonlinear methods to generate the transmitted signals. Furthermore, DP can be categorized into two subtypes: single-user precoding (referred to as Figure 9.3) which is designed to optimize

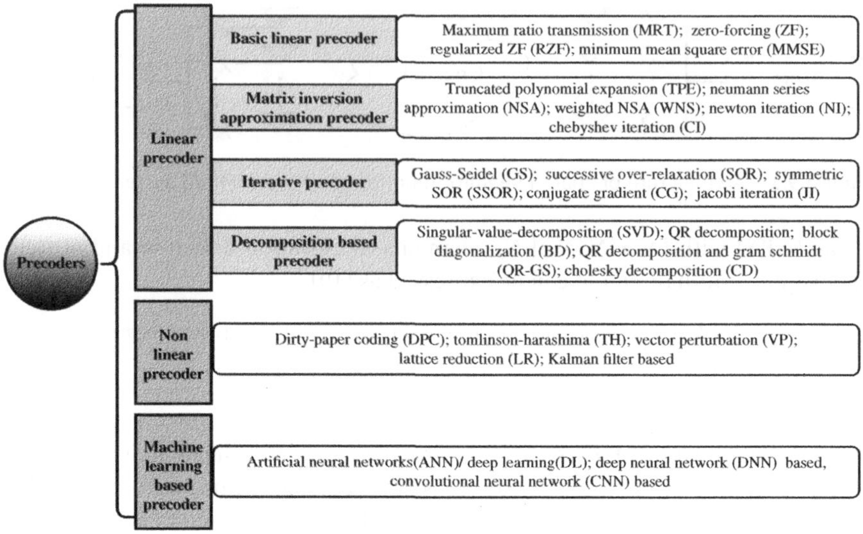

Figure 9.2 Digital precoder: algorithms.

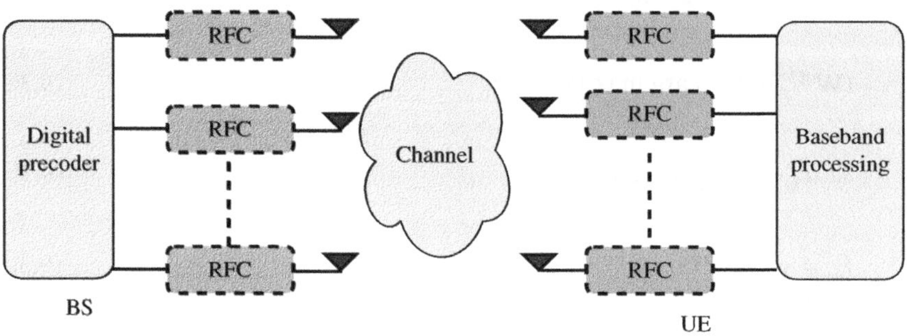

Figure 9.3 Digital precoder single user.

transmission for a single user, and MU precoding (referred to as Figure 9.4) which aims to enhance transmission for MU simultaneously.

In this section, we highlight the basic algorithms for the DPs. Let $\mathbf{H} \in \mathbb{C}^{N_T \times N_R}$ be the channel matrix between the BS and N_R single antenna UEs. The maximum ratio transmission (MRT) precoding algorithm aims to maximize the gain of the signal for a particular UE and it is also considered the counterpart to various other techniques, including matched filtering (MF) and conjugate beamforming (CB) [21]. The precoding matrix for MRT algorithm can be expressed as

$$\mathbf{F}_{\text{mrt}} = \sqrt{\beta}\,(\mathbf{H}^*), \tag{9.4}$$

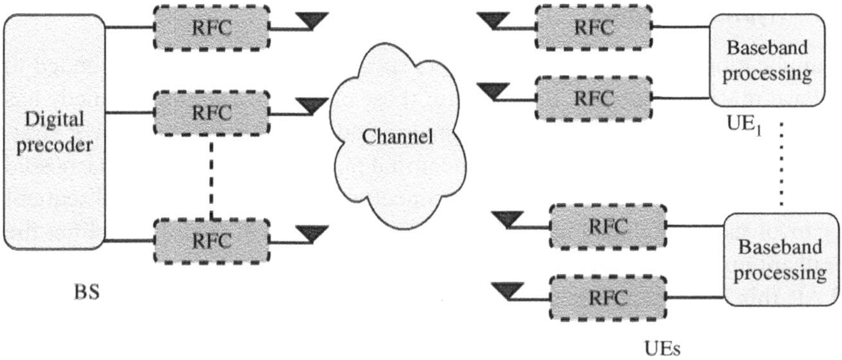

Figure 9.4 Digital precoder multiuser.

where β represents the scaling power factor and $\mathbf{H}^*$ is the complex conjugate of the channel matrix $\mathbf{H}$. The MRT algorithm typically exhibits close-to-optimal performance in scenarios where inter-user interference is negligible when compared to the noise in a given system.

The zero-forcing (ZF) algorithm is a widely used fundamental precoding technique in the MIMO systems [22]. It serves as the counterpart of channel inversion, offering a robust method to mitigate interference and enhance signal transmission. However, it is important to note that ZF may lead to noise amplification, especially in scenarios with noisy channels or limited channel state information. Therefore, some variants of ZF, such as regularized-ZF (RZF), have been developed to strike a balance between interference cancellation and noise amplification, resulting in improved performance in practical communication systems. The precoding matrix for the ZF algorithm can be presented as follows:

$$\mathbf{F}_{\text{zf}} = \sqrt{\beta}\mathbf{H}^H\left(\mathbf{H}\mathbf{H}^H\right)^{-1}. \tag{9.5}$$

Unfortunately, ZF algorithm in massive MIMO systems may not provide an optimal solution. It is only beneficial at high SNR level [21, 23].

The minimum mean square error (MMSE) algorithm is able to provide an acceptable performance under the moderate noise and interference level. It address the noise amplification issue in ZF algorithm. It is closely related to techniques such as RZF and transmit Wiener filtering:

$$\mathbf{F}_{\text{mmse}} = \sqrt{\beta}\mathbf{H}^H\left(\mathbf{H}\mathbf{H}^H + \frac{1}{\gamma}\mathbf{I}\right)^{-1}, \tag{9.6}$$

where γ denotes the SNR and $\mathbf{I} \in \mathbb{C}^{N_R \times N_R}$ is the identity matrix.

9.2.3 Hybrid Precoding

The application of DP and AP encounters specific challenges when extended to mmWave-mMIMO systems. To surmount these obstacles, a strategic remedy has been proposed in the form of hybrid analog and DP. This innovative HP strategy unfolds in two distinct stages. In the initial phase, a compact DP is harnessed to adeptly mitigate interference-related concerns. Following this, the subsequent stage involves the deployment of a large-size AP, which significantly amplifies the overall antenna array gain. By amalgamating the inherent advantages of both DP and AP, this hybrid approach emerges as a robust and pragmatic resolution tailored for mmWave massive MIMO systems.

From a structural perspective, as already stated, HP can be classified into two main types: the FC (as in Figure 9.5) and SC (as in Figure 9.6) architecture. In the FC architecture, each RFC is directly connected to every antenna through pilot assignments (PSs). This comprehensive connectivity allows for precise control over the beamforming process and provides greater flexibility in directing signals. On the other hand, in the SC architecture, each RFC is only connected to a subset of the antennas using PSs. This approach offers a tangible solution to alleviate

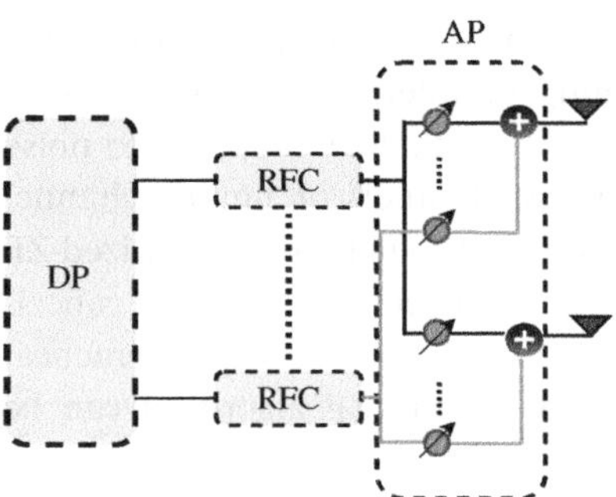

Figure 9.5 Hybrid procoding fully connected.

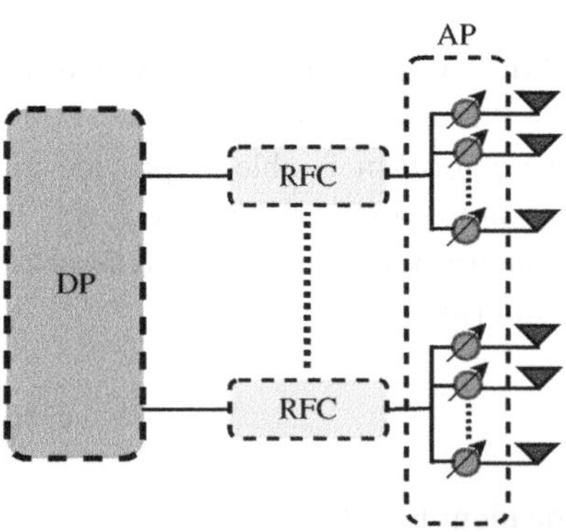

Figure 9.6 Hybrid procoding subconnected.

hardware complexity and curb power consumption, especially in systems featuring a substantial array of antennas. This quality renders it an eminently practical and viable remedy tailored to the demands of extensive mMIMO configurations.

Let the BS equipped with N_T transmit antennas with N_{RF} RFCs are utilized to provide service to K single antenna UEs [24, 25] ($N_{RF} \leq K < N_T$). More specifically, for the SC structure, each RFC is coupled to M ($= N_T/N_{RF}$) antennas. For simplification, one can consider that number of beams (D) created cannot be greater than N_{RF}, that is, $D = N_{RF}$. Under the hybrid scheme, the UEs data streams are coded through the baseband DP $\mathbf{f}_d^{BB}$ $[N_{RF} \times 1]$, $d \in [1, D]$ for the dth beam and $[N_T \times N_{RF}]$ RF AP $\mathbf{F}^{RF}$ exploiting analog PSs.

In case of a FC-HP, the AP matrix can be defined as

$$\mathbf{F}^{RF(full)} = \left[\bar{f}_1^{RF(full)}, \bar{f}_2^{RF(full)}, \ldots, \bar{f}_{N_{RF}}^{RF(full)}\right], \tag{9.7}$$

where $\bar{f}_m^{RF(full)} \in \mathbb{C}^{N_T \times 1}$ $\forall\, m = 1, 2, \ldots, N_{RF}$ is the steering vector having same amplitude $1/\sqrt{N_T}$ but with different phases [16].

For SC architecture, the AP matrix can be defined as

$$\mathbf{F}^{RF(sub)} = \begin{bmatrix} \bar{f}_1^{RF(sub)} & 0 \ \ldots\ldots & 0 \\ 0 & \bar{f}_2^{RF(sub)} & 0 \\ \vdots & \vdots & \vdots \\ 0 & 0 \ \ldots\ldots & \bar{f}_{N_{RF}}^{RF(sub)} \end{bmatrix}, \tag{9.8}$$

where $\bar{f}_m^{RF(sub)} \in \mathbb{C}^{M \times 1}$ $\forall\, m = 1, 2, \ldots, N_{RF}$ is the steering vector having same amplitude $1/\sqrt{M}$ but with different phases [26].

Now considering AP, the precoding matrix elements are subjected to satisfy $\left([\mathbf{F}^{RF}]_{:,j}\, [\mathbf{F}^{RF}]_{:,j}^H\right)_{l,l} = N_T^{-1}$ with $\left|[\mathbf{F}^{RF}]_{i,j}\right| = N_T^{-1/2}, 1 \leq i \leq N_T, 1 \leq j \leq N_{RF}$ and quantized phases: $[\mathbf{F}^{RF}]_{i,j} = (N_T)^{-1/2} \exp\left(j[\Phi]_{i,j}\right)$. And $[\Phi]_{i,j}$ is quantized as $[\Phi]_{i,j} \in \beta := \left\{\frac{2\pi n}{2^B} : n = 0, \ldots, 2^B - 1\right\}$. Apart from above constraints, the total power constraint can be realized by normalizing DP $\mathbf{f}_d^{BB}, d = 1, \ldots, D$ for the dth beam so that $\left\|\mathbf{F}^{RF} f_d^{BB}\right\|_2 = 1$ for $d = 1, \ldots, D$. The algorithm related to DP are discussed in the previous section and can be well utilized.

9.3 Related Work

This section highlights the works related to the precoding schemes for mMIMO system with brief overview on the architectural and computational complexities.

9.3.1 Linear Precoder

In this section, we provide concise overviews of various linear precoder schemes, with a specific emphasis on the following techniques: conventional basic linear precoders, matrix inversion approximation-based precoder, iterative precoder, and decomposition-based precoder.

9.3.1.1 Basic Linear Precoder

Figure 9.7 illustrates the generalized block diagrams of communication systems incorporating precoding techniques. The letter $\mathbf{P}$ represents a feedforward matrix used for linear precoding, while $\mathbf{B}$ signifies a feedback matrix utilized for linear precoding. To provide an example, when $\mathbf{B} = 0$, the generalized precoding technique functions similarly to the conventional linear precoding technique [27]. Here, $\mathbf{M}$ is the Modulo arithmetic and it is employed to regulate the average power [28]. And also $\mathbf{x}$ is the incoming information signal and $\mathbf{y}$ is the precoded signal.

Among the linear precoder schemes commonly used for mMIMO systems, ZF and MMSE stand out [29–31]. However, one significant drawback of MMSE lies in its dependency on the estimated SNR along with channel state information (CSI) and the need for power allocation among the UEs. And to address these issues, RZF algorithm [32, 33] is introduced. This approach aims to improve transmission power efficiency. Further advancements were made to enhance power efficiency by considering a complex regularization factor. In a recent investigation [34], the CRZF scheme was introduced and subjected to an in-depth performance analysis. Notably, the CRZF scheme demonstrated an equivalent computational complexity to the conventional ZF approach. This intriguing aspect suggests that additional modifications were likely incorporated to yield improved outcomes, all the while avoiding the introduction of supplementary computational load. Numerous efforts have been dedicated to the design of HP systems, with researchers exploring various schemes

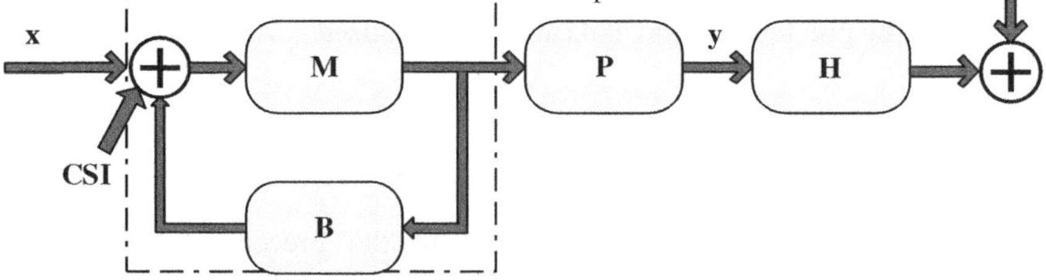

Figure 9.7 Generalized block diagram of communication systems with precoding.

such as ZF and MMSE. Additionally, the MRT scheme has been thoroughly investigated [35].

9.3.1.2 Matrix Inversion Approximation Precoder

The complexity of HP precoders in mMIMO systems, particularly due to the matrix inversion process, has raised concerns about their practical realization. Researchers have dedicated efforts to reduce this complexity, leading to the exploration of various techniques. Among them, Truncated polynomial expansion (TPE) [36] and Neumann series approximation (NSA)-based precodings [37] have been proposed and extensively studied. The NSA precoder proves advantageous when used with a small number of iterations. However, with a large number of iterations, it exhibits complexity comparable to that of the traditional ZF precoder. In mMIMO systems, NSA precoder often encounter the issue of slow convergence. To address this challenge, [38] introduce a novel approach known as the weighted NSA (WNS) precoder. Here, specific weights are meticulously designed with the aim of minimizing the error between the exact inverse and the inverse obtained through the weighted Neumann series (NS) expansion. The Newton iteration (NI) and Chebyshev iteration (CI) algorithms are employed for approximating the matrix inversion process in linear precoding for mMIMO systems through iterative computations [39, 40]. In terms of convergence speed, it has been observed that NI algorithms tend to converge more rapidly compared to the NSA algorithm [40]. To further expedite the convergence process, an innovative approach has been introduced in [41], known as the joint NI-NSA algorithm. Moreover, findings from [39] reveal that the CI algorithm demonstrates faster convergence compared to the NI algorithm. The CI algorithm achieves this accelerated convergence through the utilization of two matrix additions and three matrix multiplications.

9.3.1.3 Iterative Precoder

This limitation led researchers to seek alternative methods to decrease computational complexity. Several iterative approaches were proposed and evaluated, including Jacobi iteration (JI)-based precoding [42], Gram Schmidt (GS) precoding [43], SOR precoding [44, 45], SSOR precoding [46], weighted SSOR (WSSOR) precoding [47], and modified SOR (MSOR) [48]. These methods effectively convert the matrix inversion step into a linear equation through iterative processes. As a result, they can provide performance close to that of the ZF precoder while significantly reducing computational complexity. By adopting these iterative approaches, researchers have made substantial progress in addressing the complexity challenge associated with HP precoders for mMIMO systems.

These advancements offer promising solutions for achieving near-optimal performance while making the realization of mMIMO technology more practical and efficient. Motivated by this fact, the authors have explored the SSOR and proposed SSOR-CRZF algorithms for HP design in this chapter.

9.3.1.4 Decomposition-Based Precoder

Furthermore, decomposition-based approached also explored. Over the fundamental linear precoder algorithms like the MRT, ZF, and MMSE algorithms, it is numerically stable. Algorithms like singular value decomposition (SVD)-based approach [49], QR [50, 51], BD scheme [52, 53], modified BD scheme [54, 55], GS [50], and Cholesky decomposition (CD) [51] are tohroughly investigated. The BD algorithm delivers commendable performance, but it comes at the cost of elevated computational complexity. In this approach, every user is required to perform two SVD operations, which adds a significant computational load to the system. The QR-GS algorithm achieves performance levels akin to the BD and QR-BD algorithms but stands out due to a substantial reduction in computational complexity [50]. This reduction in computational demands makes QR-GS an attractive choice for scenarios where maintaining good performance while minimizing computational overhead is crucial.

9.3.2 Nonlinear Precoder

As a part of nonlinear approaches; lots of algorithms are investigated. The dirty-paper coding (DPC) method has outstanding performance characteristics without requiring more power on the transmission side or the sharing of CSI with the reception side. The DPC algorithm relies on infinite-length code words and necessitates complex signal processing techniques, which makes its practical implementation a tough problem [56]. Over the years, several proposals have emerged for efficient HP design, including the TH [57], Kalman filter-based approach [58], and lattice reduction (LR)-based approach [18, 59, 60].

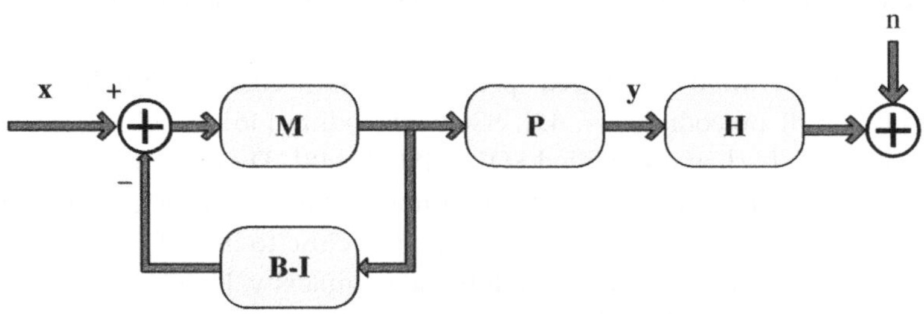

Figure 9.8 Block diagram of TH algorithm-based precoder.

The Tomlinson-Harashima (TH) precoding algorithm represents a suboptimal implementation approach derived from the DPC algorithm [61]. It essentially combines elements of the DPC algorithm with modulo arithmetic techniques to achieve its results. This amalgamation enables TH precoding to approximate the performance of DPC in a more practical and computationally feasible manner. Comparing the linear precoder, it is more complex. A simplified block diagram of a precoder based on TH algorithm is presented in Section 9.8. Further, to improve the performance, a LR-aided precoding framework in vector perturbation (VP) is introduced in [62, 63]. They explored LR-ZF, LR-successive interference cancellation (SIC), and the approximate message passing (AMP) algorithm [64, 65] to improve the overall performance of the mMIMO system, all while reducing complexity. The VP algorithm stands out by providing a full diversity order while maintaining significantly lower computational complexity when compared to the DPC algorithm.

9.3.3 Machine Learning-Based Precoder

Addressing the challenge of complexity in mMIMO systems, researchers have turned to artificial neural networks/deep learning (ANN/DL)-based algorithms [66–68] for the design of low-complexity HPs. For instance, the authors in [69] have introduced a deep neural network (DNN)-based HP framework to reduce symbol error rate (SER) and enhance SE in mmWave-mMIMO systems. In a notable contribution outlined in [70], researchers have introduced a HP technique termed "deep learning quantized phase" (DLQP) for mmWave-mMIMO systems, leveraging channel estimation. The study effectively demonstrates that the designed DLQP HP outperforms other existing methodologies, particularly those involving low-phase resolution, in terms of SE performance. Additionally, two other significant works [71, 72] have presented an innovative convolutional neural network (CNN) framework for the concurrent design of precoders and combiners. Building upon the foundations of DL, an extension of this approach is presented in [73], where authors propose a DL framework that not only enhances performance but also reduces computation time compared to traditional techniques. Furthermore, the scope of CNN-based precoding methodologies extends even to THz mMIMO systems, as highlighted in [74]. Here, an advanced scheme involving a sophisticated 1D CNN and long short-term memory (1D CNN-LSTM) architecture is introduced. This approach significantly enhances SE while simultaneously drastically reducing computational complexity, all within the framework of FC structures.

In this chapter, the authors have undertaken an exploration of the SSOR technique and have introduced the SSOR-CRZF algorithms for HP design. The primary objective of this endeavor is to enhance both SE and EE when compared to traditional linear precoders, all while maintaining a lower level of computational complexity.

9.4 Mathematical Model

This section focuses on presenting the mathematical model of the millimeter wave mMIMO-NOMA system. While the main objective of this chapter is to address the HP scheme, it also introduces a novel SSOR-CRZF DP in addition to the conventional AP. The system model considers a BS equipped with N_T transmit antennas. Under the SC structure, N^{RF} RFCs are utilized to serve K single antenna UEs. For simplicity, each RFC is connected to M antennas, resulting in a total of M antennas per RFC ($M = N_T/N^{\mathrm{RF}}$). In accordance with the findings in [24], the number of beams produced ($\mathcal{L}$) cannot exceed the number of RFCs, that is, $\mathcal{L} = N^{\mathrm{RF}}$. The system model assumes that the number of UEs served (K) is greater than N^{RF} and, therefore, requires allocating K UEs into $\mathcal{L}$ groups. In order to realize this objective, a modified K-means user clustering algorithm has been harnessed within the context of the innovative mmWave-mMIMO-NOMA framework [11]. This strategic approach has been adopted to efficiently organize UEs in the system and the UEs grouping can be denoted as $\Gamma_1, \Gamma_2, \dots, \Gamma_{\mathcal{L}}$.

Let us consider that $x_{l,k}$ represents the signal transmitted toward the kth UE, exploiting the lth beam, where ($\mathbb{E}[|x_{l,k}|^2] = 1$). It is essential to note that in the NOMA transmission scheme, the BS employs a superposition coding technique, and the UEs utilize SIC to retrieve their respective user information. Furthermore, it is presumed that the aggregate power, denoted as P, is uniformly allocated across the entirety of K UEs.

The signal at the receiver end $y_{b,k}$ corresponding to the kth UE in the lth beam $[k = 1, \dots, |S_l|]$ can be expressed as,

$$y_{l,k} = \mathbf{h}_{l,k}^H \mathbf{F}^{\mathrm{RF}} \sum_{i=1}^{\mathcal{L}} \sum_{j=1}^{|S_i|} \mathbf{f}_i^{\mathrm{BB}} \sqrt{p_{i,j}} x_{i,j} + n_{l,k}. \tag{9.9}$$

$$= \underbrace{\mathbf{h}_{l,k}^H \mathbf{F}^{\mathrm{RF}} \mathbf{f}_l^{\mathrm{BB}} \sqrt{p_l} x_{l,k}}_{\text{Desired signal}} + \underbrace{\mathbf{h}_{l,k}^H \mathbf{F}^{\mathrm{RF}} \mathbf{f}_l^{\mathrm{BB}} \left(\sum_{j=1}^{k-1} \sqrt{p_{l,j}} x_{l,j} + \sum_{j=k+1}^{|S_l|} \sqrt{p_{l,j}} x_{l,j} \right)}_{\text{Intra-beam interference}}$$

$$+ \underbrace{\mathbf{h}_{l,k}^H \mathbf{F}^{\mathrm{RF}} \sum_{i \neq l} \sum_{j=1}^{|S_i|} \mathbf{f}_i^{\mathrm{BB}} \sqrt{p_{i,j}} x_{i,j}}_{\text{Inter-beam interference}} \underbrace{+ n_{l,k}}_{\text{Noise signal}}, = \mathbf{h}_{l,k}^H \mathbf{F}^{\mathrm{RF}} \mathbf{F}^{\mathrm{BB}} \mathbf{P} \mathbf{x} + n_{l,k}, \tag{9.10}$$

where $\mathbf{F}^{RF} \in \mathbb{C}^{N_T \times N^{RF}}$ and $\mathbf{F}^{BB} = [\mathbf{f}_1^{BB}, \dots, \mathbf{f}_{\mathcal{L}}^{BB}] \in \mathbb{C}^{N_{RF} \times \mathcal{L}}$ are the precoding matrix related to AP and DP. Here, $\mathbf{h}_{l,k}^H \in \mathbb{C}^{1 \times N_T}$ denotes the channel corresponding to the kth UE in the ith beam. $\mathbf{P} = \text{diag}\{\mathbf{p}_1, \dots \mathbf{p}_{\mathcal{L}}\}$ represents the power allocation matrix. The received power corresponding to the kth UE in the lth beam is denoted by $p_{l,k}$ and $n_{l,k} \sim \mathcal{CN}(0, \sigma_n^2)$ represents the additive white Gaussian noise (AWGN) noise with zero mean and σ_n^2 variance added at the kth UE in the lth beam.

In this chapter, the authors have investigated the two-segment approach for HP design, building on the previous works as in [24, 25, 75] for references. The focus of those study lies in separately considering the design of the analog RF precoder and the digital baseband precoder. The analog RF precoder algorithm ($\mathbf{F}^{RF}$) is inspired by the work presented in [25], and while it is an essential part of the study, the authors have opted not to delve into its detailed discussion in this chapter. As a part of low-dimensional DP design, the authors have introduced a low complexity algorithm (for $\mathbf{F}^{BB}$) and the performance of this algorithm is then compared with existing DPs.

The referenced work [34] introduces the CRZF algorithm as a novel solution to improve overall system performance. This algorithm's unique attributes and capabilities distinguish it as a valuable contribution to the field. Furthermore, the specified precoding matrix, as in [11, 34] can be expressed as,

$$\mathbf{F}_{\text{crzf}} = \beta \mathbf{H}^H (\mathbf{H}\mathbf{H}^H + \alpha \mathbf{I})^{-1}, \tag{9.11}$$

where α represents the complex valued regularization parameter, as defined in [34]. As presented in Eq. (9.11), the precoding matrix for the CRZF involve matrix inversion and thereby it is unsuitable for the practical implementation. To resolve the issue, in this chapter, the authors have introduced low-complex SSOR-CRZF precoder [11]. As a part of HP scheme, SSOR-CRZF DP is implemented after the design of the AP ($\mathbf{F}^{RF}$), considering the effective channel matrix, $\grave{\mathbf{H}} (= \mathbf{H}^H \mathbf{F}^{RF})$. To initiate the process, the CRZF filtering matrix ($\mathbf{G}$) is computed as the primary step, and its formulation is articulated as follows:

$$\mathbf{G} = \left(\grave{\mathbf{H}} \grave{\mathbf{H}}^H + \alpha \mathbf{I}_{N^{RF} \times N^{RF}} \right) \tag{9.12}$$

As illustrated in [46], the transmitted signal ($\mathbf{x}$) following the application of CRZF precoding can be mathematically articulated.

$$\mathbf{x} = \mathbf{F}_{\text{crzf}} \mathbf{s} = \beta \grave{\mathbf{H}}^H \left(\grave{\mathbf{H}} \grave{\mathbf{H}}^H + \alpha \mathbf{I}_{N^{RF} \times N^{RF}} \right)^{-1} \mathbf{s} = \beta \grave{\mathbf{H}}^H \mathbf{G}^{-1} \mathbf{s} = \beta_{\text{CRZF}} \grave{\mathbf{H}}^H \mathbf{t}, \tag{9.13}$$

where $\mathbf{s}$ is the information signal.

The sole purpose of the SSOR algorithm is to produce the precoder matrix $\mathbf{t}$ without any matrix inversion process ($\mathbf{G}^{-1}$). In the SSOR method, the initial step involves decomposing the matrix $\mathbf{G}$, which can be considered as the sum of three matrices: $\mathbf{D}$, $\mathbf{L}$, and $\mathbf{L}^H$. that is, $\mathbf{G} = \mathbf{D} + \mathbf{L} + \mathbf{L}^H$. Here, $\mathbf{D}$ represents the diagonal matrix, $\mathbf{L}$ is the lower triangular matrix, and $\mathbf{L}^H$ is the upper triangular matrix of $\mathbf{G}$. The detailed algorithm is presented as follows.

Algorithm 9.1: Proposed digital precoder

Input: $K(> N^{RF})$; N^{RF}; $\mathbf{H} = [\mathbf{h}_1, \dots, \mathbf{h}_K]$; $\Gamma = [\Gamma_1, \dots, \Gamma_{\mathcal{L}}]$; $\mathbf{F}^{RF}$;N_T;
$\mathcal{L} = N^{RF}$; α; $\mathcal{I}_{iter}$

Output: $\mathbf{F}^{BB} \in \mathbb{C}^{N^{RF} \times \mathcal{L}}$ is the DP matrix

Initialize: (B)

$\grave{\mathbf{H}} = \mathbf{H}^H \mathbf{F}^{RF}$:Effective Channel after AP.

$\mathbf{G} = \left(\grave{\mathbf{H}}\grave{\mathbf{H}}^H + \alpha \mathbf{I}_{N^{RF} \times N^{RF}}\right)$; CRZF Filtering Matrix

Decomposition: $\mathbf{G}$;$\mathbf{D} = diag(Diag(\mathbf{G}))$;$\mathbf{L} = -(tril(\mathbf{G}) - \mathbf{D})$;$\mathbf{U} = -(triu(\mathbf{G}) - \mathbf{D})$

$[r, t] = size(\mathbf{G})$

$\Xi = [1 + sqrt(r/t)]^2 - 1$; $\zeta = 2/(1 + sqrt((1 - \Xi^2)))$; relaxation parameter

$Sym = I_{r,r}$

$\mathbf{Sy}_0 = zeros(r, length(Sym))$

$iter = 1$

$\mathbf{Sy}_2 = \mathbf{Sy}_0$;

while $iter \leqslant \mathcal{I}_{iter}$ **do**

  $\mathbf{Sy}_0 = \mathbf{Sy}_2$

  $\mathbf{Sy}_1 = (D - \zeta * L)^{-1} * (\zeta * Sym + ((1 - \zeta) * D + \zeta * U) * \mathbf{Sy}_0)$

  $\mathbf{Sy}_2 = (D - \zeta * U)^{-1} * (\zeta * Sym + ((1 - \zeta) * D + \zeta * L) * \mathbf{Sy}_1)$

  $iter = iter + 1$

$\hat{\mathbf{F}}^{BB} = \grave{\mathbf{H}} * Sy_2$

$\mathbf{F}^{BB} = \mathbf{0}^{N^{RF} \times K}$

$[\mathbf{F}^{BB}]_{:,[\Gamma]:,1} = \hat{\mathbf{F}}^{BB}$

for $l = 1$ *to* N^{RF} **do**

  $\Delta = [\Gamma]_{l,:}$

  $\Delta = non\ zeros(\Delta)^T$; discard the zero element

  for $n = 2$ *to* $|\Delta|$ **do**

    $[\mathbf{F}^{BB}]_{:,\Delta_n} = [\mathbf{F}^{BB}]_{:,[\Gamma]_{l,1}}$

9.5 Results

In this section, we embark on a comprehensive comparison between the newly proposed SSOR-CRZF HP and a set of traditional precoders, like, ZF, RZF, MRT, TPE, SSOR, and CRZF techniques. Our primary objective revolves around showcasing the superiority of the proposed algorithm within a scenario where a

SC structure is adopted for the HP. Furthermore, we assume that the transmitter possesses complete access to the CSI. To effectively quantify the performance of the proposed HP technique in the context of a mMIMO-NOMA system, we delve into the analysis of three pivotal metrics: SE, EE, and computational complexity. For the sake of precision and meticulous evaluation, the simulations are meticulously executed using the MATLAB platform with the parameters as in Table 9.1. Notably, the outcomes stem from an extensive set of 1000 distinct channel realizations, ensuring robust statistical significance throughout the results.

Figure 9.9 depicts a comprehensive performance evaluation of different precoding strategies, showcasing their SE profiles across varying SNR. The simulation results are obtained using the following parameters: $N_T = 32$ antennas at the BS, $N^{\text{RF}} = 8$ RFCs, $K = 10$ users, number of multipath components $= 10$, and $B = 4$ bits. Notably, the proposed SSOR-CRZF precoder exhibits remarkable superiority over other precoders designed based on the SC architecture, such as ZF, RZF, MRT, TPE, SSOR, and CRZF. In fact, the performance gap between the SSOR-CRZF and the previous SC-based precoders is quite substantial.

It is important to acknowledge that both full-digital (FD) and FC schemes outperform the SC-based precoder in terms of SE. However, the SC-SSOR-CRZF approach significantly enhances the SE performance, especially in the high SNR region, as shown in Figure 9.9. These simulation findings demonstrate the promising potential of the SSOR-CRZF HP method in achieving high SE values, even though it operates in the SC domain, which generally lags behind the FD and FC schemes in terms of SE.

Table 9.1 Simulation parameters.

Parameters	Values
Base station antennas (N_T)	32
Number of RF chains (N^{RF})	8
Number of propagation paths per cluster	10
Number of users (K)	10
Direction-of-arrival (DOA)/angle-of-arrival (AOA) phase quantization	$B = 4$
Channel model ($\mathbf{H}$)	Geometric extended Saleh–Valenzuela model
HP algorithms	MRT, ZF, RZF, TPE, SSOR,CRZF, SSOR-CRZF
HP structure	FD, FC, SC

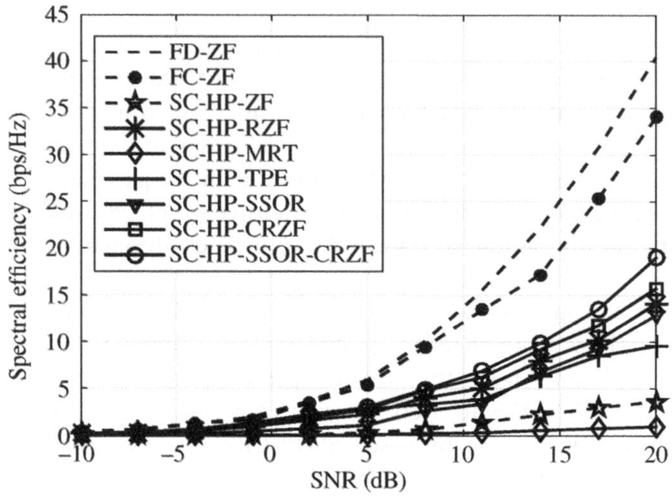

Figure 9.9 Spectral efficiency variation with SNR.

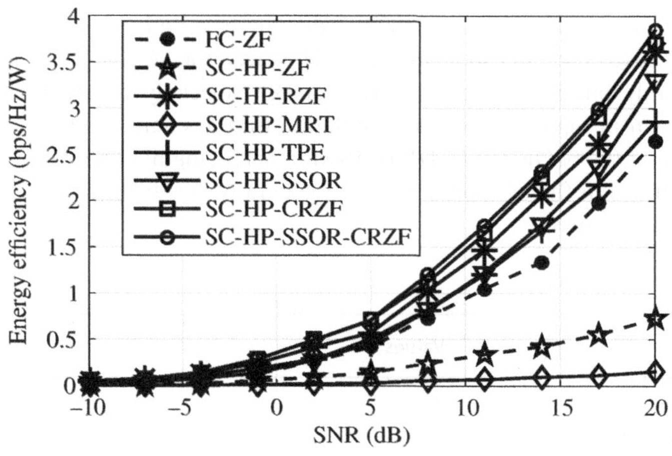

Figure 9.10 Energy efficiency variation with SNR.

The EE (bps/Hz/W) of the mmWave-mMIMO-NOMA system is displayed in Figure 9.10. The EE is calculated based on the specification as mentioned in [24]. In the FC scheme, the total number of PSs, denoted as N_{ps}, is equal to $N_T N^{RF}$. On the other hand, for the SC architecture, $N_{ps} = N_T$. The analysis assumes a maximum transmitted power of $P_T = 30$ mW, which is fixed for all precoder algorithms. Each RFC utilizes $P_{RF} = 300$ mW of power. Additionally, the power consumption associated with the PSs and baseband processing are specified as $P_{PS} = 40$ mW and $P_{BB} = 200$ mW, respectively.

The EE comparison of several precoders vs the SNR variation is shown in Figure 9.10. For this performance analysis, following parameters are considered, N_T= 32, N^{RF}=8, K =10, and B =4. As seen in Figure 9.10, it is clear that the EE is worse in a FC system when compared to other systems. This is due to the FC system's energy consumption, which is caused by the fact that the number of RFCs and BS antennas are equal. On the other hand, in a SC system, there are far fewer RFCs. Because fewer PS are used in SC-HPs than in FC-HPs, they are more energy efficient.

Figure 9.10 depicts the EE comparison of several precoders as a function of SNR variation. The analysis considers the following parameters: $N_T = 32$ antennas at the transmitter, $N^{RF} = 8$ RFCs, $K = 10$ UEs, and $B = 4$ bits. From the results presented in Figure 9.10, it is evident that the EE of the FC system is inferior when compared to other precoder architectures. This discrepancy in EE is primarily due to the higher energy consumption associated with the FC system, which arises from the fact that the number of RFCs and BS antennas are equal in FC systems. In contrast, a SC system uses substantially fewer RFCs, which results in a configuration that is more energy-efficient. The utilization of fewer PSs in SC-HPs, as opposed to FC-HPs, contributes to their improved EE. This is an important observation as it highlights the potential advantages of employing SC-based HP to enhance the EE performance in mmWave-mMIMO-NOMA systems. The findings establish the significance of considering the trade-offs between energy consumption and system performance when selecting the appropriate precoder design for practical implementations.

Complexity analysis:

Within this designated section, a comprehensive investigation into the computational complexity of the SSOR-CRZF precoder is meticulously undertaken. Moreover, a comparative analysis is presented, encompassing several existing precoding strategies like ZF, CRZF. As in case of ZF, CRZF, and SSOR-ZF, it is required to compute $\mathbf{HH}^H$, therefore, the investigation can be extended beyond the realm of $\mathbf{HH}^H$ computation. From the fundamental equations for the precoding matrix the computational complexities for the ZF, CRZF, and SSOR-CRZF can be expressed as $\mathcal{O}(2(N^{RF})^3 + (N^{RF})^2)$, $\mathcal{O}(2(N^{RF})^3 + 3(N^{RF})^2)$, and $\mathcal{O}(N^{RF} + (N^{RF})^2 + i2(N^{RF})^2)$ respectively. Therefore, it is clearly visible that the proposed SSOR-CRZF is computationally less complex.

9.6 Challenges of Hybrid Precoding and Future Research Scope

HP has received considerable attention as a prospective option for tackling the issues posed by mmWave-mMIMO systems, owing especially to its ability to reduce hardware costs and power consumption. The HP technique strategically deploys a limited number of RFC, thereby effectively curbing the energy consumption associated with mixed signal components, such as analog-to-digital converters (ADCs), and also the associated baseband processing. However, it's worth noting that a substantial portion of existing HP methodologies presume a FC architecture. This architectural choice entails an extensive employment of phase shifters, which in turn raises concerns about energy consumption. Moreover, the implementation of such an architecture necessitates a considerable number of phase shifters, consequently leading to heightened energy demands. Given the economic and energetic implications, constraints arise in accommodating a reduced number of signal mixers and ADCs, both of which exhibit notable cost and power requirements. One potential avenue to address these challenges involves the incorporation of low-resolution phase shifters. This strategic inclusion enables the reduction of hardware complexity while maintaining a level of performance that aligns with the objectives of mmWave-mMIMO systems. By thoughtfully managing the balance between architectural intricacies, energy considerations, and cost-effectiveness, the utilization of low-resolution phase shifters emerges as a viable solution in the pursuit of efficient HP for mmWave-mMIMO systems. The HP techniques have been extensively researched spanning microwave and mmWave frequencies, with current study also extending into the THz region. However, the unique attributes of the THz band, coupled with the intricacies of large-scale MIMO systems, introduce substantial challenges in the domain of THz HP design. Thus, the pursuit of practical THz HP architectures and algorithms gains significance. This shift arises from the realization that the conventional high-resolution phase shifts are still beyond the current realm of feasibility at THz band. Therefore, the development of tailored HP strategies that accommodate low-resolution components becomes an imperative in the quest for effective THz HP solutions.

In scenarios involving large-scale MIMO systems, the acquisition of CSI poses a significant challenge. Compounding this issue, the adverse impact stemming from imprecise CSI and resultant beam distortion is notably pronounced within such systems. As a result, in the field of large-scale MIMO-THz HP, it becomes very important to investigate on the designs and algorithms capable of successfully mitigating the challenges posed by imperfect CSI.

9.7 Conclusion

This scholarly exploration undertakes a comprehensive and meticulous examination of performance attributes within the intricate landscape of mmWave-mMIMO-NOMA systems. Anchored by a primary aim, the study centers on the conception and empirical validation of a HP solution, hinging upon the SSOR-CRZF methodology. This proposition unfolds within a multifaceted framework, encompassing a thorough evaluation spanning SE, EE, and the intricate realm of computational complexity. Notably, this investigation draws a sharp contrast between the proposed SSOR-CRZF HP and a repertoire of conventional precoding counterparts, including ZF, RZF, MRT, TPE, SSOR, and CRZF. This empirical analysis convincingly demonstrates the merit of the proposed SSOR-CRZF approach, which substantially outperforms the conventional precoding techniques in terms of multiple performance metrics. As a stepping stone to future research endeavors, the authors signify their intention to extend the scope of their work toward energy harvesting applications.

References

1 Yang, P., Xiao, Y., Xiao, M., and Li, S. (2019). 6G wireless communications: vision and potential techniques. *IEEE Network* 33 (4): 70–75. https://doi.org/10.1109/mnet.2019.1800418.

2 Saad, W., Bennis, M., and Chen, M. (2020). A vision of 6G wireless systems: applications, trends, technologies, and open research problems. *IEEE Network* 34 (3): 134–142. https://doi.org/10.1109/mnet.001.1900287.

3 Imoize, A.L., Adedeji, O., Tandiya, N., and Shetty, S. (2021). 6G enabled smart infrastructure for sustainable society: opportunities, challenges, and research roadmap. *Sensors* 21 (5): 1709. https://doi.org/10.3390/s21051709.

4 Imoize, A.L., Obakhena, H.I., Anyasi, F.I., and Sur, S.N. (2022). A review of energy efficiency and power control schemes in ultra-dense cell-free massive MIMO systems for sustainable 6G wireless communication. *Sustainability* 14 (17): 11100. https://doi.org/10.3390/su141711100.

5 Jiang, H. and Dai, L. (2023). Transformer-based downlink precoding design in massive MIMO systems for 5G-advanced and 6G. *Science China Information Sciences* 66 (9): 199301. https://doi.org/10.1007/s11432-023-3838-4.

6 Uwaechia, A.N., Mahyuddin, N.M., Ain, M.F. et al. (2019). On the spectral-efficiency of low-complexity and resolution hybrid precoding and combining transceivers for mmWave MIMO systems. *IEEE Access* 7: 109259–109277. https://doi.org/10.1109/access.2019.2933785.

7 Liu, Y., Yi, W., Ding, Z. et al. (2022). Developing NOMA to next generation multiple access: future vision and research opportunities. *IEEE Wireless Communications* 29 (6): 120–127. https://doi.org/10.1109/mwc.007.2100553.

8 Zhu, Z., Deng, H., Xu, F. et al. (2022). Hybrid precoding-based millimeter wave massive MIMO-NOMA systems. *Symmetry* 14 (2): 412. https://doi.org/10.3390/sym14020412.

9 Wang, B., Dai, L., Wang, Z. et al. (2017). Spectrum and energy-efficient beamspace MIMO-NOMA for millimeter-wave communications using lens antenna array. *IEEE Journal on Selected Areas in Communications* 35 (10): 2370–2382. https://doi.org/10.1109/jsac.2017.2725878.

10 Yuan, W., Kalokidou, V., Armour, S.M.D. et al. (2017). Application of non-orthogonal multiplexing to mmWave multi-user systems. *2017 IEEE 85th Vehicular Technology Conference (VTC Spring)*. IEEE. https://doi.org/10.1109/vtcspring.2017.8108678.

11 Sur, S.N., Kandar, D., Silva, A. et al. (2022). Hybrid precoding algorithm for millimeter-wave massive MIMO-NOMA systems. *Electronics* 11 (14): 2198. https://doi.org/10.3390/electronics11142198.

12 Uwaechia, A.N. and Mahyuddin, N.M. (2020). A comprehensive survey on millimeter wave communications for fifth-generation wireless networks: feasibility and challenges. *IEEE Access* 8: 62367–62414. https://doi.org/10.1109/access.2020.2984204.

13 Heath, R.W., Gonzalez-Prelcic, N., Rangan, S. et al. (2016). An overview of signal processing techniques for millimeter wave MIMO systems. *IEEE Journal of Selected Topics in Signal Processing* 10 (3): 436–453. https://doi.org/10.1109/jstsp.2016.2523924.

14 Molisch, A.F., Ratnam, V.V., Han, S. et al. (2017). Hybrid beamforming for massive MIMO: a survey. *IEEE Communications Magazine* 55 (9): 134–141. https://doi.org/10.1109/mcom.2017.1600400.

15 Du, J., Wang, Z., Zhang, Y. et al. (2021). Multi-user hybrid precoding for mmWave massive MIMO systems with sub-connected structure. *EURASIP Journal on Wireless Communications and Networking* 2021 (1): https://doi.org/10.1186/s13638-021-02031-0.

16 El Ayach, O., Rajagopal, S., Abu-Surra, S. et al. (2014). Spatially sparse precoding in millimeter wave MIMO systems. *IEEE Transactions on Wireless Communications* 13 (3): 1499–1513. https://doi.org/10.1109/twc.2014.011714.130846.

17 Sharma, P. and Sur, S.N. (2021). Performance analysis of energy-efficient hybrid precoding for massive MIMO. In: *Advances in Communication, Devices and Networking*, 415–423. Singapore: Springer. https://doi.org/10.1007/978-981-16-2911-2_43.

18 Kandar, D., Sur, S.N., Singh, A.K., and Nandi, S. (2021). Performance analysis of lattice reduction-assisted precoder for multi-user millimeter wave MIMO system. *International Journal of Communication Systems* 34 (11). https://doi.org/10.1002/dac.4853.

19 Alkhateeb, A., El Ayach, O., Leus, G., and Heath, R.W. (2014). Channel estimation and hybrid precoding for millimeter wave cellular systems. *IEEE*

Journal of Selected Topics in Signal Processing 8 (5): 831–846. https://doi.org/10.1109/jstsp.2014.2334278.

20 Osama, I., Rihan, M., Elhefnawy, M. et al. (2022). A review on precoding techniques for mm-Wave massive MIMO wireless systems. *International Journal of Communication Networks and Information Security (IJCNIS)* 14 (1): https://doi.org/10.17762/ijcnis.v14i1.5206.

21 Fatema, N., Hua, G., Xiang, Y. et al. (2018). Massive MIMO linear precoding: a survey. *IEEE Systems Journal* 12 (4): 3920–3931. https://doi.org/10.1109/jsyst.2017.2776401.

22 Mengistu, F.G. and Worku, G.A. (2023). Performance analysis of downlink massive MIMO system with precoding techniques and pilot reuse factor. *Cogent Engineering* 10 (1). https://doi.org/10.1080/23311916.2023.2210387.

23 Joham, M., Utschick, W., and Nossek, J.A. (2005). Linear transmit processing in MIMO communications systems. *IEEE Transactions on Signal Processing* 53 (8): 2700–2712. https://doi.org/10.1109/tsp.2005.850331.

24 Dai, L., Wang, B., Peng, M., and Chen, S. (2019). Hybrid precoding-based millimeter-wave massive MIMO-NOMA with simultaneous wireless information and power transfer. *IEEE Journal on Selected Areas in Communications* 37 (1): 131–141. https://doi.org/10.1109/jsac.2018.2872364.

25 Uwaechia, A.N. and Mahyuddin, N.M. (2020). Spectrum and energy efficiency optimization for hybrid precoding-based SWIPT-enabled mmWave mMIMO-NOMA systems. *IEEE Access* 8: 139994–140007. https://doi.org/10.1109/access.2020.3013305.

26 Gao, X., Dai, L., Han, S. et al. (2016). Energy-efficient hybrid analog and digital precoding for MmWave MIMO systems with large antenna arrays. *IEEE Journal on Selected Areas in Communications* 34 (4): 998–1009. https://doi.org/10.1109/jsac.2016.2549418.

27 Simeone, O., Spagnolini, U., and Bar-Ness, Y. (2003). Linear and non-linear precoding/decoding for MIMO systems using the fading correlation at the transmitter. *2003 4th IEEE Workshop on Signal Processing Advances in Wireless Communications - SPAWC 2003 (IEEE Cat. No.03EX689)*. IEEE. https://doi.org/10.1109/spawc.2003.1318911.

28 Cai, J., Rong, B., and Sun, S. (2016). A low complexity hybrid precoding scheme for massive MIMO system. *2016 16th International Symposium on Communications and Information Technologies (ISCIT)*. IEEE, https://doi.org/10.1109/iscit.2016.7751711.

29 Lyu, T.K. (2016). Capacity of multi-user MIMO systems with MMSE and ZF precoding. *2016 IEEE Conference on Computer Communications Workshops (INFOCOM WKSHPS)*. IEEE. https://doi.org/10.1109/infcomw.2016.7562264.

30 Sheikh, T.A., Bora, J., and Hussain, M.A. (2021). Capacity maximizing in massive MIMO with linear precoding for SSF and LSF channel with perfect CSI. *Digital Communications and Networks* 7 (1): 92–99. https://doi.org/10.1016/j.dcan.2019.08.002.

31 Palhares, V.M.T., Flores, A.R., and de Lamare, R.C. (2021). Robust MMSE precoding and power allocation for cell-free massive MIMO systems. *IEEE Transactions on Vehicular Technology* 70 (5): 5115–5120. https://doi.org/10.1109/tvt.2021.3072828.

32 Mehana, A.H. and Nosratinia, A. (2014). Diversity of MIMO linear precoding. *IEEE Transactions on Information Theory* 60 (2): 1019–1038. https://doi.org/10.1109/tit.2013.2289860.

33 Saleeb, B., Shehata, M., Mostafa, H., and Fahmy, Y. (2019). Performance evaluation of RZF precoding in multi-user MIMO systems. *2019 IEEE 62nd International Midwest Symposium on Circuits and Systems (MWSCAS)*. IEEE. https://doi.org/10.1109/mwscas.2019.8885118.

34 Mostafa, M., Newagy, F., and Hafez, I. (2021). Complex regularized zero forcing precoding for massive MIMO systems. *Wireless Personal Communications* 120 (1): 633–647. https://doi.org/10.1007/s11277-021-08482-4.

35 Elmagzoub, H.M. (2020). On the MMSE-based multiuser millimeter wave MIMO hybrid precoding design. *International Journal of Communication Systems* 33 (11): e4409. https://doi.org/10.1002/dac.4409.

36 Mueller, A., Kammoun, A., Björnson, E., and Debbah, M. (2016). Linear precoding based on polynomial expansion: reducing complexity in massive MIMO. *EURASIP Journal on Wireless Communications and Networking* 2016 (1). https://doi.org/10.1186/s13638-016-0546-z.

37 Prabhu, H., Rodrigues, J., Edfors, O., and Rusek, F. (2013). Approximative matrix inverse computations for very-large MIMO and applications to linear pre-coding systems. *2013 IEEE Wireless Communications and Networking Conference (WCNC)*. IEEE. https://doi.org/10.1109/wcnc.2013.6554990.

38 Nagy, B., Elsabrouty, M., and Elramly, S. (2018). Fast converging weighted Neumann series precoding for massive MIMO systems. *IEEE Wireless Communications Letters* 7 (2): 154–157. https://doi.org/10.1109/lwc.2017.2761876.

39 Zhang, C., Li, Z., Shen, L. et al. (2017). A low-complexity massive MIMO precoding algorithm based on Chebyshev iteration. *IEEE Access* 5: 22545–22551. https://doi.org/10.1109/access.2017.2760881.

40 Tang, C., Liu, C., Yuan, L., and Xing, Z. (2016). High precision low complexity matrix inversion based on Newton iteration for data detection in the massive MIMO. *IEEE Communications Letters* 20 (3): 490–493. https://doi.org/10.1109/lcomm.2015.2514281.

41 Shao, L. and Zu, Y. (2016). Joint Newton iteration and Neumann series method of convergence-accelerating matrix inversion approximation in linear precoding for massive MIMO systems. *Mathematical Problems in Engineering* 2016: 1–5. https://doi.org/10.1155/2016/1745808.

42 Minango, J. and de Almeida, C. (2018). A low-complexity linear precoding algorithm based on Jacobi method for massive MIMO systems. *2018 IEEE 87th Vehicular Technology Conference (VTC Spring)*. IEEE, https://doi.org/10.1109/vtcspring.2018.8417575.

43 Lee, W.-S., Jang, J.-Y., Ro, J.-H. et al. (2020). An efficient modified Gauss Seidel precoder for downlink massive MIMO systems. *IEEE Access* 8: 202164–202173. https://doi.org/10.1109/access.2020.3036008.

44 Albreem, M.A.M. and Vasudevan, K. (2020). Efficient hybrid linear massive MIMO detector using Gauss–Seidel and successive over-relaxation. *International Journal of Wireless Information Networks* 27 (4): 551–557. https://doi.org/10.1007/s10776-020-00493-5.

45 Albreem, M.A., Salah, W., Kumar, A. et al. (2021). Low complexity linear detectors for massive MIMO: a comparative study. *IEEE Access* 9: 45740–45753. https://doi.org/10.1109/access.2021.3065923.

46 Xie, T., Dai, L., Gao, X. et al. (2016). Low-complexity SSOR-based precoding for massive MIMO systems. *IEEE Communications Letters* 20 (4): 744–747. https://doi.org/10.1109/lcomm.2016.2525807.

47 Zhang, L. and Hu, Y. (2016). Low complexity WSSOR-based linear precoding for massive MIMO systems. *2016 7th International Conference on Cloud Computing and Big Data (CCBD)*. IEEE. https://doi.org/10.1109/ccbd.2016.033.

48 Liu, Y., Li, Y., Cheng, X. et al. (2020). Low-complexity and fast-convergence linear precoding based on modified SOR for massive MIMO systems. *Digital Signal Processing* 107: 102864. https://doi.org/10.1016/j.dsp.2020.102864.

49 Liu, X., Li, X., Cao, S. et al. (2019). Hybrid precoding for massive mmWave MIMO systems. *IEEE Access* 7: 33577–33586. https://doi.org/10.1109/access.2019.2903166.

50 Wu, J., Fang, S., Li, L., and Yang, Y. (2014). QR decomposition and gram Schmidt orthogonalization based low-complexity multi-user MIMO precoding. *10th International Conference on Wireless Communications, Networking and Mobile Computing (WiCOM 2014)*. Institution of Engineering and Technology. https://doi.org/10.1049/ic.2014.0074.

51 Shahabuddin, S., Islam, M.H., Shahabuddin, M.S. et al. (2020). Matrix decomposition for massive MIMO detection. *2020 IEEE Nordic Circuits and Systems Conference (NorCAS)*. IEEE. https://doi.org/10.1109/norcas51424.2020.9264998.

52 Ni, W. and Dong, X. (2016). Hybrid block diagonalization for massive multiuser MIMO systems. *IEEE Transactions on Communications* 64 (1): 201–211. https://doi.org/10.1109/tcomm.2015.2502954.

53 Li, A. and Masouros, C. (2017). Hybrid analog-digital millimeter-wave MU-MIMO transmission with virtual path selection. *IEEE Communications Letters* 21 (2): 438–441. https://doi.org/10.1109/lcomm.2016.2621741.

54 Crasmariu, V.-F., Arvinte, M.-O., Enescu, A.-A., and Ciochina, S. (2017). A reduced complexity multi-user massive MIMO precoding method. *2017 40th International Conference on Telecommunications and Signal Processing (TSP)*. IEEE. https://doi.org/10.1109/tsp.2017.8075963.

55 Liu, Y., Zhang, Q., He, X. et al. (2022). Spectral-efficient hybrid precoding for multi-antenna multi-user mmWave massive MIMO systems with low complexity. *EURASIP Journal on Wireless Communications and Networking* 2022 (1). https://doi.org/10.1186/s13638-022-02150-2.

56 Jacobsson, S., Durisi, G., Coldrey, M. et al. (2017). Quantized precoding for massive MU-MIMO. *IEEE Transactions on Communications* 65 (11): 4670–4684. https://doi.org/10.1109/tcomm.2017.2723000.

57 Harashima, H. and Miyakawa, H. (1972). Matched-transmission technique for channels with intersymbol interference. *IEEE Transactions on Communications* 20 (4): 774–780. https://doi.org/10.1109/tcom.1972.1091221.

58 Vizziello, A., Savazzi, P., and Chowdhury, K.R. (2018). A Kalman based hybrid precoding for multi-user millimeter wave MIMO systems. *IEEE Access* 6: 55712–55722. https://doi.org/10.1109/access.2018.2872738.

59 Halak, B., El-Hajjar, M., and Hassanein, A. (2018). Hardware efficient architecture for element-based lattice reduction aided K-best detector for MIMO systems. *Journal of Sensor and Actuator Networks* 7 (2): 22. https://doi.org/10.3390/jsan7020022.

60 Sur, S.N., Bera, R., Kumar Bhoi, A. et al. (2020). Capacity analysis of lattice reduction aided equalizers for massive MIMO systems. *Information* 11 (6): 301. https://doi.org/10.3390/info11060301.

61 Stankovic, V. and Haardt, M. (2008). Generalized design of multi-user MIMO precoding matrices. *IEEE Transactions on Wireless Communications* 7 (3): 953–961. https://doi.org/10.1109/lcomm.2008.060709.

62 Ma, Y., Yamani, A., Yi, N., and Tafazolli, R. (2016). Low-complexity MU-MIMO nonlinear precoding using degree-2 sparse vector perturbation. *IEEE Journal on Selected Areas in Communications* 34 (3): 497–509. https://doi.org/10.1109/jsac.2016.2525378.

63 Lyu, S. and Ling, C. (2019). Hybrid vector perturbation precoding: the blessing of approximate message passing. *IEEE Transactions on Signal Processing* 67 (1): 178–193. https://doi.org/10.1109/tsp.2018.2877205.

64 Chen, J.-C., Wang, C.-J., Wong, K.-K., and Wen, C.-K. (2016). Low-complexity precoding design for massive multiuser MIMO systems using approximate message passing. *IEEE Transactions on Vehicular Technology* 65 (7): 5707–5714. https://doi.org/10.1109/tvt.2015.2457450.

65 Shoukath Ali, K., Perarasi, T., Poongodi, C. et al. (2021). Approximate message passing for mmWave massive MIMO architecture using optimal hybrid precoder/combiner. *2021 Smart Technologies, Communication and Robotics (STCR)*. IEEE. https://doi.org/10.1109/stcr51658.2021.9588908.

66 Hu, Q., Cai, Y., Shi, Q. et al. (2021). Iterative algorithm induced deep-unfolding neural networks: precoding design for multiuser MIMO systems. *IEEE Transactions on Wireless Communications* 20 (2): 1394–1410. https://doi.org/10.1109/twc.2020.3033334.

67 Balatsoukas-Stimming, A., Castaneda, O., Jacobsson, S. et al. (2019). Neural-network optimized 1-bit precoding for massive MU-MIMO. *2019 IEEE 20th International Workshop on Signal Processing Advances in Wireless Communications (SPAWC)*. IEEE. https://doi.org/10.1109/spawc.2019.8815519.

68 Li, X. and Alkhateeb, A. (2019). Deep learning for direct hybrid precoding in millimeter wave massive MIMO systems. *2019 53rd Asilomar Conference on Signals, Systems, and Computers*. IEEE. https://doi.org/10.1109/ieeeconf44664.2019.9048966.

69 Huang, H., Song, Y., Yang, J. et al. (2019). Deep-learning-based millimeter-wave massive MIMO for hybrid precoding. *IEEE Transactions on Vehicular Technology* 68 (3): 3027–3032. https://doi.org/10.1109/tvt.2019.2893928.

70 Ma, W., Qi, C., Zhang, Z., and Cheng, J. (2020). Sparse channel estimation and hybrid precoding using deep learning for millimeter wave massive MIMO. *IEEE Transactions on Communications* 68 (5): 2838–2849. https://doi.org/10.1109/tcomm.2020.2974457.

71 Elbir, A.M. (2019). CNN-based precoder and combiner design in mmWave MIMO systems. *IEEE Communications Letters* 23 (7): 1240–1243. https://doi.org/10.1109/lcomm.2019.2915977.

72 Bao, X., Feng, W., Zheng, J., and Li, J. (2020). Deep CNN and equivalent channel based hybrid precoding for mmWave massive MIMO systems. *IEEE Access* 8: 19327–19335. https://doi.org/10.1109/access.2020.2967402.

73 Elbir, A.M. and Papazafeiropoulos, A.K. (2020). Hybrid precoding for multiuser millimeter wave massive MIMO systems: a deep learning approach. *IEEE Transactions on Vehicular Technology* 69 (1): 552–563. https://doi.org/10.1109/tvt.2019.2951501.

74 Murshed, R.U., Ashraf, Z.B., Hridhon, A.H. et al. (2023). A CNN-LSTM-based fusion separation deep neural network for 6G ultra-massive MIMO hybrid beamforming. *IEEE Access* 11: 38614–38630. https://doi.org/10.1109/access.2023.3266355.

75 Alkhateeb, A., Leus, G., and Heath, R.W. (2015). Limited feedback hybrid precoding for multi-user millimeter wave systems. *IEEE Transactions on Wireless Communications* 14 (11): 6481–6494. https://doi.org/10.1109/twc.2015.2455980.

10

Intelligent Reflecting Surfaces and Next-Generation Wireless Systems

Yashuai Cao[1,2], Hetong Wang[3], Tiejun Lv[3], and Wei Ni[4]

[1] *Department of Electronic Engineering, Tsinghua University, Beijing, China*
[2] *Beijing National Research Center for Information Science and Technology (BNRist), Beijing, China*
[3] *School of Information and Communication Engineering, Beijing University of Posts and Telecommunications (BUPT), Beijing, China*
[4] *School of Electrical and Data Engineering, University of Technology Sydney, Australia*

10.1 Introduction

Massive multiple-input multiple-output (MIMO) and beamforming are two advanced techniques that can improve the performance and efficiency of wireless networks [1]. However, they also pose significant challenges in terms of hardware and software complexity and cost. Besides, extremely high-energy consumption prohibits practical deployment. Thus, scalability and compatibility should be considered as the system needs to scale up to accommodate a large number of users and antennas.

From the perspective of single base station (BS), the average power consumption of fifth-generation (5G) BS can reach approximately 3–3.5 times than that of fourth-generation (4G) BS [2]. From the network perspective, the network densification deployment and massive MIMO application led to a sharp rise in power consumption of 5G systems. According to [3], 5G systems would consume 12 times more power than 4G networks. Compared to the 4T4R MIMO in 4G Remote Radio Unit (RRU), the active antenna unit (AAU) in 5G BS adopts 64T64R MIMO. With the increasing channels, the maximum power consumption of a single AAU can reach $1000 \sim 1400$W, and each BS usually consists of 3 AAUs. Meanwhile, the baseband unit (BBU) in 5G BSs also consumes more power due to the computation increment. Figure 10.1 compares the maximum power consumption of a single BS between 4G and 5G, where the maximum power consumption of AAU/RRU and BBU components in 5G BSs increases 68% compared with that in 4G. As reported in [4], the energy cost takes over $19\% \sim 23\%$

Massive MIMO for Future Wireless Communication Systems: Technology and Applications,
First Edition. Edited by Webert Montlouis and Agbotiname Lucky Imoize.

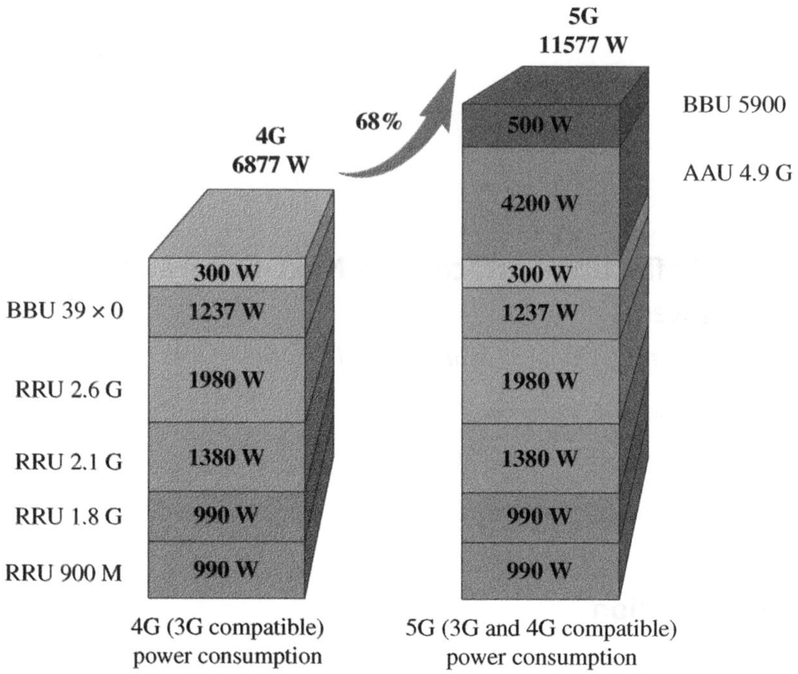

Figure 10.1 Comparison of power consumption between 4G and 5G BSs. Source: Adapted from [4].

of radio access network (RAN) operating expenses (OpEx). By 2030, the overall power consumption of 5G BSs would indirectly cause 990,404 tonnes of carbon footprint, which is unacceptable for economic and industrial development and environmental protection. Particularly for millimeter wave (mmWave) massive MIMO networks, signal blockage would inevitably increase the transmitted energy to guarantee the basic communication quality of service (QoS).

Against the background, intelligent reflecting surface (IRS) has been regarded as a promising low-cost technology for massive MIMO systems. The concept of IRS has originated from electromagnetic (EM) metamaterial technology based on the generalized Snell's law. IRS comprises massive low-cost passive meta-atoms and a smart controller, which can intentionally adjust the phase shift of reflected EM waves. The revolutionary idea of IRS is to proactively engineer radio signal propagation while traditional MIMO systems only adapt to wireless environments. Several passive meta-atoms can achieve the ability to control the propagation channel. This means a powerful received signal enhancement with low-cost passive reflective elements instead of active transmit radio frequency (RF) chains.

Table 10.1 Comparison of massive MIMO and IRS.

Technology	Massive MIMO	IRS
Hardware [5]	Several active antennas High complexity Power hungry	Massive passive meta-atoms Low cost Energy efficiency
System capacity [6]	Logarithmic relationship with average transmit power	Linear relationship with average transmit power
Operating mechanism [7]	Receive and transmit signal	Reflect signal
Full-duplex communication	RF link required	No RF link required
Beamforming [6]	Fully digital or fully analog or hybrid	Passive and smart
Power budget	Very high	Negligible

Table 10.1 compares the concept of massive MIMO with IRS. The massive MIMO adopts the operating mechanism of transmitting and receiving signals by RF chains where analog-to-digital converters (ADCs), digital-to-analog converters (DACs), power amplifiers, and various hardware are required to realize signal processing. Conversely, the IRS can passively reflect the incident signals and empower the smart radio environment to intelligently reconfigure the wireless propagation environment, which avoids introducing additional noise and leads to negligible power consumption. The optimized IRS reflection modes can achieve constructive or destructive combinations for different users, thus enhancing the overall performance gain of the communication system. Therefore, IRS can be used to promote the journey toward the next level of massive MIMO, that is, massive MIMO 2.0 [5].

In support of ubiquitous connectivity, IRS acts in the same role as the active relay in wireless networks. In particular, IRS is expected to settle the communication coverage issues [8] in high-frequency bands, for example, mmWave massive MIMO systems. Therefore, IRS can provide a new paradigm for future 6G ubiquitous connectivity. Naturally, deployments and configuration of IRSs are critical for reaping their full passive beamforming gains. Existing studies assume the rich-scattering environment between BS and IRS [9], but the low-rank BS-IRS channel should be considered when it comes to mmWave transmissions. The sparse multipath channel between the BS and IRS severely limits the degrees of freedom (DoF) of the cascaded channel provided by IRSs. The other pressing challenge arising from incorporating IRSs into existing MIMO systems is overcoming the adverse effect of practically imperfect, typically outdated, channel state information (CSI). Most existing resource allocation algorithms designed for IRS-assisted MIMO systems require the perfect and instantaneous CSI (I-CSI) to perform slot-by-slot IRS optimization, or they can hardly reap the promising gains of the IRSs. However, slot-by-slot IRS optimization can result in non-negligible overhead and computational complexity in the phase-shift configurations of IRSs [10]. In general, the phase-shift configurations of the IRSs are closely coupled with MIMO beamformers in IRS-assisted MIMO systems.

In the face of the challenges of IRS deployments and configuration overheads, it is of practical interest to investigate the channel rank improvement and low-overhead IRS reflection configuration schemes in the massive MIMO networks. The main contributions are listed as follows.

- We propose a channel rank improvement method based on a distributed IRS deployment scheme in the mmWave massive MIMO system. By analyzing the DoF of combined channels, we prove the feasibility of a distributed IRS-assisted channel rank improvement scheme. The optimization of each IRS reflection pattern can achieve signal coverage enhancement.
- We propose an alternating optimization (AO) framework under the perfect CSI scenario for slot-by-slot sum-rate maximization. In particular, we jointly optimize the active precoding at the BS and the passive reflection pattern of the IRS. Based on the equivalent transformation, we decouple the original optimization problem into two subproblems. For the active precoding problem, we develop a Lagrange multiplier method. For the reflection optimization problem, we propose a closed-form semi-definite programming (SDP) method and a low-complexity manifold optimization method to solve the IRS reflection pattern design.
- We establish a two-timescale reflection design framework for the imperfect and outdated CSI (O-CSI) scenario, where slot-by-slot optimization overhead is significantly reduced. For the small timescale, we exploit the inherent temporal correlation of the channel between consecutive time slots to suppress the performance degradation incurred by O-CSI. For the large timescale, a modified particle swarm optimization (PSO) method, namely, recursive sampling PSO (rsPSO), is proposed to solve the IRS reflection optimization with reduced complexity.

The remainder of this chapter is outlined as follows. Section 10.2 discusses the related studies of IRS-assisted MIMO systems. Section 10.3 describes the hardware architecture and work principles of IRS. In Section 10.4, we present the channel model and distributed-IRS system model, and analyze the channel rank improvement. In Section 10.5, we formulate a sum-rate maximization problem and propose an AO framework to configure the IRS reflection pattern slot-by-slot. Section 10.6 proposes a two-timescale IRS reflection pattern design strategy, where the rsPSO-based IRS configuration algorithm at a large timescale is developed. Section 10.7 provides simulation results and corresponding discussions. Sections 10.8 and 10.9 draw conclusions and identify some future works.

10.2 Related Work

Recent advancement in IRS shows the vast perspectives across signal modulation [11], relay transmission [12, 13], target sensing [14], and over-the-air (OTA) analog computation [15], as illustrated in Figure 10.2. These new functions depend strongly on the proper IRS reflection pattern optimization.

Existing IRS studies have concentrated on reflection optimization to improve spectral efficiency (SE) and energy efficiency (EE) separately in most scenarios, as summarized in Table 10.2. The authors of [9] investigate an IRS-assisted downlink multi-input single-output (MISO) communication system, which shows that the upper bound of the ergodic SE is only proportional to the numbers of IRS atoms and BS antennas but independent of to the phase-shift values in the case of Rayleigh fading. In [12, 16], the IRS-assisted multiuser MISO network

Table 10.2 Literature review of related works.

References	Scenario	Optimization objectives	Reflection pattern configuration scheme	Role of IRS	CSI
Han et al. [9]	Single-user MISO	Maximize ergodic SE	Slot-by-slot	Coverage enhancement	Perfect I-CSI
Wu and Zhang [12]	Multiuser MISO	Maximize transmit power	Slot-by-slot	Hardware cost reduction	Perfect I-CSI
Huang et al. [16]	Multiuser MISO	Maximize EE	Slot-by-slot	Coverage enhancement	Perfect I-CSI
Wang et al. [17]	mmWave MISO	Maximize received power	Slot-by-slot	Channel rank improvement	Perfect I-CSI
Zhou et al. [18]	Multiuser MISO	Minimize transmit power	Slot-by-slot	Save power consumption	Imperfect CSI
Zhao et al. [19]	Multiuser MISO	Maximize expected SE	Two-timescale	Improve channel quality	Perfect S-CSI and I-CSI

is considered to guarantee the individual QoS of each mobile user (MU). The optimized IRS reflection patterns can attain the same sum-rate performance while improving the EE by up to threefold compared to relay-aided systems. Most IRS-assisted systems assume that the propagation channel between BS and IRS is rich-scattering [9, 16]. However, for the mmWave massive MIMO scenario, sparsity and low rank are the key features of channels, thus leading to a low-rank cascaded channel matrix and small DoF. Considering the demand for multi-stream transmissions in mmWave communication networks, additional reflection paths are introduced by deploying IRSs to maximize the received signal power [17], minimize system energy consumption [20], resist jamming attacks [21], or strengthen the integrity of OTA federated learning [22]. Moreover, IRSs have also been integrated into unmanned aerial vehicles (UAVs) to provide uninterrupted air-to-ground connections [23–25], provide radio surveillance [26], or combat adversarial jamming attacks [27].

High-accuracy channel estimation is vital to reaping the promising gains of the IRS. However, it is rather challenging to estimate the IRS-involved channels because the passive IRS meta-atoms do not involve active RF chains. Although IRS-related cascaded channel estimations have been widely studied in [28–30], CSI is always imperfect due to channel estimation errors. In [18], the authors focused on robust transmissions under imperfect CSI to satisfy each user's QoS. The above-mentioned works are optimized for the reflection pattern slot by slot. To reduce the channel training overhead, a two-timescale transmission protocol is proposed in [19], where the sum-rate can be maximized by optimizing IRS and precoder with S-CSI and I-CSI. For the larger Rician factor, an S-CSI-based scheme becomes more effective because of more deterministic channels. Nevertheless, the I-CSI is generally dramatically and quickly outdated in time-variant channels. Existing studies generally overlook the temporal correlation between channel slots for IRS design.

10.3 Intelligent Reflecting Surfaces Hardware and Architectures

In this section, we first briefly describe the reflection pattern model of IRS and provide the corresponding work principle. Then, the reconfigurable impedance network is modeled to explain how IRS achieves spatially varying phase response.

To elaborately describe the array response model of passive IRS, let us first consider the two-dimensional (2D) array architecture in the three-dimensional (3D) Cartesian coordinate system as Figure 10.3. Consider a rectangular IRS of N meta-atoms placed on the xoy-plane, with N_x meta-atoms arranged along the x-axis and N_y meta-atoms arranged along the y-axis. Taking the origin O as the

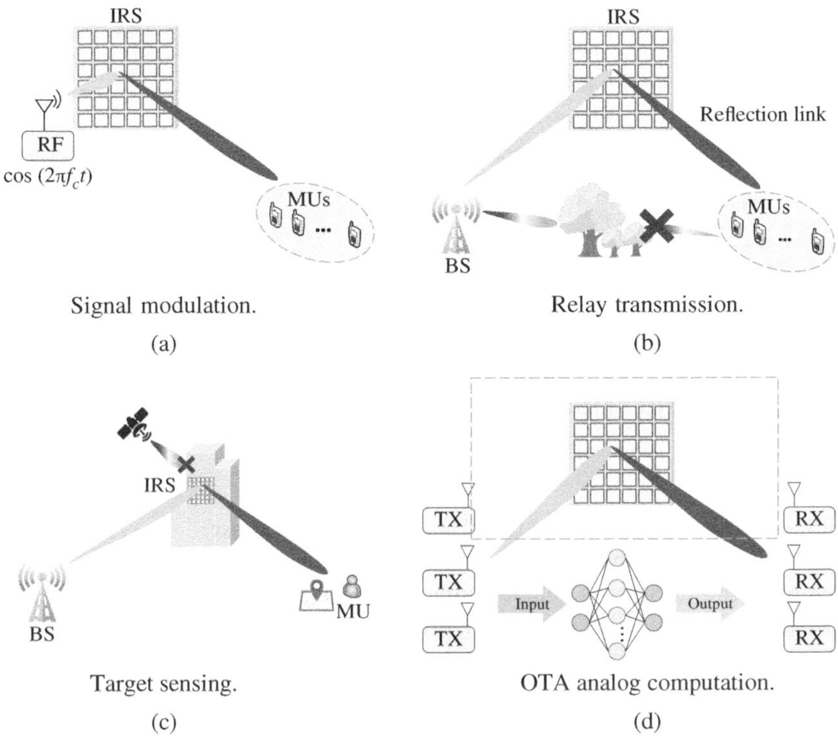

Figure 10.2 IRS-enabled new functions: (a) modulate the pure carrier and transmit information symbols by IRS reflection; (b) overcome the signal blind spot to ensure robust transmissions; (c) IRS performs beamforming and provides desirable RF propagation properties for sensing; (d) realize the specified computation tasks by controlling the wave propagation.

reference center, we define the vector pointing from O to the (x, y)th meta-atom as $\mathbf{b}_{x,y}$ with $x \in \{1, 2, \dots, N_x\}$ and $y \in \{1, 2, \dots, N_y\}$. The unit vector of the lth incident path is denoted by $\mathbf{e}_l$. Then, the phase of the signal incident upon the (x, y)th meta-atom is given by

$$\varphi^{\mathrm{i}}_{x,y} = \frac{2\pi \mathbf{b}^{\mathrm{T}}_{x,y} \mathbf{e}_l}{\lambda}, \tag{10.1}$$

where λ is the carrier wavelength. Likely, if $\mathbf{e}_l$ denotes the lth departure path, the phase of the signal departure from the (x, y)th meta-atom is

$$\varphi^{\mathrm{d}}_{x,y} = \frac{2\pi \mathbf{b}^{\mathrm{T}}_{x,y} \mathbf{e}_l}{\lambda}. \tag{10.2}$$

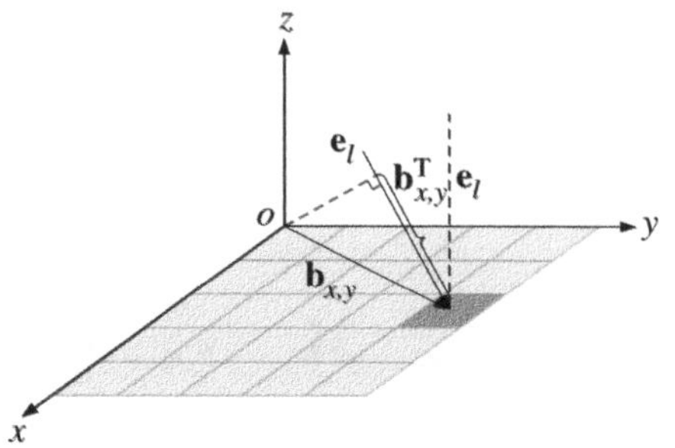

Figure 10.3 Array geometry structure of IRS.

Accordingly, the array response vector of the lth incident/departure path can be written as

$$\mathbf{a}_{\mathrm{i/d}}(\mathbf{e}_l) = \exp\left(-\mathrm{j}\left[\varphi_{1,1}^{\mathrm{i/d}}, \varphi_{2,1}^{\mathrm{i/d}}, \dots, \varphi_{N_x,1}^{\mathrm{i/d}}, \varphi_{1,2}^{\mathrm{i/d}}, \varphi_{2,2}^{\mathrm{i/d}}, \dots, \varphi_{N_x,2}^{\mathrm{i/d}}, \dots, \varphi_{1,N_y}^{\mathrm{i/d}}, \varphi_{2,N_y}^{\mathrm{i/d}}, \dots, \varphi_{N_x,N_y}^{\mathrm{i/d}}\right]^{\mathrm{T}}\right) \tag{10.3}$$

IRS can provide the functionality of spatially varying phase response. To explain this physical mechanism conveniently, we model the IRS by a single-connected reconfigurable impedance network, as shown in Figure 10.4. There are N meta-atom connected to the N-port reconfigurable impedance network. Since each port is single-connected without connecting to the other ports, each port connects to the ground through the independently reconfigurable and passive impedance. Consequently, each port can generate an independent phase shift, thus leading to the diagonal scattering matrix $\mathbf{\Phi}$ [31]:

$$[\mathbf{\Phi}]_{n,n} = \frac{Z_n - Z_0}{Z_n + Z_0} = \frac{\mathrm{j}X_n - Z_0}{\mathrm{j}X_n + Z_0} = e^{\mathrm{j}\theta_n}, \quad \forall n = 1, 2, \dots, N, \tag{10.4}$$

where θ_n is the phase response of the nth meta-atom; $\mathbf{\Phi} = \mathrm{diag}\left([e^{\theta_1}, e^{\theta_2}, \dots, e^{\theta_N}]\right)$ is a diagonal phase-shift matrix (reflection pattern) of IRS; Z_n and X_n are the reconfigurable impedance and reactance of the nth meta-atom, respectively; and Z_0 is the reference impedance.

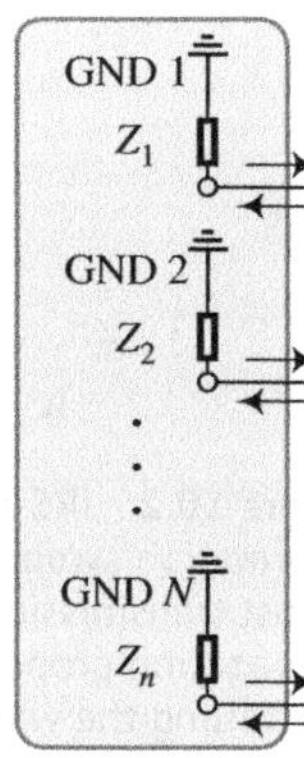

Figure 10.4 N-port single-connected reconfigurable impedance network, which explains the physical significance of the IRS reflection pattern.

Theorem 10.1 *Given the angle $\varphi_n^{\mathrm{i/d}}$ which is the nth entry of vector $\mathbf{a}_{\mathrm{i/d}}$, the array gain of the IRS from the incident wave to the departure wave is equivalent to:*

$$G_a(\mathbf{\Phi}) = \left|\mathbf{a}_{\mathrm{i}}^{\mathrm{H}}\mathbf{\Phi}\mathbf{a}_d\right|^2 = \left|\sum_{n=1}^{N} e^{\mathrm{j}(\varphi_n^{\mathrm{i}}+\theta_n-\varphi_n^{\mathrm{d}})}\right|^2. \tag{10.5}$$

The maximum array gain can be achieved by setting

$$\theta_n = \varphi_n^{\mathrm{d}} - \varphi_n^{\mathrm{i}}. \tag{10.6}$$

Proof: The array gain in (10.5) can be reframed as

$$G_a(\mathbf{\Phi}) = \left|\sum_{n=1}^{N} e^{\mathrm{j}(\varphi_n^{\mathrm{i}} - \varphi_n^{\mathrm{d}})} e^{\mathrm{j}\theta_n}\right|^2 = \left|\boldsymbol{\varphi}^{\mathrm{T}}\boldsymbol{\theta}\right|^2, \tag{10.7}$$

where $\boldsymbol{\varphi} = [e^{\mathrm{j}(\varphi_1^{\mathrm{i}} - \varphi_1^{\mathrm{d}})}, e^{\mathrm{j}(\varphi_2^{\mathrm{i}} - \varphi_2^{\mathrm{d}})}, \dots, e^{\mathrm{j}(\varphi_N^{\mathrm{i}} - \varphi_N^{\mathrm{d}})}]^{\mathrm{T}}$ and $\boldsymbol{\theta} = [e^{\mathrm{j}\theta_1}, e^{\mathrm{j}\theta_2}, \dots, e^{\mathrm{j}\theta_N}]^{\mathrm{T}}$. Defining $\varphi_n' = e^{\mathrm{j}(\varphi_n^{\mathrm{i}} - \varphi_n^{\mathrm{d}})}$, we note that

$$\left|\boldsymbol{\varphi}^{\mathrm{T}}\boldsymbol{\theta}\right|^2 \leq \left|\sum_{n=1}^{N} |\varphi_n'||\theta_n|\right|^2, \tag{10.8}$$

which is equal to satisfying the condition of $\theta_n = \arg(\varphi_n') = \varphi_n^{\mathrm{d}} - \varphi_n^{\mathrm{i}}$. ■

10.4 Intelligent Reflecting Surfaces and the Path Loss Model

In this section, we first provide the product-distance path loss model for IRS cascaded channels. Then, we build the communication models for the centralized and distributed IRS deployments in the mmWave massive MIMO system, respectively. Finally, we analyze the channel rank improvement by the distributed deployment.

IRS is regarded as an emerging and appealing 6G-enabled technology, according to its passive and intelligent reflection ability. The terminology "passive" indicates that meta-atoms only reflect the incident waves immediately without any analog and digital signal processing and time delay. As such, IRS can be regarded as intelligent scatters in the physical space to improve the propagation. Particular for future high-frequency communication systems with severe propagation attenuation and poor diffraction bottlenecks, IRS is significantly compatible with the high-frequency bands (e.g., mmWave and Terahertz) by providing uninterrupted wireless connectivity with low power consumption [32, 33].

Although IRS can bring promising reflection gain in a low-cost manner, the actual cascaded channel fading should be considered. As shown in Figure 10.5, the distances from the BS to IRS and from the IRS to MU can be denoted as d_1 and d_2, respectively. The path loss of BS-IRS and IRS-MU links through the nth meta-atom of the IRS follow $\mathbb{E}\left(|g_n|^2\right) \propto c_1 \frac{d_1}{d_0}^{-\alpha_1}$ and $\mathbb{E}\left(|h_n|^2\right) \propto c_2 \frac{d_2}{d_0}^{-\alpha_2}$, where c_1 (c_2) and α_1 (α_2) represent the path loss value at reference distance d_0 and the path loss exponent, respectively. As a result, the path loss of the IRS cascaded link can be denoted as $P_r \propto \frac{1}{d_1^{\alpha_1} d_2^{\alpha_2}}$, which is inversely proportional to product-distance

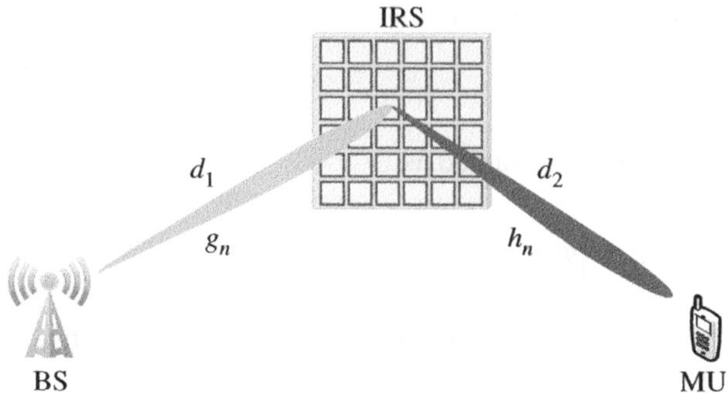

Figure 10.5 Product-distance path loss model in IRS-assisted mmWave system.

instead of the sum-distance. Consequently, the product-distance path loss effect would lead to double fading attenuation [34], which should be properly compensated by the IRS reflection gains.

Consider an IRS-assisted downlink mmWave system, as shown in Figure 10.6. The BS is equipped with a uniform linear array (ULA) composed of N_t antenna elements. With the assistance of the IRS, K single-antenna MUs are served. Two deployment schemes of the IRS are provided: centralized and distributed. In a typical technical scenario of the IRS, the direct mmWave links between BS and MUs are severely blocked by obstacles, while the MUs are located in the hotspot area served by the IRS. Thus, the BS can only communicate with the MUs via the complementary links offered by IRS since the mmWave links between BS and MUs are highly susceptible to environmental blockages and dynamics.

For the centralized deployment scheme, we assume a single IRS is placed near the hotspot. Based on the discussion in Section 10.3, we can directly obtain the communication model as:

$$\mathbf{y} = \mathbf{H}^{\mathsf{H}}\boldsymbol{\Phi}^{\mathsf{H}}\mathbf{G}\sum_{k=1}^{K}\mathbf{w}_k s_k + \mathbf{n}, \tag{10.9}$$

where $\mathbf{H} = [\mathbf{h}_1, \mathbf{h}_2, \dots, \mathbf{h}_K]$ collects the channels of all MUs, $\mathbf{G}$ is the channel between the BS and IRS, $\mathbf{W} = [\mathbf{w}_1, \mathbf{w}_2, \dots, \mathbf{w}_K]$ is the precoding matrix at the BS, $\mathbf{s} = [s_1, s_2, \dots, s_K]^{\mathsf{T}}$ collects the intended signals of MUs, and $\mathbf{n}$ is the complex Gaussian noise of $\mathcal{CN}(0, \sigma^2\mathbf{I})$.

According to the widely used Saleh–Valenzuela channel model [35] for mmWave communications, $\mathbf{G}$ can be modeled as

$$\mathbf{G} = \sum_{l=0}^{L} \nu^{(l)}\mathbf{a}_{\mathrm{B}}(\mathbf{e}_l)\mathbf{a}_{\mathrm{I}}^{\mathsf{H}}(\mathbf{e}_l), \tag{10.10}$$

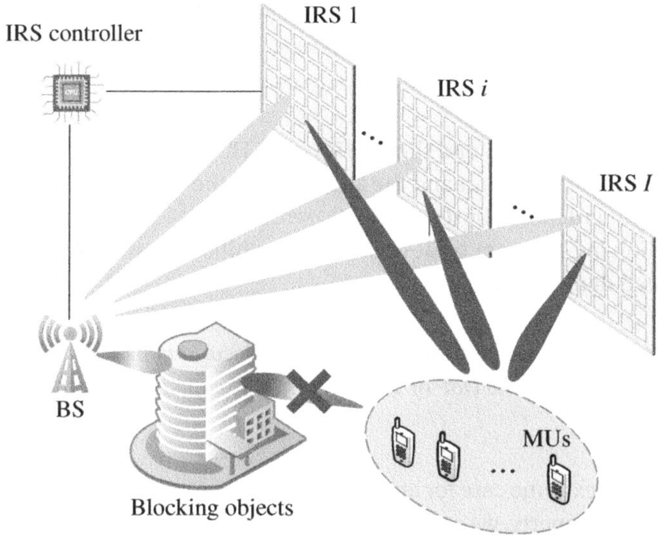

Figure 10.6 IRS-assisted downlink mmWave massive MIMO system.

where L denotes the number of non-line-of-sight (NLoS) paths, $l = 0$ represents the line-of-sight (LoS) path, $v^{(l)}$ is the complex gain of the lth path, and $\mathbf{a}_{\mathrm{B}}$ and $\mathbf{a}_{\mathrm{I}}$ are the array steering vectors of the BS and IRS.

Typically, the IRS is densely distributed in the hotspot spaces, which gives rise to a high probability of LoS propagation. Due to the severe path loss, the transmit power of 2 or more reflections can be ignored so that only LoS is considered [12]. Similarly, the channel between the IRS and the kth MU is given by

$$\mathbf{h}_k = v_k \mathbf{a}_{\mathrm{U}} \left(\mathbf{e}_k\right), \forall k = 1, 2, \ldots, K, \tag{10.11}$$

where v_k is the channel gain, and $\mathbf{a}_{\mathrm{U}}$ is the array steering vector of IRS.

However, mmWave channels generally consist of only a small number of dominant multipath components and typically exhibit 3–5 paths in realistic environments [36], while the scattering at sub-6 GHz is relatively rich. Therefore, the virtual LoS channel created by the IRS can hardly enjoy the spatially multiplexed benefits. Note that the maximum number of signal streams carried by the virtual LoS channel is determined by the rank of the cascaded channel matrix $\mathbf{H}^{\mathrm{H}}\boldsymbol{\Phi}^{\mathrm{H}}\mathbf{G}$. When rank($\mathbf{G}$) $< K$, the multiuser communication system does not work.

For this reason, we propose a distributed deployment scheme to achieve channel rank improvement. Specifically, we adopt I IRS units to serve K MUs, and each IRS unit is composed of N meta-atoms. These distributed IRS units are managed by a smart controller, which exchanges CSI and coordinates the reflection pattern for all IRS units.

Then, the received signal at the kth MU can be given by

$$\mathbf{y} = \sum_{i=1}^{I} \mathbf{H}_i^{\mathsf{H}} \mathbf{\Phi}_i^{\mathsf{H}} \mathbf{G}_i \sum_{k=1}^{K} \mathbf{w}_k s_k + \mathbf{n}, \tag{10.12}$$

where $\mathbf{G}_i$ is the channel between BS and the ith IRS unit, $\mathbf{H}_i = [\mathbf{h}_{i,1}, \mathbf{h}_{i,2}, \dots, \mathbf{h}_{i,K}]$ is the channel between the ith IRS unit and the kth MU, and the reflection pattern matrix of the ith IRS unit is denoted by $\mathbf{\Phi}_i = \mathrm{diag}(\boldsymbol{\theta}_i)$ with $\boldsymbol{\theta}_i = [e^{\theta_{i,1}}, e^{\theta_{i,2}}, \dots, e^{\theta_{i,N}}]^{\mathsf{T}}$.

We provide the following theorem to describe the feasibility of rank improvement by the distributed deployment.

Theorem 10.2 *There always exist multiple deployment locations of IRSs which can be carefully chosen to provide the DoF of K by creating cascaded channel matrices* $\tilde{\mathbf{H}}_i = \mathbf{H}_i^{\mathsf{H}} \mathbf{\Phi}_i^{\mathsf{H}} \mathbf{G}_i$.

Proof: Let us consider an extreme case for any cascaded channel matrix $\tilde{\mathbf{H}}_i$ of rank one. Due to the rank-one property, we can decompose $\tilde{\mathbf{H}}_i$ as $\tilde{\mathbf{H}}_i = \mathbf{a}_i \mathbf{b}_i^{\mathsf{T}}$. Since array steering vectors $\mathbf{a}_i$ and $\mathbf{b}_i$ are determined by their particular angles, it is easy to find multiple array steering vectors with independent angles, thus leading to

$$\sum_{i=1}^{I} \tilde{\mathbf{H}}_i = \sum_{i=1}^{I} \mathbf{a}_i \mathbf{b}_i^{\mathsf{T}} = \begin{bmatrix} \mathbf{a}_1 & \mathbf{a}_2 & \cdots & \mathbf{a}_I \end{bmatrix} \begin{bmatrix} \mathbf{b}_1^{\mathsf{T}} \\ \mathbf{b}_2^{\mathsf{T}} \\ \vdots \\ \mathbf{b}_I^{\mathsf{T}} \end{bmatrix}. \tag{10.13}$$

For the ease of illustration, we define $\mathbf{A} = [\mathbf{a}_1, \mathbf{a}_2, \dots, \mathbf{a}_I]$ and $\mathbf{B} = [\mathbf{b}_1, \mathbf{b}_2, \dots, \mathbf{b}_I]$. If matrices $\mathbf{A}$ and $\mathbf{B}$ are constructed by independent angles, we have $\mathrm{rank}(\mathbf{A}) = I$ and $\mathrm{rank}(\mathbf{B}) = I$. According to Sylvester's rank inequality, summing these cascaded channel matrices yields

$$\mathrm{rank}(\sum_{i=1}^{I} \tilde{\mathbf{H}}_i) = \mathrm{rank}(\mathbf{A}\mathbf{B}^{\mathsf{T}}) \geq \mathrm{rank}(\mathbf{A}) + \mathrm{rank}(\mathbf{B}) - I = I. \tag{10.14}$$

This indicates that only $I \geq K$ IRSs deployed independently can provide the DoF of K under the rank-one case. For the higher-rank case, it may require a small I. ■

10.5 IRS-Empowered Slot Scheduling and Cost-Efficient Reflection Optimization

In this section, we study a slot-by-slot IRS refection design scheme to maximize the system sum rate. With the perfect CSI, a joint precoding and reflection pattern design problem is established, where the precoding and reflection pattern are optimized in each time slot. Then, we develop an AO framework to solve the optimization problem. Two optimization algorithms are proposed to solve the reflection pattern design problem. One is the semidefinite relaxation (SDR) method in a closed-form update manner, and the other is the unit-circle manifold optimization (UCMO) method.

10.5.1 Slot-by-Slot Reflection Pattern Design

To achieve the sum-rate maximization in the considered downlink mmWave network, the precoder at the BS and the reflection pattern of distributed IRSs should be jointly designed. The individual rate of the kth MU is given by

$$R_k = \log_2(1 + \gamma_k), \tag{10.15}$$

where the signal-to-interface-noise (SINR) of the kth MU is computed by

$$\gamma_k = \frac{\left|\sum_{i=1}^{I} \mathbf{h}_{i,k}^{\mathsf{H}} \mathbf{\Phi}_i^{\mathsf{H}} \mathbf{G}_i \mathbf{w}_k\right|^2}{\sum_{j=1, j\neq k}^{K} \left|\sum_{i=1}^{I} \mathbf{h}_{i,k}^{\mathsf{H}} \mathbf{\Phi}_i^{\mathsf{H}} \mathbf{G}_i \mathbf{w}_j\right|^2 + \sigma^2}. \tag{10.16}$$

The resulting sum-rate maximization problem is formulated as

$$\max_{\mathbf{W}, \{\mathbf{\Phi}\}_{i=1}^{I}} f_1(\mathbf{W}, \{\mathbf{\Phi}\}_{i=1}^{I}) = \sum_{k=1}^{K} \omega_k R_k \tag{10.17a}$$

$$\text{s.t. } \operatorname{tr}\left(\mathbf{W}\mathbf{W}^{\mathsf{H}}\right) \leq P_{\max}, \tag{10.17b}$$

$$\theta_{i,n} \in [0, 2\pi), \quad \forall i, \forall n, \tag{10.17c}$$

where the weight ω_k is the required service priority of the kth MU.

It is observed that Problem (10.17) is a weighted sum-rate maximization problem. Due to the sum-logarithm form and complicated coupling relationship between $\mathbf{W}$ and $\{\mathbf{\Phi}\}_{i=1}^{I}$, it is difficult to optimize the precoder and reflection patterns simultaneously. To address this issue, we examine the transformation steps for iteratively optimizing $\mathbf{W}$ and $\{\mathbf{\Phi}\}_{i=1}^{I}$. First, we constitute an equivalent optimization problem to (10.17) by introducing auxiliary variables $\boldsymbol{\alpha} = \left[\alpha_1, \alpha_2, \dots, \alpha_K\right]^{\mathsf{T}}$ for SINR. Hence, the objective (10.17a) can be transformed as

$$f_2(\mathbf{W}, \{\mathbf{\Phi}\}_{i=1}^{I}, \boldsymbol{\alpha}) = \max_{\mathbf{W}, \{\mathbf{\Phi}\}_{i=1}^{I}, \boldsymbol{\alpha}} \frac{1}{\ln 2} \sum_{k=1}^{K} \omega_k \ln(1 + \alpha_k) - \omega_k \alpha_k + \frac{\omega_k (1 + \alpha_k) \gamma_k}{1 + \gamma_k}. \tag{10.18}$$

The equivalence between (10.18) and (10.17) is confirmed by calculating the optimal α_k for f_2, given fixed $\mathbf{W}$ and $\{\mathbf{\Phi}\}_{i=1}^{I}$. By setting $\frac{\partial f_2}{\partial \alpha_k} = 0$, the optimal solution for α_k is $\alpha_k^{\text{opt}} = \gamma_k$. This means that optimizing the α_k equals to optimizing Problem (10.18), while optimizing $\mathbf{W}$ and $\{\mathbf{\Phi}\}_{i=1}^{I}$ without need of dealing with the logarithm.

Based on the above discussions, given the fixed α_k in each iteration, only the last term in (10.18) is involved in optimizing $\mathbf{W}$ and $\{\mathbf{\Phi}\}_{i=1}^{I}$ due to the SINR function of γ_k. From (10.18), the joint precoder and reflection pattern optimization problem is recast to

$$\max_{\mathbf{W},\{\mathbf{\Phi}\}_{i=1}^{I}} f_3(\mathbf{W},\{\mathbf{\Phi}\}_{i=1}^{I}) = \sum_{k=1}^{K} \frac{\omega_k(1+\alpha_k)\gamma_k}{1+\gamma_k}$$
$$\text{s.t. } (10.17\text{b}), (10.17\text{c}). \tag{10.19}$$

Now, we transform the joint precoder and reflection pattern optimization problem into a multi-ratio fractional programming (MRFP) form.

To solve the precoding matrix $\mathbf{W}$ when given fixed IRS reflection patterns $\{\mathbf{\Phi}_i\}_{i=1}^{I}$, we first define $\tilde{\mathbf{h}}_k^{\mathsf{H}} = \sum_{i=1}^{I} \mathbf{h}_{i,k}^{\mathsf{H}} \mathbf{\Phi}_i^{\mathsf{H}} \mathbf{G}_i$ for later convenience. Substituting $\tilde{\mathbf{h}}_k$ into (10.16), Problem (10.19) can be rewritten as

$$\max_{\mathbf{W}} f_4(\mathbf{W}) = \sum_{k=1}^{K} \frac{\tilde{\omega}_k \left|\tilde{\mathbf{h}}_k^{\mathsf{H}} \mathbf{w}_k\right|^2}{\sum_{j=1}^{K} \left|\tilde{\mathbf{h}}_k^{\mathsf{H}} \mathbf{w}_j\right|^2 + \sigma^2}$$
$$\text{s.t. } (10.17\text{b}), \tag{10.20}$$

where $\tilde{\omega}_k = \omega_k(1+\alpha_k)$. Problem (10.20) is an MRFP problem. Based on the quadratic transform [37], we reformulate (10.20) as

$$f_5(\mathbf{W},\boldsymbol{\beta}) = \sum_{k=1}^{K} 2\sqrt{\tilde{\omega}_k} \text{Re}\{\beta_k^* \tilde{\mathbf{h}}_k^{\mathsf{H}} \mathbf{w}_k\} - |\beta_k|^2 \left(\sum_{j=1}^{K} \left|\tilde{\mathbf{h}}_k^{\mathsf{H}} \mathbf{w}_j\right|^2 + \sigma^2 \right), \tag{10.21}$$

where $\boldsymbol{\beta} = [\beta_1, \beta_2, \ldots, \beta_K]^{\mathsf{T}}$ is an auxiliary vector by the quadratic transform. Next, we can directly obtain the optimal $\boldsymbol{\beta}$ by setting $\partial f_5/\partial \beta_k = 0$.

Then, we adopt the Lagrange multiplier method to address the power constraint (10.17a) to obtain the optimal $\mathbf{w}_k$, as given by

$$\mathbf{w}_k^{\text{opt}} = \sqrt{\tilde{\omega}_k}\beta_k \left(\mu \mathbf{I}_N + \sum_{i=1}^{K} |\beta_i|^2 \tilde{\mathbf{h}}_i \tilde{\mathbf{h}}_i^{\mathsf{H}} \right)^{-1} \tilde{\mathbf{h}}_k, \tag{10.22}$$

where $\mu \geq 0$ is the Lagrange multiplier.

To facilitate the optimization of reflection patterns, we adopt the following identity:

$$\sum_{i=1}^{I} \mathbf{h}_{i,k}^{\mathsf{H}} \mathbf{\Phi}_i^{\mathsf{H}} = \sum_{i=1}^{I} \theta_i^{\mathsf{H}} \text{diag}\left(\mathbf{h}_{i,k}^{\mathsf{H}}\right). \tag{10.23}$$

Given $\boldsymbol{\alpha}$ and $\mathbf{W}$, substituting (10.23) into (10.19) yields

$$\max_{\theta_i} f_6(\theta_i) = \sum_{k=1}^{K} \frac{\tilde{\omega}_k \left| \sum_{i=1}^{I} \theta_i^{\mathsf{H}} \mathbf{v}_{i,k,k} \right|^2}{\sum_{j=1}^{K} \left| \sum_{i=1}^{I} \theta_i^{\mathsf{H}} \mathbf{v}_{i,k,j} \right|^2 + \sigma^2}$$

$$\text{s.t. (10.17c),} \tag{10.24}$$

where $\mathbf{v}_{i,k,j} = \text{diag}\left(\mathbf{h}_{i,k}^{\mathsf{H}}\right) \mathbf{G}_i \mathbf{w}_j$. Problem (10.24) can be further reduced to a compact form, as given by

$$\max_{\tilde{\theta}} \quad f_7(\tilde{\theta}) = \sum_{k=1}^{K} \frac{\tilde{\omega}_k \left| \tilde{\theta}^{\mathsf{H}} \tilde{\mathbf{v}}_{k,k} \right|^2}{\sum\limits_{j=1}^{K} \left| \tilde{\theta}^{\mathsf{H}} \tilde{\mathbf{v}}_{k,j} \right|^2 + \sigma^2}. \tag{10.25}$$

where $\tilde{\theta} = \text{vec}\,([\theta_1, \theta_2, \dots, \theta_I])$ and $\tilde{\mathbf{v}}_{k,j} = \text{vec}\,([\mathbf{v}_{1,k,j}, \mathbf{v}_{2,k,j}, \dots, \mathbf{v}_{I,k,j}])$.

Note that Problem (10.25) is also an MRFP, but with non-convex unit-modulus constraints (10.17c). To get rid of the fractional form, applying the quadratic transform to (10.25) yields

$$\max_{\tilde{\theta}, \rho} \quad f_8(\tilde{\theta}, \rho) = \sum_{k=1}^{K} 2\sqrt{\tilde{\omega}_k} \text{Re}\left\{ \rho_k^* \tilde{\theta}^{\mathsf{H}} \tilde{\mathbf{v}}_{k,k} \right\} - |\rho_k|^2 \left(\sum_{j=1}^{K} \left| \tilde{\theta}^{\mathsf{H}} \tilde{\mathbf{v}}_{k,j} \right|^2 + \sigma^2 \right), \tag{10.26}$$

where ρ_k refers to the auxiliary variable of quadratic transform. Then, the optimal ρ_k can be directly obtained by setting $\frac{\partial f_8}{\rho_k} = 0$.

Based on the Lagrange multiplier method, optimizing $\tilde{\theta}$ reduces to

$$\max_{\tilde{\theta}} \quad f_9(\tilde{\theta}) = -\tilde{\theta}^{\mathsf{H}} \mathbf{A} \tilde{\theta} + 2\text{Re}\{\tilde{\theta}^{\mathsf{H}} \mathbf{b}\} - \sum_{k=1}^{K} |\rho_k|^2 \sigma^2, \tag{10.27}$$

where $\mathbf{A} = \sum_{k=1}^{K} \left(|\rho_k|^2 \sum_{j=1}^{K} \tilde{\mathbf{v}}_{k,j} \tilde{\mathbf{v}}_{k,j}^{\mathsf{H}} \right)$ and $\mathbf{b} = \sum_{k=1}^{K} \sqrt{\tilde{\omega}_k} \rho_k^* \tilde{\mathbf{v}}_{k,k}$. Since $\mathbf{A}$ is a positive-definite matrix, Problem (10.27) becomes a quadratic constraint quadratic programming (QCQP). To tackle the unit-modulus (10.17c), we relax it into a convex constraint $|\theta_n| \leq 1, \forall n = 1, 2, \dots, N$. Therefore, we rewrite (10.27) as

$$\max_{\tilde{\theta}} f_{10}(\tilde{\theta}) = -\tilde{\theta}^{\mathsf{H}} \mathbf{A} \tilde{\theta} + 2\text{Re}\{\tilde{\theta}^{\mathsf{H}} \mathbf{b}\}, \tag{10.28a}$$

$$\text{s.t.}\ |\theta_n|^2 \leq 1, \quad \forall n = 1, 2, \dots, N, \tag{10.28b}$$

According to Slater's condition, it is easy to check that the strong duality holds. Hence, the Lagrangian of (10.28) can be written as

$$f_{\mathcal{L}}(\tilde{\theta}, \zeta) = f_{10}(\tilde{\theta}) - \sum_{n=1}^{N} \zeta_n \left(|\theta_n|^2 - 1 \right), \tag{10.29}$$

where ζ_k denotes the dual variable for constraints (10.28b). It is noticed that $f_{\mathcal{L}}(\tilde{\boldsymbol{\theta}}, \zeta)$ is concave with respect to $\tilde{\boldsymbol{\theta}}$. Then, the dual transformation of (10.29) is given by

$$\min_{\zeta} f_D(\zeta) = \sup_{\tilde{\boldsymbol{\theta}}} \quad f_{\mathcal{L}}(\tilde{\boldsymbol{\theta}}, \zeta), \tag{10.30a}$$

$$\text{s.t. } \zeta_n \geq 0, \quad \forall n = 1, 2, \dots, N, \tag{10.30b}$$

where $f_D(\zeta)$ is the Lagrange dual function. To get the dual objective in the form that only depends on ζ, we attain the optimal $\tilde{\boldsymbol{\theta}}$ by setting $\partial f_{\mathcal{L}} / \partial \tilde{\boldsymbol{\theta}} = 0$. Thus, we have

$$\tilde{\theta}^{\text{opt}} = \left(\mathbf{A} + \text{diag}\,(\zeta)\right)^{-1} \mathbf{b} = \mathbf{D}(\zeta)\mathbf{b}, \tag{10.31}$$

where $\mathbf{D}(\zeta) = \left(\mathbf{A} + \text{diag}\,(\zeta)\right)^{-1}$. The optimal reflection patterns require the value of ζ. The following lemma reveals the property of optimal ζ.

Lemma 10.1 *The optimal ζ for Problem (10.30) can be given by*

$$\zeta^{\text{opt}} = \left\{\zeta_n \geq 0 : [\mathbf{D}(\zeta)\mathbf{b}] \circ [\mathbf{D}(\zeta)\mathbf{b}]^* = \mathbf{1}\right\}, \tag{10.32}$$

which is exactly equivalent to $|\theta_n|^2 = 1$, that is, unit-modulus phase-shift constraints.

Proof: According to the chain rule in the matrix derivatives, we have

$$\begin{aligned}\frac{\partial f_D}{\partial \zeta_n} &= 1 - \text{tr}\left[\mathbf{D}(\zeta)\mathbf{b}\mathbf{b}^{\mathsf{H}}\mathbf{D}(\zeta)\frac{\partial\left(\mathbf{A} + \text{diag}\,(\zeta)\right)}{\partial \zeta_n}\right] \\ &= 1 - \left[\mathbf{D}(\zeta)\mathbf{b}\mathbf{b}^{\mathsf{H}}\mathbf{D}(\zeta)\right]_{n,n}.\end{aligned} \tag{10.33}$$

Forcing (10.33) to be zero yields

$$[\mathbf{D}(\zeta)\mathbf{b}] \odot [\mathbf{D}(\zeta)\mathbf{b}]^* = \tilde{\boldsymbol{\theta}} \circ \tilde{\boldsymbol{\theta}}^{\mathsf{H}} = \mathbf{1}. \tag{10.34}$$

This completes our proof. ■

Substituting (10.31) into (10.30), we have

$$f_D(\zeta) = \mathbf{b}^{\mathsf{H}}\mathbf{D}(\zeta)\mathbf{b} + \text{tr}\left(\text{diag}\,(\zeta)\right). \tag{10.35}$$

Notably, based on the Schur complement, (10.35) can be recast as an SDP form:

$$\max_{\zeta, \epsilon} \quad \epsilon - \text{tr}(\text{diag}\,(\zeta)), \tag{10.36a}$$

$$\text{s.t.} \quad \begin{bmatrix}\mathbf{A} + \text{diag}\,(\zeta) & \mathbf{b} \\ \mathbf{b}^{\mathsf{H}} & -\epsilon\end{bmatrix} \succcurlyeq \mathbf{0}, \tag{10.36b}$$

which can be efficiently solved by convex tools such as CVX.

Algorithm 10.1: The Proposed AO Framework

Initialization: Set feasible values of $\{\mathbf{W}^{(0)}, \mathbf{\Phi}_i^{(0)}\}$ and iteration index $t = 0$.

repeat
- Set $t \leftarrow t + 1$;
- Update $\alpha_k^{(t)} = \gamma_k$ by (10.16);
- Update $\mathbf{w}_k^{(t)}$ by (10.22);
- Update $\zeta^{(t)}$ by solving Problem (10.36) to obtain $\tilde{\theta}^{(t)}$ by using (10.31);
- With given $\tilde{\theta}^{(t)}$, update $\mathbf{\Phi}_i^{(t)}$;

until *The function* (10.18) *converges*;

The details of the developed AO framework in this subsection are summarized in Algorithm 10.1.

We provide the following proposition to further illustrate the convergence of the proposed AO algorithm.

Proposition 10.1 *The problem in* (10.18) *converges when the AO algorithm is adopted.*

Proof: For ease of illustration, we introduce a variable t as the iteration index in Algorithm 10.1. Since the optimum solution can be attained at each iteration, we have

$$\begin{aligned} f_1(\mathbf{W}^{(t+1)}, \mathbf{\Phi}_i^{(t+1)}) &= f_2(\mathbf{W}^{(t+1)}, \mathbf{\Phi}_i^{(t+1)}, \alpha^{(t+1)}) \\ &\geq f_2(\mathbf{W}^{(t+1)}, \mathbf{\Phi}_i^{(t+1)}, \alpha^{(t)}) \\ &\geq f_2(\mathbf{W}^{(t)}, \mathbf{\Phi}_i^{(t+1)}, \alpha^{(t)}) \\ &\geq f_2(\mathbf{W}^{(t)}, \mathbf{\Phi}_i^{(t)}, \alpha^{(t)}) \\ &= f_1(\mathbf{W}^{(t)}, \mathbf{\Phi}_i^{(t)}). \end{aligned} \tag{10.37}$$

Note that the weighted sum-rate maximization problem is bounded above due to the power constraints. Hence, the original objective f_1 is monotonically non-decreasing after each iteration. ∎

10.5.2 Low-Complexity Reflection Pattern Algorithm

Although the relaxation-based SDP method in Algorithm 10.1 is a straightforward solution to the reflection pattern optimization, unit-modulus constraints are generally converted into quadratic terms. By suppressing the rank-one constraint, Problem (10.28) becomes SDR form, typically solved by eigen-decomposition. However, the SDR problem suffers from a critical drawback: the number of optimization variables increases quadratically with the number of IRS elements. For this reason, we develop a UCMO method that can directly deal with the unit-modulus constraints.

In light of the special geometry of the constraint $|\theta_n| = 1$, we resort to Riemannian-Geometric optimization tools [38]. The feasible region of Problem (10.27) constitutes a unit-circle complex manifold. The manifold representation can provide a relatively concise form without relaxation. The principle of Riemannian manifold optimization extends the gradient descent (GD) from Euclidean space to a Riemannian manifold space. Thus, the original constrained optimization problem can be transformed into an unconstrained optimization problem and optimized by Riemannian GD.

Referring to Problem (10.28), the unit-modulus constrained optimization can be given by

$$\max_{\tilde{\theta}} f_{10}(\tilde{\theta}) = -\tilde{\theta}^{\mathsf{H}}\mathbf{A}\tilde{\theta} + 2\mathrm{Re}\{\tilde{\theta}^{\mathsf{H}}\mathbf{b}\}, \tag{10.38a}$$

$$\text{s.t. } |\theta_n|^2 = 1, \quad \forall n = 1, 2, \ldots, N. \tag{10.38b}$$

By constructing the smooth manifold structure on mapping space, Problem (10.38) can be reformulated as:

$$\min_{\tilde{\theta}\in \mathcal{S}^N} f_{\mathrm{R}}(\tilde{\theta}) = -\tilde{\theta}^{\mathsf{H}}\mathbf{A}\tilde{\theta} + 2\mathrm{Re}\left(\tilde{\theta}^{\mathsf{H}}\mathbf{b}\right), \tag{10.39}$$

where $\mathcal{S}^N$ denotes the manifold space defined by the unit-modulus constraints, that is,

$$\mathcal{S}^N = \left\{\tilde{\theta} \in \mathbb{C}^{N\times 1} : |\theta_1| = |\theta_2| = \cdots = |\theta_N| = 1\right\}, \tag{10.40}$$

where $\mathcal{S} = \left\{\theta_n \in \mathbb{C} : \theta_n\theta_n^* = \mathrm{Re}\{\theta_n\}^2 + \mathrm{Im}\{\theta_n\}^2 = 1\right\}$ is known as a complex circle and can be viewed as a sub-manifold of $\mathbb{C}$. The search space $\mathcal{S}^N$ is the product of N complex circles. Referred to as a unit-circle complex manifold, the search space $\mathcal{S}^N$ is a sub-manifold of $\mathbb{C}^{N\times 1}$.

The developed UCMO algorithm performs GD on the unit-circle complex manifold. The GD on the Riemannian manifold is similar to that in the Euclidean space. It consists of two phases: finding the descent direction of the current solution by computing the negative Riemannian gradient and decreasing the value of the objective function via line search [39]. The Riemannian gradient of $f_{\mathrm{R}}(\tilde{\theta})$ at the current iteration point $\tilde{\theta}^{(t)} \in \mathcal{S}^N$ is the projection of the search direction in the Euclidean space onto the tangent space $\mathcal{T}_{\tilde{\theta}^{(t)}}\mathcal{S}^N$, as given by

$$\mathcal{T}_{\tilde{\theta}^{(t)}}\mathcal{S}^N = \left\{\boldsymbol{\eta} \in \mathbb{C}^{(N+1)\times 1} : \mathrm{Re}\{\boldsymbol{\eta}^* \circ \tilde{\theta}^{(t)}\} = \mathbf{0}\right\}. \tag{10.41}$$

Based on the projection operator, the Riemannian gradient of $f_{\mathrm{R}}(\tilde{\boldsymbol{\theta}}^{(t)})$ is obtained as

$$\begin{aligned}\nabla_{\mathcal{S}^N} f_{\mathrm{R}}(\tilde{\boldsymbol{\theta}}^{(t)}) &= \mathrm{Proj}_{\mathcal{T}_{\tilde{\theta}^{(t)}}\mathcal{S}^N}\left(\nabla f_{\mathrm{R}}(\tilde{\boldsymbol{\theta}}^{(t)})\right) \\ &= \nabla f_{\mathrm{R}}(\tilde{\boldsymbol{\theta}}^{(t)}) - \mathrm{Re}\{\nabla f_{\mathrm{R}}(\tilde{\boldsymbol{\theta}}^{(t)})^* \circ f_{\mathrm{R}}(\tilde{\boldsymbol{\theta}}^{(t)})\} \circ f_{\mathrm{R}}(\tilde{\boldsymbol{\theta}}^{(t)}),\end{aligned} \tag{10.42}$$

where $\nabla f_{\mathrm{R}}(\tilde{\boldsymbol{\theta}}^{(t)})$ is the Euclidean gradient of $f_{\mathrm{R}}(\tilde{\boldsymbol{\theta}}^{(t)})$ at $\tilde{\boldsymbol{\theta}}^{(t)}$, as given by

$$\nabla f_{\mathrm{R}}(\tilde{\boldsymbol{\theta}}^{(t)}) = -2\mathbf{A}\tilde{\boldsymbol{\theta}}^{(t)} + 2\mathbf{b}. \tag{10.43}$$

As a result, the current point $\tilde{\boldsymbol{\theta}}^{(t)}$ in the tangent space $\mathcal{T}_{\tilde{\theta}^{(t)}}\mathcal{S}^N$ is updated as

$$\tilde{\boldsymbol{\theta}}'^{(t)} = \tilde{\boldsymbol{\theta}}^{(t)} - \zeta \nabla_{\mathcal{S}^N} f_{\mathrm{R}}(\tilde{\boldsymbol{\theta}}^{(t)}), \tag{10.44}$$

where $\zeta > 0$ is a carefully chosen constant step size.[1] Since $\tilde{\boldsymbol{\theta}}'^{(t)}$ is still in the tangent space $\mathcal{T}_{\theta^{(t)}}\mathcal{S}^{N+1}$ but may not be on the manifold $\mathcal{S}^N$, the Retraction mapping operation should be applied to move the point $\tilde{\boldsymbol{\theta}}'^{(t)}$ back onto the manifold $\mathcal{S}^N$. Finally, the point $\tilde{\boldsymbol{\theta}}^{(t+1)}$ after using the Retraction mapping is given by

$$\tilde{\boldsymbol{\theta}}^{(t+1)} = \mathrm{Ret}_{\tilde{\theta}^{(t)}}\left(-\zeta \nabla_{\mathcal{S}^N} f_{\mathrm{R}}(\tilde{\boldsymbol{\theta}}^{(t)})\right) = \frac{\tilde{\boldsymbol{\theta}}^{(t)} - \zeta \nabla_{\mathcal{S}^N} f_{\mathrm{R}}(\tilde{\boldsymbol{\theta}}^{(t)})}{\| \tilde{\boldsymbol{\theta}}^{(t)} - \zeta \nabla_{\mathcal{S}^N} f_{\mathrm{R}}(\tilde{\boldsymbol{\theta}}^{(t)}) \|}. \tag{10.45}$$

The above operations are illustrated in Figure 10.7 and summarized in Algorithm 10.2.

We compare the computational complexity between the SDR and the developed UCMO algorithms in reflection pattern optimization. The SDR presented

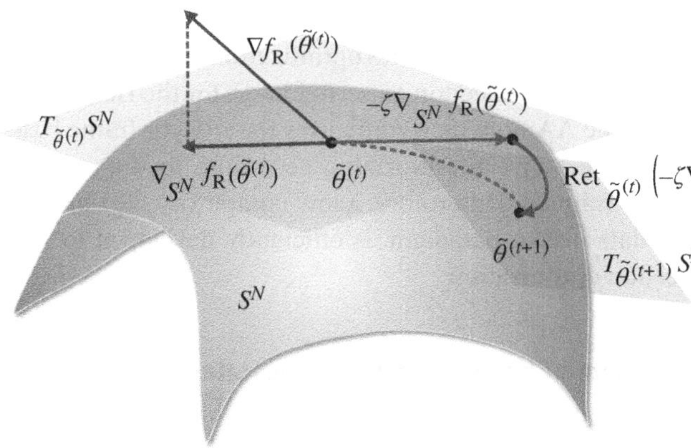

Figure 10.7 An illustration of GD on Riemannian manifolds.

1 To ensure stability and convergence of the UCMO algorithm, the step size ζ should be selected to satisfy $\zeta \leq 1/\lambda_{\mathbf{A}}$ where $\lambda_{\mathbf{A}}$ represents the largest eigenvalue of the matrix $\mathbf{A}$ in Problem (10.39). This optimization problem can be solved by leveraging the Manopt toolbox [40].

Algorithm 10.2: The Proposed UCMO Algorithm for Reflection Pattern Optimization

Initialization: Set feasible values of $\tilde{\theta}^{(0)}$ and iteration index $t = 0$.
Output : Reflection pattern $\tilde{\theta}$.

repeat
  Set $t \leftarrow t + 1$;
  Calculate the Euclidean gradient $\nabla f_{\mathrm{R}}(\tilde{\theta}^{(t)})$ at $\tilde{\theta}^{(t)}$ using (10.43);
  Construct the tangent space $\mathcal{T}_{\theta^{(t)}} \mathcal{S}^N$ and calculate the current Riemannian gradient $\nabla_{\mathcal{S}^N} f_{\mathrm{R}}(\tilde{\theta}^{(t)})$ using (10.45);
  Perform GD algorithm over the current tangent space using (10.44);
  Update $\tilde{\theta}^{(t+1)}$ using the Retraction mapping operator according to (10.45);
until *The value of* $|f_{\mathrm{R}}(\tilde{\theta}^{(t)}) - f_{\mathrm{R}}(\tilde{\theta}^{(t-1)})|$ *in* (10.39) *converges*;

in Section 10.5.1 incurs a high complexity of $\mathcal{O}(N^{6.5})$ [41] per iteration, since it expands the variable dimension. For the proposed UCMO algorithm described in Algorithm 10.2, the complexity is dominated by the evaluation of the Riemannian GD, which is $\mathcal{O}(N^2)$ per iteration [42]. To this end, the proposed UCMO algorithm is order of magnitude lower than the SDR-based alternative.

10.6 Two-Timescale Reflection Pattern Design

Precise CSI, real-time reflection pattern computing and control, and low-overhead configuration signaling are key challenges for IRS optimization. Therefore, in this section, we develop a two-timescale reflection pattern scheme for the IRS-assisted MIMO system to maximize the AASR, which configures the IRS per frame based on the S-CSI and updates the active beamforming according to the available I-CSI slot by slot during the frame. Furthermore, a low-complexity PSO method, which is inspired by the data-driven paradigm, is efficiently developed to solve the large-timescale reflection optimization.

10.6.1 Temporally Correlated Channel and Two-Timescale Transmission Strategy

We consider a generic IRS-assisted MIMO communication system, where the BS is equipped with N_{t} transmit antennas of ULA, and MU is equipped with K receive antennas of ULA. The IRS is a uniform rectangular array (URA) consisting of $N = N_x N_y$ meta-atoms, with N_x meta-atoms per row and N_y meta-atoms per column. The direct channel from the BS to the MU is denoted by $\mathbf{H}_{\mathrm{d}}$, and the

BS-IRS and IRS-MU channels are denoted by $\mathbf{G}$ and $\mathbf{H}_r$, respectively. For the brevity of notation, we define $\mathcal{H} = \{\mathbf{G}, \mathbf{H}_d, \mathbf{H}_r\}$ as the full channel ensemble.

Typically, IRS is deployed to achieve the LoS propagation between the BS and the IRS. So, the rank-one BS-IRS channel is assumed. Generally, the locations of IRSs are known to the BS and so are the BS-IRS channel, which is thus assumed to be time-invariant due to the stationarity of both the BS and IRS. The BS-IRS channel is given by

$$\mathbf{G} = \sqrt{L_{br}}\mathbf{a}(\phi)\mathbf{b}^{H}(\phi_e, \phi_a) \in \mathbb{C}^{N_t \times N}, \tag{10.46}$$

where L_{br} is the large-scale fading, ϕ is the angle-of-departure (AoD) at the BS, ϕ_e(or ϕ_a) is the elevation (or azimuth) angles-of-arrival (AoA) at the IRS, $\mathbf{a}(\phi)$ is the array steering vector of the BS, and $\mathbf{b}(\phi_e, \phi_a)$ is the array steering vector of the IRS. The array steering vectors are defined in the same way as (10.3).

Due to the mobility of MUs, the IRS-MU and BS-MU channels are multi-path and fast time-varying. To illustrate, we focus on a time frame within which the S-CSI of all links remains unchanged. The time frame is divided evenly into T time slots. The channel coefficients of the links, that is, I-CSI, do not change during a time slot. The NLoS channel coefficient matrices are correlated across different time slots. At time slot t, the IRS-MU channel can be given

$$\mathbf{H}_r[t] = \sqrt{L_{ur}}\left(\kappa\mathbf{H}_r^{LoS} + \sqrt{1-\kappa^2}\mathbf{H}_r^{NLoS}[t]\right) \in \mathbb{C}^{N\times K}, \tag{10.47}$$

where L_{ur} accounts for large-scale fading; κ is the Rician fading-related factor; $\mathbf{H}_r^{LoS}$ and $\mathbf{H}_r^{NLoS} \sim \mathcal{CN}(0, \mathbf{I})$ are the deterministic LoS component and the NLoS component, respectively. Likewise, the BS-UE channel can be given by

$$\mathbf{H}_d[t] = \sqrt{L_{bu}}\left(\kappa\mathbf{H}_d^{LoS} + \sqrt{1-\kappa^2}\mathbf{H}_d^{NLoS}[t]\right) \in \mathbb{C}^{N_t\times K}. \tag{10.48}$$

Herein, we use the first-order auto-regressive process [43] to characterize temporal-variant NLoS channel as

$$\mathbf{H}^{NLoS}[t] = \rho\mathbf{H}^{NLoS}[t-\tau] + \mathbf{\Delta}[t], \tag{10.49}$$

where $\mathbf{\Delta}[t] \sim \mathcal{CN}\left(0, (1-\rho^2)\mathbf{I}\right)$ yields the complex Gaussian distribution. Following Jakes' model [44], the temporal correlation coefficient is $\rho = J_0(2\pi\overline{f}_d)$, where $J_0(\cdot)$ is the zeroth order Bessel function of the first kind; $\overline{f}_d = f_d\tau$ is the normalized Doppler frequency; f_d is the maximum Doppler frequency; and τ is the delay between the uplink channel estimation and downlink transmission.

In time-varying scenarios, the BS first estimates the CSI based on the uplink pilots sent by the MU. Then, according to the estimated CSI, the BS computes the beamformer for downlink transmissions. When used for downlink transmissions, the CSI estimated in the uplink can expire by τ slots. Thus, the channel reciprocity may not hold in a practical time-division duplex (TDD) system.

In the considered IRS-assisted MIMO system, the BS transmits $M \leq N_t$ data streams to the MU. At the tth slot, the received signal at the MU is given by

$$\mathbf{y}[t] = \mathbf{W}_r^H \underbrace{\left(\mathbf{H}_d^H[t] + \mathbf{H}_r^H[t]\mathbf{\Theta}^H[t]\mathbf{G}^H[t]\right)}_{\text{effective channel}} \mathbf{W}_t\mathbf{P}[t]\mathbf{x}[t] + \mathbf{n}[t]$$
$$= \mathbf{W}_r^H \tilde{\mathbf{H}}^H[t]\mathbf{W}_t\mathbf{P}[t]\mathbf{x}[t] + \mathbf{n}[t], \qquad (10.50)$$

where $\mathbf{x} \in \mathbb{C}^{M\times 1}$ is the transmit signal, $\mathbf{n} \sim \mathcal{CN}(0, \sigma^2\mathbf{I}_K)$ is the additive Gaussian noise, $\mathbf{W}_t$ is the transmit beamformer at the BS, $\mathbf{W}_r$ is the receive beamformer at the MU, $\mathbf{P} = \mathrm{diag}\left\{[\sqrt{P_1}, \sqrt{P_2}, \dots, \sqrt{P_M}]\right\} \in \mathbb{C}^{M\times M}$ collects the transmit powers allocated to data streams, $\mathbf{\Theta} = \mathrm{diag}\{\exp(j\theta)\} \in \mathbb{C}^{N\times N}$ is the reflection pattern matrix of the IRS, and $\tilde{\mathbf{H}} \triangleq \mathbf{H}_d + \mathbf{G}\mathbf{\Theta}\mathbf{H}_r \in \mathbb{C}^{N_t\times K}$ is defined as *effective channel.*

It is not straightforward to estimate the effective channel $\tilde{\mathbf{H}}$, because part of it can be augmented by the IRS. The passive meta-atoms are made up of positive-intrinsic-negative (PIN)-diodes and do not involve active RF chains or any signal processing. For this reason, cascaded channel estimation is widely adopted to estimate the effective channels [30]. However, as for the slot-by-slot reflection configuration of the IRS, the cascaded channel estimation needs to be frequently conducted every time the IRS is reconfigured. In consequence, unacceptably large training overhead and delay would be required.

To this end, we develop a two-timescale beamforming strategy, configuring the IRS once per frame based on the S-CSI and updating the active beamformers in every slot based on available CSI, which may be outdated. As illustrated in Figure 10.8, the S-CSI parameters including $\{L_{ur}, L_{bu}\}$ and $\mathbf{H}_{r/d}^{\mathrm{LoS}}$ are evaluated at the beginning of a frame based on the S-CSI estimation approach [45, 46]. Based on the S-CSI, the smart controller optimizes the IRS reflection patterns for the entire frame. Given the large-timescale IRS reflection pattern in a frame, the small-timescale beamformers can be conducted slot by slot throughout the frame.

Next, we analyze the small-timescale beamforming schemes in the case of time-varying channels. As shown in Figure 10.9a and b, both BS and MU directly

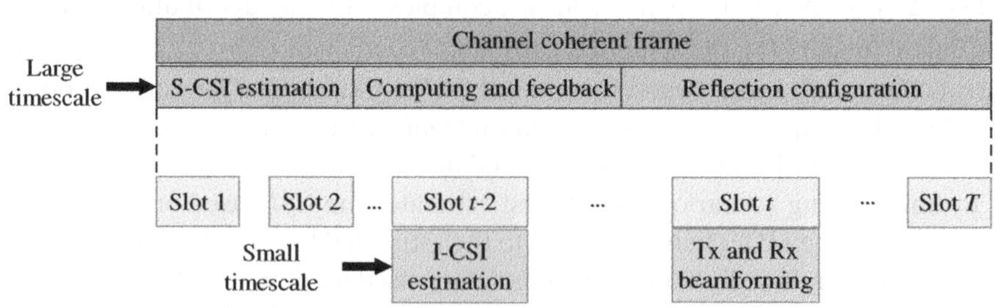

Figure 10.8 Pictorial illustration of the proposed two-timescale beamforming strategy.

apply singular value decomposition (SVD) beamformers generated from the O-CSI, that is, $\tilde{\mathbf{H}}[t-\tau] = \mathbf{U}[t-\tau]\boldsymbol{\Lambda}[t-\tau]\mathbf{V}^{\mathsf{H}}[t-\tau]$. The mismatch between beamformers and actual channel ($\mathbf{V}[t-\tau]$ and $\tilde{\mathbf{H}}[t]$) can lead to non-negligible inter-stream interference, deteriorating the rate performance. This is because mismatched unitary matrices cannot fully parallelize subchannels.

For this reason, we adopt a hybrid SVD and zero-forcing (ZF) beamforming scheme and extend it under power constraints to optimize the power allocation based on the O-CSI. As shown in Figure 10.9c, the BS still performs the SVD beamformer $\mathbf{W}_{\mathrm{t}} = \mathbf{V}[t-\tau]$ based on the O-CSI but sends the downlink pilots before the transmit beamformer, instead of sending pilots before the wireless channel after the transmit beamformer. Then, the MU conducts the ZF beamformer based on the combined information of the channel and transmit beamformer in the current slot.

Since both pilots and symbols are precoded and transmitted through the equivalent combined channel, $\check{\mathbf{H}}[t] \triangleq \mathbf{V}^{\mathsf{H}}[t-1]\tilde{\mathbf{H}}[t] \in \mathbb{C}^{M\times K}$. The MU can estimate $\check{\mathbf{H}}[t]$ based on the precoded pilots. Based on the estimated combined channel, that is, $\check{\mathbf{H}}$, the MU detects the data symbols using the following ZF beamformer:

$$\mathbf{W}_{\mathrm{r}} = \left[\left(\tilde{\mathbf{H}}^{\mathsf{H}}[t]\mathbf{V}[t-\tau]\right)^{\mathsf{H}}\tilde{\mathbf{H}}^{\mathsf{H}}[t]\mathbf{V}[t-\tau]\right]^{-1}\left(\tilde{\mathbf{H}}^{\mathsf{H}}[t]\mathbf{V}[t-\tau]\right)^{\mathsf{H}}$$

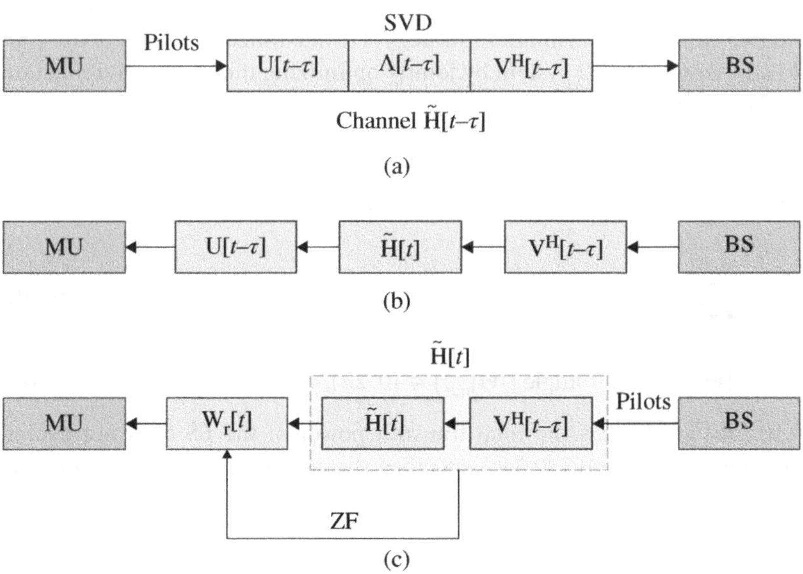

Figure 10.9 The small-timescale beamforming procedure: (a) uplink channel estimation; (b) conventional SVD-based downlink transmission; and (c) hybrid SVD and ZF beamforming-based downlink transmission.

$$= \left(\check{\mathbf{H}}[t]\check{\mathbf{H}}^{\mathsf{H}}[t]\right)^{-1}\check{\mathbf{H}}[t]. \tag{10.51}$$

The received signal is obtained as

$$\begin{aligned} \mathbf{y}[t] &= \mathbf{W}_{\mathrm{r}}^{\mathsf{H}}\left(\tilde{\mathbf{H}}^{\mathsf{H}}[t]\mathbf{V}[t-\tau]\mathbf{P}[t]\mathbf{x}[t] + \mathbf{n}[t]\right) \\ &= \mathbf{P}[t]\mathbf{x}[t] + \left(\check{\mathbf{H}}[t]\check{\mathbf{H}}^{\mathsf{H}}[t]\right)^{-1}\check{\mathbf{H}}[t]\mathbf{n}[t]. \end{aligned} \tag{10.52}$$

The resulting SINR of the mth data stream is given by

$$\gamma_m[t] = \frac{P_m[t]}{\sigma^2\left[\left(\check{\mathbf{H}}[t]\check{\mathbf{H}}^{\mathsf{H}}[t]\right)^{-1}\right]_{(m,m)}}. \tag{10.53}$$

This beamforming scheme can efficiently eliminate the inter-stream interference resulting from the SVD beamforming based on the O-CSI. The SVD precoding with water-filling power allocation is also known to be optimal under I-CSI [47]. It is still likely to produce effective parallel streams based on the O-CSI by taking advantage of the temporal correlation of the auto-regressive channels. To this end, the small-timescale SVD-ZF beamforming strategy is increasingly closer to the optimum as the channel changes more slowly and undergoes stronger temporal correlation. The strategy becomes optimal when the channel stops changing.

With the hybrid SVD and ZF scheme in the small-timescale, the goal of the proposed two-timescale transmission strategy is to maximize the AASR of the considered IRS-assisted MIMO system by jointly optimizing the long-term reflection pattern of the IRS and the per-slot power allocation of the BS. This problem is formulated as

$$(\mathrm{P1}): \max_{\mathbf{\Theta}} \mathbb{E}\left\{\max_{\mathbf{P}} \sum_{m=1}^{M} \log_2\left[1 + \gamma_m(\mathbf{\Theta}, \mathbf{P}[t])\right]\right\} \tag{10.54a}$$

$$\text{s.t.} \quad \sum_{m=1}^{M} P_m = P_{\text{tot}}, \quad P_m \geq 0, \forall m, \tag{10.54b}$$

$$\left|[\mathbf{\Theta}]_{n,n}\right| = 1, \quad \text{angle}\left([\mathbf{\Theta}]_{n,n}\right) \in [0, 2\pi), \forall n. \tag{10.54c}$$

where (10.54b) guarantees the total transmit power of the BS does not exceed its power budget P_{tot}, and (10.54c) specifies the unit-modulus nature and the phase-shift range of the IRS meta-atoms.

10.6.2 Reflection Pattern Design with Statistical CSI

We reformulate Problem (P1) as a joint design of power allocation at the BS and reflection configuration at the IRS. Then, we provide the solutions for large- and small-timescale designs. We first rewrite the SINR of the mth data stream at slot t, $\gamma_m(P_m[t], \boldsymbol{\Theta}, \mathcal{H}[t-\tau])$, as

$$\gamma_m(P_m[t], \boldsymbol{\Theta}, \mathcal{H}[t-\tau]) = \frac{P_m[t]}{\sigma^2\left[\left\{\check{\mathbf{H}}[t]\check{\mathbf{H}}^{\mathsf{H}}[t]\right\}^{-1}\right]_{(m,m)}} = \frac{P_m[t]}{\sigma^2\left[\mathbf{R}^{-1}[t]\right]_{(m,m)}} = \frac{P_m[t]}{\sigma^2 f_m(\boldsymbol{\theta})[t]}, \tag{10.55}$$

where $\mathbf{R}[t] = \check{\mathbf{H}}[t]\check{\mathbf{H}}^{\mathsf{H}}[t]$ is the channel covariance, $f_m(\boldsymbol{\theta})[t] = \left[\mathbf{R}^{-1}[t]\right]_{(m,m)}$ is the mth diagonal entry of $\mathbf{R}^{-1}$. Problem (P1) can be rewritten as

$$(\text{P1'}): \max_{\theta} \mathbb{E}\left\{\max_{\mathbf{P}} \sum_{m=1}^{M} \log_2\left[1 + \frac{P_m[t]}{\sigma^2 f_m(\boldsymbol{\theta})[t]}\right]\right\} \tag{10.56a}$$

$$\text{s.t. } \mathbf{p} \succcurlyeq 0, \quad \mathbf{1}^{\mathsf{T}}\mathbf{p} = P_{\text{tot}}, \tag{10.56b}$$

$$\boldsymbol{\Theta} = \text{diag}\left\{e^{j\theta}\right\}, \theta_n \in [0, 2\pi), \forall n = 1, 2, \dots, N, \tag{10.56c}$$

where $\mathbf{p} = [P_1, \dots, P_M]^{\mathsf{T}}$ collects the transmit powers of M data streams.

Problem (P1') is a stochastic optimization problem because of the per-slot optimization of the power allocation at the small timescale. At each slot, the power allocation is based on the available O-CSI, with the constant IRS reflection pattern applied at the current frame. On the other hand, the optimization of the IRS configuration needs to farsightedly maximize the AASR over random channel samples and potential per-slot power allocations throughout the frame. Although the stochastic optimization problem is converted into a deterministic non-convex optimization problem, minimizing each diagonal entry of the inverse of the channel covariance, that is, $\left[\mathbf{R}^{-1}\right]_{(m,m)}$, is still arduous.

Conversely, the small-timescale power allocation is a deterministic optimization problem. Given the IRS reflection pattern of the large timescale, with the hybrid SVD and ZF scheme, Problem (P1) is reformulated as

$$(\text{P2}): \quad \max_{\mathbf{p}} \sum_{m=1}^{M} \log\left(1 + \frac{P_m[t] f_m^{-1}(\boldsymbol{\theta})[t]}{\sigma^2}\right) \quad \text{s.t. (10.56b)}.$$

Problem (P2) is convex with respect to $\mathbf{p}$ and can be solved by the classic water-filling algorithm. The optimal small-timescale power allocation is given by

$$P_m^{\text{opt}}[t] = \left[\frac{1}{\mu \ln 2} - \sigma^2 f_m(\boldsymbol{\theta})[t]\right]^{+}. \tag{10.57}$$

where μ is the Lagrange multiplier and $[x]^+ = \max\{x, 0\}$. The resulting per-slot achievable rate is

$$\sum_{m=1}^{M} \log_2\left(1 + \frac{P_m[t]}{\sigma^2 f_m(\boldsymbol{\theta})[t]}\right) = \sum_{m=1}^{M} \log\left(1 + \frac{1}{\sigma^2}\left[\frac{1}{\mu f_m(\boldsymbol{\theta})[t]\ln 2} - \sigma^2\right]^+\right). \tag{10.58}$$

The per-frame IRS reflection optimization is a stochastic optimization problem due to the expectation operation in the objective of the Problem (P1'). To circumvent this impasse, we first transform the objective in a deterministic form by taking the average rate of B random channel samples to approximate the expectation in the objective. Suppose that the bth sample of the IRS-related time-variant channels in the tth time slot is $\mathcal{H}^b[t] = \{\mathbf{H}_{\mathrm{d}}^b[t-\tau], \mathbf{H}_{\mathrm{r}}^b[t-\tau]\}$, accounting for the O-CSI at the BS. The AASR of the mth data-stream is approximated by

$$\bar{r}_m\left(P_m^{\mathrm{opt}}(\boldsymbol{\theta}, \{\mathcal{H}^b\}_{b=1}^B), \boldsymbol{\theta}\right) = \frac{\sum_{t=1}^{T}\sum_{b=1}^{B} \log_2\left[1 + \gamma_m\left(P_m[t], \boldsymbol{\theta}, \mathcal{H}^b[t-\tau]\right)\right]}{TB}. \tag{10.59}$$

The large-timescale passive beamforming problem is cast as

$$(\text{P3}): \quad \max_{\boldsymbol{\theta}} \sum_{m=1}^{M} \bar{r}_m\left(P_m^{\mathrm{opt}}(\boldsymbol{\theta}, \{\mathcal{H}^b\}_{b=1}^B), \boldsymbol{\theta}\right) \quad \text{s.t. (10.56c)},$$

Apart from the non-convex constant-modulus constraint, the power allocation variables are closely coupled with the reflection patterns of the IRS. Moreover, maximizing the AASR involves minimizing the diagonal entries, that is, $\left[\mathbf{R}^{-1}\right]_{(m,m)}$, which makes it difficult to optimize $\boldsymbol{\theta}$ since $\left[\mathbf{R}^{-1}\right]_{(m,m)}$ cannot be written as an explicit function of $\boldsymbol{\theta}$. Conventional convex optimization or convexification methods cannot be applied to solve the Problem (P3).

We propose a modified PSO framework to solve the problem (P3). As a widely used meta-heuristic bionic optimization algorithm [48], PSO can find the optimal solution with little prior information. It enjoys the benefits of fewer parameters, simple calculation implementation, and fast convergence. In PSO, the position of a particle stands for a potential solution. The fitness function of a particle is typically defined to be the optimization objective. We define a possible reflection vector $\boldsymbol{\theta}$ as the position of a particle. The fitness function can be taken as the objective of (P3), as given by

$$J(\boldsymbol{\theta}) = \sum_{m=1}^{M} \bar{r}_m\left(P_m^{\mathrm{opt}}(\boldsymbol{\theta}, \{\mathcal{H}^b\}_{b=1}^B), \boldsymbol{\theta}\right). \tag{10.60}$$

For each particle $\boldsymbol{\theta}$, the optimal power solution can be uniquely found according to (10.57). This guarantees the uniqueness of the fitness function in (10.60). There are P particles utilized to search for the optimal reflection pattern in the constrained search space, and each is assigned a velocity at every iteration. The set of particle positions is denoted by $\mathcal{L}_P = \{\theta_1, \theta_2, \dots, \theta_P\}$. A particle is assessed based on its fitness function in every iteration to determine whether the current position implies a good solution. The particles record their best positions ever found.

The global optimal position is selected from the best positions of all particles. Let $\hat{\theta}_p^{(i)}$ and $\hat{\theta}^{(i)}$ denote the best position of the pth particle and the global optimal position of all particles at the ith iteration, respectively. The pth particle updates its position according to its velocity $\mathbf{v}_p^{(i)}$ at the ith iteration. The newly updated position is used to determine the best position of this particle, if the fitness value of the previous best position is lower than that of the new position. Otherwise, the best position of this particle remains unchanged. After one round of the fitness value evaluation, the global optimal position of all particles is accordingly determined. Both the velocity and the position of the pth particle at the ith iteration are updated by

$$\mathbf{v}_p^{(i)} = w\mathbf{v}_p^{(i-1)} + c_1\varepsilon_1(\hat{\theta}_p^{(i-1)} - \theta_p^{(i-1)}) + c_2\varepsilon_2(\hat{\theta}^{(i-1)} - \theta_p^{(i-1)}), \tag{10.61}$$

$$\theta_p^{(i)} = \theta_p^{(i-1)} + \mathbf{v}_p^{(i)}, \tag{10.62}$$

where w is the non-negative inertia weight of a particle; c_1 and c_2 denote cognitive and social scaling factors, respectively; and ε_1 and ε_2 are two independent random variables with a uniform distribution in (0, 1). The other control parameters, such as the swarm size P and iteration number I_{iter}, are critical for the convergence rate and global search performance. Once the parameters updated in (10.62) are outside $(-\pi, \pi]$, their boundary values are taken.

Each particle requires a notable amount of calculations for fitness evaluation per iteration since the fitness function requires a large number of random channel samples. It can be computationally prohibitive to take large-scale channel samples (which are four-dimensional matrices with size of $N_t \times K \times B \times P$) to obtain a fitness function value per iteration, potentially hindering the PSO convergence.

We develop a rsPSO algorithm that is able to substantially reduce the total fitness evaluation complexity while reaping satisfactory AASR performance. Specifically, we introduce a recursive sampling surrogate function to replace the fitness function in (10.60). We partition all B random channel samples into N_B batches with $B_s = B/N_B$ samples per batch. At each iteration, the mini-batch sampling surrogate function is given by

$$J^{(i)}(\theta) = \overline{\mu}^{(i)} J^{(i-1)} + \mu^{(i)} \sum_{m=1}^{M} \sum_{t=1}^{T} \sum_{B=(i-1)B_s+1}^{iB_s} \frac{\log_2\left[1 + \gamma_m\left(P_m[t], \theta, \mathcal{H}^b[t-\tau]\right)\right]}{TB_s}, \tag{10.63}$$

where $\overline{\mu}^{(i)} = 1 - \mu^{(i)}$, $\mu^{(i)}$ denotes a decay weight coefficient associated with the new sampling of the AASR. To guarantee more accurate sampling approximation as i increases in the stochastic optimization, $\mu^{(i)} = i^{-0.2}$ is commonly used according to the diminishing stepsize rules [49].

The proposed rsPSO algorithm is summarized in Algorithm 10.3. Given its use of random channel samples, the algorithm is *data-driven* in spirit. Given the S-CSI estimated at the beginning of a frame, the BS can generate the L_B channel samples to approximate the future channels in the frame used for downlink transmissions. Recall that there are T slots in the frame. The BS can generate the non-line-of-sight (NLoS) components of $\frac{B}{T}$ channel samples, and therefore, we set $T = I_{\text{iter}}$ without loss of generality. By repeating this process from $t = 1$ to T, we can obtain B time-varying channel samples to approximate CSI measurements of the upcoming T slots.

Finally, we analyze the computational complexity of the proposed rsPSO algorithm measured by floating point operations per second (FLOPS). According to (10.55), the operations dominating the computational complexity of each particle include a matrix addition, four matrix multiplications, and a matrix inversion per channel sample per iteration, requiring $F_1 = 4\left(N_t + K\right)N^2 + 4MN_tK + 4M^2K + (4M^3 + M^2 + M)$ FLOPS [50]. The overall computational complexity of the algorithm is $P \times B_s \times F_1$ FLOPS, according to (10.63). In contrast, a direct use of all B channel samples to evaluate the fitness function per iteration using (10.59) incurs $P \times B_s \times F_1$ FLOPS, which is N_B times the complexity of the new rsPSO algorithm since $B = N_B B_s$.

10.7 Results and Discussions

To demonstrate the potential of using IRSs for MIMO networks, in this section, we provide numerical results to verify the effectiveness and benefits of the above-proposed transmission architecture and resource allocation. In particular, the BS is located in the origin point, and MUs are uniformly and randomly distributed in a circle centered at (40 m, 0 m) with a radius of 10 m. The total transmit power is $P_{\max} = 30$ dBm.

First, we investigate the passive beamforming gains of IRS in the presence of direct BS-MU links. Figure 10.10 shows the impact of the number of BS antennas M and IRS elements N on the achievable sum-rate. The sum-rate of the proposed algorithm with the IRS exhibits superiority over the scheme without the IRS. The achievable sum-rate grows as M and N increase. We see that the achievable sum rate with N grows faster than that with M. More importantly, this implies that the higher rate gain can be achieved by low-cost passive beamforming of properly deployed IRSs instead of active but expensive massive MIMO antennas.

Algorithm 10.3: The Proposed rsPSO algorithm

Step 1: Swarm initialization
  Initialize the PSO control parameters: $\{P, w, c_1, c_2, \varepsilon_1, \varepsilon_2, I_{\text{iter}}\}$.
  Initialize the position and velocity : $\{\boldsymbol{\theta}_p^{(0)}\}_{p=1}^{P}, \{\mathbf{v}_p^{(0)}\}_{p=1}^{P}$.
  Generate B outdated channel samples according to the S-CSI and first-order AR model;
  Separate the total channel samples into N_B mini-batches with each B_s samples;
  For each particle $\boldsymbol{\theta}_p^{(0)}$, obtain the associated optimal powers $\mathbf{p}_p^{(0)}$ by using (10.57);
  Evaluate the fitness value in (10.63) of each particle using the first mini-batch of samples, and obtain the initial best position set $\{\hat{\boldsymbol{\theta}}_p\}_{p=1}^{P}$;
  Find the global optimal position $\hat{\boldsymbol{\theta}} = \arg\max_{\hat{\boldsymbol{\theta}}}\{J^{(0)}(\hat{\boldsymbol{\theta}}_1), J^{(0)}(\hat{\boldsymbol{\theta}}_2), \cdots, J^{(0)}(\hat{\boldsymbol{\theta}}_P)\}$;
end
Step 2: Iterative search
  for $i = 1 : I_{\text{iter}}$ **do**
    Update particle velocity $\{\mathbf{v}_p^{(i)}\}_{p=1}^{P}$ and position $\{\boldsymbol{\theta}_p^{(i)}\}_{p=1}^{P}$ by using (10.61) and (10.62);
    for $p = 1 : P$ **do**
      Given particle $\boldsymbol{\theta}_p^{(i)}$, calculate the optimal powers $\mathbf{p}_p^{(i)}$ according to (10.57);
      Evaluate the fitness value of particle $\boldsymbol{\theta}_p^{(i)}$ over i^{th} mini-batch of samples by using (10.63);
      if $J^{(i)} > J^{(i-1)}$ **then**
        $\hat{\boldsymbol{\theta}}_p = \boldsymbol{\theta}_p^{(i)}$;
      else
        $\hat{\boldsymbol{\theta}}_p = \boldsymbol{\theta}_p^{(i-1)}$;
      Update $\hat{\boldsymbol{\theta}}_p = \arg\max_{\hat{\boldsymbol{\theta}}}\{J^{(i)}(\hat{\boldsymbol{\theta}}_1), J^{(i)}(\hat{\boldsymbol{\theta}}_2), \cdots, J^{(i)}(\hat{\boldsymbol{\theta}}_P)\}$;
end

Therefore, the IRS can be viewed as a promising candidate for massive MIMO 2.0 technology.

Considering that the reflective elements cannot be implemented with the infinite resolution values in practice, the set of the low-resolution phase shifts at each element is given by

$$\mathcal{F}_d = \left\{ \theta_{g,m} = e^{j\varphi_{g,m}} \,|\, \varphi_{g,m} \in \left\{ \frac{2\pi i}{2^B} \right\}_{i=0}^{2^B-1} \right\}, \tag{10.64}$$

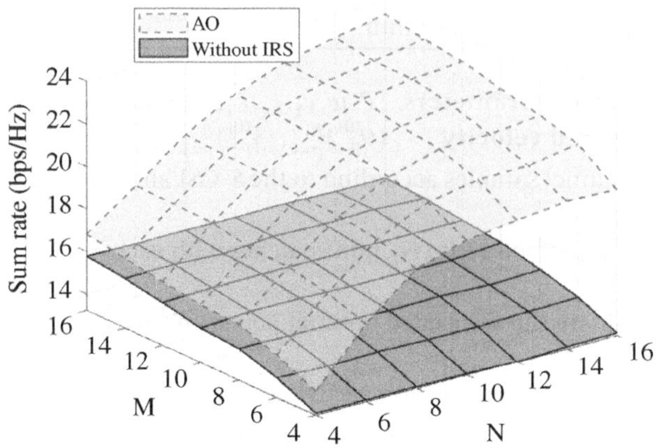

Figure 10.10 Achievable sum-rate vs the numbers of BS antennas M and IRS elements N.

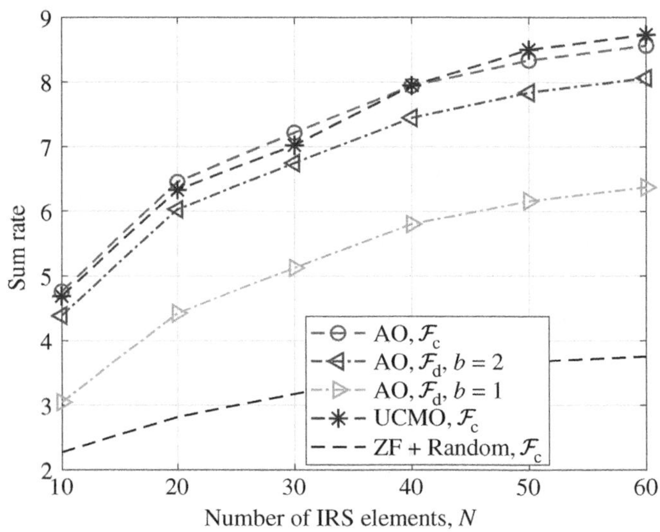

Figure 10.11 Achievable sum-rate vs the IRS elements N of different methods.

where B denotes the phase resolution in number of bits, and $B \to \infty$ corresponds to the continuous phase-shift set $\mathcal{F}_c$. In Figure 10.11, we plot the sum-rate vs the number of reflecting elements under the single-IRS system. ZF precoding with a random reflection pattern method is selected as the baseline. We see that the proposed AO and UCMO algorithms outperform the baseline method in terms of sum-rate. Besides the performance, the complexity superiority of the proposed

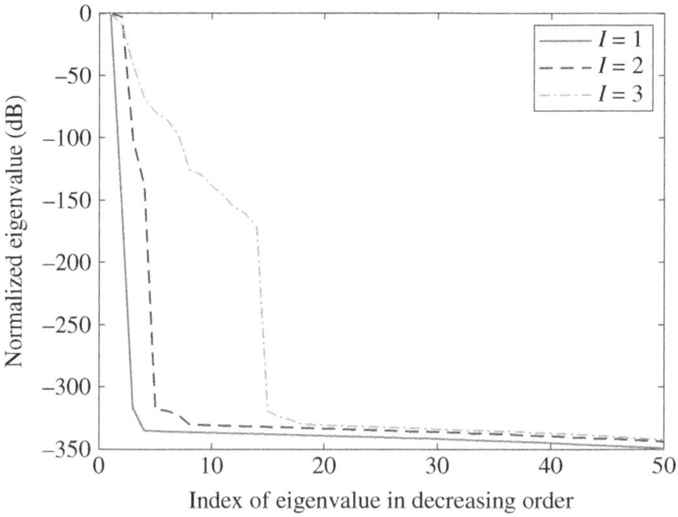

Figure 10.12 Normalized eigenvalues of the channel correlation matrix vs the number of IRS units.

UCMO approaches over the SDR method. In addition, the 1-bit phase-shifter still stably precedes the baseline method, while the 2-bit phase-shifter has an overwhelming advantage.

In Figure 10.12, we assess the impact of distributed IRS deployment on the channel rank. To characterize the channel DoF, we adopt the normalized eigenvalues of the channel correlation $\tilde{\mathbf{H}}\tilde{\mathbf{H}}^{\mathsf{H}}$ as the indicator. For single-IRS cases where only IRS unit 1 is activated, only 2–3 valid independent channels can carry signal streams. This is due to the sparse BS-IRS channel, which can serve up to 3 users. The number of valid independent channels is expanded for distributed IRS cases with $I > 1$. This confirms the channel rank improvement of the distributed IRS scheme.

Finally, we validate the two-timescale reflection optimization method by considering time-varying MIMO channels. Figure 10.13 presents the AASRs achieved by the different schemes against the temporal correlation coefficient ρ. A larger temporal correlation coefficient ρ indicates a smaller time delay τ, thus resulting in a smaller channel error between consecutive slots. In the extreme case where the O-CSI is independent of the I-CSI, the precoding matrices of the schemes are independent of the I-CSI and equivalent to random precoding schemes. We see that more severe O-CSI leads to a lower AASR. This confirms the importance of exploiting the channel correlation between consecutive time slots to improve the AASR performance. Under the nearly independent channel conditions, that is, $\rho \to 0$, the SVD precoder based on the O-CSI is, in essence, an arbitrary precoder. Nevertheless, the proposed SVD-ZF-based scheme can still outperform the

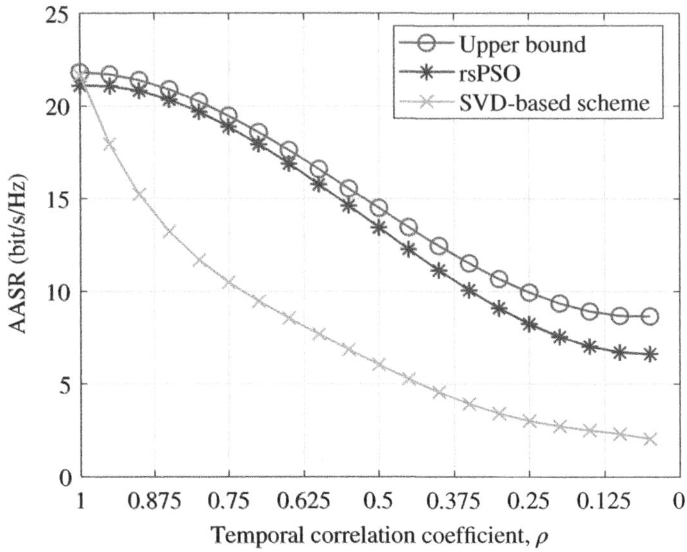

Figure 10.13 AASR against temporal correlation coefficient ρ.

SVD-based scheme based on the O-CSI because of its use of the ZF detector to suppress inter-stream interference based on the I-CSI at the UE. In the case of $\rho = 1$, the CSI remains unchanged and never outdated throughout all time slots. It is seen that the conventional SVD scheme can achieve the same AASR as the upper bound since SVD precoding is known to be optimal under the I-CSI.

10.8 Conclusion

In this chapter, slot-by-slot and two-timescale reflection pattern design schemes are designed. First, the distributed IRS-assisted mmWave massive MIMO system is built, which can overcome the channel blockage with passive meta-atoms and improve channel rank to enhance the signal coverage and channel DoF. Then, slot-by-slot IRS reflection pattern optimization is formulated to reap the maximum system weighted sum-rate in the considered mmWave system. To maximize the sum-rate, the precoder at the BS and the reflection pattern of the distributed IRS are jointly optimized in each time slot. The AO framework is developed, where the popular SDP algorithm and the low-complexity UCMO algorithm are proposed to obtain the reflection pattern, respectively. Next, to avoid the heavy overhead configuration signaling and frequently cascaded channel estimation in time-variant MIMO channels, a two-timescale reflection pattern optimization strategy is devised to maximize the long-term AASR of the IRS-assisted

MIMO system. Specifically, the small-timescale transmit beamformer and power allocation at the BS are updated in every slot with the SVD-ZF scheme to mitigate the performance deterioration resulting from outdated I-CSI. An rsPSO algorithm is proposed to solve the large-timescale IRS reflection pattern efficiently. The proposed rsPSO algorithm can dramatically reduce the total fitness evaluation complexity with light sampling of random channels.

10.9 Future Work

Recent research in the realm of IRS-assisted wireless networks has indicated the important role of the IRS in the convergence of communication, sensing and computing (CSC) for future mobile systems. However, the study domain of the IRS-assisted CSC system is still in its infancy. This section highlights possible research directions for the future investigation of the IRS.

- For the multi-cell multi-IRS scenario, each cellular network is deployed by a distinct network operator. BSs or cell-edge MUs may be interfered with by some IRSs from other operators, as IRSs cannot realize strictly narrow-band response. Thus, interference management between different operators is a vital research direction for IRS-assisted communications.
- Some studies on IRS-enabled modulation and precoding schemes have initially shown the potential of wave-based computing [11, 51]. Holographic IRS can generate desired beams or signals by controlling the radiation amplitude and phase of the EM wave. The resulting computing ability may not meet the complicated tasks for the current low-cost phase-shift-only IRS. For this reason, active IRS or few-layer IRS can provide more wave-control DoFs.
- Since IRSs can provide virtual LoS links, facilitating the sensing functions for the integrated sensing and communication (ISAC) network. Moreover, the near-field channel may be necessary for the extremely large IRS, and a near-field LoS channel can be rank-sufficient due to the spherical wavefront. As near-field channel involves both distance and angle information, they can benefit both sensing and computing more. This implies that the related information-theoretic limits analysis and near-field range extension are worth noting in future ISAC directions.

References

1 Lu, L., Li, G.Y., Swindlehurst, A.L. et al. (2014). An overview of massive MIMO: benefits and challenges. *IEEE Journal on Selected Topics in Signal Processing* 8 (5): 742–758.

2 Fall, M., Balboul, Y., Fattah, M. et al. (2023). Towards sustainable 5G networks: a proposed coordination solution for macro and pico cells to optimize energy efficiency. *IEEE Access* 11: 50794–50804.

3 Han, S. and Bian, S. (2020). Energy-efficient 5G for a greener future. *Nature Electronics* 3 (4): 182–184.

4 Shurdi, O., Ruci, L., Biberaj, A., and Mesi, G. (2021). 5G energy efficiency overview. *European Scientific Journal* 17 (3): 315–327.

5 Sanguinetti, L., Björnson, E., and Hoydis, J. (2020). Toward massive MIMO 2.0: understanding spatial correlation, interference suppression, and pilot contamination. *IEEE Transactions on Communications* 68 (1): 232–257.

6 Ning, B., Chen, Z., Chen, W. et al. (2021). Terahertz multi-user massive MIMO with intelligent reflecting surface: beam training and hybrid beamforming. *IEEE Transactions on Vehicular Technology* 70 (2): 1376–1393.

7 Hu, S., Rusek, F., and Edfors, O. (2017). The potential of using large antenna arrays on intelligent surfaces. *2017 IEEE 85th vehicular technology conference (VTC Spring)*, 1–6. Sydney, Australia: IEEE.

8 Cao, Y., Lv, T., and Ni, W. (2020). Intelligent reflecting surface aided multi-user mmWave communications for coverage enhancement. *2020 IEEE 31st Annual International Symposium on Personal, Indoor and Mobile Radio Communications*, 1–6. London, UK: IEEE.

9 Han, Y., Tang, W., Jin, S. et al. (2019). Large intelligent surface-assisted wireless communication exploiting statistical CSI. *IEEE Transactions on Vehicular Technology* 68 (8): 8238–8242.

10 Cao, Y., Lv, T., and Ni, W. (2022). Two-timescale optimization for intelligent reflecting surface-assisted MIMO transmission in fast-changing channels. *IEEE Transactions on Wireless Communications* 21 (12): 10424–10437.

11 Liu, R., Li, H., Li, M., and Liu, Q. (2019). Symbol-level precoding design for intelligent reflecting surface assisted multi-user MIMO systems. *2019 11th International Conference on Wireless Communications and Signal Processing (WCSP)*, 1–6. Xi'an, China: IEEE.

12 Wu, Q. and Zhang, R. (2019). Intelligent reflecting surface enhanced wireless network via joint active and passive beamforming. *IEEE Transactions on Wireless Communications* 18 (11): 5394–5409.

13 Cao, Y., Lv, T., Ni, W., and Lin, Z. (2021). Sum-rate maximization for multi-reconfigurable intelligent surface-assisted device-to-device communications. *IEEE Transactions on Communications* 69 (11): 7283–7296.

14 Hu, J., Zhang, H., Bian, K. et al. (2021). MetaSensing: Intelligent metasurface assisted RF 3D sensing by deep reinforcement learning. *IEEE Journal on Selected Areas in Communications* 39 (7): 2182–2197.

15 Sanchez, S.G., Reus-Muns, G., Bocanegra, C. et al. (2022). AirNN: Over-the-air computation for neural networks via reconfigurable intelligent surfaces. *IEEE/ACM Transactions on Networking*, early access. https://doi.org/10.1109/TNET.2022.3225883.

16 Huang, C., Zappone, A., Alexandropoulos, G.C. et al. (2019). Reconfigurable intelligent surfaces for energy efficiency in wireless communication. *IEEE Transactions on Wireless Communications* 18 (8): 4157–4170.

17 Wang, P., Fang, J., Yuan, X. et al. (2020). Intelligent reflecting surface-assisted millimeter wave communications: joint active and passive precoding design. *IEEE Transactions on Vehicular Technology* 69 (12): 14960–14973.

18 Zhou, G., Pan, C., Ren, H. et al. (2020). Robust beamforming design for intelligent reflecting surface aided MISO communication systems. *IEEE Wireless Communications Letters* 9 (10): 1658–1662.

19 Zhao, M.-M., Wu, Q., Zhao, M.-J., and Zhang, R. (2021). Intelligent reflecting surface enhanced wireless networks: two-timescale beamforming optimization. *IEEE Transactions on Wireless Communications* 20 (1): 2–17.

20 Sun, C., Ni, W., Bu, Z., and Wang, X. (2022). Energy minimization for intelligent reflecting surface-assisted mobile edge computing. *IEEE Transactions on Wireless Communications* 21 (8): 6329–6344.

21 Yuan, X., Hu, S., Ni, W. et al. (2023). Joint user, channel, modulation-coding selection, and RIS configuration for jamming resistance in multiuser OFDMA systems. *IEEE Transactions on Communications* 71 (3): 1631–1645.

22 Zheng, J., Tian, H., Ni, W. et al. (2022). Balancing accuracy and integrity for reconfigurable intelligent surface-aided over-the-air federated learning. *IEEE Transactions on Wireless Communications* 21 (12): 10964–10980.

23 Eskandari, M., Savkin, A.V., and Ni, W. (2023). Consensus-based autonomous navigation of a team of RIS-equipped UAVs for LoS wireless communication with mobile nodes in high-density areas. *IEEE Transactions on Automation Science and Engineering* 20 (2): 923–935.

24 Eskandari, M., Huang, H., Savkin, A.V., and Ni, W. (2023). Model predictive control-based 3D navigation of a RIS-equipped UAV for LoS wireless communication with a ground intelligent vehicle. *IEEE Transactions on Intelligent Vehicles* 8 (3): 2371–2384.

25 Savkin, A.V., Huang, C., and Ni, W. (2023). Collision-free 3-D navigation of a UAV team for optimal data collection in Internet-of-Things networks with reconfigurable intelligent surfaces. *IEEE Systems Journal* 17 (3): 4070–4077.

26 Yuan, X., Hu, S., Ni, W. et al. (2023). Deep reinforcement learning-driven reconfigurable intelligent surface-assisted radio surveillance with a fixed-wing UAV. *IEEE Transactions on Information Forensics and Security* 18: 4546–4560.

27 Hu, S., Yuan, X., Ni, W. et al. (2023). RIS-assisted jamming rejection and path planning for UAV-borne IoT platform: a new deep reinforcement learning framework. *EEE Internet of Things Journal* 10 (22): 20162–20173.

28 Nadeem, Q.U.A., Alwazani, H., Kammoun, A. et al. (2020). Intelligent reflecting surface-assisted multi-user MISO communication: channel estimation and beamforming design. *IEEE Open Journal of the Communications Society* 1: 661–680.

29 Wang, Z., Liu, L., and Cui, S. (2020). Channel estimation for intelligent reflecting surface assisted multiuser communications: framework, algorithms, and analysis. *IEEE Transactions on Wireless Communications* 19 (10): 6607–6620.

30 Liu, H., Yuan, X., Zhang, Y.-J.A., and IEEE, J. (2020). Matrix-calibration-based cascaded channel estimation for reconfigurable intelligent surface assisted multiuser MIMO. *IEEE Journal on Selected Areas in Communications* 38 (11): 2621–2636.

31 Shen, S., Clerckx, B., and Murch, R. (2022). Modeling and architecture design of reconfigurable intelligent surfaces using scattering parameter network analysis. *IEEE Transactions on Wireless Communications* 21 (2): 1229–1243.

32 Cao, Y., Lv, T., Lin, Z., and Ni, W. (2021). Delay-constrained joint power control, user detection and passive beamforming in intelligent reflecting surface-assisted uplink mmWave system. *IEEE Transactions on Cognitive Communications and Networking* 7 (2): 482–495.

33 Ahmad, S., Khan, S., Khan, K.S. et al. (2023). Resource allocation for IRS-assisted networks: a deep reinforcement learning approach. *EEE Communications Standards Magazine* 7 (3): 48–55.

34 Etcibaşı, A.Y. and Aktaş, E. (2023). Coverage analysis of IRS-aided millimeter-wave networks: a practical approach. *IEEE Transactions on Wireless Communications*, early access. https://doi.org/10.1109/TWC.2023.3310664.

35 Akdeniz, M.R., Liu, Y., Samimi, M.K. et al. (2014). Millimeter wave channel modeling and cellular capacity evaluation. *IEEE Journal on Selected Areas in Communications* 32 (6): 1164–1179.

36 Alkhateeb, A., El Ayach, O., Leus, G., and Heath, R.W. (2014). Channel estimation and hybrid precoding for millimeter wave cellular systems. *IEEE Journal on Selected Topics in Signal Processing* 8 (5): 831–846.

37 Shen, K. and Yu, W. (2018). Fractional programming for communication systems—Part I: Power control and beamforming. *IEEE Transactions on Signal Processing* 66 (10): 2616–2630.

38 Absil, P.-A., Mahony, R., and Sepulchre, R. (2008). *Optimization Algorithms on Matrix Manifolds*. Princeton, NJ: Princeton University Press.

39 Chen, J. (2018). Manifold optimization approach for data detection in massive multiuser MIMO systems. *IEEE Transactions on Vehicular Technology* 67 (4): 3652–3657.

40 Boumal, N., Mishra, B., Absil, P.-A., and Sepulchre, R. (2014). Manopt, a Matlab toolbox for optimization on manifolds. *The Journal of Machine Learning Research* 15 (1): 1455–1459.

41 Wang, J., Liao, X., and Liu, Y. (2021). Joint active and passive beamforming optimization for multigroup multicast system aided by intelligent reflecting surface. *IET Communications* 15 (4): 642–652.

42 Alhujaili, K., Monga, V., and Rangaswamy, M. (2019). Transmit MIMO radar beampattern design via optimization on the complex circle manifold. *IEEE Transactions on Signal Processing* 67 (13): 3561–3575.

43 Mehanna, O. and Sidiropoulos, N.D. (2014). Channel tracking and transmit beamforming with frugal feedback. *IEEE Transactions on Signal Processing* 62 (24): 6402–6413.

44 Ge, Y., Zhang, W., Gao, F. et al. (2019). Beamforming network optimization for reducing channel time variation in high-mobility massive MIMO. *IEEE Transactions on Communications* 67 (10): 6781–6795.

45 Mu, X., Liu, Y., Guo, L. et al. (2021). Intelligent reflecting surface enhanced indoor robot path planning: a radio map-based approach *IEEE Transactions on Wireless Communications* 20 (7): 4732–4747.

46 Zhao, M.-M., Liu, A., Wan, Y., and Zhang, R. (2021). Two-timescale beamforming optimization for intelligent reflecting surface aided multiuser communication with QoS constraints. *IEEE Transactions on Wireless Communications* 20 (9): 6179–6194.

47 Eldad, P. and Robert, S. (2008). *Next Generation Wireless LANs Throughput Robustness and Reliability in 802.11 n*. Cambridge: Cambridge University Press.

48 Jian, R., Chen, Y., Liu, Z., and Xia, Y. (2020). Hybrid precoding for multiuser massive MIMO systems based on MMSE-PSO. *Wireless Networks* 26 (2): 1291–1299.

49 Yang, Y., Scutari, G., Palomar, D.P., and Pesavento, M. (2016). A parallel decomposition method for nonconvex stochastic multi-agent optimization problems. *IEEE Transactions on Signal Processing* 64 (11): 2949–2964.

50 Hei, Y.-q., Li, X.-H., Yi, K.-C., and Li, X. (2009). Multiuser scheduling in downlink MIMO systems using particle swarm optimization. *2009 IEEE Wireless Communications and Networking Conference*, 1–5. Budapest, Hungary: IEEE.

51 An, J., Xu, C., Ng, D.W.K. et al. (2023). Stacked intelligent metasurfaces for efficient holographic MIMO communications in 6G. *IEEE Journal on Selected Areas in Communications* 41 (8): 2380–2396.

11

ABER Performance Evaluation of RIS-Aided Millimeter Wave Massive MIMO System Under 3GPP 5G Channels

Vishnu Vardhan Gudla[1], Vinoth Babu Kumaravelu[2], Anjana B. S.[2], Poongundran Selvaprabhu[2], Nivetha Baskar[2], Helen Sheeba John Kennedy[2], Samarendra Nath Sur[3], Webert Montlouis[4], Agbotiname Lucky Imoize[5,6], and Arthi Murugadass[7]

[1]*Department of Electronics and Communication Engineering, Aditya Engineering College, Andhra Pradesh, Surampalem, India*
[2]*Department of Communication Engineering, School of Electronics Engineering, Vellore Institute of Technology, Vellore, Tamil Nadu, India*
[3]*Department of Electronics and Communication Engineering, Sikkim Manipal Institute of Technology, Sikkim Manipal University, Sikkim, India*
[4]*Applied Physics Laboratory (APL) and Electrical and Computer Engineering Department, Johns Hopkins University, Baltimore, MD, USA*
[5]*Department of Electrical and Electronics Engineering, Faculty of Engineering, University of Lagos, Nigeria*
[6]*Department of Electrical Engineering and Information Technology, Ruhr University, Bochum, Germany*
[7]*School of Computer Science and Engineering, Vellore Institute of Technology, Chennai, Tamil Nadu, India*

11.1 Introduction

Fourth-generation (4G) achieves 15 b/s/Hz in peak spectral efficiency, while fifth-generation (5G) doubles that demand to around 30 b/s/Hz. The anticipated sixth-generation (6G) surpasses both with a remarkable 100 b/s/Hz, indicating a threefold increase over 5G [1–3]. These improvements, driven by innovations like higher frequency bands and advanced techniques, highlight the evolution toward higher data rates and superior network performance across wireless generations. In terms of energy efficiency, 5G is expected to be 100 times more energy efficient than 4G, while 6G aims for a 10 times improvement over 5G. Technological advancements, such as beamforming and dynamic spectrum sharing, enhance data transmission rates and reduce latency. These features also improve energy efficiency by allowing devices to enter low-power states faster, thus ultimately making newer generation networks more sustainable than their predecessors.

Massive MIMO for Future Wireless Communication Systems: Technology and Applications,
First Edition. Edited by Webert Montlouis and Agbotiname Lucky Imoize.

11.1.1 Massive MIMO (M-MIMO)

Massive multiple-input and multiple-output (MIMO) is a cutting-edge wireless communication technology designed to significantly enhance the performance of wireless networks. In a massive MIMO (M-MIMO) system, a base station (BS) is equipped with an exceptionally large number of antennas, potentially in the order of hundreds or even thousands [4]. The M-MIMO-assisted communication is pictorially represented in Figure 11.1.

The fundamental idea behind M-MIMO is to exploit the spatial domain more effectively. By deploying a massive array of antennas, the system can simultaneously serve multiple users with spatially focused beams, greatly increasing spectral efficiency and overall system capacity [5]. M-MIMO technology can transmit multiple data streams concurrently by utilizing the spatial dimension, significantly increasing data throughput and network capacity. This technology uses advanced signal processing techniques, like beamforming, to focus radio frequency (RF) signals directly toward the intended user devices [6]. Meanwhile, it enhances the signal strength and quality of the network. This technology helps mitigate interference, improving the overall reliability and performance of the wireless network. Despite the increased number of antennas, M-MIMO systems are designed to be energy-efficient, contributing to a more sustainable and eco-friendly communication infrastructure [7]. Through its ability to support numerous spatial connections, M-MIMO enhances spectral efficiency that makes more effective use of available frequency bands. Moreover, it is a crucial technology contributing to the evolution of wireless communication networks, providing higher data rates, improved reliability, and better radio spectrum utilization.

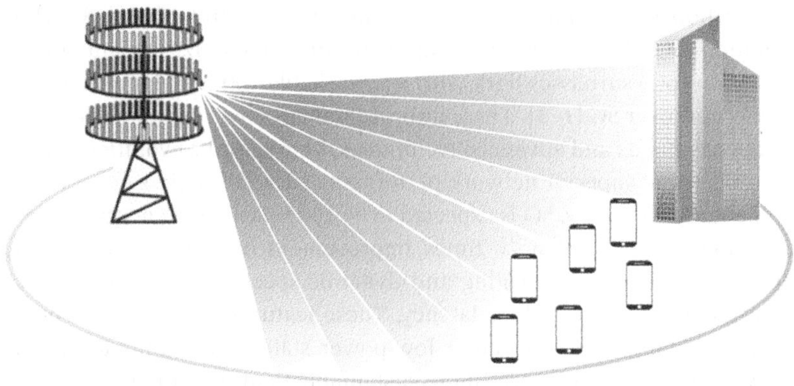

Figure 11.1 Pictorial representation of M-MIMO.

The key characteristics and advantages of M-MIMO include:

- **Spatial multiplexing**: M-MIMO leverages spatial multiplexing, allowing multiple users to be served on the same time–frequency resources through the use of highly directional beams.
- **Increased spectral efficiency**: With a large number of antennas, M-MIMO can spatially separate users, enabling more simultaneous connections and significantly boosting spectral efficiency.
- **Improved energy efficiency**: Despite having a higher number of antennas, M-MIMO can achieve better energy efficiency compared to traditional MIMO systems, as it can focus energy where it is needed.
- **Reduced interference**: Precise beamforming in M-MIMO helps minimize interference between users, leading to improved signal quality and system performance.
- **Enhanced coverage**: M-MIMO can enhance coverage and reliability, particularly in challenging environments such as urban areas with high user density.

While M-MIMO offers remarkable advantages, challenges include increased system complexity of signal processing and hardware requirements [8]. As the number of antennas grows significantly, the system faces challenges related to channel estimation, signal processing algorithms, and the need for a large number of RF chains. This complexity can result in higher energy consumption, increased computational demands, and higher implementation costs. Additionally, M-MIMO systems may experience challenges in mobility scenarios, as tracking fast-changing channels with a large number of antennas can be computationally intensive. This required the need for advanced signal processing algorithms and addressing potential issues related to hardware implementation. Nonetheless, as a technology at the forefront of 5GB, M-MIMO holds great promise for revolutionizing wireless communication networks.

11.1.2 RIS

RIS stands for "reconfigurable intelligent surfaces." It is a technology that has gained prominence in wireless communication systems, particularly in the context of 5GB. An RIS comprises numerous small and individually controllable elements, such as antennas or reflecting elements [9]. These surfaces consist of a large number of closely spaced, passive elements that can dynamically adjust their electromagnetic (EM) properties in response to the changing RF environment. The RIS-assisted communication is pictorially represented in Figure 11.2.

The reflecting surface can also be adjusted to control the reflected waves' direction, phase, and amplitude, which enables various applications in wireless communications, such as beamforming and signal enhancement [10]. Moreover, it is a

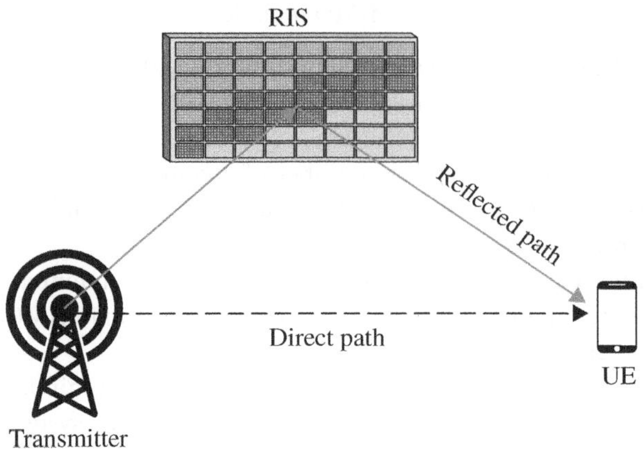

Figure 11.2 RIS-assisted communication.

technology that uses programmable materials to adapt the behavior of RF signals. These reflecting elements can be adjusted in real time to manipulate the characteristics of radio waves. By intelligently managing EM waves, RIS has the potential to address coverage issues, enhance spectral efficiency, and contribute to the overall performance of advanced wireless networks [11]. The key aspects of RIS include:

- **Passive elements**: The surface is composed of passive elements, such as metamaterials or programmable reflecting elements, which do not require a direct power supply. They respond to incoming signals by altering their reflective or refractive properties.
- **Dynamic adjustments**: RIS can dynamically reconfigure the reflection and scattering of EM waves. By adjusting the phase and amplitude of the reflected signals, RIS can optimize the wireless channel and improve signal quality.
- **Beamforming and signal enhancement**: RIS technology enables precise beamforming, allowing for the redirection and focusing of signals. This leads to improved signal strength, reduced interference, and enhanced communication reliability.
- **Interference management**: RIS can be employed to mitigate interference in wireless communication networks. By intelligently modifying the propagation environment, RIS helps create favorable conditions for communication.
- **Energy efficiency**: Since RIS elements are passive and do not require a continuous power supply, this technology has the potential to enhance energy efficiency in wireless networks.

RIS finds applications in various scenarios, including indoor wireless communication, outdoor networks, and Internet of Things (IoT) deployments. They can be utilized to optimize coverage, increase data rates, and enhance overall network performance [12]. The drawback of RIS is the potential complexity in optimizing their configurations. Achieving optimal settings for RIS elements to enhance wireless communication can be computationally intensive and may require sophisticated algorithms, making practical implementation challenging. Additionally, RIS deployment can face challenges in terms of power consumption and cost, impacting their feasibility in certain applications. As research and development in this field progress, RIS is expected to play a crucial role in shaping the future of wireless communication systems.

11.1.3 Millimeter Waves

Millimeter waves (mmWave) refer to RF in the EM spectrum with wavelengths ranging from 1 to 10 mm and frequencies between 30 and 300 GHz. This portion of the spectrum holds immense potential for various applications, particularly in wireless communication and technology. The high frequency of mmWave enables the transmission of large amounts of data, making it a key player in the evolution of wireless networks, including the 5GB [13]. The shorter wavelength allows for more antennas to be packed into smaller spaces, contributing to the development of advanced communication systems like beamforming and M-MIMO. Figure 11.3

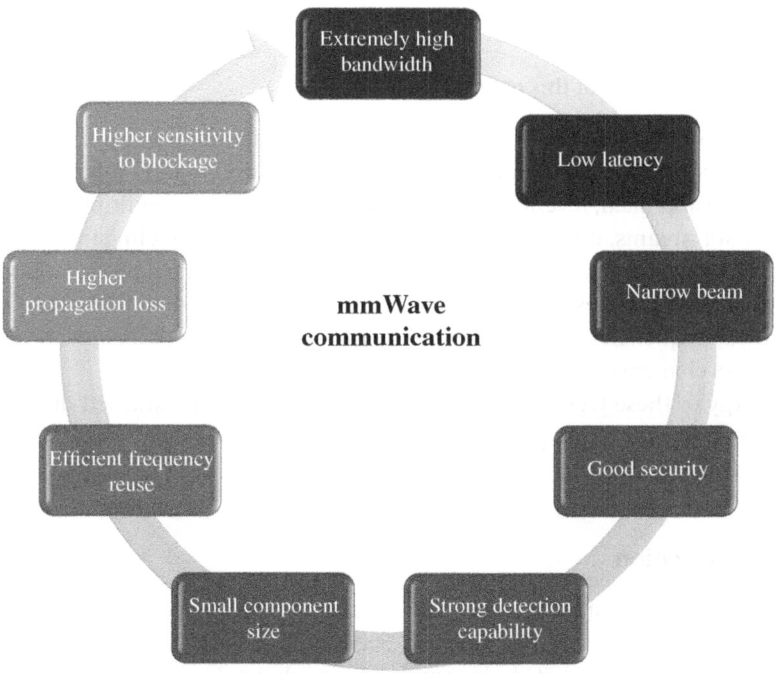

Figure 11.3 Benefits and challenges of mmWave communication.

provides a visual representation of the benefits and challenges that come with mmWave communication.

While mmWave technology offers significant data transfer rates and low latency, it faces challenges such as limited propagation range and susceptibility to atmospheric absorption and obstacles like buildings [14]. The presence of obstacles and signal attenuation in the mmWave band significantly impacts network service quality. Despite these challenges, the potential for high-speed, low-latency communication makes mmWave an exciting frontier in the realm of wireless technology.

Innovations like M-MIMO and RIS offer a compelling solution to address these issues and aim to enhance the spectral efficiency and overall performance of mmWave systems. RIS has individual reflecting elements that independently manipulate the phase shifts and enable tailored adjustments to the wireless propagation environment [15]. This solution proves especially beneficial in scenarios where obstacles disrupt the line-of-sight (LoS) path, thus allowing RIS deployment to create alternative LoS paths between the transmitter and receiver. Such technology holds promise for enhancing the performance of mmWave communications by optimizing the wireless channel in real-time.

11.1.4 Contributions

The major contributions of this chapter are:

- The novelty of this proposed chapter lies in the integration of advanced techniques, including M-MIMO, mmWave, and RIS. This combination represents a cutting-edge approach, leveraging the benefits of M-MIMO for multiple simultaneous data streams, mmWave for higher-frequency bands enabling increased data capacity, and RIS for intelligent reflection to optimize signal propagation.
- This study demonstrates the successful outcomes in both indoor and outdoor scenarios at frequencies of 28 and 73 GHz by varying numbers of transmit antennas and RIS elements.
- The synergy of these techniques introduces a novel and sophisticated dimension to the proposed work, promising enhanced performance and efficiency in wireless communication.

11.1.5 Organization

The structure of this chapter is organized as follows: The related works on RIS, mmWave, and M-MIMO are discussed in Section 11.2. Sections 11.3 and 11.4 demonstrate the system and 5G indoor and outdoor channel models respectively. Section 11.5 introduces the average bit error rate (ABER) performance evaluation and its corresponding discussions. Finally, the summary and conclusion of the chapter are discussed in Section 11.6.

11.2 Related Work

This section discusses the existing works on mmWave, M-MIMO, and RIS, highlighting their findings, merits, and limitations. Also, it emphasizes the prior works that integrate these key technologies.

11.2.1 Prior Works on mmWave M-MIMO System

Ai et al. [16] suggested the research gaps by investigating M-MIMO channels at an mmWave frequency (around 26 GHz), an area often overlooked in previous studies. This work conducts an indoor measurement campaign with 64 and 128 array elements, extracting key metrics. Using a calibrated ray-tracing simulator, extensive simulations with 1024 antenna elements provide insights into mean delay and root mean square (RMS) delay variation. Notably, the study observes a good agreement between measurement and simulation results, contributing valuable understanding to mmWave M-MIMO behavior. The study by Sun and Qi [17] explores analog beamforming and combining in mmWave M-MIMO communications. Using a hierarchical codebook, the research investigates weighted sum rate maximization by jointly considering channel training duration and received signal-to-noise ratio (SNR). Simulation results reveal the suggested algorithm's superiority over multi-sectional search, achieving almost double the weighted sum rate in scenarios with 35 symbols per block and a channel SNR of 20 dB.

Elbir and Papazafeiropoulos [18] suggested a convolutional neural network (CNN)-based deep learning (DL) framework for hybrid precoding design in multi-user (MU) mmWave MIMO systems. The CNN takes user channel matrices as input and outputs analog precoders and combiners. Comparative experiments show that CNN-based precoding outperforms existing hybrid precoding methods, including optimization-based and DL-based approaches like multilayer perceptron (MLP), by demonstrating superior performance and reduced computation time. Song et al. [19] introduced a channel estimation technique for mmWave M-MIMO systems, targeting beam squint reduction. While demonstrating low estimation error with a small number of subcarriers, the method is sensitive to parameter initialization and low SNR. It is designed for LoS scenarios and can perform poorly at indoors. The simulation results align well with theoretical expressions, highlighting the method's effectiveness and limitations under specific constraints.

The research by Srinivas and Borugadda [20] focuses on maximizing energy and spectral efficiency in an M-MIMO system by selecting optimal RF chains through a hybrid model. The optimized hybrid model named differential evolution firefly selection model, evaluated against existing approaches that prove highly effective in enhancing system efficiency for transmitting the data from transmitter to receiver. Du et al. [21] discuss the CNN-based dynamic partially connected (DPC) algorithm, which employs a hybrid precoding approach for DPC structures in mmWave M-MIMO systems. It claims higher energy and spectral efficiency based on simulation results, suggesting potential advancements in wireless communication technology.

11.2.2 Prior Works on mmWave RIS System

He et al. [22] introduced RIS for mmwave MIMO systems to control EM waves. The suggested adaptive phase shifter system improves data rate and positioning accuracy without using active sensors in the RIS. It is based on mobile station feedback and hierarchical codebooks. Despite low SNR, simulation findings indicate better performance than random design approaches and convergences to an exhaustive search technique. Pradhan et al. [23] suggested an alternating optimization method for jointly constructing the hybrid precoding and phase shifters in the RIS-assisted MU mmWave communication network. The element-wise constant modulus necessitates analog precoding and phase shifters are handled by the gradient-projection method. Finally, analytical proof ensures convergence and numerical examples that validate the performance gain of the suggested design.

Nemati et al. [24] devised an innovative RIS-empowered wall to enhance outdoor-to-indoor mmWave communication, utilizing a metasurface design with chipless RF identification (RFID) sensors for a seamless signal transition from outdoor-to-indoor spaces. The controlled RFID sensors enable adaptive passive beamforming, and mathematical analysis validates the system's feasibility and assesses blockage effects. Wang et al. [25] explore extending traditional beamforming training methods to RIS-aided mmWave systems, introducing a new multidirectional approach that should improve the performance under low SNR and reduce training overhead compared to exhaustive search schemes. Finally, open research issues in RIS-aided beamforming training are also addressed.

Feng et al. [26] suggested an innovative RIS-aided mmWave wireless network system featuring twin resolution with hybrid and passive beamforming architecture. This paper introduced a greedy method and complex circuit manifold-based strategy for configuring connections and upgrading RIS elements that show improved performance, as evidenced by simulation results. This recommended structure achieves a favorable trade-off between power consumption and bandwidth efficiency. The work by Chen et al. [27] tackles channel estimation in

RIS-aided mmWave MIMO systems, recommending a protocol for directly estimating cascaded channels by leveraging their inherent sparsity. The sparse channel matrix minimizes training overhead using compressive sensing. The paper also explores optimizing the training sequence for RIS reflection coefficients. Simulation results finally confirm the efficiency of the recommended estimator.

11.2.3 Prior Works on RIS-Aided mmWave M-MIMO System

Ying et al. [28] discuss the potential of RIS for energy-efficient wireless environments in smart cities. They introduced a novel geometric mean decomposition-based beamforming for RIS-aided mmWave hybrid MIMO systems. This approach enhances ABER performance without complex bit/power allocation and treats multiple spatial domain streams equally. Simulation results demonstrate superior ABER performance compared to conventional methods. Zhou et al. [29] present a cascaded channel estimation method for RIS-assisted mmWave M-MIMO systems. Employing angle-of-departure (AoD) and angle-of-arrival (AoA), the channel gain and angle estimation are determined using the least squares (LS) method with a lower pilot overhead. Finally, the simulation results show superior normalized mean square error (NMSE) performance compared to orthogonal matching pursuit (OMP)-based techniques.

Yildirim et al. [30] introduced an RIS-assisted angular-based hybrid beamforming (AB-HBF) system for hybrid mmWave M-MIMO. This study explores various RIS-assisted AB-HBF architectures to optimize achievable rate performance that aims for harmonious collaboration between RIS and HBF technologies. Numerical results demonstrate a substantial increase in achievable rates, even without a LoS channel. The findings suggest that integrating RISs in AB-HBF systems holds the potential for enhanced reliability and flexibility in system design. The work by Wang et al. [31] focuses on cascaded channel estimation in a mmWave M-MIMO system with an RIS and few-bit analog-to-digital converter (ADC). The bilinear generalized approximate message passing (BiG-AMP) technique and Bayesian optimum estimation framework are employed to handle the low-rank cascaded channel estimation as a matrix completion problem. Simulation results demonstrate accurate cascaded channel estimation for the RIS-assisted mmWave M-MIMO communication system. Singh et al. [32] describe a novel technique for optimizing transmit power and active RF chains in RIS-assisted mmWave MU-MIMO downlink systems focusing on maximizing energy efficiency. This technique formulates an optimization problem and provides closed-form solutions by demonstrating its efficiency through simulation results. This work addresses significant challenges and contributes to improving the energy efficiency of RIS-assisted systems.

The work by Wang et al. [33] focuses on joint beamforming in point-to-point RIS-aided mmWave MIMO-orthogonal frequency division multiplexing (OFDM) systems. This work introduces a novel singular value decomposition approximation-based method and creates a mathematical connection between singular values and reflection coefficients. Simulation results illustrate its effectiveness in crafting a favorable propagation channel with minor conditions, demonstrating superior performance compared to existing methods. These results represent a significant advancement in beamforming optimization for RIS-assisted systems.

11.3 System Model

Figures 11.4 and 11.5 showcase the RIS-assisted mmWave M-MIMO system. The system consists of N_T transmit antennas at the BS, N_R receive antennas at the user equipment (UE), and an RIS equipped with N reflective elements. These figures illustrate both indoor and outdoor communication scenarios where the BS communicates with the UE through the RIS. It is assumed that there is no direct LoS link between the BS and the UE, and communication occurs solely through the RIS.

The received signal at the UE reflected from the RIS is expressed as

$$\underbrace{\mathbf{r}}_{N_R\times 1} = \underbrace{\mathbf{G}\boldsymbol{\phi}\mathbf{H}}_{N_R\times N_T} \underbrace{\mathbf{x}}_{N_T\times 1} + \underbrace{\mathbf{w}}_{N_R\times 1}, \tag{11.1}$$

where $\mathbf{G}\boldsymbol{\phi}\mathbf{H}$ is the complete overall channel between BS and UE. $\mathbf{H}$ is a $N \times N_T$ channel matrix corresponding to the BS-RIS link, and $\mathbf{G}$ is a $N_R \times N$ channel

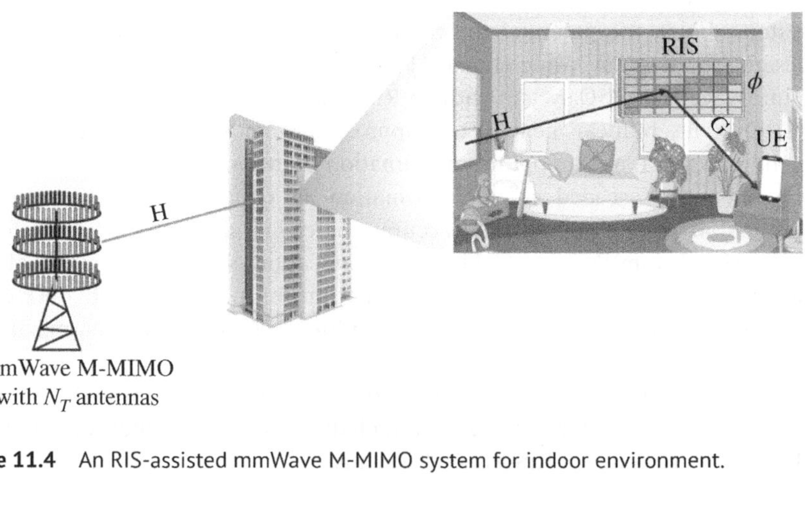

Figure 11.4 An RIS-assisted mmWave M-MIMO system for indoor environment.

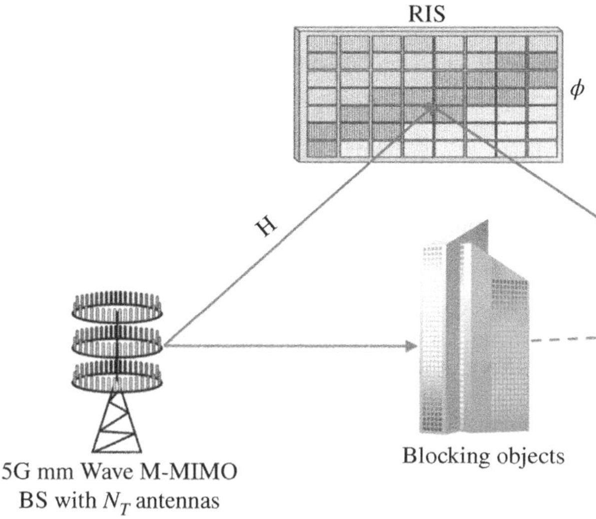

Figure 11.5 An RIS-assisted mmWave M-MIMO system for outdoor environment.

matrix corresponding to the RIS-UE link. $\boldsymbol{\phi} = \text{diag}\left(\left[e^{j\phi_1}, e^{j\phi_2}, \dots, e^{j\phi_N}\right]\right)$ is a matrix capturing phase shifts to be introduced at the RIS elements. $\mathbf{x}$ is transmit signal vector and $\mathbf{w}$ is the noise vector with mean 0 and variance N_0. We are using a diversity mode for MIMO transmission.

To provide a more explicit explanation, let us consider that UE with $N_R = 1$. Equation (11.1) can then be further simplified as

$$r = \underbrace{\mathbf{g}\boldsymbol{\phi}\mathbf{H}}_{1\times N_T} \underbrace{\mathbf{x}}_{N_T\times 1} + w. \tag{11.2}$$

Here, we can consider the individual elements of channel matrices as $g_m = \beta_m e^{-j\psi_m}$ and $h_{m,n} = \alpha_{m,n} e^{-j\theta_{m,n}}$, $m \in \{1, 2, \dots, N\}$, $n \in \{1, 2, \dots, N_T\}$. Due to diversity mode, the same symbol (x) is transmitted from all the antennas with average energy E_s. Hence, (11.2) can be written as

$$r = \left(\sum_{m=1}^{N} g_m \left[\sum_{n=1}^{N_T} h_{m,n} \right] e^{j\phi_m} \right) x + w. \tag{11.3}$$

Let $\sum_{n=1}^{N_T} h_{m,n} = h_m^s = \alpha_m^s e^{-j\theta_m^s}$. Substituting in (11.3) results

$$r = \left(\sum_{m=1}^{N} g_m h_m^s e^{j\phi_m} \right) x + w. \tag{11.4}$$

The instantaneous SNR at the UE can be composed as

$$\chi = \frac{\left|\sum_{m=1}^{N} \beta_m \alpha_m^s e^{j(\phi_m - \psi_m - \theta_m^s)}\right|^2 E_s}{N_o}. \tag{11.5}$$

To optimize the RIS phase shift responses for maximum instantaneous SNR at the UE receiver, channel state information (CSI) availability at the RIS is assumed and considered in selecting the optimal phase shift responses. The optimal phases are given as follows:

$$\phi_m = \psi_m + \theta_m^s,\ m \in \{1, 2, \dots, N\}. \tag{11.6}$$

The maximized instantaneous SNR at the receiver UE is expressed as

$$\chi_{\max} = \frac{\left|\sum_{m=1}^{N} \beta_m \alpha_m^s\right|^2 E_s}{N_o}. \tag{11.7}$$

The transmitted symbol can be detected using the optimal maximum likelihood (ML) detector, assuming perfect CSI knowledge at the UE.

$$\hat{x} = \underset{x}{\text{argmin}} \left| r - \left(\sum_{m=1}^{N} g_m h_m^s e^{j\phi_m} \right) x \right|^2 \tag{11.8}$$

11.4 Channel Model

This section presents channel models for RIS-aided mmWave M-MIMO systems for indoor and outdoor environments using the widely adopted 3GPP clustered statistical channel model [34].

11.4.1 Indoor Physical Channel Model

This section shows the channel model for InH-indoor office environments at 28 and 73 GHz mmWave frequencies. The models for the BS-RIS and RIS-UE channels are described in the subsections below.

11.4.1.1 BS-RIS Channel

Let us consider that P scatterers/reflectors are available between BS and RIS. These scatterers are grouped into Q clusters each having S_q sub-rays for $q = 1, 2, \dots, Q$, i.e., $P = \sum_{q=1}^{Q} S_q$. A cluster typically comprises a group of sub-rays that share similar spatial and/or temporal characteristics. The $N \times N_T$ BS-RIS channel vector is defined by considering array responses and path attenuations as follows [35, 36]:

$$\mathbf{H} = \gamma \sum_{q=1}^{Q} \sum_{s=1}^{S_q} \beta_{q,s} \sqrt{G_e(\theta_{q,s}^{\text{BS-RIS}}) L_{q,s}^{\text{BS-RIS}}}\, \mathbf{a}\left(\phi_{q,s}^{\text{BS-RIS}}, \theta_{q,s}^{\text{BS-RIS}}\right) \mathbf{a}^T\left(\phi_{q,s}^{\text{BS}}, \theta_{q,s}^{\text{BS}}\right) + \mathbf{H}_{\text{LoS}} \tag{11.9}$$

where $\gamma = \sqrt{\frac{1}{\sum_{q=1}^{Q} S_q}}$ indicate the scaling factor, $\beta_{q,s}$ denote the complex Gaussian distributed small-scale path gain associated with sth sub-ray in qth cluster. $G_e\left(\theta_{q,s}^{\text{BS-RIS}}\right)$ is the RIS element gain and $L_{q,s}^{\text{BS-RIS}}$ is the large-scale path attenuation associated with sth sub-ray in qth cluster. $\mathbf{a}\left(\phi_{q,s}^{\text{BS-RIS}}, \theta_{q,s}^{\text{BS-RIS}}\right)$ and $\mathbf{a}\left(\phi_{q,s}^{\text{BS}}, \theta_{q,s}^{\text{BS}}\right)$ are the array response vectors for the elevation arrival angles $\theta_{q,s}^{\text{BS-RIS}}, \theta_{q,s}^{\text{BS}}$ and azimuth arrival angle $\phi_{q,s}^{\text{BS-RIS}}, \phi_{q,s}^{\text{BS}}$ respectively. $\mathbf{H}_{\text{LoS}}$ is the LoS component. The process to calculate the array response vectors is explained in detail in [35, 37].

In general, Poisson distribution models the number of mmWave communication clusters $Q \sim \mathcal{P}(\lambda)$. Here λ is the variance, which depends on operating frequency [38]. The propagation environment must contain at least one cluster, $Q = \max\{1, \mathcal{P}(\lambda)\}$. In a given cluster, $\phi_{q,s}^{\text{BS-RIS}}$ and $\phi_{q,s}^{\text{BS}}$ are Laplacian distributed with mean $\mathcal{U}\left[-\frac{\pi}{2}, \frac{\pi}{2}\right]$. Similarly, the $\theta_{q,s}^{\text{BS-RIS}}$ and $\theta_{q,s}^{\text{BS}}$ are conditionally Laplacian with a mean $\mathcal{U}\left[-\frac{\pi}{4}, \frac{\pi}{4}\right]$. The expression for path loss $L_{q,s}^{\text{BS-RIS}}$ for (q, s)th path can be found in [35, 39]

The $\mathbf{H}_{\text{LoS}}$ can be formulated as

$$\mathbf{H}_{\text{LoS}} = I_h\left(d_{\text{BS-RIS}}\right) \sqrt{G_e(\theta_{\text{LoS}}^{\text{BS-RIS}}) L_{\text{LoS}}^{\text{BS-RIS}}}\, e^{j\eta} \mathbf{a}(\phi_{\text{LoS}}^{\text{BS-RIS}}, \theta_{\text{LoS}}^{\text{BS-RIS}})\, \mathbf{a}^T(\phi_{\text{LoS}}^{\text{BS}}, \theta_{\text{LoS}}^{\text{BS}}) \tag{11.10}$$

where $I_h\left(d_{\text{BS-RIS}}\right)$ is the Bernoulli distributed random variable with values $\{0,1\}$ that indicate the presence of LoS link between BS and RIS separated by a distance of $d_{\text{BS-RIS}}$. $\eta \sim \mathcal{U}[0, 2\pi]$ is a random phase term. $L_{\text{LoS}}^{\text{BS-RIS}}$ indicates attenuation of the LoS link for the separation distance $d_{\text{BS-RIS}}$ [35, 39]. $G_e\left(\theta_{\text{LoS}}^{\text{BS-RIS}}\right)$ is the RIS element gain for the LoS link between BS and RIS. $\mathbf{a}\left(\phi_{\text{LoS}}^{\text{BS-RIS}}, \theta_{\text{LoS}}^{\text{BS-RIS}}\right)$ and $\mathbf{a}\left(\phi_{\text{LoS}}^{\text{BS}}, \theta_{\text{LoS}}^{\text{BS}}\right)$ are the corresponding array response vectors for angles $\phi_{\text{LoS}}^{\text{BS-RIS}}, \theta_{\text{LoS}}^{\text{BS-RIS}}$ and $\phi_{\text{LoS}}^{\text{BS}}, \theta_{\text{LoS}}^{\text{BS}}$, respectively.

11.4.1.2 RIS-UE Channel

For indoor environments, RIS and UE are closer. Hence, only a direct LoS link is present between them. The RIS-UE LoS channel $\mathbf{g}$ is obtained by recalculating array response vectors for corresponding azimuth and elevation angles $\phi_{UE}^{\text{RIS-}UE}, \theta_{UE}^{\text{RIS-}UE}$ as steps discussed by Basar et al. [35]. The vector $\mathbf{g}$ can be formulated as follows:

$$\mathbf{g} = \sqrt{G_e\left(\theta_{UE}^{\text{RIS}-UE}\right) L_{\text{LoS}}^{\text{RIS}-UE}}\, e^{j\eta} \mathbf{a}\left(\phi_{UE}^{\text{RIS}-UE}, \theta_{UE}^{\text{RIS}-UE}\right) \tag{11.11}$$

11.4.2 Outdoor Physical Channel Model

This section presents the channel model for outdoor UMi street Canyon environments at 28 and 73 GHz mmWave frequencies. The channel model developed for indoor environments can be extended to outdoor environments by varying the path loss parameters as per the 3GPP clustered model. When extending the indoor channel model to outdoor environments, the critical distinction is the channel between the RIS and the UE.

In the outdoor setting, this channel may experience small-scale fading, similar to indoor environments, but with the additional characteristic of having a random number of distinctive clusters. Distance and environmental circumstances fluctuate outdoors, and the LoS dominance assumption may not hold.

The channel vector **g** for outdoor scenario can be generally formulated as [35, 39]

$$\mathbf{g} = \tilde{\gamma} \sum_{q=1}^{\tilde{Q}} \sum_{s=1}^{\tilde{S}_q} \beta_{c,s} \sqrt{G_e\left(\theta_{q,s}^{\text{RIS}-UE}\right) L_{\text{q,s}}^{\text{RIS}-UE}}\, \mathbf{a}\left(\phi_{q,s}^{\text{RIS}-UE}, \theta_{q,s}^{\text{RIS}-UE}\right) + \mathbf{g}_{\text{LoS}}. \tag{11.12}$$

Here, $\tilde{Q}$ and $\tilde{S}_q$ denote cluster count and sub-rays per cluster for the RIS-UE link, respectively. $\tilde{\gamma} = \sqrt{\frac{1}{\sum_{q=1}^{\tilde{Q}} \tilde{S}_q}}$ is the normalization factor. $G_e\left(\theta_{q,s}^{\text{RIS}-UE}\right)$ is the RIS element gain and $L_{q,s}^{\text{RIS}-UE}$ is the large-scale attenuation in the direction of $(q, s)^{th}$ scatterer. $\mathbf{a}\left(\phi_{q,s}^{\text{RIS}-UE}, \theta_{q,s}^{\text{RIS}-UE}\right)$ is the array response vector for the associated azimuth and elevation angles. The path loss parameters for indoor and outdoor environments are briefly described in [35, 39].

11.5 Simulations and Discussions

The simulation setup details are provided in Table 11.1. Algorithm 11.1 outlines the procedure used to simulate the performance curves. For simplicity, the InH-indoor office and UMi-street canyon outdoor environments are just called indoor and outdoor, respectively. Figure 11.6 depicts the ABER performance of RIS-aided 1×16 MIMO in indoor at 28 GHz with varying N. To attain the ABER demand, $N = 16$, $N = 64$, and $N = 256$ systems require an SNR of ~75, ~71, and ~68 dB respectively. The $N = 256$ system has an SNR gain of ~7 dB compared to the $N = 16$ system. Therefore, when N increases, a significant SNR gain is observed.

Table 11.1 Simulation parameters.

Parameters	Values
Operating frequencies (GHz)	28 and 73
MIMO setups $(N_R \times N_T)$	$1 \times 4, 1 \times 16, 1 \times 64$
Reflecting components (N)	16, 64, 256
Total channel realizations	10^5
Desired ABER	10^{-4}
Constellation order	2
Channel models	3GPP TR 38.901 V16.1.0 5G
Channel environments	InH- indoor office (indoor), UMi-street canyon outdoor (outdoor)
Simulation tools	MATLAB R2023a and SimRIS

Algorithm 11.1

Input: $N_R \times N_T$, N, operating frequency.
Output: ABER.
Step 1: Generate random bits and perform constellation mapping.
Step 2: Generate 3GPP InH and UMi channels using SimRIS channel Simulator for the required number of realizations.
Step 3: Optimize the phase of RIS elements with the help of generated channel matrices as per (11.6).
Step 4: Transmit the symbols from the M-MIMO with the aid of RIS through 3GPP InH and UMi channels to the receivers.
Step 5: At the receiver, detect the transmitted bits using ML detector and compute bit error rate (BER).
Step 6: Repeat BER calculations for different channel realizations and compute ABER for every SNR.

Figure 11.7 shows the ABER performance of RIS-aided MIMO in indoor at 28 GHz with $N = 64$ and varying the N_T. To reach the ABER demand, $N_T = 4$, $N_T = 16$, and $N_T = 64$ systems require an SNR of ~79, ~71, and ~67 dB respectively. The SNR gain increases with an increase in N_T. Using MIMO configurations in diversity mode increases diversity gains as N_T increases. For example, the $N_T = 64$ system has a gain of ~12 dB compared to $N_T = 4$ system.

The ABER performance of RIS-aided 1×16 MIMO in outdoor at 28 GHz with varying N is shown in Figure 11.8. To achieve the ABER demand, $N = 16$, $N = 64$,

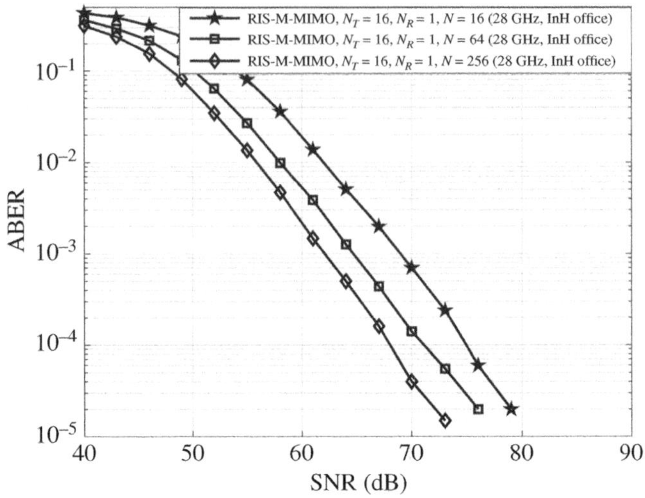

Figure 11.6 ABER performance of RIS-aided 1 × 16 MIMO in indoor at 28 GHz with varying N.

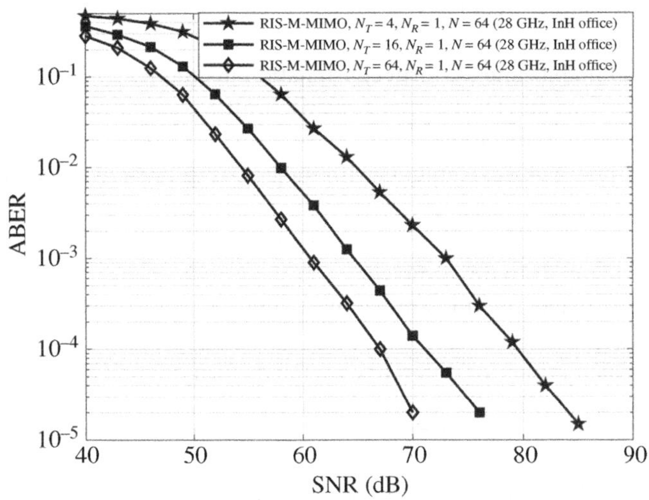

Figure 11.7 ABER performance of RIS-aided MIMO in indoor at 28 GHz with $N = 64$ and varying N_T.

and $N = 256$ systems require an SNR of ~79, ~73, and ~71dB, respectively. The SNR gain increases with N. $N = 256$ system has a gain of ~8 dB over $N = 16$ system. For $N = 256$ systems, the SNR required for outdoor is ~71 dB whereas for indoor is ~68 dB. The path loss in mmWave communication is higher in outdoor

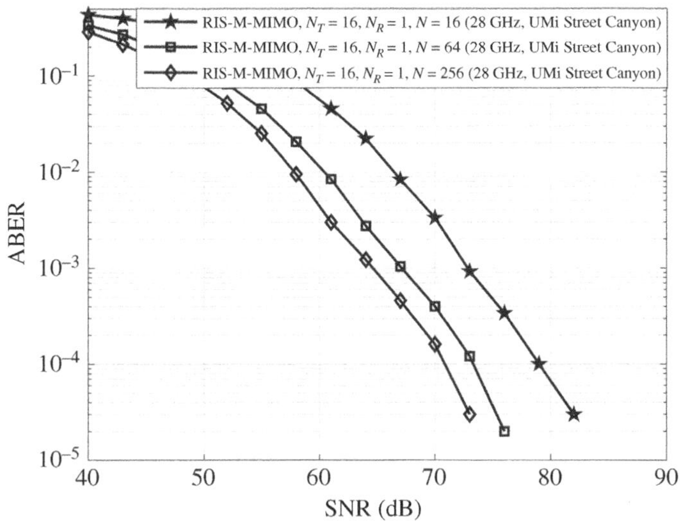

Figure 11.8 ABER performance of RIS-aided 1 × 16 MIMO in outdoor at 28 GHz with varying *N*.

Table 11.2 Comparison of the SNR requirements of RIS-aided 1 × 16 MIMO indoors and outdoors at 28 GHz with varying *N*.

Number of reflector elements (*N*)	28 GHz		SNR loss (dB) due to outdoor environment
	Indoor	Outdoor	
16	~75 dB	~79 dB	~4
64	~71 dB	~73 dB	~2
256	~68 dB	~71 dB	~3

than indoor environments. Therefore, the ABER performance in the indoor is better than in the outdoor environment. Table 11.2 compares the SNR requirements of RIS-aided 1 × 16 MIMO indoors and outdoors at 28 GHz with varying *N*.

Figure 11.9 shows the ABER performance of RIS-aided 1 × 16 MIMO in indoor at 73 GHz with varying *N*. To reach the ABER demand, $N = 16$, $N = 64$, and $N = 256$ systems need an SNR of ~84, ~79, and ~76 dB, respectively. The $N = 256$ system has an SNR gain of ~8 dB over $N = 16$ system. The SNR gain increases with *N*. In Figures 11.6 and 11.9, it is clear that the increase in frequency degrades the ABER performance of the system due to path loss. For $N = 256$, the SNR required to reach the ABER demand is ~76 dB for 73 GHz, whereas ~68 dB for 28 GHz.

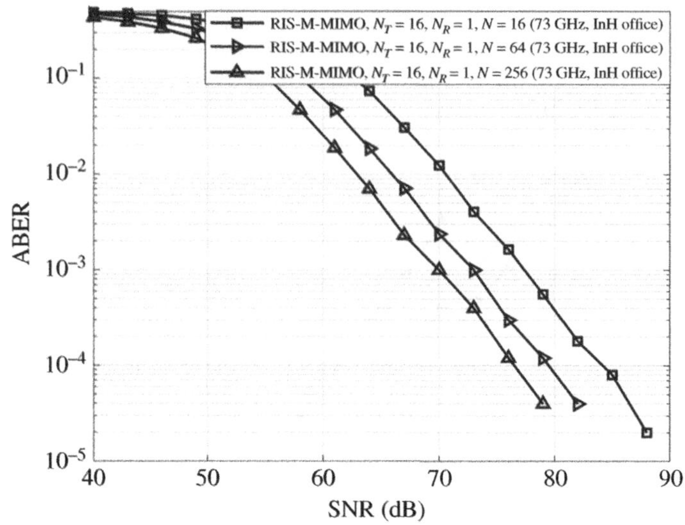

Figure 11.9 ABER performance of RIS-aided 1 × 16 MIMO in indoor at 73 GHz with varying *N*.

Table 11.3 Comparison of the SNR requirements of RIS-aided 1 × 16 MIMO indoors at 28 and 73 GHz with varying *N*.

Number of reflector elements (*N*)	Indoor		SNR loss (dB) due to higher frequency
	28 GHz	73 GHz	
16	~75 dB	~84 dB	~9
64	~71 dB	~79 dB	~8
256	~68 dB	~76 dB	~8

When mmWave with higher frequency travel through the atmosphere, they experience significant attenuation due to oxygen, moisture, and rainfall absorption. As a result, the point-to-point propagation distance for 73 GHz is relatively short, and beyond a certain distance, the signal will weaken, leading to a higher error rate compared to 28 GHz. Table 11.3 compares the SNR requirements of RIS-aided 1 × 16 MIMO for indoor at 28 and 73 GHz with varying *N*.

Figure 11.10 depicts the ABER performance of RIS with 64 reflector elements in indoor at 73 GHz and varying N_T. To attain the ABER demand, $N_T = 4$, $N_T = 16$, and $N_T = 64$ systems require an SNR of ~87, ~79, and ~76 dB, respectively. The $N_T = 64$ system has a gain of ~11 dB over $N_T = 4$ system. Therefore, when the N_T

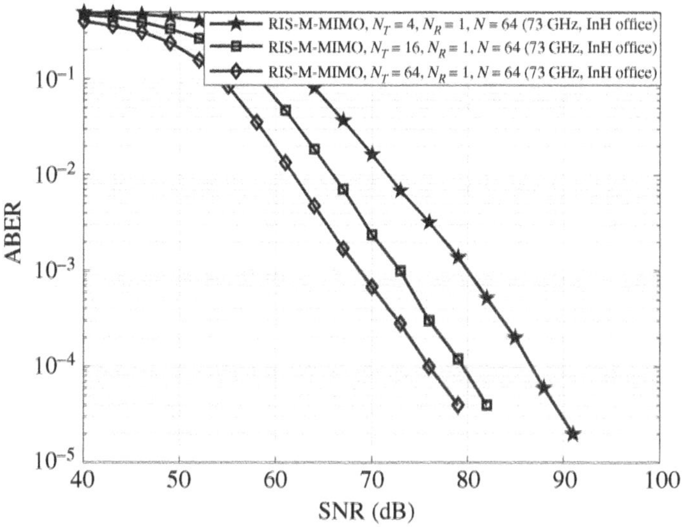

Figure 11.10 ABER performance of RIS-MIMO in indoor at 28 GHz with $N = 64$ and varying N_T.

Table 11.4 Comparison of the SNR requirements of $N = 64$ RIS-aided MIMO indoors at 28 and 73 GHz with varying N_T.

Number of transmitter antennas (N_T)	Indoor		SNR loss (dB) due to higher frequency
	28 GHz	73 GHz	
4	~79 dB	~87 dB	~8
16	~71 dB	~79 dB	~8
64	~67 dB	~76 dB	~9

increases, there is a significant SNR gain. Table 11.4 compares the SNR requirements of $N = 64$ RIS-aided MIMO for indoor environment at 28 and 73 GHz with varying N_T.

11.6 Conclusions

RIS-aided M-MIMO for mmWave communications is proposed in this chapter. We evaluated the system performance for indoor and outdoor environments for 28 and 73 GHz mmWave frequencies in 3GPP TR 38.901 V16.1.0 5G channel models using a SimRIS channel simulator. The ABER performance of the proposed system is evaluated for indoor and outdoor scenarios by changing N and N_T. According to simulation results, the ABER performance of 28 GHz is more outstanding than 73 GHz by at least ~7 dB. Increasing the operating frequency leads to performance degradation due to path loss, and the path loss in mmWave communication is more incredible in outdoor than indoor environments. The indoor environment has an SNR gain of at least ~2 dB over the outdoor environment. As we move from 28 to 73 GHz, the achievable data rate can be increased at the cost of degraded ABER performance. However, the performance degradation of the system can be compensated by increasing both N and N_T in outdoor communications. In contrast, in indoor communications, system performance can be increased by increasing only N. Due to the significant merits, the proposed system model can be better suited for 5GB communications. As a future work, the infinite resolution RIS in the proposed system can be replaced with finite resolution RIS. The performance of the proposed system can be explored in other mmWave operating frequencies.

References

1. Imoize, A.L., Adedeji, O., Tandiya, N., and Shetty, S. (2021). 6G enabled smart infrastructure for sustainable society: opportunities, challenges, and research roadmap. *Sensors* 21 (5): 1709.
2. Chowdhury, M.Z., Shahjalal, M., Ahmed, S., and Jang, Y.M. (2020). 6G wireless communication systems: applications, requirements, technologies, challenges, and research directions. *IEEE Open Journal of the Communications Society* 1: 957–975.
3. De Alwis, C., Kalla, A., Pham, Q.-V. et al. (2021). Survey on 6G frontiers: trends, applications, requirements, technologies and future research. *IEEE Open Journal of the Communications Society* 2: 836–886.
4. Albreem, M.A., Juntti, M., and Shahabuddin, S. (2019). Massive MIMO detection techniques: a survey. *IEEE Communications Surveys & Tutorials* 21 (4): 3109–3132.
5. Chataut, R. and Akl, R. (2020). Massive MIMO systems for 5G and beyond networks—overview, recent trends, challenges, and future research direction. *Sensors* 20 (10): 2753.
6. Albreem, M.A., Al Habbash, A.H., Abu-Hudrouss, A.M., and Ikki, S.S. (2021). Overview of precoding techniques for massive MIMO. *IEEE Access* 9: 60764–60801.
7. Borges, D., Montezuma, P., Dinis, R., and Beko, M. (2021). Massive MIMO techniques for 5G and beyond—opportunities and challenges. *Electronics* 10 (14): 1667.

8 Björnson, E. and Sanguinetti, L. (2020). Scalable cell-free massive MIMO systems. *IEEE Transactions on Communications* 68 (7): 4247–4261.

9 ElMossallamy, M.A., Zhang, H., Song, L. et al. (2020). Reconfigurable intelligent surfaces for wireless communications: principles, challenges, and opportunities. *IEEE Transactions on Cognitive Communications and Networking* 6 (3): 990–1002.

10 Liu, Y., Liu, X., Mu, X. et al. (2021). Reconfigurable intelligent surfaces: principles and opportunities. *IEEE Communications Surveys & Tutorials* 23 (3): 1546–1577.

11 Trichopoulos, G.C., Theofanopoulos, P., Kashyap, B. et al. (2022). Design and evaluation of reconfigurable intelligent surfaces in real-world environment. *IEEE Open Journal of the Communications Society* 3: 462–474.

12 Pan, C., Ren, H., Wang, K. et al. (2021). Reconfigurable intelligent surfaces for 6G systems: principles, applications, and research directions. *IEEE Communications Magazine* 59 (6): 14–20.

13 Hassan, K., Masarra, M., Zwingelstein, M., and Dayoub, I. (2020). Channel estimation techniques for millimeter-wave communication systems: achievements and challenges. *IEEE Open Journal of the Communications Society* 1: 1336–1363.

14 Nor, A.M., Halunga, S., and Fratu, O. (2022). Survey on positioning information assisted mmWave beamforming training. *Ad Hoc Networks* 135: 102947.

15 Wang, P., Fang, J., Duan, H., and Li, H. (2020). Compressed channel estimation for intelligent reflecting surface-assisted millimeter wave systems. *IEEE Signal Processing Letters* 27: 905–909.

16 Ai, B., Guan, K., He, R. et al. (2017). On indoor millimeter wave massive MIMO channels: measurement and simulation. *IEEE Journal on Selected Areas in Communications* 35 (7): 1678–1690.

17 Sun, Y. and Qi, C. (2017). Weighted sum-rate maximization for analog beamforming and combining in millimeter wave massive MIMO communications. *IEEE Communications Letters* 21 (8): 1883–1886.

18 Elbir, A.M. and Papazafeiropoulos, A.K. (2019). Hybrid precoding for multiuser millimeter wave massive MIMO systems: a deep learning approach. *IEEE Transactions on Vehicular Technology* 69 (1): 552–563.

19 Song, Y., Gong, Z., Chen, Y., and Li, C. (2022). Efficient channel estimation for wideband millimeter wave massive MIMO systems with beam squint. *IEEE Transactions on Communications* 70 (5): 3421–3435.

20 Srinivas, C.V.V.S. and Borugadda, S. (2023). RF chain selection using hybrid optimization with precoding in mm-wave massive MIMO systems. *Wireless Personal Communications* 131 (3): 1997–2017.

21 Du, R., Li, T., Li, X., and Liu, F. (2023). CNN-DPC algorithm for hybrid precoding in millimeter-wave massive MIMO systems. *Wireless Networks* 29 (6): 2447–2456.

22 He, J., Wymeersch, H., Sanguanpuak, T. et al. (2020). Adaptive beamforming design for mmWave RIS-aided joint localization and communication. *2020 IEEE Wireless Communications and Networking Conference Workshops (WCNCW)*, 1–6. IEEE.

23 Pradhan, C., Li, A., Song, L. et al. (2020). Hybrid precoding design for reconfigurable intelligent surface aided mmWave communication systems. *IEEE Wireless Communications Letters* 9 (7): 1041–1045.

24 Nemati, M., Maham, B., Pokhrel, S.R., and Choi, J. (2021). Modeling RIS empowered outdoor-to-indoor communication in mmWave cellular networks. *IEEE Transactions on Communications* 69 (11): 7837–7850.

25 Wang, P., Fang, J., Zhang, W. et al. (2022). Beam training and alignment for RIS-assisted millimeter-wave systems: state of the art and beyond. *IEEE Wireless Communications* 29 (6): 64–71.

26 Feng, C., Shen, W., An, J., and Hanzo, L. (2022). Joint hybrid and passive RIS-assisted beamforming for mmWave MIMO systems relying on dynamically configured subarrays. *IEEE Internet of Things Journal* 9 (15): 13913–13926.

27 Chen, J., Liang, Y.-C., Cheng, H.V., Yu, W. (2023). Channel estimation for reconfigurable intelligent surface aided multi-user mmWave MIMO systems. *IEEE Transactions on Wireless Communications* 22 (10): 6853–6869.

28 Ying, K., Gao, Z., Lyu, S. et al. (2020). GMD-based hybrid beamforming for large reconfigurable intelligent surface assisted millimeter-wave massive MIMO. *IEEE Access* 8: 19530–19539.

29 Zhou, G., Pan, C., Ren, H., and Wang, K. (2021). Channel estimation for RIS-aided millimeter-wave massive MIMO systems. *2021 55th Asilomar Conference on Signals, Systems, and Computers*, 698–703. IEEE.

30 Yildirim, I., Koc, A., Basar, E., and Le-Ngoc, T. (2022). RIS-aided angular-based hybrid beamforming design in mmWave massive MIMO systems. *GLOBECOM 2022-2022 IEEE Global Communications Conference*, 5267–5272. IEEE.

31 Wang, R., Ren, H., Pan, C. et al. (2023). Channel estimation for RIS-aided mmWave massive MIMO system using few-bit ADCs. *IEEE Communications Letters* 27 (3): 961–965.

32 Singh, J., Srivastava, S., Yadav, S.P. et al. (2023). Energy efficiency optimization in reconfigurable intelligent surface aided hybrid multiuser mmwave MIMO systems. *IEEE Open Journal of Vehicular Technology* 4: 581–589.

33 Wang, H., Fang, J., and Li, H. (2023). Joint beamforming and channel reconfiguration for RIS-assisted millimeter wave massive MIMO-OFDM systems. *IEEE Transactions on Vehicular Technology* 72 (6): 7627–7638.

34 3GPP Radio Access Network Working Group (2018). Study on Channel Model for Frequencies from 0.5 to 100 GHz (release 15). *Technical report, 3GPP TR 38.901*.

35 Basar, E., Yildirim, I., and Kilinc, F. (2021). Indoor and outdoor physical channel modeling and efficient positioning for reconfigurable intelligent surfaces in mmWave bands. *IEEE Transactions on Communications* 69 (12): 8600–8611.

36 Basar, E. and Yildirim, I. (2021). Reconfigurable intelligent surfaces for future wireless networks: a channel modeling perspective. *IEEE Wireless Communications* 28 (3): 108–114.

37 Balanis, C.A. (2016). *Antenna Theory: Analysis and Design*. Wiley.

38 Hemadeh, I.A., Satyanarayana, K., El-Hajjar, M., and Hanzo, L. (2017). Millimeter-wave communications: physical channel models, design considerations, antenna constructions, and link-budget. *IEEE Communications Surveys & Tutorials* 20 (2): 870–913.

39 NTT Docomo (2016). 5G Channel Model for Bands up to 100 GHz. Technical report.

12

Massive MIMO for Non-terrestrial Wireless Communication Systems

Unwana Macaulay Ekpe[1], *Agbotiname Lucky Imoize*[2], *and Webert Montlouis*[3]

[1] *Department of Electrical and Electronic Engineering, Faculty of Engineering, Akwa Ibom State University, Ikot Akpaden, Nigeria*
[2] *Department of Electrical and Electronics Engineering, Faculty of Engineering, University of Lagos, Akoka, Lagos, Nigeria*
[3] *Johns Hopkins University, Baltimore, MD, USA*

12.1 Introduction

A non-terrestrial wireless communication system refers to a network or infrastructure that facilitates wireless communication beyond Earth's surface. This type of system is crucial for various applications in space exploration, satellite communication, lunar missions, and potentially interplanetary or deep space missions. This chapter will therefore discuss the important components of non-terrestrial networks (NTNs). Special emphasis will be placed on the role of satellites, the signal processing and the security requirements for such networks, and the recent efforts that have been made to standardize the components and protocols of such networks.

Shown in Figure 12.1 is a simplified depiction of a NTN, which is envisioned to be a hybrid network of networks with varying levels of integration between the spaceborne, airborne, and terrestrial-based components. Such a network is capable of short-range, wide-area, and global communications using satellites with regenerative or bent-pipe (transparent) payloads, airborne vehicles such as unmanned aerial systems (UASs) and high-altitude platforms (HAPs), and ground infrastructure which includes user equipment (UE), 5G core network (5GCN), next-generation Node B (gNB), and NTN gateways and relays. The ground components provide feeder links and service links to the space and airborne platforms. Also included as part of NTNs are inter-satellite links (ISLs) and inter-platform links (IPLs), which allow for connections to be more efficiently

Massive MIMO for Future Wireless Communication Systems: Technology and Applications, First Edition. Edited by Webert Montlouis and Agbotiname Lucky Imoize.

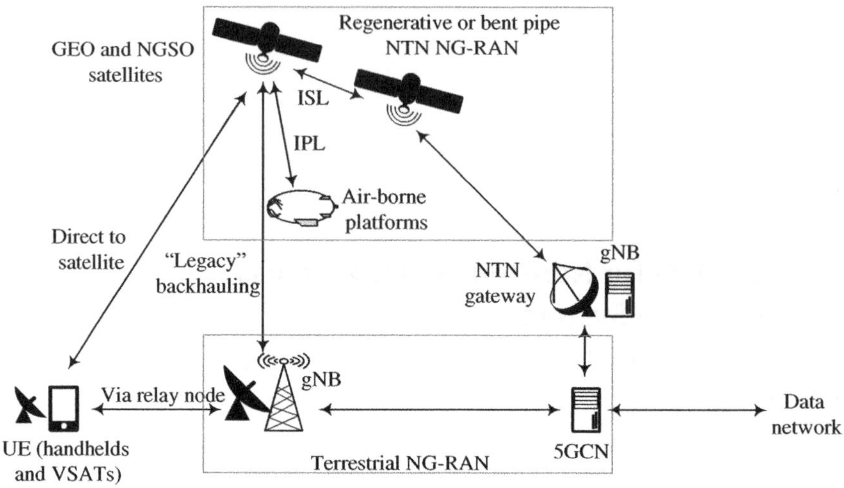

Figure 12.1 Non-terrestrial network showing space, air, and ground components.

routed from source to destination. The key components and concerns for the operation of NTNs are described in the following subsections.

12.1.1 Satellite Networks

Satellites are perhaps the most important component of NTNs. Historically, traditional satellite operators such as Inmarsat and Intelsat made use of just a few large Geostationary Earth Orbit (GEO) satellites to deliver fixed satellite services (FSS), broadcast satellite services (BSS), and/or mobile satellite services (MSS). However, with the recent increase in demand for low latency services on a global scale, there is now a compelling business case for the operation of very large constellations of Low Earth Orbit (LEO) satellites. Examples of such satellite constellations currently in operation include Starlink and OneWeb.

From a communications point of view, satellite networks comprise a space segment and a ground segment. The satellite is the only space segment component and its orbital parameters and telecommunication functions have to be continuously monitored by a satellite control center (SCC) and a network control center (NCC) respectively. While the SCC uses telemetry to track, monitor, and control the satellite's altitude, attitude, fuel supply, battery state of charge, and solar power supply, the NCC is responsible for radio resource management (choice of coding, modulation, multiplexing, and multiple access schemes) to guarantee service provision to authorized ground-based users within and outside the satellite network. Both the SCC and NCC are ground-based infrastructure and for effective support of non-GEO satellite constellations, their physical locations have to be dispersed to give them continuous view of the satellites as they process in their orbits.

Satellite constellations can therefore be viewed as "communications networks in the sky" and such constellations depend on ISLs for networking. ISLs may be intra-plane, inter-plane, or inter-orbit. While some constellations use the microwave band of frequencies for ISLs, others use the optical frequencies to achieve high-bandwidth networking. Efficient routing via ISLs has now made it possible for satellite networks to support low-latency applications such as autonomous aerial and ground vehicles, remote medical procedures, gaming, and financial transactions.

Apart from the GEO and LEO earlier mentioned, other orbits used by satellite networks include the Medium Earth Orbit (MEO) and the Highly Elliptic Orbit (HEO). It is therefore worth noting that traditional satellite networks have hitherto operated independently of terrestrial systems. However, with the adoption of 5G and the evolution toward 6G, there is now a unique opportunity for satellite and terrestrial operators to build hybrid or fully integrated networks. One of the advantages of hybrid and integrated networks is that geographical, physical, and atmospheric obstacles which often affect traditional terrestrial communication systems would no longer be a problem. Therefore, with such networked systems, it would be possible to achieve service ubiquity, service continuity, and scalability on a global scale.

12.1.2 Spectrum Utilization and Interference Management

Operating radio-based systems such as satellite communication networks can cause harmful interference to other systems. Therefore, spectrum should be carefully managed to guarantee an acceptable quality of service for existing systems and also permit access to new and emerging systems. The International Telecommunications Union (ITU) is responsible for regulating the use of radio spectrum worldwide and working with national spectrum administrations, it has allocated to satellite communication systems the exclusive and shared use of spectrum in the L, S, C, X, Ku, K, and Ka-bands. Within this context, the ITU convenes the World Radiocommunications Conference (WRC) every four years to ensure that the most advantageous set of frequencies are allocated for use by satellite services and while these frequencies are being used, interference to other systems must be minimized.

As much as practically feasible, spectrum allocation is done in such a manner to allow for reasonable compatibility between different systems. This allows network operators in the same or adjacent territories to coordinate their spectrum usage. Coordination refers to the process whereby different network operators agree on the acceptability or otherwise of interference to their systems. Also, the use of advanced technologies to minimize interference and improve spectral efficiency is strongly encouraged. For example, some satellites now have multi-spot beams and employ the fractional frequency reuse technique. This technique divides up allocated frequencies such that adjacent spot beams do not use the same set of frequencies. More recently, cognitive radio has been proposed to improve spectrum utilization in cases where two or more independent, hybrid, or integrated terrestrial and NTNs share the same frequency bands.

12.1.3 Security and Reliability of Service

Our dependence on wireless communication systems will be further deepened when NTNs are fully integrated into future wireless communication systems. The provision of critical and essential services in the civil and military domains would increasingly rely on wireless communication networks. It is therefore important that such networks be made as secure as possible and traditional methods of improving security usually rely on strong encryption and authentication mechanisms. For terrestrial networks, this has so far presented acceptable levels of protection against eavesdroppers and has routinely prevented unauthorized and rogue agents from taking undue advantage of sensitive information.

However, with the emergence of NTNs and the requirements for service ubiquity, service continuity, and scalability, such networks are prone to dynamic changes in their topology. Therefore, existing security mechanisms may need to be reviewed. Also, based on the stringent power efficiency and continuity of service requirements for NTNs, some open security questions have recently emerged. For example, to conserve power, a UE may not have to be awake all the time. Also, in high mobility scenarios, a UE experiences a temporary loss of service while switching from terrestrial coverage to non-terrestrial coverage. In such cases, there is a need to comprehensively define how the data stored and/or forwarded to UEs that are temporarily unavailable can be protected. It is also necessary to ensure that authentic UE location information and other private and confidential information are not tampered with during transmission.

In terms of reliability of service, there is a need for robust redundancies to be built into the non-terrestrial components of future communication networks as these components are usually not easily accessible for repairs or replacement. For example, it is necessary to have in-orbit and on-ground spare satellites ready to be deployed into position at the shortest notice to cover for failed satellites. Also, interim corrective measures may include the deployment of intelligent algorithms to re-route data from severed ISLs and the activation of in-built redundant and failover mechanisms to ensure continuity of service.

12.1.4 Energy Efficiency

Energy efficiency is a very important consideration for the space and ground segments of NTNs. The main space segment component is the satellite, and the power supply available to the satellite is limited by the duration of insolation hours, the size of deployed solar arrays, and the storage capacity of backup batteries. It is therefore vital that satellites should have very high levels of energy efficiency to maximize their operational lifespan in orbit. Conversely, the majority of the ground segment components are UE and these have been traditionally designed to operate in an energy-consuming always-on mode. Fortunately, the are emerging research efforts to devise power-saving measures and subsequently improve the energy efficiency of UEs deployed in NTNs. With the anticipated explosion in their numbers, improving the energy efficiency of UEs should ensure that they do not adversely contribute to elevated greenhouse emissions.

12.1.5 Standards and Interoperability of Service

Standardization ensures that different equipment manufactured by independent vendors for the same system are compatible while interoperability considerations allow for collaborative service provision between operators of different communication systems. These are very important considerations for NTNs and the 3rd Generation Partnership Project (3GPP) has been the main driver for standardization activities. 3GPP is a global initiative that brings together seven telecommunication standards development organizations in Asia, Europe, and North America. The partner organizations include the Association of Radio Industries and Businesses (ARIB), Alliance for Telecommunications Industry Solutions (ATIS), China Communications Standards Association (CCSA), European Telecommunications Standards Institute (ETSI), Telecommunications Standards Development Society, India (TSDSI), Telecommunications Technology Association (TTA), and Telecommunications Technology Committee (TTC), and they set the standards for interoperability in radio access, core network and service capability. In particular, 3GPP's Technical Specification Group on Radio Access Networks (TSG RAN) works in close cooperation with ITU's Radiocommunication Sector's (ITU-R) Working Party 5D (WP 5D) to create a roadmap for the development and standardization of new radio interfaces, spectrum allocation, and spectrum regulation.

12.1.6 Key Contributions and Organization of the Chapter

It is important to note that non-terrestrial wireless communication systems are pivotal in providing reliable global connectivity, enhancing space exploration, and advancing scientific research. These systems will enable humans to communicate and share information in environments where traditional terrestrial communication infrastructure is unavailable or impractical. These systems will become even more integral in bringing about a more highly connected world as technology evolves. This chapter has therefore made the following contributions to the body of knowledge.

(a) An overview of signal processing techniques for the NTN environment is presented in Section 12.3.
(b) A massive MIMO system model for NTNs and the necessary signal processing techniques to compensate for the effects of Doppler and propagation delay have been explained using easy-to-understand terminology. Details of this can be found in Sections 12.3.1 and 12.3.2.
(c) Five major linear precoding techniques that can be applied in the downlink channels of single-cell and multi-cell NTNs have been presented. See Sections 12.4 and 12.5.
(d) The security requirements for key aspects of NTNs have been briefly explained in Section 12.6.
(e) Recent standardization activities covering time and frequency synchronization, random access, and the modifications necessary to improve the reliability of service have been presented in Section 12.7.

In presenting the above contributions, this chapter follows the following logical arrangement. Section 12.2 describes the role of satellites in future non-terrestrial wireless networks while Section 12.3 provides an overview of NTN-based massive MIMO and the basic signal processing requirements. Section 12.4 presents linear precoding techniques for a single-cell massive MIMO NTN system while Section 12.5 extends the precoding analysis to the multi-cell scenario. Section 12.6 discusses the security considerations for satellite-based NTNs. Section 12.7 presents standards and interoperability requirements for hybrid and integrated NTNs and Section 12.8 concludes the chapter and makes some recommendations for further study.

12.2 The Role of Satellites in Future Non-terrestrial Networks

Satellites are one of the most important components of NTNs. Their function is to overcome geographical and physical obstacles and provide service to areas that cannot be economically covered by terrestrial networks. As depicted in Figure 12.2, such areas include isolated or remote places (aircraft, maritime vessels, oil rigs, deserts, and mountainous areas) and underserved suburban and rural communities. Satellites are also expected to improve service reliability anywhere on Earth by providing alternative backhaul links for human-centric and machine-to-machine (M2M) communications in a scalable manner.

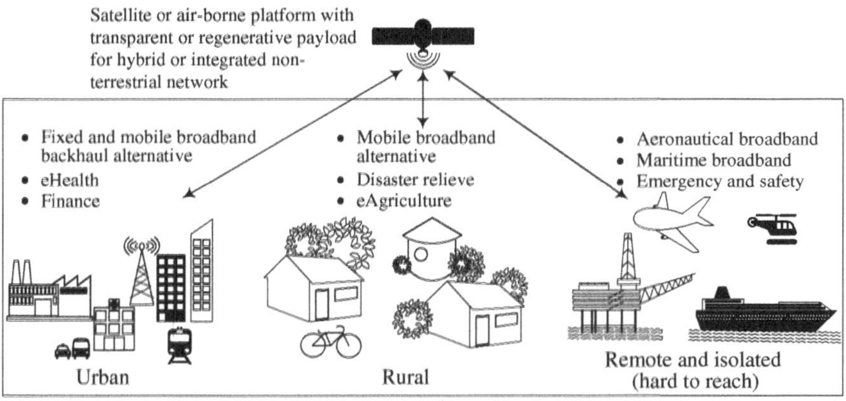

Figure 12.2 Role of satellites in non-terrestrial networks.

Table 12.1 Non-terrestrial platforms.

Platform	Type	Altitude	Coverage area (cell/footprint)
Spaceborne	GEO satellite	35,786 km	Earth-fixed global beam and/or regional spot beams of 200–3500 km diameter
	MEO satellite	7000–25,000 km	Earth-fixed or satellite-fixed moving cells of 100–1000 km diameter
	LEO satellite	300–2000 km	Earth-fixed or satellite-fixed moving cells of 100–1000 km diameter
	HEO satellite	1500–39,000 km	Earth-fixed regional and beams of variable diameter
Airborne	HAP	20 km	Earth-fixed cells of 5–200 km diameter
	UAS	8–50 km	

Table 12.1 shows the different deployment options for NTNs. The two categories of non-terrestrial platforms are spaceborne and airborne platforms. Satellites are the only spaceborne platforms while types of airborne platforms include HAPs and UAS. UAS include tethered or untethered lighter-than-air types and heavier-than-air vehicles. The altitude of the platforms shall range from 8 km for airborne platforms to more than 35,000 km for GEO and HEO satellites. In the network of the future, satellites will complement rather than compete with terrestrial and other non-terrestrial platforms. The subsections that follow therefore present satellite architectures and their key performance indicators, the applications of different satellite architectures, and the frequency bands reserved for satellite services.

12.2.1 Satellite Architectures and Key Performance Indicators

Satellite architectures for NTNs will have satellites operating in LEO, MEO, GEO, and HEO, forming a multilayer network with ISLs and links to HAPs and UASs. GEO is a circular orbit on the equatorial plane at an altitude of 35,786 km and an orbital period of 24 hours. These characteristics ensure that a GEO satellite appears to be fixed in the sky, with an unchanging elevation and azimuth when viewed by an Earth-bound UE. As shown in Table 12.1, the coverage area or footprint diameter of a GEO satellite's main beam or spot beam can range from 200 to 3500 km. MEO and LEO are both circular orbits and their altitude range is respectively from 7000 and 25,000 km and 300 to 2000 km. HEO is designed to serve high or low-latitude locations since their orbital parameters enable them to experience apsidal dwell around apogee ranging from 12 to 24 hours. According to [1], the key performance indicators (KPIs) for 6G NTN systems would have to exceed those set for 5G. This includes the use of sub 6 GHz, mmWave, and THz frequency bands (above 300 GHz), peak data rate of up to 1 Tbps, connected UE/device density of up to 10 million/km^2, and latency as low as 0.1 ms, among several other ambitious targets.

12.2.2 Applications of the Satellite Architecture

There are two major application modes for satellites within the NTN. These modes are currently being standardized and they include when the satellites are equipped with transparent or bent-pipe payloads, and when the satellite payloads are fully regenerative. These application modes are shown in Figure 12.3a–c. The transparent architecture has been the traditional mode of operation of satellites until recently. In the case of Figure 12.3a, the satellite's functionality within the network is limited to radio RF filtering, frequency conversion, and amplification. The satellite therefore only relays the New Radio signal (*NR-Uu*) between the UE and gNB. Satellites with regenerative payloads can effectively implement some (Figure 12.3b) or all the functions (Figure 12.3c) of a terrestrial-based next-generation base station. The functions include demodulation, decoding, switching and/or routing, re-encoding, and re-modulation. Depending on the functions carried out by the regenerative satellite, the feeder link can be either an F1 interface (with control plane and user plane separation, see Ref. [2]) or the N1/N2/N3 interface. The N1, N2, and N3 interfaces, respectively, represent reference points between the UE and the access and mobility management function (AMF), the radio access network (RAN) and the AMF, and the RAN and the user plane functions (UPFs). N6 represents the reference point between the UPF and a data network. A detailed description of this can be found in [3].

Another way in which the applications of satellites in future NTN architectures are classified is based on the type of access granted to Earth-bound UEs. This

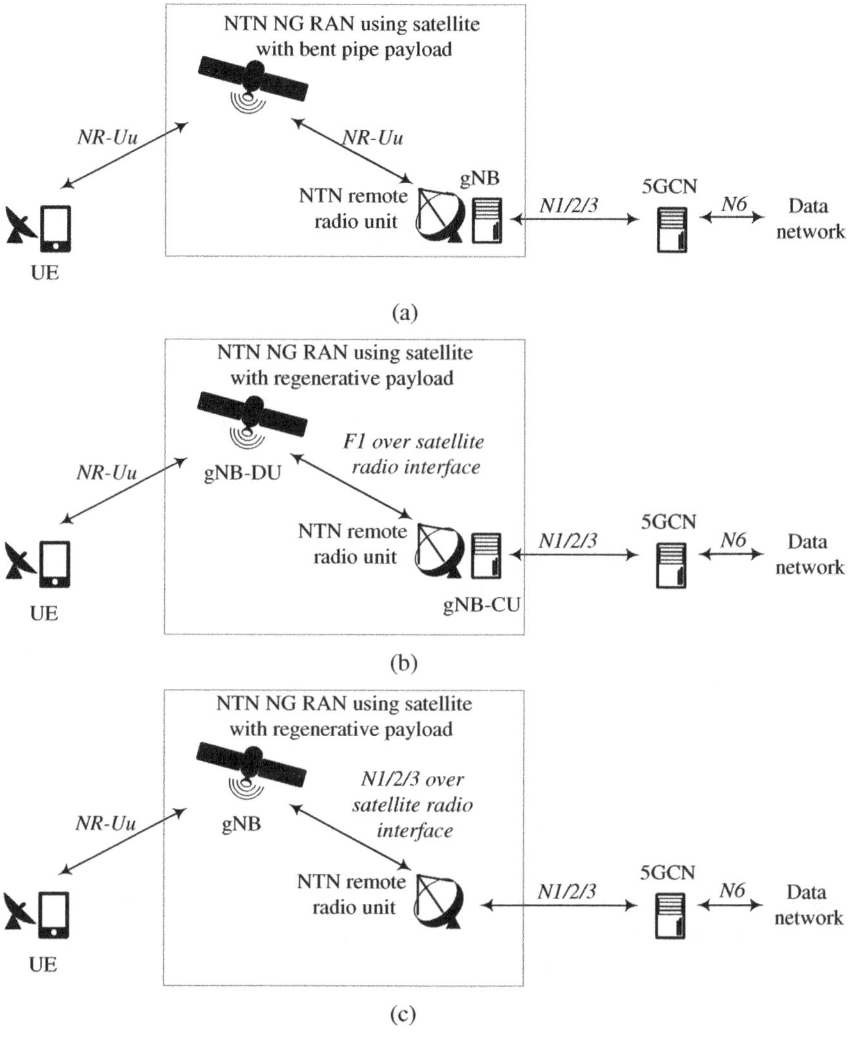

Figure 12.3 Applications of satellites in non-terrestrial networks showing, (a) when the satellite has a bent-pipe payload, (b) when the satellite has a distributed unit of gNode B, and (c) full functionality gNode B.

gives rise to direct-access and indirect-access architectures. As the name implies, direct-access UEs have enough transmit power to enable them to directly access the satellite. For indirect access, Earth-bound relay terminals aggregate the signals emanating from UEs and other parts of the terrestrial network and then serve as satellite backhaul links. The key applications of satellites in the future network will enable service provision across new industry verticals including telemedicine; agriculture; finance; aerial, maritime, and terrestrial automotive and transportation systems; mission-critical applications; media and entertainment; energy; and public safety. According to [4], NTNs will make possible new human-centric applications such as augmented reality, virtual reality, and immersive mixed reality. To meet the service expectations in terms of energy and spectral efficiency requirements, new spectrum allocations will be needed and such networks will have to rely on artificial intelligence and advanced network virtualization functions.

12.2.3 Frequency Bands of Operation

Table 12.2 shows the frequency bands allocated for the operation of satellite communication services. Some of the bands are shared with other services and this creates a potential for harmful interference. Careful coordination must therefore be carried out to ensure that the different systems in the shared bands can operate within acceptable limits. Recently, the spectrum coordination problem has been compounded by the introduction of very high-throughput GEO satellites and the emergence of very large constellations of LEO satellites.

It is worth mentioning that the choice of spectrum use is determined by the targeted service type and propagation characteristics. For example, mobile satellite operators use the L and S bands for their service links because of the favorable antenna and propagation characteristics. The radiation pattern of handheld UE antennas allows for user mobility and flexibility of UE orientation.

Table 12.2 Frequency band allocations for satellite communication services.

Band	Downlink (satellite to Earth) frequency (GHz)	Uplink (Earth to satellite) frequency (GHz)
L band	1.53–1.559	1.6265–1.6605
S band	2.50–2.655	2.655–2.69
C band	3.40–4.20 4.50–4.80	5.725–7.075
X band	7.25–7.75	7.90–8.40
Ku band	10.70–12.75	12.75–13.25 14.0–14.8
K band	18.1–21.2	17.3–18.1
Ka/Q/V bands	37.5–40.5	27.0–31.0 42.5–43.5 47.2–50.2 50.4–51.4

However, fixed and BSS operators mainly use the higher-frequency C, Ku, K, and Ka bands and can take advantage of narrow beamwidth high gain antennas even though such antennas may require regular steering in some cases. The newly introduced very high throughput satellites (vTHSs) will require new spectrum and at present, there are no exclusive satellite bands above the Ka band. Fortunately, cognitive database-driven techniques have been shown in the literature [5] to be capable of improving the spectrum sharing and spectrum efficiency prospects of future NTNs.

12.3 Signal Processing for Non-terrestrial Network-Based Massive MIMO Systems

Signal processing for massive MIMO majorly involves vector and matrix multiplication and inversion. To explain these operations, we start by presenting a schematic of a generic massive MIMO system in Figure 12.4 and noting that the central requirement of massive MIMO is the assumption that the number of transmit antennas, m on the non-terrestrial platform shall always be more than the number of single receive antennas, k, on the terrestrial/maritime/aeronautical-bound UEs. It is also assumed that in the converged future wireless communication system, NTNs will provide service to UEs in hard-to-reach areas

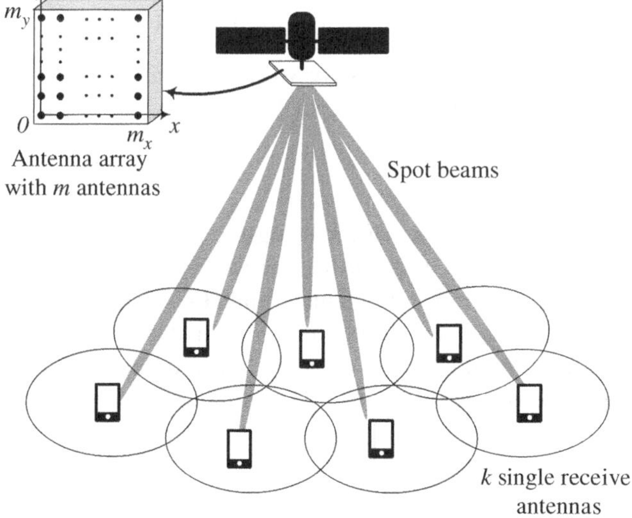

Figure 12.4 The massive MIMO NTN communication channel showing m transmit antennas and k receive antennas.

characterized by sparse user population, low user density, and relatively high latency as compared with terrestrial networks [6].

The channel propagation characteristics of NTNs are expected to have relatively high and variable Doppler shifts and Doppler spreads, and significant propagation delay. These channel impediments must be compensated for through signal processing. Also, once non-terrestrial platforms are launched into their orbital positions, they become physically inaccessible for repairs and servicing. To overcome this drawback, design and engineering compromises are usually made to trade-off onboard signal processing performance with satellite lifetime reliability based on available power generation and consumption profile, among other variables.

The huge distances and relatively high mobility between communicating terminals in NTNs make it difficult for the transmitting non-terrestrial platform to use the channel reciprocity principle to extract up-to-date channel state information (CSI) from the received UE signals. Notwithstanding, some theoretical studies still assume the availability of instantaneous CSI (iCSI), while other studies such as [7] have shown that the use of statistical CSI (sCSI) and low complexity linear precoding techniques transceivers can give performance that approaches theoretical limits in certain scenarios. To appreciate the scale of the signal processing requirements, the next two subsections describe the NTN massive MIMO system model and the signal processing requirements for Doppler shift, Doppler spread, and propagation delay compensation.

12.3.1 Non-terrestrial Network Massive MIMO System Model

The overall input–output relationship of the massive MIMO system shown can be represented by a simplified time-invariant frequency-flat fading channel model given as:

$$\boldsymbol{y} = \mathbf{H}\boldsymbol{x} + \boldsymbol{n}, \tag{12.1}$$

where $\boldsymbol{y}$ is a $k \times 1$ received signal vector, with k being the number of receive antennas belonging to k distinct users. $\boldsymbol{x}$ is a $m \times 1$ normalized transmit signal vector, with m being the number of transmit antennas, and $\mathbf{H}$ is a $k \times m$ channel coefficient matrix representing the channel attenuation terms between receive antenna k and transmit antenna m. $\boldsymbol{n}$ indicates an additive white Gaussian noise component at each of the receive antennas. Since $\mathbf{H}$ is a matrix, it is generally represented as:

$$\mathbf{H} = \begin{bmatrix} h_{11} & h_{12} & \cdots & h_{1m} \\ h_{21} & h_{22} & \cdots & h_{2m} \\ \vdots & \vdots & \ddots & \vdots \\ h_{k1} & h_{k2} & \cdots & h_{km} \end{bmatrix}. \tag{12.2}$$

In the non-terrestrial environment where there is a good line of sight (LOS) between the transmit and the receive terminals and where there are few intervening objects, it is more beneficial to implement multiuser beamforming. In beamforming, the signal directed at a specific user is transmitted through multiple antenna elements but with different phase weightings applied to each of the antenna elements. This steers the created beam toward the desired direction while placing nulls in the direction of interferers. Beamforming can be achieved with either passive or active components and the NTN components may have either transparent or regenerative payloads. The ground components can be handheld UE, very small aperture type terminals (VSAT), or phased array antennas mounted on fixed or moving platforms as shown in Figure 12.2.

The transceiver architecture and the attendant signal processing requirements dictate how much of the advantages of massive MIMO can be harnessed. In the case of semi-passive payloads at the non-terrestrial platform, beamforming can be achieved using RIS-based techniques [8], while for regenerative payloads, precoding techniques have been shown in the literature to give good performance [9]. However, before the precoding techniques can be applied, the effects of Doppler and delay must be taken care of.

12.3.2 Compensation for Doppler and Propagation Delay

In the NTN environment depicted in Figure 12.1, since the Earth-bound UEs are separated by several wavelengths, it is fair to assume that the respective propagation channels from the non-terrestrial platform to the UEs are uncorrelated. Based on this assumption and following the ray tracing-based model of [7], the complex baseband downlink channel between the non-terrestrial platform and UE, k, at time instant, t, and operating frequency, f, can be represented by

$$\mathbf{h}_k(t,f) = \sum_{p=0}^{P_k-1} h_{k,p} \exp(\bar{j}2\pi[t\nu_{k,p} - f\tau_{k,p}]).\mathbf{v}_{k,p} + \mathbf{n} \in \mathbb{C}^{M\times 1}, \tag{12.3}$$

where $\mathbb{C}^{M\times 1}$ represents an $M \times 1$ dimensional complex-valued vector space, $\bar{j} = \sqrt{1}$ and P_k represents the number of propagation paths from the non-terrestrial platform to the UE while $h_{k,\,p}$, $\nu_{k,\,p}$, $\tau_{k,\,p}$ and $\mathbf{v}_{k,\,p}$, respectively, represent channel impulse response, Doppler shift, propagation delay, and the non-terrestrial platform's downlink array response vector associated with path propagation path p. $\mathbf{n}$ represents additive white Gaussian noise.

The channel model presented in Eq. (12.3) is applicable over periods where the relative position of the non-terrestrial platform and the UE do not significantly change. If this period is exceeded, the channel parameters will need to be updated. While many of the characteristics of the NTN environment are similar to that of terrestrial networks, they exhibit dissimilar Doppler and propagation delay characteristics and these must be compensated for.

12.3.2.1 Doppler

For the NTN system, the scattering environment only exists within the vicinity of the UE [10] and this environment is usually stationary within the time interval of interest. Therefore, the Doppler shift, $\nu_{k,p}$ related to the propagation path, p of UE k comprises two independent shifts that are respectively attributed to the motion of the non-terrestrial platform and the motion of the UE [11]. However, due to the high altitude of the non-terrestrial platform, the Doppler shift experienced by a particular UE is the same irrespective of the different signal propagation paths. Also, the Doppler shift due to the motion of a UE is characteristically different for different propagation paths and this leads to the phenomena of Doppler spread.

12.3.2.2 Propagation Delay

The relatively huge distance between the non-terrestrial platform and UEs ensures that the propagation delay is much larger than for typical terrestrial systems. However, the maximum and minimum propagation delays would give rise to a delay spread that would be in many cases smaller than that of terrestrial systems.

12.3.2.3 Compensation Techniques

One of the practical techniques to compensate for the effects of Doppler and propagation delay is to use properly selected orthogonal frequency division multiplexing (OFDM) parameters for the wideband massive MIMO non-terrestrial system [7]. For example, if N_{SC} is the number of OFDM subcarriers and N_{CP} is the cyclic prefix samples and T_S is the system sampling interval, then the OFDM symbol length, $T_{\text{SL}} = N_{\text{SC}}T_S$ while the cyclic prefix length, $T_{\text{CP}} = N_{\text{CP}}T_S$. Following the expression given in [7], the OFDM signal transmitted by the non-terrestrial platform can be expressed as

$$\mathbf{x}_l(t) = \sum_{n=0}^{N_{\text{sc}}-1} \mathbf{x}_{l,n} \exp\left(\bar{j}2\pi \frac{n}{T_{\text{SL}}} t\right), -T_{\text{CP}} \leq t - l(T_{\text{CP}} + T_{\text{SL}}) < T_{\text{SL}}, \tag{12.4}$$

where l represents the OFDM symbol duration. The signal received by UE k is then given as

$$\mathbf{y}_{k,l}(t) = \int_{-\infty}^{\infty} [\mathbf{h}_k(t,\tau)]^T \mathbf{x}_l(t-\tau)d\tau + \mathbf{n}, \tag{12.5}$$

where $\mathbf{h}_k(t, \tau)$ is the inverse Fourier transform of $\mathbf{h}_k(t, f)$ given in Eq. (12.3). Compensation for the effects of Doppler and propagation delay is achieved through frequency and time synchronization, particularly by making the synchronized Doppler shift equal to the Doppler shift resulting from the motion of the non-terrestrial platform ($\nu_k^{\text{syn}} = \nu_{k,p}^{\text{NTN}}$). Also, the synchronized delay is made equal to the minimum delay ($\tau_k^{\text{syn}} = \tau_k^{\min}$) [7]. Applying these to the received signal results in

$$\mathbf{y}_{k,l}^{\text{syn}}(t) = \mathbf{y}_{k,l}\left(t + \tau_k^{\text{syn}}\right)\exp\left(\bar{j}2\pi\left[t + \tau_k^{\text{syn}}\right]\nu_k^{\text{syn}}\right) + \mathbf{n}. \tag{12.6}$$

Using Eq. (12.6), the corresponding signal dispersion in the frequency (Doppler) and time (delay) domains can be significantly reduced by selecting adequate OFDM parameters that effectively mitigate the effects of inter-symbol and inter-carrier interference. Therefore, the demodulated signal received by UE k over subcarrier n of the OFDM symbol is given as

$$\mathbf{y}_{k,l,n}(t) = (\mathbf{h}_{k,l,n})^T\mathbf{x}_{l,n} + \mathbf{n}, \tag{12.7}$$

where $\mathbf{h}_{k,l,n}$ is the downlink channel of UE k over symbol l and subcarrier n, which in turn is given in [7] as

$$\mathbf{h}_{k,l,n}(t) = \mathbf{v}_k h_{k,l,n} \in \mathbb{C}^{M\times 1}, \tag{12.8}$$

where $h_{k,l,n} = h_k(l[T_{\text{SL}} + T_{\text{CP}}], n/T_{\text{SL}})$.

Note that the received signal presented in (12.7) is accurate only if the delay and Doppler compensation are adequate. Having presented the basic signal processing requirements, the next section provides details of specific linear and nonlinear precoding techniques for NTN-based massive MIMO transceivers when used in single-cell, multi-cell, and cell-free configurations.

12.4 Single-Cell Massive MIMO Linear Precoding Techniques

The single-cell massive MIMO system refers to a single non-terrestrial platform equipped with a large number of antennas that serve multiple single-antenna UEs. The precoding techniques treated in this section include matched filtering, zero forcing (ZF), eigenvalue-based beamforming (EVB), truncated polynomial expansion (TPE), and phased zero forcing (PZF). These techniques only apply to single-cell systems but with appropriate extensions in terms of feedback and sharing of channel information, they can be applied in multi-cell systems.

12.4.1 Matched Filtering

At the front of most linear receivers is a filter that is matched to the convolution of the channel coefficient, $h_{k,m}(t,\tau)$, and the transmitted symbol, $x(t)$. Here, t represents time and τ indicates delay. The main function of such a filter is to limit the amount of noise and interference that is sampled from the channel. At the transmit link end, MF entails the application of precoder weights to the transmit signal to enforce channel orthogonality. However, in the LOS conditions of the NTN environment, channel orthogonality can only be achieved when the number of transmit antennas, m, is far more than the number of receive antennas, k [12].

Dropping the time and delay notations, the matched filter precoder weights can be represented as

$$\mathbf{W}_{\text{MF}} = \sqrt{\alpha}\mathbf{H}^*, \tag{12.9}$$

where α represents a scaling factor used in normalizing the signal power while $\mathbf{H}^*$ is the conjugate transpose of the downlink channel matrix. The matched filtered received signal vector is therefore given as

$$\mathbf{y}_{\text{MF}} = \sqrt{\alpha}\mathbf{H}^*\mathbf{H}\boldsymbol{x} + \boldsymbol{n}. \tag{12.10}$$

As m increases in number, the term $\mathbf{H}^*\mathbf{H}$ starts to become a diagonal matrix and this leads to optimal filtering. However, if m reduces in number, the interference experienced by the UEs increases. In such cases, a better precoding technique would be needed.

12.4.2 Zero Forcing

ZF is a simple linear channel precoding technique. It operates on the assumption that the massive MIMO channel matrix, $\mathbf{H}$, is fully invertible and the product of the received signal and the coefficient of the inverted channel must fulfill the criteria:

$$[\mathbf{W}_{\text{ZF}}(t)\boldsymbol{y}(t)]_{t=iT_s} = \begin{cases} 1, & i = 0, \\ 0, & \text{otherwise}, \end{cases} \tag{12.11}$$

$$\mathbf{W}_{\text{ZF}} = \sqrt{\alpha}\mathbf{H}^*\mathbf{H}^+, \tag{12.12}$$

where $\mathbf{W}_{\text{ZF}}(t)$ is a $k \times m$ matrix of ZF complex weights, $\boldsymbol{y}(t)$ represents the received signal vector, $\mathbf{H}^+$ is the Moore–Penrose pseudo-inversed channel matrix, and T_s represents the duration of the received symbol. Equation (12.11) means that the received signal vector is forced to zero at all instants except when $t = 0$. The expression also assumes that the channel is narrowband. If the channel happens to be wideband, multiple taps are needed and $\mathbf{W}_{\text{ZF}}$ for each of the taps are scaled with respect to their level of contribution to the overall received signal power.

The Moore–Penrose pseudo-inversion process is given as

$$\mathbf{H}^+ = (\mathbf{H}\mathbf{H}^*)^{-1} \tag{12.13}$$

while the received signal vector after ZF precoding is represented by

$$\mathbf{y}_{\text{ZF}} = \sqrt{\alpha}\mathbf{H}\mathbf{H}^*\mathbf{H}^+\boldsymbol{x} + \boldsymbol{n}. \tag{12.14}$$

ZF can achieve optimal performance in removing interference but while inverting the channel, it enhances background channel noise. However, despite the noise enhancement drawback, ZF's simplicity makes it well suited for the largely LOS propagation characteristic of the NTN since noise in such channels is usually minimal.

12.4.3 Eigenvalue-Based Beamforming

EVB is a precoding technique referred to in some texts as regularized zero forcing (RZF) [7], [13]. The EVB precoder weights are given by:

$$\mathbf{W}_{\text{EVB}} = \sqrt{\alpha}\mathbf{H}^*(\mathbf{H}\mathbf{H}^* + \mathbf{X} + \lambda\mathbf{I}_{\boldsymbol{K}})^{-1}. \tag{12.15}$$

It can be observed in Eq. (12.15) that EVB uses both MF and ZF precoding weights which are made to remain fully invertible by adding a non-negative Hermitian matrix, $\mathbf{X}$, and a $k \times k$ identity matrix, $\mathbf{I}_K$, which has been regularized by λ. These additions ensure that channel orthogonality is maintained, noise is not enhanced and the mean square error (MSE) between the transmitted and received symbols is minimized under different channel fading conditions. The received signal vector for EVB is given as

$$\mathbf{y}_{\text{EVB}} = \sqrt{\alpha}\mathbf{H}\mathbf{H}^*(\mathbf{H}\mathbf{H}^* + \mathbf{X} + \lambda\mathbf{I}_{\boldsymbol{K}})^{-1}\boldsymbol{x} + \boldsymbol{n}. \tag{12.16}$$

EVB is a more complex signal processing procedure than MF and ZF as it may have to invert matrices whose dimensions are very large. It is also very complex because it aims to support transmit beamforming from the non-terrestrial platform to the multiple UEs at all times by oscillating between MF and ZF modes. For example, if $\mathbf{X} = 0$ and $\lambda \to \infty$, $\mathbf{W}_{\text{EVB}} = \mathbf{W}_{\text{MF}}$. Alternatively, if $\mathbf{X} = 0$ and $\lambda \to 0$, $\mathbf{W}_{\text{EVB}} = \mathbf{W}_{\text{ZF}}$.

12.4.4 Truncated Polynomial Expansion

The TPE-based precoding reduces the computational complexity of EVB by replacing the matrix inversion with a TPE. The TPE precoding weights are given as

$$\mathbf{W}_{\text{TPE}} = \sum_{j=0}^{J-1} \omega_j(\mathbf{H}\mathbf{H}^*)^j\mathbf{H}, \tag{12.17}$$

where ω_j represents a jth-order precoder polynomial that allows for a smooth transition between the low complexity MF and the high complexity EVB precoding. For example, setting $j = 1$ gives $\mathbf{W}_{\text{TPE}} = \omega_0\mathbf{H}$, which is equivalent to MF precoding. Alternatively, increasing the value of j improves the performance of TPE at the expense of increased computational complexity. Therefore, TPE signal processing comprises several simple matrix–vector multiplications based on the value of j and its complexity can approach that of the EVB precoder.

12.4.5 Phased Zero Forcing

PZF is a low-complexity hybrid precoding technique proposed in [14] for multiuser massive MIMO systems. Here, the precoder matrix is divided into radio frequency and baseband submatrices, **R** and **B**, respectively. While the order of $\mathbf{R} = m \times k$, the order of $\mathbf{B} = k \times k$. Recall that m represents the number of NTN-based transmit antenna elements while k represents the number of receive antennas and for massive MIMO to achieve orthogonality in the NTN environment, $m > k$. Therefore, the signal processing complexity of PZF is determined based on the dimension-reduced square matrix **B**. Following the notation given in [14], the precoding weights of PZF are therefore given as

$$\mathbf{W}_{\mathrm{PZF}} \triangleq \mathbf{B} = \tilde{\mathbf{H}}^{\boldsymbol{H}}(\tilde{\mathbf{H}}\tilde{\mathbf{H}}^{\boldsymbol{H}})^{-1}\boldsymbol{\Lambda}, \tag{12.18}$$

where $\tilde{\mathbf{H}} \triangleq \mathbf{HR}$ and $\boldsymbol{\Lambda}$ is a $k \times k$ positive diagonal matrix whose function is to normalize the column power. The PZF received signal vector is therefore given as

$$\mathbf{y}_{\mathrm{PZF}} = \mathbf{HRB}\boldsymbol{x} + \boldsymbol{n}. \tag{12.19}$$

In summary, while the lowest complexity linear precoding techniques are MF and PZF, ZF has medium complexity and the complexity of TPE is scalable and can approach that of the highest complexity EVB. Signal processing can be performed using general-purpose computers and/or dedicated hardware such as field programmable gate arrays (FPGAs). The next section extends the single-cell analysis to multicell scenarios.

12.5 Multi-cell Massive MIMO Linear Precoding Techniques

The single-cell massive MIMO linear precoding techniques discussed in the previous section can also be applied in the multi-cell environment. The multi-cell environment has several linked and multilayered single-cell systems that can be operated in a cooperative or noncooperative manner. In the multi-cell environment, the non-terrestrial platforms can be at different altitudes and they may project different and overlapping cells that provide services either cooperatively or noncooperatively. In the noncooperative multi-cell scenario, TPE is less complex to implement and can achieve higher throughput than EVB [9]. On the other hand, cooperative multi-cell massive MIMO must be supported by backhaul infrastructure and would require increased signal processing in terms of both complexity and cost. The following subsections present the model of the downlink multi-cell scenario and the signal processing requirements of the major forms of cooperative and noncooperative multi-cell linear precoding techniques.

12.5.1 Model of the Downlink Multi-cell Scenario

The model of the downlink multi-cell scenario as shown in Figure 12.5 is developed from the assumption that there are L number of non-terrestrial platforms that generate j number of cells. Each platform has m antennas that serve k users and the signal received by the kth user in the jth cell is given as

$$\mathbf{y}_{j,k} = \sum_{l=1}^{L} \mathbf{h}_{l,j,k}\mathbf{W}_l\mathbf{x}_l + \mathbf{n}_{j,k}, \tag{12.20}$$

where $\mathbf{h}_{l,j,k}$ is an $m \times k$ vector representing the channel between the lth non-terrestrial platform and the kth user in the jth cell, $\mathbf{W}_l$ is an $m \times k$ precoding matrix for lth cell, $\mathbf{x}_l$ is an $m \times 1$ transmit signal vector from the non-terrestrial platform serving the lth cell, and $\mathbf{n}_{j,k}$ is a zero mean complex valued additive white Gaussian noise component.

Taking into account the interference from the multiple cells, the received signal vector in the jth cell is

$$\mathbf{y}_j = \sum_{l=1}^{L} \mathbf{H}_{l,j}\mathbf{W}_l\mathbf{x}_l + \mathbf{n}_j, \tag{12.21}$$

$$\mathbf{y}_j = \mathbf{H}_{j,j}\mathbf{W}_j\mathbf{x}_j + \mathbf{n}_j + \sum_{l=1, l\neq j}^{L} \mathbf{H}_{l,j}\mathbf{W}_l\mathbf{x}_l. \tag{12.22}$$

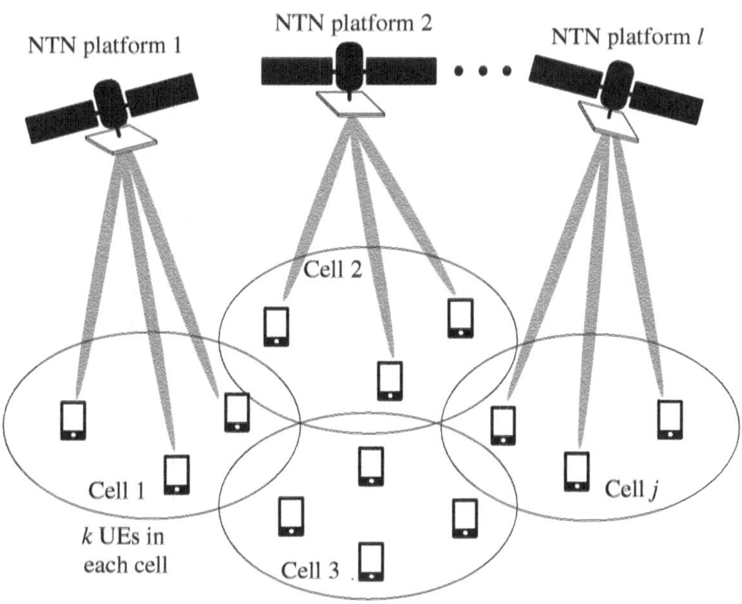

Figure 12.5 Noncooperative multi-cell downlink channels for massive MIMO.

The left portion of the right-hand side of Eq. (12.22) is the wanted signal and is equivalent to the single-cell model while the right portion is the inter-cell interference. Based on the developed model, at least two types of noncooperative multi-cell precoding techniques are applicable. These are H-infinity precoding and precoding for the maximum signal-to-interference ratio.

12.5.2 H-Infinity Precoding

The H-infinity method is a noncooperative precoding technique that aims to find a precoding matrix, $\mathbf{W}_j$ such that the ratio between the estimated error in the signal received in cell j (after precoding) and the total interference resulting from all other cells is less than a prescribed bound, p_b. It has been reported in [15] that H-infinity can achieve better performance than ZF and EVB precoding techniques and its objective function is given as

$$\sup_{\mathbf{H}_{l,j}\mathbf{W}_l + \mathbf{n}_j} \frac{\|\mathbf{H}_{j,j}\mathbf{W}_j\mathbf{x}_j - \mathbf{x}_j\|^2{}_w}{\left\|\sum_{l=1,\ l\neq j}^{L}\mathbf{H}_{l,j}\mathbf{W}_l\mathbf{x}_l + \mathbf{n}_j\right\|^2} < p_b, \tag{12.23}$$

where sup represents supremum and $\mathbf{W}_j$ must always be adjusted such that the ratio is always less than p_b.

12.5.3 Precoding for Maximum Signal-to-Interference and Noise Ratio

This precoding algorithm aims to maximize the ratio of the signal to inter-cell interference and noise (SINR) and keep it above a given threshold while at the same time minimizing the total transmit power. Here, it is assumed that the maximum power available to each non-terrestrial platform is P, and ϕ is the utilization rate (with value 0–1) of the power, subject to the condition that the signal-to-noise plus interference ratio of user k in cell j ($\text{SINR}_{j,k}$) is greater than a threshold, $\gamma_{j,k}$. This precoding algorithm aims to maximize ϕ while minimizing the total power ϕP. Therefore, its objective function is expressed in [16] as

$$\min_{\phi,\mathbf{W}_{j,k}} \phi P, \ \text{s.t.} \sum_{k=1}^{K} |\mathbf{W}_{j,k}|^2 \leq \phi P \quad \text{and} \quad \text{SINR}_{j,k} \geq \gamma_{j,k}. \tag{12.24}$$

It has been shown in [16] that this precoding algorithm outperforms single-cell massive MIMO MF precoding even in the presence of severe pilot contamination. However, to achieve this performance, the scheme assumes the presence of accurate CSI which is used by the individual cells without cooperation.

12.5.4 Partial Cooperation

In partial cooperation, non-terrestrial platforms share the CSI of the active users and make a compromise between achieving a reasonable load on the backhaul and attaining acceptable performance gains. Figure 12.6 depicts the downlink channels for a cooperative massive MIMO NTN system. Observe that the main difference between the noncooperative mode shown in Figure 12.5 and the cooperative mode is the presence of ISLs. These links facilitate the sharing of CSI.

The shared CSI in partial cooperation can be used in designing joint precoding matrices for downlink transmission. However, the computational complexity and backhaul signaling requirements make practical implementation difficult. An exception to this is when a heuristic strategy such as the distributed virtual SINR (DV-SINR) maximizing technique is adopted. This technique aims at producing precoder weights that maximize beamforming in all cells and for all receive antennas. According to [17], this is achieved by solving the following equation:

$$\mathbf{W}_{j,k}^{\text{DV-SINR}} = \underset{\|\mathbf{W}\|^2}{\arg\max} \frac{\left|\mathbf{h}_{j,k}^{H}\mathbf{w}\right|^2}{\frac{\sigma_k^2}{P_j} + \sum_{\bar{k}\neq k}\left|\mathbf{h}_{j,\bar{k}}^{H}\mathbf{W}\right|^2}, \tag{12.25}$$

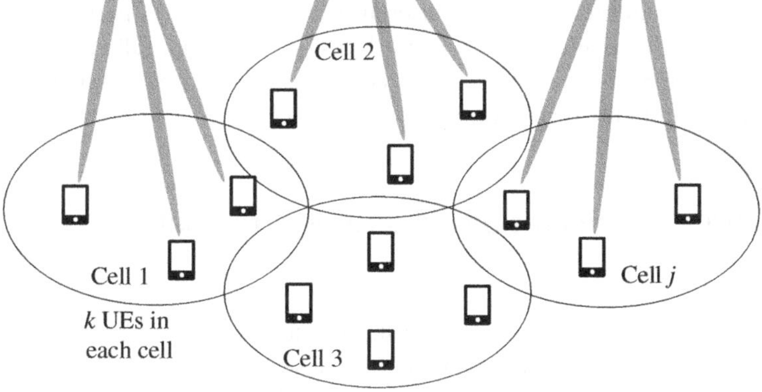

Figure 12.6 Downlink channels of a massive MIMO non-terrestrial network.

where $||\mathbf{W}||^2 = 1$, σ_k^2 is the variance of the power allocation coefficients, P_j, $\mathbf{h}_{j,k}^H$ is the Hermitian transpose of the channel coefficient, which must be real valued to ensure that the signals arriving at a given receiver from different non-terrestrial platforms would constructively combine. In essence, DV-SINR performs a trade-off between maximizing the useful signal power at a target user terminal and generating harmful interference at other terminals.

Other attempts at reducing the computational complexity and backhaul signaling requirements have resulted in different types of partial precoding schemes. While some of such schemes propose grouping non-terrestrial platforms into small clusters for ease of coordination and joint precoding, others propose multistage precoding techniques to improve spectral efficiency, limit inter-cell interference, and improve the SINR of end users.

12.5.5 Full Cooperation

In full cooperation, non-terrestrial platforms share CSI and these are then used in designing joint precoding matrices for downlink transmission. Here, the transmit antennas are effectively pooled to jointly serve UEs. Adapting the expression given in [9, 18] to suit the case of non-terrestrial platforms in full cooperation using ZF precoders, we have the total power minimization constraint expressed as

$$\min_{p \geq 0} \sum_{l=1}^{n} \sum_{m=1}^{K} p_{l,m}; \quad \text{s.t.} \left\| \sum_{l=1}^{n} \frac{\sqrt{p_{l,m}}}{||f_{l,m}||} \right\|^2 \geq \gamma_m, \quad l \leq m \leq K, \tag{12.26}$$

where $p_{l,m}$ is the transmit power at the lth non-terrestrial platform directed at the mth UE, γ_m is the target SINR and $f_{l,m}$ is the normalized transmit precoding weights ($w_{l,m}$).

Some of the issues that have to be dealt with in the fully cooperative NTN system include pilot contamination and interference, unavailability of perfect CSI, and limitation in the throughput of backhaul links. In summary, fully cooperative massive MIMO for NTN systems is highly complex and may not be practically feasible to implement in the near future.

12.6 Security Considerations in a Non-terrestrial Network

Similar to the security architecture of terrestrial systems given in [19], NTNs should have at least six defined security domains. These are shown in Figure 12.7, and they include (I) the network access domain, (II) the internal network security domain, (III) the user domain, (IV) the application domain, (V)

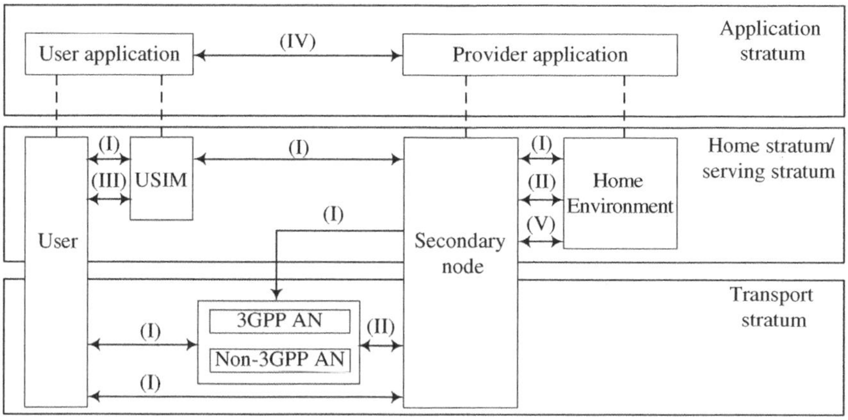

Figure 12.7 Overview of the security architecture for non-terrestrial networks.

services-based architecture (SBA) domain, and a domain that deals with visibility and configurability of security features (not shown).

In the network access domain, we deal with the set of features that enable a UE to be authenticated and to securely access the NTN, while the internal network domain covers the set of security features that enable nodes within a network to securely exchange signaling and user data. User domain security is concerned with the features that give a subscriber secure access to communicating equipment, while application domain security allows applications to securely exchange information between end users and providers. The SBA domain includes security features such as network function registration, discovery, and authorization, and these enable secure intra- and inter-network communication as well as protection for service-based interfaces. The security visibility and configurability domain deal with the protocols that allow a user to be informed whether a security feature is active or not.

The subsection that follows explains the common security requirements applicable in all but the application domain and presents them as they are generally applicable at the UE, gNB, and for AMF. Security requirements in the core network are treated in Section 12.6.2.

12.6.1 Requirements at the UE, gNB, and AMF

The AMF is a control plane component that terminates the interface from both 3GPP and non-3GPP access networks. It also manages access control and mobility. According to [19], there must be support for ciphering and integrity protection of both signaling data and user data exchanged between the UE and the gNB and at the AMF. This requirement is currently applicable to terrestrial systems and will undoubtedly be extended to non-terrestrial systems based on the fact that only transparent satellite payloads are currently being considered [6]. The ciphering algorithms supported include the New radio Encryption Algorithm (NEA0), 128-NEA1, 128-NEA2, and 128-NEA3.

Depending on prevailing regulations in the country of use, confidentiality protection of user data and signaling data between UE and gNB, and at the AMF, shall be optional. It is, however, recommended that there must be an in-built capability for integrity protection and replay protection of user data and signaling data exchanged between the UE and gNB. This is especially important for UEs that may not be awake all the time and for UEs that may be temporarily out of coverage areas. The supported integrity protection algorithms include the NIA0, 128-NIQ1, 128-NIA2, and 128-NIA3. It is however not mandatory to use the integrity protection algorithms, and NIA0 is expressly forbidden in certain instances since it adds signal processing overheads at the UE and gNB but with no security benefits.

It is recommended that the UE shall be capable of secure storage and processing of subscription credentials using tamper-resistant hardware components while the gNB shall be configured such that attackers would not be able to modify their software settings via local or remote access. Additionally, the UE should be able to support subscriber privacy and therefore, not allow SUbscription Permanent Identifier (SUPI) data to be transferred in clear text over the RAN. For the gNB, any part of it that stores or processes user plane and control plane data in clear text must be protected from physical attacks, and management traffic sent between central and distributed gNB units must have integrity, confidentiality, and replay protection.

12.6.2 Requirements in the Core Network

The core network is the heart of the NTN as it aggregates data traffic, communicates with the UEs, and securely delivers essential services among other functions. Since NTNs are essentially a network of networks, it is recommended that mobile network operators (MNOs) and satellite network operators (SNOs) whose networks are integrated must subdivide their networks into trust zones. Subnetworks of different operators should be assumed to lie in different trust zones and messages that are not end-to-end protected and happen to cross trust boundaries must follow certain protocols. For example, network function service requests and responses should support mutual authentication between service providers and service consumers.

To guarantee secure service registration, discovery, and authorization in the core network, all incoming messages must be validated and internal core information such as the data network name (DNN) must not be sent outside of the operator's domain. End-to-end confidentiality and/or integrity must be supported to allow destination network elements to be able to determine the authenticity of the source network sending specific information.

12.7 Standards and Interoperability Requirements in a Non-terrestrial Network

As earlier introduced in Section 12.1.5, standardization allows for equipment from different vendors, operators, and networks to be used in a global NTN of networks. So far, 3GPP has driven the research to standardize the air interface and other aspects of the NTN. These efforts were initiated right from when 5G was introduced and as reported in [6], the channel model for NTNs has already been established, the possible deployment scenarios defined, and the potential impacts of NTN components on present 5G NR networks well studied. Other current 3GPP research works are focused on determining the necessary enhancements to the 5G physical layer and upper layer protocol stack specifications to enable them to accommodate the unique characteristics of the NTN. Some of the specific standardization efforts are detailed as follows.

12.7.1 Solution to the Uplink Time Synchronization Problem

One of the major challenges of integrating non-terrestrial with terrestrial networks has to do with large and time-varying propagation delay, the high velocity of satellites (especially LEO satellites), and the resultant effect on transmission timing and synchronization. A study item in 3GPP Release 17 [20] proposed a synchronization approach based on the periodic transmission of predefined system information block (SIB) messages using the synchronization signal (SS)/physical broadcasting channel (PBCH). This method makes it easy for UEs to be synchronized during downlink transmission. However, for the uplink, the difference in the distance from the non-terrestrial platform to individual UEs (even in the same cell) creates a potentially huge interference problem due to arrival time mismatch. To overcome this, different timing advances (TAs) have to be applied to the uplink signals of each UE to ensure that the uplink frames received at the non-terrestrial platform are aligned with the downlink frames. The TA is expressed in [21] as

$$T_{\text{TA}} = (N_{\text{TA}} + N_{\text{TA,offset}})T_C, \tag{12.27}$$

where N_{TA} denotes a closed-loop TA, $N_{\text{TA,offset}}$ is an offset applied to take care of different subcarrier spacings (SCS) and frequency bands, and T_C is a basic time unit for 5G NR systems. A simple solution to the TA problem would be to extend the range of N_{TA} to accommodate the propagation delay. However, this solution will introduce huge signaling overheads based on the fact that N_{TA} would need to be frequently sent to the UE to maintain the required uplink timing.

An implementable solution with lower signaling overhead is to separate the full propagation delay into two parts. The first part is a delay that is common to all UEs within the same cell and is computed using a single reference point, while the second part is a UE-specific delay based on the displacement from the reference point. The standards indicate that the reference point-based propagation delay can be computed from

$$T_{\text{TA,common}} = \frac{d_{\text{ref}}}{c}, \tag{12.28}$$

where d_{ref} denotes the distance between the satellite and the reference point, and c is the velocity of light. A depiction of this is shown in Figure 12.8. Also, the UE-specific delay can be computed using

$$T_{\text{TA,UE}} = \frac{d_{\text{UE}} - d_{\text{ref}}}{c}, \tag{12.29}$$

where d_{UE} is the distance between the non-terrestrial platform and the UE. Substituting (12.29) into (12.27) gives

$$T_{\text{TA}} = (N_{\text{TA}} + N_{\text{TA,offset}} + N_{\text{TA,common}} + N_{\text{TA,UE}})T_C, \tag{12.30}$$

where $N_{\text{TA,common}} = 2T_{\text{TA,common}}$ and $N_{\text{TA,UE}} = 2T_{\text{TA,UE}}$. It is recommended that signaling overhead can be reduced by broadcasting only $N_{\text{TA,common}}$ since it is the same for all UEs in the same cell.

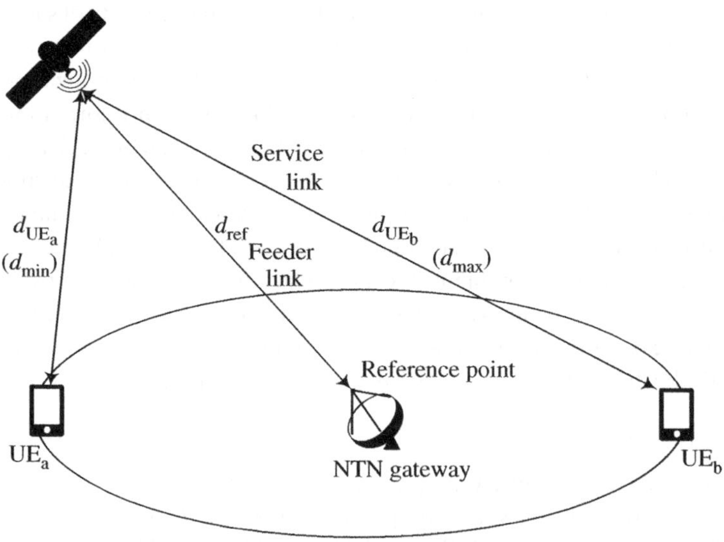

Figure 12.8 Reference and UE propagation delay.

It is important to note that the standards assume that UEs are equipped with Global Navigation Satellite System (GNSS) chips to enable them to calculate their $T_{TA,UE}$ based on the knowledge of their positions and the ephemeris of the non-terrestrial platform as supplied by the network.

12.7.2 Considerations for Random Access

The channel random access procedure is based on the UE first transmitting a preamble to the gNB using the physical random access channel (PRACH). The preamble comprises a cycle prefix with duration, T_{CP}, a Zadoff-Chu sequence, T_{seq}, and a guard period, GP as shown in Figure 12.9. To take care of UEs that experience relatively longer propagation delay such as those at the cell edge, T_{CP} and T_{seq} are sized as follows.

$$T_{CP} = \frac{2(d_{max} - d_{min})}{c} + \sigma_{DS},$$
$$T_{seq} = \frac{2(d_{max} - d_{min})}{c} + \sigma_{DS}, \tag{12.31}$$

where d_{max} and d_{min}, respectively, represent the maximum and minimum distances from the UE to the non-terrestrial platform as shown in Figure 12.8. Since the common signaling delay, $T_{TA,common}$ is broadcast to all UEs, the design of T_{CP} and T_{seq} only considers the difference between d_{max} and d_{min}.

The current formats for random access in 5G NR are suitable for cell sizes generated by all airborne platforms. However, for satellites, the difference between d_{max} and d_{min} can be very large and this would cause T_{CP} and T_{seq} to be unreasonably long, thereby increasing the signaling overhead. Fortunately, if the UEs can determine their location based on inbuilt GNSS chips and the supplied satellite ephemeris data, this problem can be easily solved and the existing preamble format reused without any modification. Two types of random access procedures have been specified in [22] and work is ongoing to optimize them for the GNSS-capable UEs that would be used in NTNs.

12.7.3 Enhancements to Uplink Frequency Synchronization

Doppler effects are known to create frequency synchronization problems and as discussed in Section 12.3.2, these effects must be compensated for. The two types

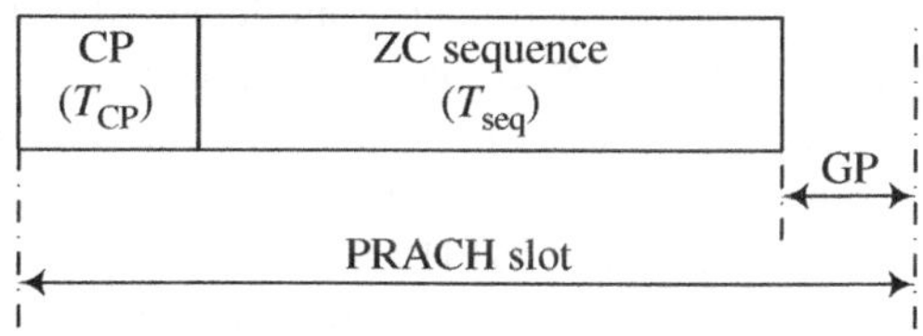

Figure 12.9 Preamble for random access in non-terrestrial networks.

of Doppler effects include service link frequency shifts resulting from the relative velocity between the satellite and the UE, and the feeder link frequency shifts caused by the satellite transponder.

The frequency shift in the service link, especially for LEO satellites, can result in frequency offsets larger than the SCS. To compensate for this, the UE uses state vector-based and orbital element-based satellite ephemeris data as supplied by the NTN control center to compute the necessary frequency adjustments. The ephemeris data used are the satellite's orientation and velocity as determined using an Earth-centered, Earth-fixed reference frame, and the six orbital parameters being the semi-major axis, eccentricity, argument of periapsis, longitude of ascending node, inclination and mean anomaly. Using these data and the received downlink frequencies, the UE makes the necessary adjustments to its local frequency reference and applies the required compensation to the uplink transmission frequencies. The frequency shift in the feeder link is taken care of by the non-terrestrial gateway or the satellite transponder as necessary.

12.7.4 Hybrid Automatic Repeat Request Enhancements

The Hybrid Automatic Repeat reQuest (HARQ) is a feedback-based physical layer mechanism used in terrestrial 5G NR systems to increase the reliability of data transmission. While HARQ achieves high Block Error Rate (BLER) targets, it pays a latency penalty as a result of the increased round trip time (RTT). Since RTT comprises propagation time delay (T_p), transmission time (T_{slot}), and hardware processing time (T_{proc}), it is obvious that RTT will have a comparatively larger value for non-terrestrial systems due to the huge distances between the non-terrestrial platform and the UE. In 3GPP Release 15 [6], some enhancements have been proposed to the legacy HARQ NR specifications to make it more suitable for NTNs. These enhancements are detailed below.

The minimum number of HARQ processes ($N_{\text{HARQ,min}}$) has been increased beyond 16 and this can be computed using the following formula:

$$N_{\text{HARQ,min}} \geq \frac{T_{\text{HARQ}}}{T_{\text{slot}}}, \tag{12.32}$$

where T_{HARQ} is the time duration between the initial transmission of one transport block (TB) and the beginning of the second TB. Within this period, decoding of the corresponding acknowledgment (ACK) or negative acknowledgment (NACK) message must have been completed. The timing diagram for a single HARQ process in a NTN employing a transparent satellite is shown in Figure 12.10. Observe in the diagram that T_{slot} is the time it takes to transmit a TB while T_{proc} is the hardware processing time. Note that T_{slot} has a value of 1 ms for a reference numerology with 15 kHz SCS. For other numerologies with larger SCS (i.e., $2^k \times 15$ kHz), $N_{\text{HARQ,min}}$ can be scaled by 2^k.

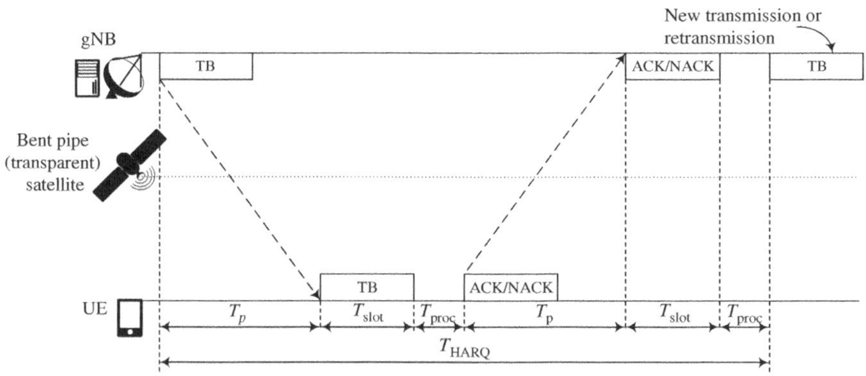

Figure 12.10 Timing diagram for a single HARQ process with a transparent satellite.

Based on Figure 12.10, it can be deduced that a more detailed expression for $N_{HARQ,min}$ is

$$N_{HARQ,min} = \frac{2(T_p + T_{slot} + T_{proc})}{T_{slot}}. \tag{12.33}$$

The computed values of $N_{HARQ,min}$ for a LEO satellite are 50, while the values for MEO and GEO/HEO satellites are, respectively, 180 and 600. According to [6], it is currently feasible to implement $N_{HARQ,min}$ for LEO satellites while that of MEO, GEO, and HEO satellites need to be further studied, especially the aspects that have to do with the potential impact on the size of the TB.

Another enhancement is the recommendation to disable HARQ feedback on a semi-statistical basis using radio resource control signaling. The standards recommend that the choice of enabling or disabling HARQ should be configured on a per UE and per HARQ process basis. Depending on the type of HARQ codebooks used, there are specific recommendations on how the gNB can be informed of the current configuration (whether HARQ feedback has been disabled or not). Recent simulation works such as [23] show that using 16 HARQ processes without feedback can have similar resource efficiencies as that of 32 HARQ processes in LEO satellite-based NTNs.

12.8 Conclusion and Recommendations

The NTN of the future will be capable of short-range, wide-area, and global coverage using airborne and spaceborne platforms. Since satellites will play a very prominent role in such networks, this chapter has described satellite network architectures and the spectrum utilization issues that will arise from the operation of such networks. The chapter has also provided details on the need for security and reliable communications and the ongoing standardization work championed by 3GPP to ensure interoperability of service for the different components of NTNs.

Building on the above foundation, the chapter treated the signal processing techniques necessary for implementing massive MIMO in the non-terrestrial environment. These techniques are based on linear precoding methods and have been found in the literature to be appropriate for the envisaged single-cell and multi-cell massive MIMO implementation scenarios. The linear precoding techniques presented in order of increasing complexity include MF, ZF, EVB, TPE, PZF, H-infinity precoding, and precoding for maximum SINR. Models of beamforming coefficients and power minimization constraints were, respectively, presented for situations where there is no cooperation between cells and when there is cooperation between the cells formed by NTN platforms.

Using the already established framework for terrestrial systems, specific security requirements on the UE, gNB, and the core network were detailed. The details reveal that there is a need for privacy and confidentiality of signaling and user data, and these can be actualized using various kinds of ciphering and integrity protection algorithms. The chapter also presented some solutions to uplink time and frequency synchronization problems in NTNs. Given that the propagation characteristics of the non-terrestrial environment are different from the terrestrial scenario, considerations for random access and the need to modify the implementation of the standard hybrid automatic repeat request protocol were presented.

So far, standardization efforts have been limited to transparent (bent-pipe) satellite payloads. It would therefore be interesting for work to commence on signal processing for satellites with full onboard signal processing capabilities. Even though NTNs have inherited many of their security features from existing terrestrial systems, there is a need for a new direction in the set of security protocols and architectures. Also, specific enablers for satellite access in new frequency bands to serve both stationary and mobile platforms need to be agreed upon and worked on.

References

1 Slalmi, A., Chaibi, H., Chehri, A. et al. (2021). Towards 6G: understanding network requirements and key performance indicators. *Transactions on Emerging Telecommunication Technologies* 32 (3): 4201–4220.

2 ETSI (August 2018). 5G; F1 General Aspects and Principles. *Release 15, 3GPP Technical Specification 38.470 v 15.2.0.*

3 ETSI (September 2018). 5G; System Architecture for the 5G System. *Release 15, 3GPP Technical Specification 23.501 v15.3.0.*

4 Evans, B. (2022). 6G satellite communications. *27th Asia Pacific Conference on Communications,* Jeju Island, Korea.

5 Kokkinen, H., Piemontese, A., Kulacz, L. et al. (2023). Coverage and interference in co-channel spectrum sharing between terrestrial and satellite networks. *IEEE Aerospace Conference*, Big Sky, MT, USA.

6 3GPP (September 2020). Study on New Radio (NR) to Support Non-terrestrial Networks (Release 15) v15.4.0. *3GPP, Technical Report 38.811.*

7 You, L., Li, K.-X., Wang, J. et al. (2020). Massive MIMO transmission for LEO satellite communications. *IEEE Journal on Selected Areas in Communications* 38 (8): 1851–1865.

8 Tekbiyik, K., Kurt, G.K., Ekti, A.R., and Yanikomeroglu, A.H. (2022). Reconfigurable intelligent surfaces empowered THz communication in LEO satellite networks. *IEEE Access* 10: 121957–121696.

9 Fatema, N., Hua, G., Xiang, Y. et al. (2018). Massive MIMO linear precoding: a survey. *IEEE Systems Journal* 12 (4): 3920–3931.

10 Ekpe, U.M. (2012). Modelling and measurement analysis of the satellite MIMO radio channel. PhD thesis. University of Surrey.

11 Perez Fontan, F. and Marino Espineira, P. (2008). *Modeling the Wireless Propagation Channel: A Simulation Approach with MATLAB.* Chichester: John Wiley & Sons.

12 Zhu, J., Schober, R., and Bhargava, V.K. (2016). Linear precoding of data and artificial noise in secure massive MIMO systems. *IEEE Transactions on Wireless Communications* 15 (3): 2245–2261.

13 Muller, A., Couillet, R., Bjornson, E. et al. (2015). Interference-aware RZF precoding for multicell downlink systems. *IEEE Transactions on Signal Processing* 63 (15): 3959–3973.

14 Liang, L., Xu, W., and Dong, X. (2014). Low-complexity hybrid precoding in massive multiuser MIMO systems. *IEEE Wireless Communications Letters* 3 (6): 653–656.

15 Xu, P., Wang, J., Wang, J., and Qi, F. (2014). Multi-cell H-inf precoding in massive MIMO systems. *2014 IEEE International Conference on Communications*, Sydney, Australia.

16 Jing, J. and Zheng, X. (2014). A downlink max-SINR precoding for massive MIMO system. *International Journal of Future Generation Communication and Networking* 7 (3): 107–116.

17 Bjornson, E., Zakhour, R., Gesbert, D., and Ottersten, B. (2010). Cooperative multicell precoding: rate region characterization and distributed strategies with instantaneous and statistical CSI. *IEEE Transactions on Signal Processing* 58 (8): 4298–4310.

18 Kang, J., Kang, J., Lee, N. et al. (2013). Minimizing transmit power for cooperative multicell system with massive MIMO. *10th Annual IEEE Consumer Communications and Networking Conference*, Las Vegas.

19 3GPP (October 2023). Release 17, Security Architecture and Procedures for 5G System. *3GPP, Technical Specification 33.501 v17.11.1.*

20 3GPP (January 2023). Release 17 Description; Summary of Rel-17 Work Items. *3GPP, Technical Report 21.917 v17.0.1.*

21 3GPP (June 2022). Release 16, NR; Physical Channels and Modulation. *3GPP Technical Specification 38.211 v16.10.0.*

22 3GPP (September 2023). Release 16, NR; Physical Layer Procedures for Control. *3GPP Technical Specification 38.213 v16.15.0.*

23 Cioni, S., Lin, X., Chamaillard, B. et al. (2022). Physical layer enhancements in 5G-NR for direct access via satellite systems. *International Journal of Satellite Communications and Networking* 41 (3): 262–275.

13

Artificial Intelligence and Machine Learning for Channel Estimation in Massive MIMO Wireless Communication Systems

Thiruvengadam Sundarrajan Jayaraman[1], Velmurugan Periyakarupan Gurusamy Sivabalan[1], Vinoth Babu Kumaravelu[2], Agbotiname Lucky Imoize[3,4], and Helen Sheeba John Kennedy[2]

[1]*Department of Electronics and Communication Engineering, Thiagarajar College of Engineering, Madurai, Tamil Nadu, India*
[2]*Department of Communication Engineering, School of Electronics Engineering, Vellore Institute of Technology, Vellore, Tamil Nadu, India*
[3]*Department of Electrical and Electronics Engineering, Faculty of Engineering, University of Lagos, Lagos, Nigeria*
[4]*Department of Electrical Engineering and Information Technology, Ruhr University, Bochum, Germany*

13.1 Introduction

In the ever-evolving landscape of wireless communication, the advent of massive multiple input-multiple output (MIMO) technology is a revolutionary stride toward realizing the ambitious goals of fifth-generation (5G) and future network generations. At its core, massive MIMO is an advanced iteration of traditional MIMO technology, which offers substantial improvements in data rates and network capacity [1, 2]. Massive MIMO, a technological knight in shining armor, wielding an arsenal of hundreds, even thousands of antennas at a base station (BS). It represents a complex interplay among antennas, radios, and the spectrum, all governed by intricate algorithms. These algorithms play a crucial role in directing and molding the radio waves, maximizing the efficiency of signal transmission within the given spectrum [3]. This sophisticated management enables a larger number of devices to simultaneously utilize the same spectral resources without encroaching on each other's signals. As a result, this technology fosters a more equitable wireless environment, where data is distributed more fairly among users, significantly reducing the struggle for bandwidth in high-demand situations. It is designed to meet the burgeoning demands for higher data rates and capacity, brought about by the explosion of connected

Massive MIMO for Future Wireless Communication Systems: Technology and Applications,
First Edition. Edited by Webert Montlouis and Agbotiname Lucky Imoize.

devices and the ever-growing appetite for data-intensive applications like virtual reality, ultra-high-definition video streaming, and the Internet of Things (IoT). In this context, massive MIMO is not just a technological innovation; it is a critical enabler that will underpin the smart, interconnected world of tomorrow.

However, its efficiency heavily relies on accurate channel estimation, which is a challenging task due to the system's complexity. With potentially hundreds or thousands of antennas at each BS, the sheer volume of channel data that needs to be estimated grows significantly. This increase in scale leads to more complex channel matrices, demanding higher computational resources for accurate estimation.

In massive MIMO, the same set of pilot signals is often reused across different cells to keep the pilot overhead manageable. This reuse can lead to pilot contamination, where signals from users in different cells interfere with each other [4]. This interference distorts the channel estimates, leading to a degradation in performance. The allocation of pilot sequences in a massive MIMO system is a critical factor. Efficient allocation can minimize interference and improve channel estimation accuracy, but achieving this efficiency is challenging due to the dynamic nature of wireless environments and user mobility. The processing of signals for channel estimation in massive MIMO requires sophisticated algorithms that can handle high-dimensional data efficiently. Developing these algorithms is challenging, especially when aiming to balance accuracy with computational feasibility.

Artificial intelligence (AI) has become increasingly important in wireless communication systems, driving significant advancements and innovations [5]. The complexity of modern networks, along with the surge in data traffic and the diversity of services, makes AI an invaluable tool in managing and optimizing wireless communications. AI is not just an add-on but a fundamental component in modern wireless communication systems. Its ability to learn, predict, and adapt makes it essential for handling the complexity and demands of today's and future network environments. As wireless technology continues to evolve, the integration of AI will become even more ingrained, leading to smarter, more efficient, and more user-centric networks.

A considerable amount of recent research has been dedicated to exploring key issues in wireless networking through the lens of various machine learning (ML) techniques [6]. These studies primarily concentrate on areas such as managing and sharing spectrum resources, allocating network resources efficiently, predicting fading channels, and forecasting network traffic patterns. This research underscores the growing importance and effectiveness of ML applications in tackling fundamental challenges in the realm of wireless communications.

Implementing centralized ML (CML) approaches in large-scale systems remains problematic due to several key issues. These include scalability challenges, the increased communication overhead, vulnerability to a single point of failure, concerns related to privacy and security, as well as issues with latency and the necessity for real-time processing. These factors collectively pose significant hurdles in the effective application of traditional ML techniques in expansive network environments.

Decentralized ML (DML) represents a paradigm shift in the way we approach data processing and algorithmic learning in complex systems [7]. Unlike traditional centralized ML models where data is aggregated and processed in a single location, DML distributes the learning process across multiple nodes or devices. This approach is increasingly relevant in an era marked by vast volumes of data generated at the edge of networks, such as in IoT devices, smartphones, and sensors.

At the heart of DML is the concept of bringing the algorithm to the data, rather than the conventional method of bringing data to the algorithm. One of the most prominent examples of DML is federated learning (FL), where multiple devices collaboratively learn a shared model while keeping the data localized [8]. This approach not only preserves privacy but also allows for a diverse dataset collected from different sources, leading to potentially more robust and generalized models.

13.1.1 Key Contributions of the Chapter

The following are the significant contributions of this chapter:

- Comprehensive review on LS and MMSE-based channel estimation in various configurations of wireless communication system in both narrowband (flat fading) and wideband (frequency-selective fading) environments.
- Detailed analysis on the impact of channel uncertainty (CU) and pilot contamination on channel estimation performance in massive MIMO system.
- Summary on AI and ML-based approaches for channel estimation in massive MIMO system.
- Developing FL-based channel estimation in massive MIMO from the comprehensive review on LS and MMSE-based channel estimation.
- Study on protection of privacy of user data in cell-free massive MIMO using FL.

13.1.2 Chapter Organization

Section 13.2 presents the related works on the various channel estimation techniques in MIMO and massive MIMO system. Section 13.3 describes the detailed analysis of LS and MMSE estimation in various configurations of massive MIMO system in both flat fading and frequency-selective fading environment. Section 13.4 presents the essential of AI and ML-based approaches for channel estimation in massive MIMO system. Section 13.5 discusses the framework for channel estimation in cell-free massive MIMO system using FL. Section 13.6 explains the importance and protection of privacy of user data in cell-free massive MIMO using FL. Section 13.7 explains the results and discussion of channel estimation using the propagation model described in 3GPP TS 38.101-4 version 16.3.0 Release 16. Conclusion and future direction of channel estimation in massive MIMO system are described in Section 13.8.

13.2 Related Work

This section discusses various existing channel estimation techniques for massive MIMO systems. These techniques include linear estimation, channel estimation with pilot contamination, sparse estimation, ML, deep learning (DL), and FL-based estimation schemes.

13.2.1 Linear Estimation Techniques

Carrera et al. [9] compared the bit error rate (BER) and error vector magnitude of different channel estimators applied to maximal ratio combining (MRC) and regularized zero-forcing (RZF) receivers. The methodology uses a massive MIMO channel and a 5G-new radio (NR) frame structure. The authors compare LS, MMSE, and the Kalman filter (KF) channel estimators. Results show that MRC receivers achieve maximum performance with KF estimators, especially at low signal-to-noise ratio (SNR) scenarios, while RZF receivers achieve maximum performance with LS estimation.

13.2.2 Sparse Estimation Techniques

Ma et al. [10] investigated a channel estimation scheme based on training sequence (TS) design and optimization in structured compressive sensing (SCS) for massive MIMO in cellular networks. The scheme optimizes block coherence using auto-coherence and cross-coherence as key merit factors. The suggested TS design and optimization method reduce block coherence in the sensing matrix, outperforming conventional TS designs in recovery probability, mean square error (MSE), and BER and approaching the Cramer–Rao lower bound. Qin et al. [11] examined time-varying channels in a MIMO-orthogonal frequency division multiplexing (OFDM) system model. Under sparsity in the delay domain and strong correlation in the spatial domain, they go through sparse channel estimation. The basis expansion method (BEM) models time-varying and spatially correlated systems. This method reduces the number of channel coefficients to be estimated. The problem is then represented as a compressive sensing problem and solved using the orthogonal matching pursuit (OMP) approach.

In [12], a massive uplink device communication scenario is examined. They focus on a BS connected to several devices that generate irregular traffic. The study aims to determine the achievable rate and cost of active user detection and channel estimation in a two-phase grant-free access scheme. The authors also explore an asymptotic regime where the BS has multiple antennas to support concurrent broadcasting devices. Albataineh et al. [13] presented a novel channel estimation technique based on compressive sensing theory to minimize pilot overhead in massive MIMO systems. The normalized mean square error (NMSE) technique is suggested, and it combines compressive sampling matching and sparsity adaptive matching pursuit techniques, allowing for sparse signal distribution. Simulation results show the suggested method outperforms conventional compressive sensing algorithms in NMSE performance and reduces computational complexity, resulting in significant performance gains and reduced pilot overhead.

Sun et al. [14] presented an interference cancellation (IC)-based channel estimation for time-shifted pilot (TSP) in massive MIMO systems with finite antennas. The TSP's effectiveness is limited by channel estimation errors (CEE), which are dominated by inter-group data interference. An IC-TSP is suggested to reduce the mean squared CEE in finite antenna massive MIMO systems, allowing dominant inter-group data interference to be cancelled based on BS cooperation. The paper evaluates the additional overhead and discusses the impact of sectorization and compressed sensing-based BS-BS channel estimations.

13.2.3 Estimation in the Presence of Pilot Contamination

De Figueiredo et al. [15] introduced a simple channel estimator for massive MIMO systems with pilot contamination in their work. The estimator is designed to address performance problems when there is moderate to heavy pilot contamination, and it constructs an approximate analytical expression for the MSE. Simulation results indicate that the estimator performs asymptotically and is comparable to the minimal MSE estimator regarding the number of antennas and multipath coefficients. De Figueiredo et al. [16] presented a practical channel estimator for multi-cell multi-user massive MIMO time division duplex (TDD) systems with pilot contamination considering flat Rayleigh fading channels. The estimator addresses performance under moderate to solid pilot contamination without cross-cell large-scale channel coefficients knowledge. It performs asymptotically and is similar to the minimum MSE estimator concerning the number of antennas. An approximate analytical MSE expression is also derived.

To minimize the weighted MSE in channel estimation for multi-cell massive MIMO systems, Shen et al. [17] suggested a coordinated pilot design strategy employing fractional programming. The effect of pilot contamination is decoupled from the MSE formulation using the quadratic transform. The approach is extended to correlated Rayleigh fading by considering the max–min fairness of data rates with orthogonal pilots. Large-scale massive MIMO has potential in future mobile communications but can lead to near-field and spatial nonstationary channel conditions. Han et al. [18] addressed the channel estimation problem by suggesting subarray-based and scatterer-based channel estimation methods. The subarray-based method provides accurate channel estimation results with low complexity, while the scatterer-based method precisely places scatterers.

13.2.4 AI-Based Estimation Techniques

He et al. [19] discussed the learned approximate message passing (AMP) based on the denoising convolutional neural networks (CNN) algorithm to predict the beam space channel. The millimeter wave (mmWave) massive MIMO system is considered, and an analysis of state evolution is carried out to identify the asymptotic effect of the work. Results show improved NMSE over different systems. In [20], a massive MIMO system based on mmWave and the application of deep CNN for complexity reduction in channel estimation is discussed. Cost reduction is addressed for the system by introducing three different CNN methods. The methods leverage the system's spatial and temporal correlation and reduce the pilot overhead. The method's accuracy is analyzed and verified by comparing the results with the MMSE channel estimator.

Yuan et al. [21] suggested a channel estimation method for massive MIMO wireless systems resolved through the ML approach. The time-based channel correlation property is used to design a TDD scheme to predict the channel state information (CSI), which minimizes the system complexity of pilot symbol-based schemes. The suggested scheme includes a CNN and autoregressive (AR) predictor, which significantly improves prediction quality, which is ensured by comparing it with other optimal channel estimation schemes.

Demir and Björnson [22] designed and trained two neural networks using the deep feedforward method to calculate the variance in the presence of distortion and channel estimation at the BS. The paper considers the nonlinearity caused by hardware issues of user devices and BSs in the uplink side of the massive MIMO system. The suggested work outperforms the linear MMSE (LMMSE) estimator in the Rician fading channel for different SNRs, which is closer to the practical scenario. For analysis, different order nonlinearity is also derived in the suggested work, and the result shows the improvement in BER and spectral efficiency corresponding to varying numbers of users.

In [23], a massive MIMO system in which channel estimation is performed using a smaller pilot length, which is assumed to be less than the number of antennas used for transmission is discussed. The estimation process consists of two sections: pilot-based and data based. The neural network, which is two layered in nature, is designed to perform the pilot-based analysis. An iterative method is adapted to improve the accuracy of estimation. Results show better performance of the suggested method over conventional schemes. Hirose et al. [24] discussed a massive MIMO system, where the effect of imperfect pilot reception is studied, and the two DL-based methods are suggested. The first one considers a connection-oriented neural network for channel prediction. To improve accuracy, the second method has adopted the CNN. Estimation based on covariance is considered for comparison purposes, and the suggested work shows improvement in terms of MSE. Sabeti et al. [25] implemented a denoising CNN to minimize channel distortion in massive MIMO systems. The OFDM-based system estimates the blind channel for multi-user scenarios by utilizing the capability of the DL approach. The massive MIMO channel is asymptotically orthogonal, and this orthogonality is used in prediction. The improvement in BER and sum throughput over the other data-aided methods are analyzed in the results.

In [26], a modified version of the AMP method for channel estimation based on learned DNN called Gaussian mixture LAMP (GM-LAMP) is suggested. The beam space channel provides the sparsity, which is utilized in the algorithm. A massive mmWave-based MIMO system is considered for analysis. Network training is performed offline, and the estimation process is done online. A metric NMSE is used for performance evaluation, and the method provides better results than conventional AMP, even with lower pilot overhead. Zhang et al. [27] suggested an AMP algorithm based on denoising CNN for the mmWave massive MIMO system. The suggested algorithm performs learning of channel characteristics and eliminates noise impacts. The results show that the algorithm minimizes the NMSE and maximizes the average sum rate.

Jeon et al. [28] presented an algorithm using the reinforcement learning method, a data-aided approach. The data was collected using the LMMSE method. The data accuracy in the suggested massive MIMO system is ensured through the Markov process. The suggested algorithm plays a crucial role in the selection of received symbols. The method suggested reduces the detection error, specifically MSE and provides a sufficient reduction in block error compared to the conventional LMMSE scheme. Melgar et al. [29] have suggested a DNN-based channel prediction method for a massive MIMO system. They have conducted time and frequency-based parameter analysis to ensure the accuracy of the suggested method. The CSI prediction is made through multi-dimensional linear interpolation. The accuracy of estimation results is analyzed based on the reduced MSE.

In [30], an FL scheme for a massive MIMO system, which improves energy efficiency through different learning designs is recommended. The asynchronous design deals with the synchronization issues in the system compared to the synchronous design. A session-oriented design is also provided, which controls uplink and downlink through separate sessions of FL. The optimization process controls the assignment of time and complexity of computation, through which minimum energy utilization can be achieved. Results show that the session-oriented system outperforms the other suggested systems and conventional schemes.

13.3 Channel Estimation Using LS and MMSE Estimators

13.3.1 Massive MIMO System Model

In this section, a C-cells multicarrier massive MIMO system is considered with N_{SC} pilot subcarriers and $K_j, j = 1, 2 \ldots, C$ user equipment (UE) in jth BS with $N_{BS_j}, j = 1, 2 \ldots, C$ antennas. Each BS is equipped with $N_{BS_j}, j = 1, 2 \ldots, C$ antennas and $N_{BS_j} \gg K_j$. It is assumed that massive MIMO system operates in synchronous TDD protocol. The users in each cell transmit N_P pilot symbols in uplink transmission and then each BS estimates the channel information of its users.

13.3.1.1 Channel Estimation in SISO System in Flat Fading Environment

In single input-single output (SISO) system, both UE and BS are equipped with single antenna. In flat fading environment, the channel h between UE and BS is modeled as Rayleigh distributed with mean zero and unit variance. The channel h is estimated by transmitting the pilot symbol sequence s_i, $i = 1, 2, \ldots, N_P$. The system model is shown in Figure 13.1. The received signal at the BS is given by,

$$y_i = \sqrt{E_s} h s_i + n_i, \ i = 1, 2, \ldots, N_P, \tag{13.1}$$

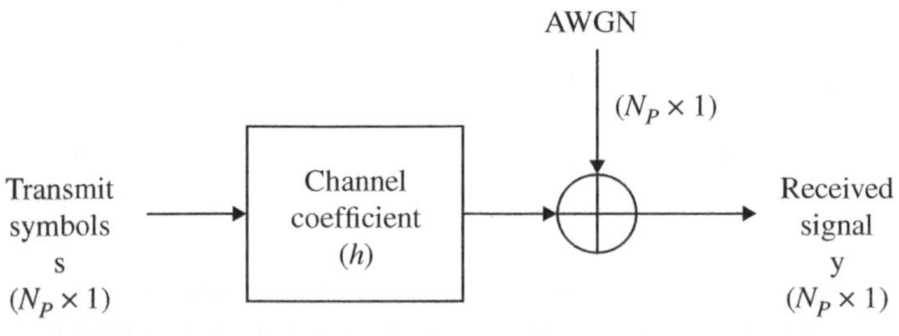

Figure 13.1 SISO system.

where E_s is the energy of the pilot symbol and n_i is modeled as additive white Gaussian noise (AWGN) with mean zero and variance N_0. The characteristic of channel h does not change during the training and actual data transmission. Assuming that the pilot sequence s_i, $i = 1, 2, \ldots, N_P$ is known at the receiver and the channel h is estimated at the BS.

The received signal in Eq. (13.1) is written in vector form as,

$$\mathbf{y} = \sqrt{E_s} h\mathbf{s} + \mathbf{n}, \tag{13.2}$$

where $\mathbf{y}$ is $N_P \times 1$ received signal vector, $\mathbf{s}$ is $N_P \times 1$ pilot sequence vector, and $\mathbf{n}$ is $N_P \times 1$ noise vector.

The LS error criterion is $J_{\text{LS}}(h) = \left(\mathbf{y} - \sqrt{E_s} h\mathbf{s}\right)^H \left(\mathbf{y} - \sqrt{E_s} h\mathbf{s}\right)$. The minimization of LS error leads to the linear LS estimate of channel h. It is given by,

$$\hat{h}_{\text{LS}} = \left(\mathbf{s}^H \mathbf{s}\right)^{-1} \mathbf{s}^H \mathbf{y} \tag{13.3}$$

The minimum LS error is given by, $J_{\text{LS,min}}(h) = \mathbf{y}^H \left(\mathbf{y} - \sqrt{E_s} \hat{h}\mathbf{s}\right)$.

The linear MMSE estimate of channel h is given by

$$\hat{h}_{\text{MMSE}} = \mathbf{c}_{hy} \mathbf{C}_{yy}^{-1} \mathbf{y}, \tag{13.4}$$

where $\mathbf{c}_{hy} = E\left[h\mathbf{y}^{\text{T}}\right]$ is $1 \times N_P$ the cross-correlation vector between channel h and the received signal vector $\mathbf{y}$. The $N_P \times N_P$ auto-correlation matrix $\mathbf{C}_{yy}$ of the received signal vector $\mathbf{y}$ is given by,

$$\mathbf{C}_{yy} = E\left[\mathbf{y}\mathbf{y}^{\text{T}}\right] = \left[\mathbf{R}_{SS} + N_0 \mathbf{I}_{N_P}\right]^{-1}, \tag{13.5}$$

where $\mathbf{I}_{N_P}$ is the $N_P \times N_P$ identity matrix and $\mathbf{R}_{SS} = E\left[\mathbf{s}\mathbf{s}^H\right]$. Substituting the values of $\mathbf{c}_{hy}$ and $\mathbf{C}_{yy}$ in Eq. (13.4), the MMSE estimate of the channel is given by,

$$\hat{h}_{\text{MMSE}} = \mathbf{s}^H \left[\mathbf{R}_{SS} + N_0 \mathbf{I}_{N_P}\right]^{-1} \mathbf{y}. \tag{13.6}$$

The NMSE performance of both LS and MMSE estimators in the estimation of channel is shown in Figure 13.2 for the pilot sequence lengths $N_P = 10$ and $N_P = 100$. It is observed that, for the pilot sequence length of $N_P = 10$, the SNR required for both LS and MMSE error of 10^{-3} is ~20 dB. The SNR requirement is reduced to ~10 dB when the pilot sequence length is increased to $N_P = 100$. When the length of the pilot sequence N_P is less, it is clearly observed that MMSE estimator outperforms LS estimator in low SNR region. However, with higher number of samples $N_P = 100$, both estimators have similar NMSE performance in channel estimation for SISO configuration.

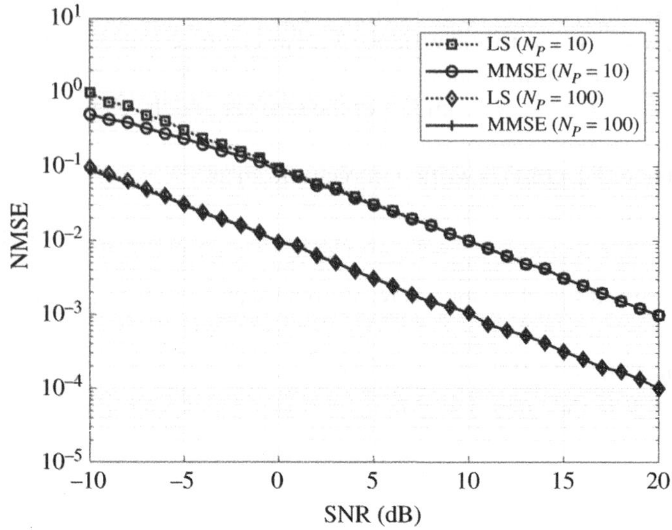

Figure 13.2 NMSE performance of LS and MMSE channel estimation in SISO configuration.

13.3.1.2 Channel Estimation in Single Input-Multiple Output (SIMO) System with Single UE

The single input-multiple output (SIMO) system model with N_{BS} antennas at BS and single UE equipped with single antenna is shown in Figure 13.3. In flat fading environment, the channels between the UE and antennas at BS are modeled as $N_{\mathrm{BS}} \times 1$ channel vector $\mathbf{h} = \left[h_1, h_2, \dots, h_{N_{BS}}\right]^{\mathrm{T}}$. The channel vector is estimated by transmitting pilot symbols s_i, $i = 1, 2, \dots, N_P$. For each pilot symbol, the $N_{\mathrm{BS}} \times 1$ received signal vector at the BS is given by,

$$\mathbf{y}_i^{(p)} = \sqrt{E_s}\mathbf{h}^{(p)} s_i + \mathbf{n}_i^{(p)}, \quad i = 1, 2, \dots, N_P, \; p = 1, 2 \dots, N_{\mathrm{BS}}. \tag{13.7}$$

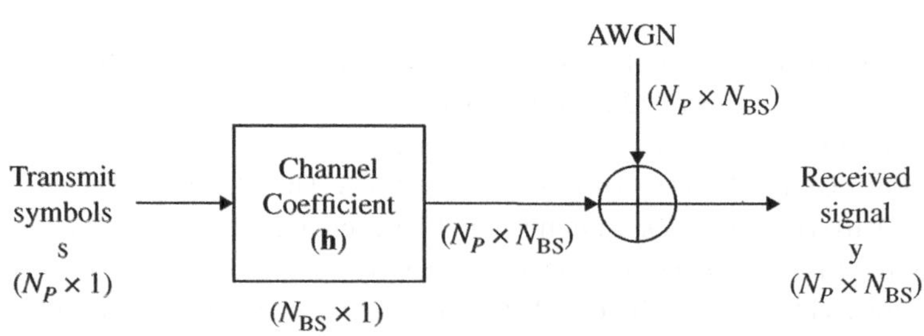

Figure 13.3 SIMO system with single UE.

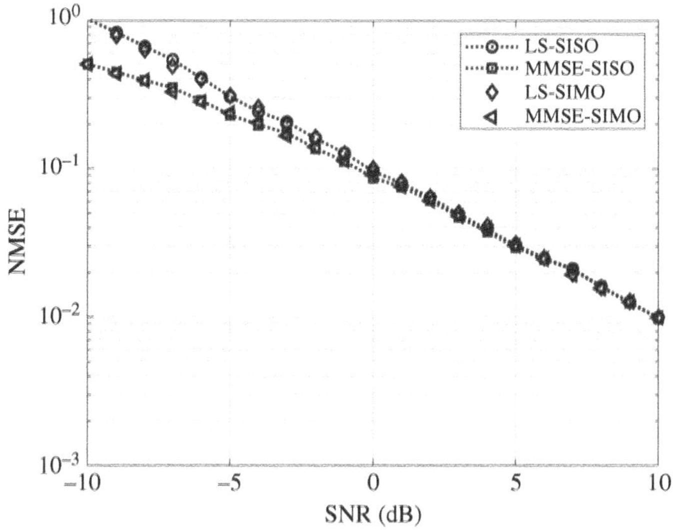

Figure 13.4 NMSE performance of LS and MMSE estimators in SIMO system with $N_{\text{BS}} = 4$ antennas at BS.

The received signal vectors at each observation are stacked at the BS and the resultant $N_P \times N_{\text{BS}}$ matrix is given by

$$\mathbf{Y} = \left[\mathbf{y}^{(1)}, \mathbf{y}^{(2)}, \ldots, \mathbf{y}^{(N_P)}\right]. \tag{13.8}$$

The size of the pilot sequence vector $\mathbf{s}$ is $N_P \times 1$. The LS estimate of the $N_{\text{BS}} \times 1$ channel vector $\mathbf{h}$ is given by

$$\hat{\mathbf{h}}_{\text{LS}} = \left(\mathbf{s}^H\mathbf{s}\right)^{-1}\mathbf{s}^H\mathbf{Y}. \tag{13.9}$$

The MMSE estimate of the $N_{\text{BS}} \times 1$ channel vector $\mathbf{h}$ is given by

$$\hat{\mathbf{h}}_{\text{MMSE}} = \mathbf{s}^H\left[\mathbf{R}_{SS} + N_0\mathbf{I}_{N_P}\right]^{-1}\mathbf{Y}. \tag{13.10}$$

The NMSE performance of both LS and MMSE estimators in the SIMO system with single UE and N_{BS} antennas at the BS is shown in Figure 13.4 for $N_P = 10$ and $N_{\text{BS}} = 4$. In the SIMO configuration also, it is observed that the MMSE estimate outperforms the LS estimate in low SNR regions and the NMSE performance converges at high SNR levels.

13.3.1.3 Channel Estimation in Single Input-Multiple Output (SIMO) System with *K* Single Antenna UEs

The SIMO system model with N_{BS} antennas at BS and K single antenna UEs is shown in Figure 13.5. The $K \times N_{\text{BS}}$ channel matrix between K single antenna UEs

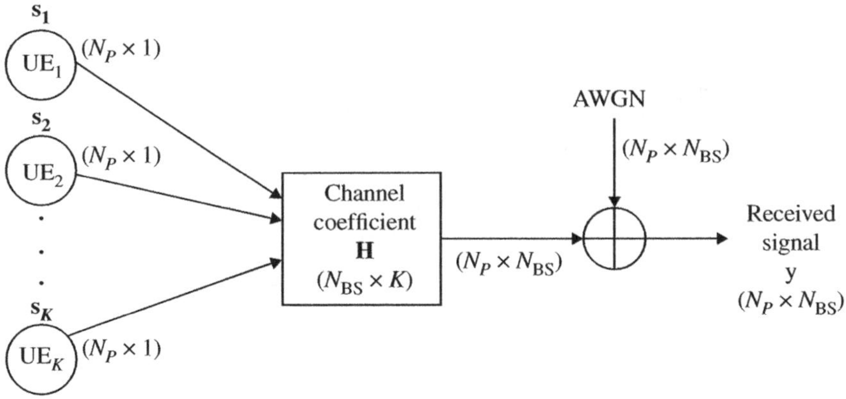

Figure 13.5 SIMO system model with N_{BS} antennas at BS and K single antenna UEs.

and N_{BS} antennas at BS is given by,

$$\mathbf{H} = \begin{bmatrix} h_1^1 & h_2^1 & \cdots & h_K^1 \\ h_1^2 & h_2^2 & \cdots & h_K^2 \\ \vdots & \vdots & \vdots & \vdots \\ h_1^{N_{BS}} & h_2^{N_{BS}} & \cdots & h_K^{N_{BS}} \end{bmatrix}. \tag{13.11}$$

The coefficients of the $N_{BS} \times K$ channel matrix $\mathbf{H}$ can be estimated at the BS, if the $N_P \times 1$ pilot sequence vectors $\mathbf{s}_1, \mathbf{s}_2, \ldots, \mathbf{s}_K$ of the K UEs are orthogonal to each other. The orthogonal pilot sequences are generated using Hadamard matrix since the dot product between any two distinct rows or distinct columns is zero. The $N_P \times K$ pilot symbol matrix is $\mathbf{S} = [\mathbf{s}_1, \mathbf{s}_2, \ldots, \mathbf{s}_K]$.

The received signal at pth antenna port of BS is given by,

$$y_i^p = \sqrt{E_s}\sum_{k=1}^{K} h_k^p s_{k,i} + n_i^p, \quad p = 1, 2, \ldots, N_{BS};\ i = 1, 2, \ldots, N_P. \tag{13.12}$$

The $N_{BS} \times 1$ received signal vector at the BS is given by,

$$\mathbf{y}_{BS,i} = \sqrt{E_s}\mathbf{HS}(k, :) + \mathbf{w}_{N_{BS},i},\ i = 1, 2, \ldots, N_P. \tag{13.13}$$

The received signal vector is stacked together in $N_P \times N_{BS}$ matrix form as,

$$\mathbf{Y}_{BS} = \left[\mathbf{y}_{BS,1}^T, \mathbf{y}_{BS,2}^T, \ldots, \mathbf{y}_{BS,N_P}^T\right]. \tag{13.14}$$

The LS estimate of $1 \times N_{BS}$ channel vector for kth user is given by,

$$\hat{\mathbf{h}}_{k,LS} = \left(\mathbf{s}_k^H \mathbf{s}_k\right)^{-1} \mathbf{s}_k^H \mathbf{Y}_{BS},\ k = 1, 2, \ldots, K. \tag{13.15}$$

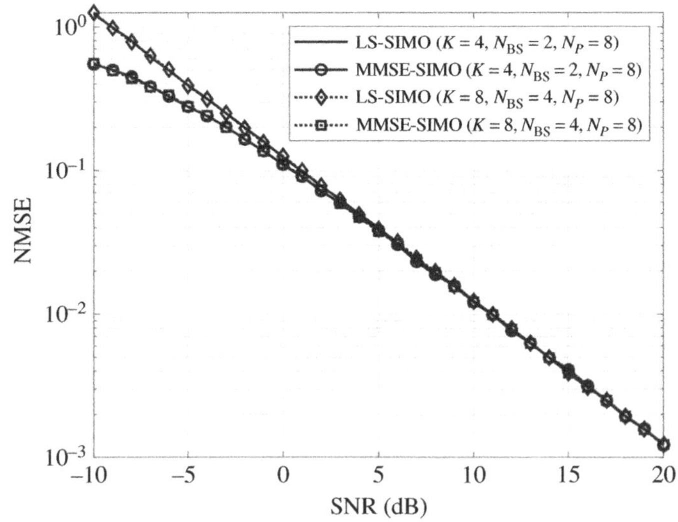

Figure 13.6 NMSE performance of LS and MMSE estimators for different N_{BS}, and K by fixing $N_P = 8$.

The MMSE estimate of $1 \times N_{BS}$ channel vector for kth user is given by,

$$\hat{\mathbf{h}}_{k,\text{MMSE}} = \mathbf{s}_k \left[\mathbf{R}_{s_k s_k} + N_0 \mathbf{I}_{N_P}\right]^{-1} \mathbf{Y}_{\text{BS}},\ k = 1, 2 \ldots, K. \tag{13.16}$$

The NMSE for LS and MMSE algorithm with ($K = 4, N_{BS} = 2, N_P = 8$) and ($K = 8, N_{BS} = 4, N_P = 8$) is shown in Figure 13.6. Similarly the NMSE for LS and MMSE with ($K = 4, N_{BS} = 8, N_P = 8$) and ($K = 4, N_{BS} = 32, N_P = 8$) is shown in Figure 13.7. In both figures, the number of observation N_P is fixed. It is observed that MMSE algorithm outperforms the LS method in low SNR regions, and their performance converges at high SNR regions in the SIMO with multiple UE configuration. Since the channel impulse response is independent and identically distributed (i.i.d.), the LS and MMSE produce the same performance for different N_{BS} and K configurations.

The NMSE for LS and MMSE algorithm with ($K = 4, N_{BS} = 8, N_P = 8$) and ($K = 4, N_{BS} = 8, N_P = 16$) is shown in Figure 13.8. It is observed that if the size of the pilot symbol vector is doubled, ~3 dB improvement in SNR is achieved for both LS and MMSE estimators.

13.3.1.4 Channel Estimation in Multi-Cell Massive MIMO System in Flat Fading Environment

Consider a multi-cell massive MIMO system as in Figure 13.9 with C cells, K single antenna UE and a BS with N_{BS} antennas at each cell. Let h_{jlkp} represent the

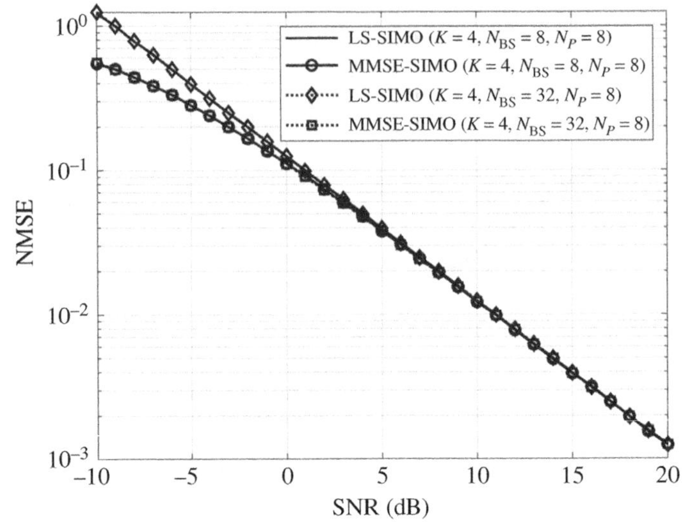

Figure 13.7 NMSE performance of LS and MMSE estimate of channel in SIMO system with $N_{BS} = 8{,}32$ and $K = 4, 8$ and $N_p = 8$.

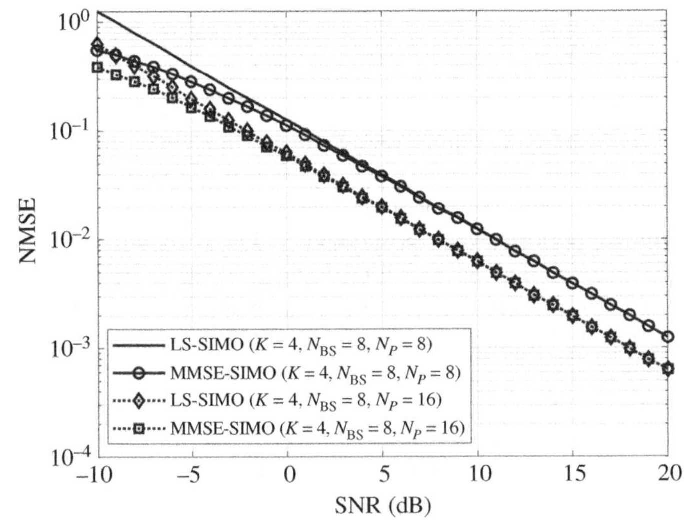

Figure 13.8 Comparison of LS and MMSE for different observation interval N_p by fixing K and N_{BS}.

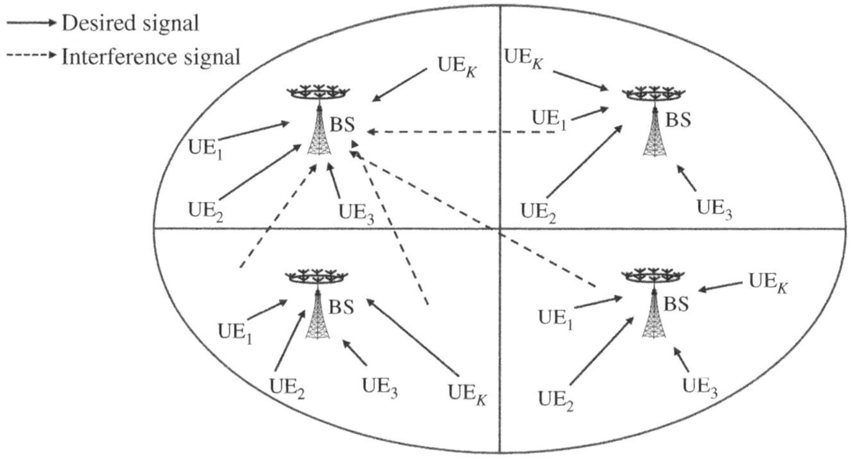

Figure 13.9 Multi-cell massive MIMO system.

channel between user k in cell l to the pth antenna of the BS in cell j in flat fading environment. The channel coefficients are modeled as i.i.d. Rayleigh distributed with zero mean and unit variance. The overall channel between jth BS and K UEs in lth cell is represented by $N_{\text{BS}} \times K$ channel matrix and denoted by $\mathbf{H}_{jl}$. In this channel matrix, the kth column $\mathbf{h}_{jlk} = [h_{jlk1}, \dots, h_{jlkN_{\text{BS}}}]^{\text{T}}$ represents the gains of the channel from user k in cell l to BS j. For example, for $j = 1, l = 1$, the $N_{\text{BS}} \times K$ channel matrix is given by,

$$\mathbf{H}_{11} = \begin{bmatrix} h_{1111} & h_{1121} & \cdots & h_{11K1} \\ h_{1112} & h_{1122} & \cdots & h_{11K2} \\ \vdots & \vdots & \vdots & \vdots \\ h_{111N_{\text{BS}}} & h_{112N_{\text{BS}}} & \cdots & h_{111KN_{\text{BS}}} \end{bmatrix}. \tag{13.17}$$

The received signal at pth antenna of the BS in the jth cell is represented by

$$y_{ji}^{p} = \sqrt{E_s} \sum_{l=1}^{C} \sum_{k=1}^{K} h_{jlk}^{p} s_{k,i} + n_i^p, \quad p = 1, 2, \dots, N_{\text{BS}},\ i = 1, 2, \cdots, N_P, j = 1, \dots, C. \tag{13.18}$$

The $N_P \times K$ pilot symbol matrix is $\mathbf{S} = \left[\mathbf{s}_1, \mathbf{s}_2, \dots, \mathbf{s}_K\right]$. In the matrix form, the $N_{\text{BS}} \times N_P$ received signal matrix at jth BS is given by,

$$\mathbf{Y}_j = \sqrt{E_s} \mathbf{H}_j^H \mathbf{S} + \mathbf{W}_j, j = 1, 2, \dots, C, \tag{13.19}$$

where $\boldsymbol{W}_j$ is $N_{\text{BS}} \times N_P$ noise matrix with i.i.d. elements with $\mathcal{CN}(0, 1)$.

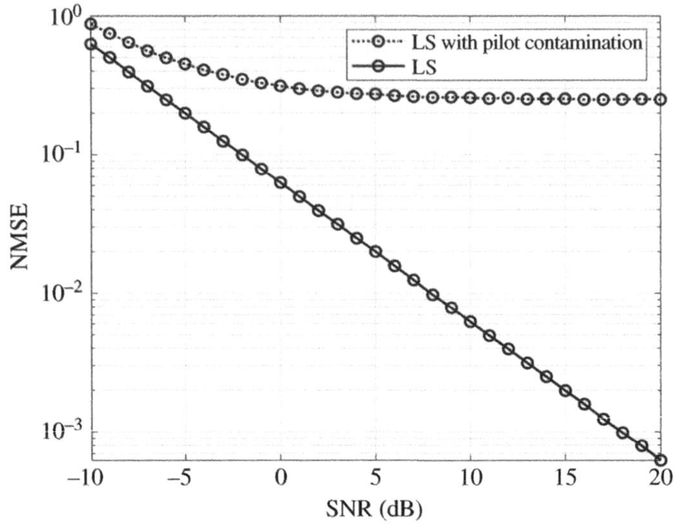

Figure 13.10 NMSE performance of LS estimator in multiuser pilot contamination in massive MIMO system.

The LS estimate of $1 \times N_{\text{BS}}$ channel vector for kth user is given by,

$$\hat{\mathbf{h}}_{k,\text{LS}} = \left(\mathbf{s}_k^H \mathbf{s}_k\right)^{-1} \mathbf{s}_k^H \mathbf{Y}_j,\ k = 1, 2, \ldots, K. \tag{13.20}$$

The MMSE estimate of $1 \times N_{\text{BS}}$ channel vector for kth user is given by,

$$\hat{\mathbf{h}}_{k,\text{MMSE}} = \mathbf{s}_k \left[\mathbf{R}_{\mathbf{s}_k\mathbf{s}_k} + N_0 \mathbf{I}_{N_P}\right]^{-1} \mathbf{Y}_j,\ k = 1, 2 \ldots, K. \tag{13.21}$$

If the pilot symbols of K users are represented by a $N_P \times K$ matrix $\mathbf{S}$ with the orthogonality property $E\left[\mathbf{S}\mathbf{S}^H\right] = \mathbf{I}_{N_P}$, the NMSE error performance is much similar to single cell multiple user massive MIMO system in flat fading environment. However, the channel estimation error in multi-cell massive MIMO system becomes higher if there is interference from other cells due to the reuse of the pilot symbol sequence. This is shown in Figure 13.10.

13.3.1.5 Channel Estimation in Frequency-Selective Fading Environment – OFDM

The channel estimation in the SISO system with L significant paths is considered in this section. The channel coefficients in multipath environment is extended to frequency-selective fading environment with L tap channel coefficient is shown in Figure 13.11.

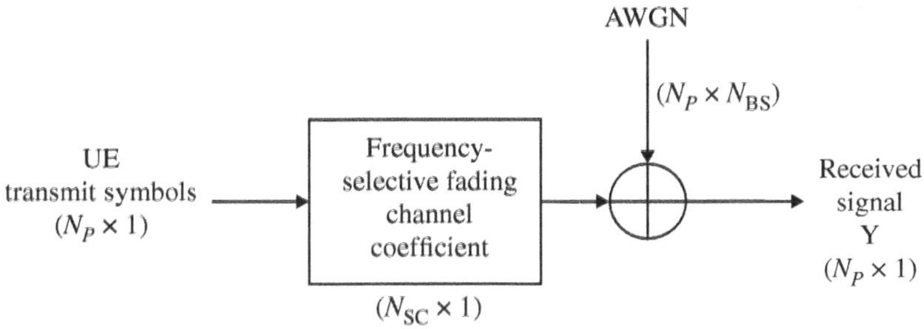

Figure 13.11 SISO system with frequency-selective fading environment.

Let $\sim$ be the variables in frequency domain. In the frequency domain, the received signal at the BS is given by,

$$\tilde{y}_i[l] = \sqrt{E_s}\tilde{h}[l]\tilde{s}[l] + \tilde{n}_i[l], \quad i = 1, 2, \ldots, N_P, \, l = 1, 2, \ldots, N_{SC}. \tag{13.22}$$

The $N_{SC} \times 1$ frequency domain received signal vector at the BS is given by,

$$\tilde{\mathbf{y}}_i = \sqrt{E_s}\widetilde{\mathbf{hs}} + \tilde{\mathbf{n}}_i, \, i = 1, 2, \ldots, N_P. \tag{13.23}$$

In the matrix form, the $N_P \times N_{SC}$ frequency domain received signal matrix at the BS is given by,

$$\widetilde{\mathbf{Y}} = \left[\widetilde{\mathbf{y}}_1^{\mathrm{T}}, \widetilde{\mathbf{y}}_2^{\mathrm{T}}, \ldots, \widetilde{\mathbf{y}}_{N_P}^{\mathrm{T}}\right]. \tag{13.24}$$

The LS estimate of the channel coefficients in frequency domain is given by

$$\hat{\mathbf{h}}_{\mathrm{LS}} = \left(\widetilde{\mathbf{s}}^H\widetilde{\mathbf{s}}\right)^{-1}\widetilde{\mathbf{s}}^H\widetilde{\mathbf{Y}}. \tag{13.25}$$

The MMSE estimate of channel coefficients in frequency domain is given by

$$\hat{\mathbf{h}}_{\mathrm{MMSE}} = \widetilde{\mathbf{s}}^H\left[\mathbf{R}_{\tilde{s}\tilde{s}} + N_0\mathbf{I}_{N_P}\right]^{-1}\widetilde{\mathbf{Y}}. \tag{13.26}$$

The comparison of LS and MMSE algorithms for the SISO system with frequency-selective fading environment by varying different pilot sequence length ($N_P = 32, N_P = 64$) is shown in Figure 13.12. Both algorithms exhibit better error performance when the pilot sequence length is increased. The frequency-selective fading environment is extended to SIMO system with multiple antennas at the BS and it is shown in Figure 13.13.

The received signal in frequency domain at the multiple antenna configuration of BS is given by,

$$\tilde{y}_i^p[l] = \sqrt{E_s}\tilde{h}^p[l]\tilde{s}[l] + \tilde{n}_i^p[l], \quad l = 1, 2, \ldots, N_{SC}, \, p = 1, 2, \ldots, N_{BS}, \; i = 1, 2, \ldots, N_P. \tag{13.27}$$

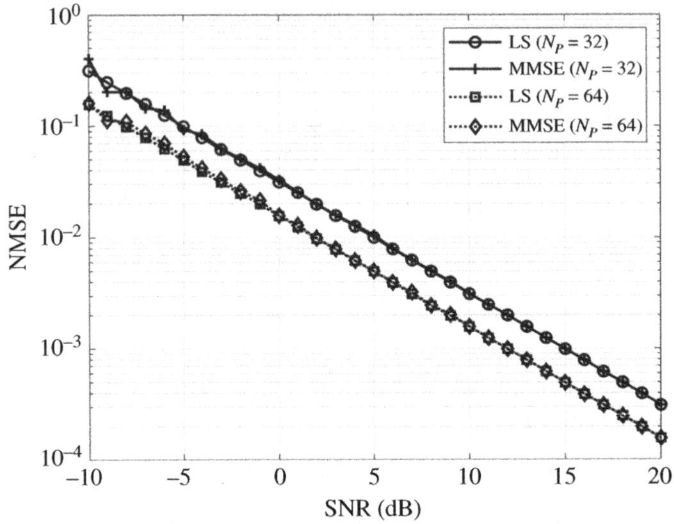

Figure 13.12 NMSE performance of LS and MMSE estimators in SISO system in frequency-selective fading environment.

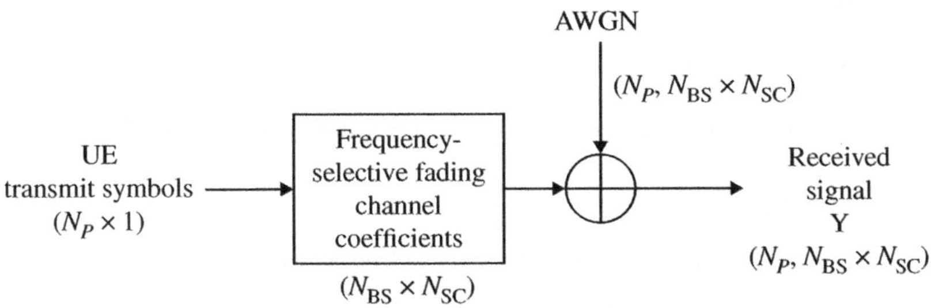

Figure 13.13 SIMO system in frequency-selective fading environment with single UE and multiple antennas at BS.

In the vector form, the received signal at the BS for SIMO system is given by,

$$\tilde{\mathbf{y}}_i^p = \sqrt{E_s}\tilde{\mathbf{h}}^p \odot \tilde{\mathbf{s}} + \tilde{\mathbf{n}}_i^p, \quad p = 1, 2, \ldots, N_{BS}, \; i = 1, 2, \ldots, N_P, \tag{13.28}$$

where $\odot$ denotes element by element multiplication.

In the matrix form, the $N_P \times N_{SC}$ received signal matrix at the pth port of the BS is given by,

$$\tilde{\mathbf{Y}}^p = \left[\tilde{\mathbf{y}}_1^{p^T}, \tilde{\mathbf{y}}_2^{p^T}, \ldots, \tilde{\mathbf{y}}_{N_P}^{p^T}\right], \; p = 1, 2, \ldots, N_{BS}. \tag{13.29}$$

The estimated channel vector in frequency domain at the BS for SIMO system using LS method is given by,

$$\hat{\mathbf{h}}_{\text{LS}}^{P}(l) = \frac{\widetilde{\mathbf{S}}^{H}\overline{\mathbf{Y}}^{p}(:,l)}{\widetilde{\mathbf{S}}^{H}\widetilde{\mathbf{S}}}, \; l = 1, 2, \ldots, N_{\text{SC}}, \; p = 1, 2, \ldots, N_{\text{BS}}. \tag{13.30}$$

The MMSE estimate of channel coefficients in frequency domain is given by

$$\widetilde{\mathbf{h}}_{\text{MMSE}}^{P}(l) = \widetilde{\mathbf{s}}^{H}\left[\mathbf{R}_{\tilde{s}\tilde{s}} + N_0\mathbf{I}_{N_P}\right]^{-1}\tilde{Y}^{p}(:,l), \quad l = 1, 2, \ldots, N_{\text{SC}}, \; p = 1, 2, \ldots N_{\text{BS}}. \tag{13.31}$$

The NMSE performance for LS and MMSE estimators in wideband channel with $K = 1$, $N_{\text{BS}} = 2$, $N_{\text{SC}} = 4$ and $N_P = 16, 32$ is shown in Figure 13.14. Both LS and MMSE estimator in SIMO configuration with 1 UE and 2 antennas at BS perform better for more observation interval. Compared to LS, MMSE performs better in low SNR region. The frequency-selective fading environment is further extended to K users and N_{BS} antennas at the BS are shown in Figure 13.15.

The received signal in frequency domain at the BS with K UEs is given by,

$$\tilde{y}_i^p[l] = \sqrt{E_s}\sum_{k=1}^{K}\tilde{h}_k^p[l]\,\tilde{s}_k[l] + \tilde{n}_i^p[l], \; l = 1, \ldots, N_{\text{SC}}, \; p = 1, \ldots, N_{\text{BS}}, \; i = 1, \ldots, N_P. \tag{13.32}$$

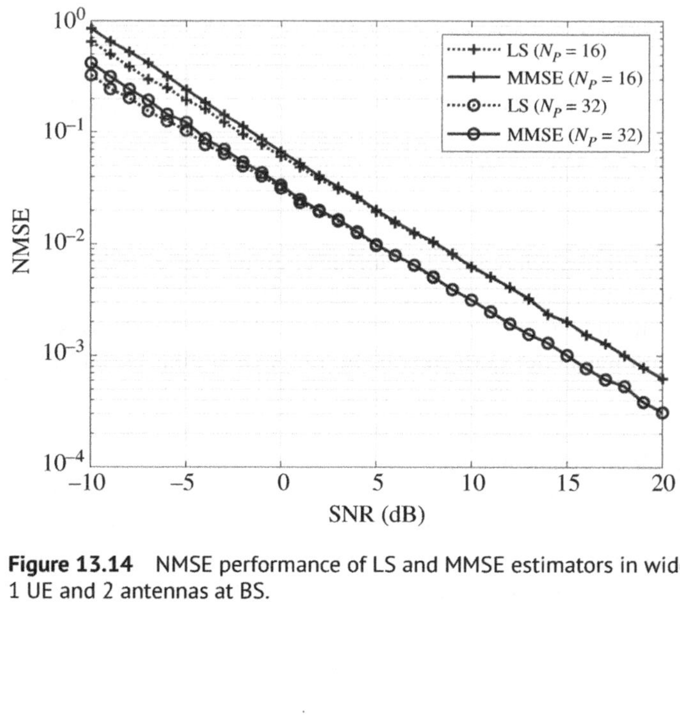

Figure 13.14 NMSE performance of LS and MMSE estimators in wideband channel with 1 UE and 2 antennas at BS.

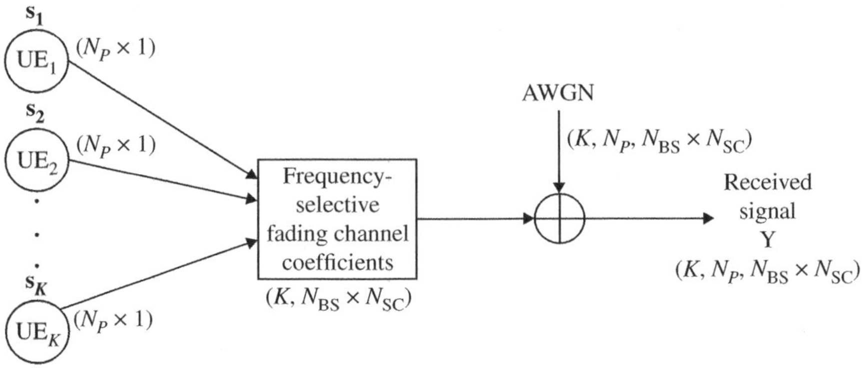

Figure 13.15 SIMO system with frequency-selective fading environment with K UEs and N_{BS} antennas at BS.

In the vector form, the received signal is given by,

$$\widetilde{\mathbf{y}}_i^p = \sqrt{E_s}\sum_{k=1}^{K}\widetilde{\mathbf{h}}_k^p\widetilde{\mathbf{s}}_{k,i} + \widetilde{\mathbf{n}}_i^p, \quad p = 1,\ldots,N_{BS},\ i = 1,\ldots,N_P. \tag{13.33}$$

The estimated channel coefficient between kth UE and pth antenna port at lth subcarrier at the BS for SIMO system using LS method is given by,

$$\widetilde{\mathbf{h}}_{k,LS}^P(l) = \frac{\widetilde{\mathbf{s}}_k^T\widetilde{\mathbf{Y}}^p(:,l)}{\widetilde{\mathbf{s}}_k^T\widetilde{\mathbf{s}}_k}, \quad l = 1,2,\ldots,N_{SC},\ p = 1,2,\ldots,N_{BS},\ k = 1,2,\ldots,K. \tag{13.34}$$

The MMSE estimate of channel coefficients in frequency domain is given by

$$\widetilde{\mathbf{h}}_{MMSE}^P(l) = \widetilde{\mathbf{s}}^H\left[\mathbf{R}_{\widetilde{s}\widetilde{s}} + N_0\mathbf{I}_{N_P}\right]^{-1}\widetilde{\mathbf{Y}}^p(:,l), \quad l = 1,2,\ldots,N_{SC},\ p = 1,2,\ldots,N_{BS}. \tag{13.35}$$

LS and MMSE error in frequency-selective fading channel with $K = 2$, $N_{BS} = 2$ for different N_p is shown in Figure 13.16. It is observed that both LS and MMSE estimator give SNR improvement for every doubling of pilot sequence observation. MMSE algorithm has better performance in low SNR similar to flat fading environment.

13.3.1.6 Impact of CU on Uplink Sum Rate

The effect of CU on the uplink sum rate performance of the massive MIMO system is examined in this section. The effective massive MIMO BS should have more antennas (N_{BS}) than the number of users (K). The uplink side's channel estimate is simpler than the downlink side's due to the reduced number of antennas.

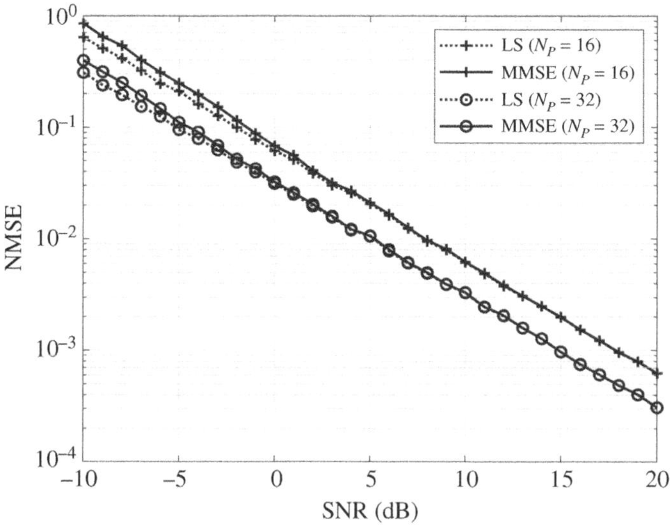

Figure 13.16 NMSE error in frequency-selective fading channel with 2 UE and 2 antennas at BS for different pilot sequence observations N_p.

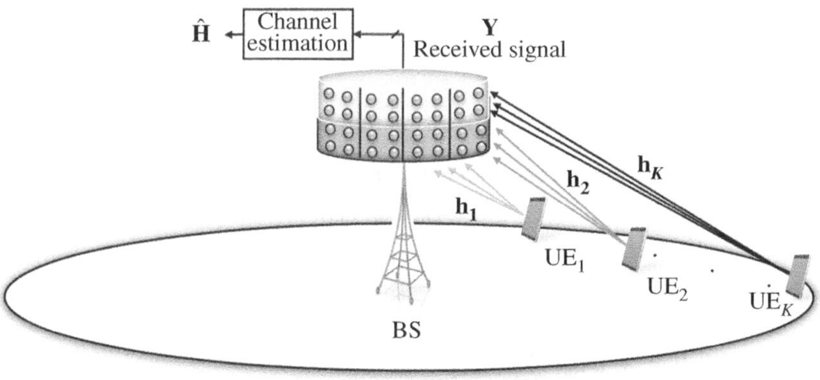

Figure 13.17 Uplink channel estimation of a massive MIMO system.

Since TDD system uses the same frequency channels for uplink and downlink, the uplink channel can be estimated. The channel for the downlink can be identified with the channel reciprocity property. The uplink model of channel estimation in a massive MIMO system is depicted in Figure 13.17.

Assume K number of single-antenna uplink users in a massive MIMO system. Hence, $N_p \geq K$ pilot transmissions are required for channel estimation. The pilot power is

$$P_p = KP_u, \tag{13.36}$$

where P_u is the user power per transmission. The pilot matrix $(\mathbf{S})_{K\times K}$ should be an orthogonal matrix, that is,

$$\mathbf{SS}^H = \mathbf{I}. \tag{13.37}$$

The received pilot matrix at massive MIMO BS is

$$\mathbf{Y}_{N_{\text{BS}}\times K} = \mathbf{H}_{N_{\text{BS}}\times K}\sqrt{P_p}\,\mathbf{S}_{K\times K} + \mathbf{N}_{N_{\text{BS}}\times K}, \tag{13.38}$$

where $\mathbf{N}$ is the noise matrix with elements i.i.d. and follows $\mathcal{CN}(0,1)$. $\mathbf{H}$ is $N_{\text{BS}} \times K$ channel matrix between K users and BS. The LS channel estimate is given by

$$\hat{\mathbf{H}} = \mathbf{Y}\frac{1}{\sqrt{P_p}}\mathbf{S}^{\mathbf{H}}. \tag{13.39}$$

Substituting (13.38) into (13.39),

$$\hat{\mathbf{H}} = \left(\mathbf{H}\sqrt{P_p}\,\mathbf{S} + \mathbf{N}\right)\frac{1}{\sqrt{P_p}}\mathbf{S}^{\mathbf{H}}. \tag{13.40}$$

Substituting (13.37) in (13.40),

$$\hat{\mathbf{H}} = \mathbf{H} + \frac{1}{\sqrt{P_p}}\mathbf{N}\,\mathbf{S}^{\mathbf{H}} = \mathbf{H} + \mathbf{E}. \tag{13.41}$$

Let $\mathbf{E} = \frac{1}{\sqrt{P_p}}\mathbf{NS}^{\mathbf{H}}$ is the channel estimation error and its elements follow $\mathcal{CN}\left(0, \frac{1}{KP_u}\right)$. The estimated channel of the qth user is $\hat{\mathbf{h}}_q = \mathbf{h}_q + \mathbf{e}_q$ which is i.i.d. Gaussian having variance $\left(\beta_q + \frac{1}{KP_u}\right)$. Here β_q is the large-scale fading coefficient associated with qth user and BS. After estimating the channel, the users transmit actual data. The signal received at the BS is

$$\mathbf{y} = \sqrt{P_u}\mathbf{h}_q x_q + \sqrt{P_u}\sum_{k=1,k\neq q}^{K}\mathbf{h}_k x_k + \mathbf{n}. \tag{13.42}$$

At the massive MIMO BS, estimated channel is used for matched filter-based beamforming or MRC. For example, to detect qth user,

$$r_q = \hat{\mathbf{h}}_q^H\mathbf{y}. \tag{13.43}$$

Substituting (13.42) into (13.43)

$$r_q = \hat{\mathbf{h}}_q^H\left(\sqrt{P_u}\mathbf{h}_q x_q + \sqrt{P_u}\sum_{k=1,k\neq q}^{K}\mathbf{h}_k x_k + \mathbf{n}\right). \tag{13.44}$$

Expression (13.44) can be represented as

$$r_q = \sqrt{P_u}\hat{\mathbf{h}}_q^H \mathbf{h}_q x_q + \sqrt{P_u} \sum_{k=1,k\neq q}^{K} \hat{\mathbf{h}}_q^H \mathbf{h}_k x_k + \hat{\mathbf{h}}_q^H \mathbf{n}. \tag{13.45}$$

Substituting $\hat{\mathbf{h}}_q = \mathbf{h}_q + \mathbf{e}_q$ in (13.45)

$$r_q = \sqrt{P_u}\|\mathbf{h}_q\|^2 x_q + \sqrt{P_u}\mathbf{e}_q^H \mathbf{h}_q x_q + \sqrt{P_u} \sum_{k=1,k\neq q}^{K} \hat{\mathbf{h}}_q^H \mathbf{h}_k x_k + \hat{\mathbf{h}}_q^H \mathbf{n}. \tag{13.46}$$

The last term in (13.46) $\hat{\mathbf{h}}_q^H \mathbf{n} \sim \mathcal{CN}\left(0, \|\hat{\mathbf{h}}_q\|^2\right)$. The signal-to-interference plus noise ratio (SINR) of qth user is

$$\gamma_q = \frac{P_u\|\mathbf{h}_q\|^4}{P_u E\{|\mathbf{e}_q{}^H \mathbf{h}_q|^2\} + P_u \sum_{k=1,k\neq q}^{K} E|\hat{\mathbf{h}}_q^H \mathbf{h}_k|^2 + \|\hat{\boldsymbol{h}}_q\|^2}. \tag{13.47}$$

Dividing the numerator and the denominator of (13.47) by $\|\mathbf{h}_q\|^2$ results in

$$\gamma_q = \frac{P_u\|\mathbf{h}_q\|^2}{\frac{1}{K} + P_u \sum_{k=1,k\neq q}^{K} E\left\{\frac{|\hat{\mathbf{h}}_q^H \mathbf{h}_k|^2}{\|\mathbf{h}_k\|^2}\right\} + \frac{\|\hat{\mathbf{h}}_q\|^2}{\|\mathbf{h}_q\|^2}}. \tag{13.48}$$

Modifying (13.48) as

$$\gamma_q = \frac{P_u\|\mathbf{h}_q\|^2}{\frac{1}{K} + P_u \sum_{k=1,k\neq q}^{K} \frac{E\left\{\left|\frac{\hat{\mathbf{h}}_q^H \mathbf{h}_k}{\sqrt{N_{BS}}}\right|^2\right\}}{\frac{\|\mathbf{h}_q\|^2}{N_{BS}}} + \frac{\frac{\|\hat{\mathbf{h}}_q\|^2}{N_{BS}}}{\frac{\|\mathbf{h}_q\|^2}{N_{BS}}}}. \tag{13.49}$$

For larger values of N_{BS}, as per law of large numbers,

$$\frac{\hat{\mathbf{h}}_q^H \mathbf{h}_k}{\sqrt{N_{\text{BS}}}} \to \mathcal{CN}\left(0, \left(\beta_q + \frac{1}{KP_u}\right)\beta_k\right), \tag{13.50}$$

$$\frac{\|\mathbf{h}_q\|^2}{N_{\text{BS}}} \to \beta_q, \tag{13.51}$$

$$\frac{\|\mathbf{h}_q\|^2}{N_{\text{BS}}} \to \left(\beta_q + \frac{1}{KP_u}\right). \tag{13.52}$$

Substituting (13.50)–(13.52) in (13.49)

$$\gamma_q = \frac{P_u\|\mathbf{h}_q\|^2}{\frac{1}{K} + P_u \sum_{k=1,k\neq q}^{K} \frac{\left(\beta_q + \frac{1}{KP_u}\right)\beta_k}{\beta_q} + \frac{\left(\beta_q + \frac{1}{KP_u}\right)}{\beta_q}}. \tag{13.53}$$

Considering the power scaling $P_u = \frac{E_u}{N_{\text{BS}}}$, the SINR becomes

$$\gamma_q = \frac{E_u \frac{\|\mathbf{h}_q\|^2}{N_{\text{BS}}}}{\frac{1}{K} + \frac{E_u}{N_{\text{BS}}} \sum_{k=1,k\neq q}^{K} \frac{\left(\beta_q + \frac{N_{\text{BS}}}{KE_u}\right)\beta_k}{\beta_q} + \left(\frac{\beta_q + \frac{N_{\text{BS}}}{KE_u}}{\beta_q}\right)}. \tag{13.54}$$

When the number of BS antennas is very high, that is, $N_{\text{BS}} \to \infty$, the second term in the denominator, that is, $\frac{E_u}{N_{\text{BS}}} \sum_{k=1,k\neq q}^{K} \frac{\left(\beta_q + \frac{N_{\text{BS}}}{KE_u}\right)\beta_k}{\beta_q} \to 0$ and the third term $\left(\frac{\beta_q + \frac{N_{\text{BS}}}{KE_u}}{\beta_q}\right) \to \infty$. Hence, $\gamma_q \to 0$. This is due to the increase of channel estimation error with the number of BS antennas. Considering the power scaling $P_u = \frac{E_u}{\sqrt{N_{\text{BS}}}}$, the SINR becomes

$$\gamma_q = \frac{E_u \frac{\|\mathbf{h}_q\|^2}{\sqrt{N_{\text{BS}}}}}{\frac{1}{K} + \frac{E_u}{\sqrt{N_{\text{BS}}}} \sum_{k=1,k\neq q}^{K} \frac{\left(\beta_q + \frac{\sqrt{N_{\text{BS}}}}{KE_u}\right)\beta_k}{\beta_q} + \left(\frac{\beta_q + \frac{\sqrt{N_{\text{BS}}}}{KE_u}}{\beta_q}\right)}. \tag{13.55}$$

Simplifying (13.55) gives

$$\gamma_q = \frac{E_u \frac{\|\mathbf{h}_q\|^2}{\sqrt{N_{\text{BS}}}}}{\frac{1}{K} + \sum_{k=1,k\neq q}^{K} \frac{\beta_k}{K\beta_q} + 1 + \frac{\sqrt{N_{\text{BS}}}}{KE_u\beta_q}}. \tag{13.56}$$

For larger values of N_{BS}, the term $\frac{\sqrt{N_{\text{BS}}}}{KE_u\beta_q}$ dominates. Hence

$$\gamma_q \approx \frac{E_u \frac{\|\mathbf{h}_q\|^2}{\sqrt{N_{\text{BS}}}}}{\frac{\sqrt{N_{\text{BS}}}}{KE_u\beta_q}} = E_u^2 \frac{\|\mathbf{h}_q\|^2}{N_{\text{BS}}} K\beta_q. \tag{13.57}$$

As per (13.51)

$$\gamma_q \to KE_u^2\beta_q^2. \tag{13.58}$$

This is the upper bound of SINR achieved by qth user when there are very large number of BS antennas. The uplink rate achievable by the qth user of massive MIMO system is

$$R_q = \log_2\left(1 + \gamma_q\right). \tag{13.59}$$

The uplink sum rate of massive MIMO system is

$$R = \sum_{k=1}^{K} R_k. \tag{13.60}$$

13.4 AI and ML-Based Approaches for Channel Estimation in Massive MIMO System

The integration of AI and ML into channel estimation for massive MIMO systems marks a significant advancement in the field of wireless communications. Massive MIMO systems employ a large number of antennas at the BS, promising significant improvements in capacity and energy efficiency. However, accurately estimating the complex channels between those antennas and a multitude of user devices is crucial for achieving these benefits. Traditional methods struggle with the high dimensionality and dynamic nature of these channels, making AI and ML-based approaches increasingly important. Here's how AI and ML shine in channel estimation for massive MIMO system.

- **Improved accuracy:** Conventional linear models may not accurately capture the complex characteristics of wireless channels in diverse environments. AI algorithms can learn complex relationships between pilot signals and channel response, leading to more accurate estimates than traditional methods. DL architectures can capture nonlinear patterns and spatial–temporal correlations in the channels, further enhancing accuracy.
- **Reduced complexity:** Massive MIMO systems involve a large number of antennas at both transmitter and receiver ends, leading to high-dimensional channel matrices. Traditional methods for channel estimation can become computationally intensive and less efficient in such scenarios. AI-based methods can often achieve comparable accuracy with simpler models compared to traditional approaches, reducing computational complexity at both the BS and user devices. This is particularly important for energy-constrained devices and real-time applications.
- **Faster adaptation:** Wireless channel conditions can vary rapidly due to mobility, fading, interference, and changing propagation conditions. AI algorithms can adapt to these changes in real time by learning from new data, leading to more robust and reliable communication. This is vital for dynamic environments like vehicular networks and high-speed communication.
- **Scalability:** As the number of users and antennas increases, traditional methods become cumbersome. AI algorithms can efficiently handle large-scale scenarios and provides a versatile tool for channel estimation across various deployment scenarios.

Here are some specific examples of AI and ML methods used for channel estimation in massive MIMO:

- **Deep neural networks:** CNNs and recurrent neural networks have been successfully applied to estimate the channel from pilot signals and other contextual information.

- **Graph neural networks:** These networks can effectively capture the spatial relationships between user devices and antennas, leading to improved channel estimates.
- **Bayesian learning:** This approach allows incorporating prior knowledge about the channel into the estimation process, further enhancing accuracy.
- DML with FL is a powerful approach that combines the principles of decentralized data processing with the collaborative aspects of ML. It is designed to address several challenges associated with traditional CML methods, particularly in terms of privacy, data security, and efficiency.

13.4.1 FL in Channel Estimation

Here is an explanation of how FL works in channel estimation.

- **Training data collection:** Each user device (e.g., smartphone, IoT sensor) collects local channel data based on received pilot signals. This data stays on the device, preserving privacy and reducing communication overhead.
- **Model training:** A shared model (e.g., a neural network) is initialized at a central server (e.g., BS). The server sends the model to participating devices. Each device trains the model on its local channel data. Model updates (e.g., weights, gradients) are sent back to the server, not raw data. The server aggregates these updates to improve the global model. This process repeats for multiple rounds, with the model becoming more accurate with each round.
- **Prediction:** The final trained model is used for channel estimation on new data at each device. Devices can locally estimate the channel conditions without relying on constant communication with the server.

13.4.2 Advantages and Disadvantages of FL

In this section, merits and demerits of FL are discussed.

13.4.2.1 Advantages

- **Privacy:** Raw data stays on devices, protecting user privacy.
- **Communication efficiency:** Only model updates are exchanged, reducing communication overhead, especially in large-scale networks.
- **Personalization:** Models can adapt to individual device characteristics and channel conditions.
- **Collaborative learning:** Devices learn from each other's data without sharing it directly.
- **Scalability:** Can handle large numbers of devices efficiently.

Table 13.1 Comparison of CML and DML.

S.no.	Feature	Federated learning (DML)	Machine learning (CML)
1.	Data storage	Local devices	Central server
2.	Communication	Model updates	Full data transfer
3.	Privacy	High	Low
4.	Scalability	High	Low
5.	Convergence speed	Slower	Faster
6.	Device requirements	Diverse hardware	Homogenous hardware
7.	Control	Limited	High

13.4.2.2 Disadvantages

- **Slow convergence:** Model training can be slower due to distributed nature and communication constraints.
- **Heterogeneity:** Devices may have varying computational capabilities and data quality, affecting model convergence.
- **Security:** Model updates need protection against adversarial attacks and data poisoning.
- **Limited control:** Central server has less direct control over training process compared to centralized learning.

In summary, FL offers a promising approach to channel estimation in wireless systems, balancing privacy, communication efficiency, and performance [31]. However, addressing its challenges is crucial for its successful deployment in real-world scenarios. A comparison of FL (DML) and existing ML (CML) model is provided in Table 13.1.

13.5 FL in Cell-Free Massive MIMO: An Overview

In massive MIMO system, ML has demonstrated superior spectral efficiency and reduced computational demands. This improvement is particularly evident in tasks such as channel estimation, hybrid beamforming, and determining the angle of arrival (AoA).

In the field of ML, the problem of channel estimation is often addressed by training a neural network with datasets collected by individual users. This training allows the network to learn a complex mapping from the input, typically the received pilot signals, to the output, which is the channel data. Historically, most research has focused on CML approaches, where all input–output data pairs are sent to the BS for the training of the model. After training, the BS distributes the model parameters to the users, enabling them to estimate the channel by inputting their received pilot data into the model. However, a significant drawback of this method is the substantial communication overhead incurred by transferring the entire dataset from the users to the BS.

Recently, to address the issue of high communication overhead associated with CML methods, FL has been introduced in DML. In FL systems, rather than transmitting the entire dataset, only the updates of the model, specifically the gradients of the model parameters, are sent. This significantly reduces the communication overhead involved.

13.5.1 FL-Based Model Training

In the context of training a model based on CML for channel estimation in massive MIMO systems, the BS executes the training of the overarching neural network. This process involves gathering datasets, denoted as $\{\mathbf{s}_k\}_{k\epsilon K}$, from each UE across the set K. In SIMO with K single antenna UE in flat fading environment, the $N_P \times K$ pilot symbol matrix is $\mathbf{S} = [\mathbf{s}_1, \mathbf{s}_2, \dots, \mathbf{s}_K]$. The received signal at pth antenna port of BS is given by,

$$y_i^p = \sqrt{E_s}\sum_{k=1}^{K} h_k^p s_{k,i} + n_i^p, \quad p = 1, 2, \dots, N_{\text{BS}};\ i = 1, 2, \dots, N_P. \tag{13.61}$$

In CML, $\mathcal{L}(\boldsymbol{\phi})$ is the loss function defined as,

$$\mathcal{L}(\boldsymbol{\phi}) = \frac{1}{KN_P}\sum_{i=1}^{KN_P} \left(f\left(\mathcal{S}_i | \boldsymbol{\phi}\right) - \mathcal{Y}_i\right)^2, \tag{13.62}$$

where $\mathcal{S} = vec(\mathbf{S})$ and $\mathcal{Y}_i$ is the $KN_P \times 1$ vector at each antenna in the BS. The training is performed at the BS by solving the optimization problem:

$$\begin{aligned} &\underset{\boldsymbol{\phi}}{\text{minimize}} \quad \mathcal{L}(\boldsymbol{\phi}) \\ &\text{subject to: } f\left(\mathcal{S}_i | \boldsymbol{\phi}\right) = \mathcal{Y}_i, i = 1, 2, \dots, KN_P. \end{aligned} \tag{13.63}$$

$\mathcal{S}^k$ is the local dataset at kth user, with input–output pairs $\left(\mathcal{S}_i^{(k)} \mathcal{Y}_i^{(k)}\right)$, $i = 1, 2, \dots, N_P$, $\boldsymbol{\phi}$ is the learnable parameters.

In DML, the training is performed at each UE. In SIMO with K single antenna UE in flat fading environment, the received signal at the kth user in downlink is given by,

$$y_{k,i} = \sqrt{E_s}\sum_{p=1}^{N_{\text{BS}}}\sum_{k=1}^{K} h_k^p s_{k,i} + n_{k,i}, \ i = 1, 2, \dots, N_P. \tag{13.64}$$

The optimization problem at each UE is defined as,

$$\begin{aligned} &\underset{\phi}{\text{minimize}} \ \overline{\mathcal{L}}(\phi) = \frac{1}{K}\sum_{k=1}^{K}\mathcal{L}_k(\phi) \\ &\text{subject to } f\left(S_i^{(k)} \middle| \phi\right) = \mathcal{Y}_i^{(k)}, i = 1, 2, \ldots, N_p, k\epsilon K, \end{aligned} \tag{13.65}$$

where $\mathcal{L}_k(\phi) = \frac{1}{N_p}\sum_{i=1}^{S_k}\left(f\left(S_i^{(k)}\middle|\phi\right) - \mathcal{Y}_i^{(k)}\right)^2$.

In order to solve both optimization problem, gradient descent (GD) is tied-up and the problem is solved iteratively. In CML, the gradient is updated with whole dataset as $g(\phi_t) = \nabla\mathcal{L}(\phi_t)$ at the BS and the update parameter is given by,

$$\phi_{t+1} = \phi_t - \mu g(\phi_t), \tag{13.66}$$

where μ is the step-size or learning rate.

In DML, each user update's its gradient individually as $g_k(\phi_t) = \nabla\mathcal{L}_k(\phi_t)$ to solve the optimization problem then sends it to the BS, the update parameter is given as,

$$\phi_{k,t+1} = \phi_t - \mu\frac{1}{K}\sum_{k=1}^{K}g_k(\phi_t). \tag{13.67}$$

13.6 Protection of Privacy of User Data in Cell-Free Massive MIMO Using FL

The rapid expansion in UEs and the use of multiple antennas at both BS and UEs pose significant challenges in channel estimation at the BS, particularly in the context of cell-free massive MIMO environments [32, 33]. These environments are characterized by their handling of immense data volumes, often reaching petabytes, within various resource-limited settings. DML emerges as a collaborative learning approach that addresses these challenges effectively, enabling learning without the need to share data with centralized servers.

However, DML is increasingly vulnerable to a range of emerging threats, particularly during the initial stages of training, learning, and inference. These threats, which include sophisticated attacks like backdoor and bit-flip, can disrupt or misdirect the learning models, compromising the trustworthiness and reliability of the outcomes for subsequent tasks.

Looking ahead, the future of DML lies in developing and implementing robust strategies for both attack prevention and defense, specifically within the realm of FL. This approach is crucial, especially when dealing with limited and varied non-i.i.d. data in nonstandard and adversarial environments. Such settings are particularly relevant to edge AI infrastructures.

To achieve these goals, it is essential to integrate unique features into FL algorithms. These include robust training methods, model restoration techniques, secure learning and inference protocols, and fault tolerance capabilities that can withstand failures and attacks. Moreover, the development of resilient, fair, and robust models is crucial. The ambition is not only to validate these advancements in classical nonstandard settings but also to apply them effectively in constrained environments like the IoT and robotic arms, where they can deliver significant benefits.

13.7 Results and Discussions

The performance of the proposed system model is validated by considering the propagation conditions specified in third-generation partnership project (3GPP) technical specification (TS) document 3GPP TS 38.101-4 version 16.3.0 Release 16 [34]. This establishes the minimum performance requirements for 5G-NR UE. The multipath propagation conditions consist of the following parts:

- A tapped delay-line (TDL) delay profile is defined by several taps placed at predetermined positions on a sampling grid. This profile is further described by its root mean square (r.m.s.) delay spread (DS) and the greatest delay covered by the taps.
- A set of channel model parameters encompasses the delay profile and the Doppler spectrum, which is defined by a traditional spectrum shape and the highest Doppler frequency.
- Different propagation models are considered for FR1 (below 6 GHz) and FR2 (above 6 GHz).

Multipath fading channel characteristics are described by its power delay profiles (PDP). The channel parameters, namely, mean excess delay, r.m.s. DS, and coherence bandwidth, are used to compare different multipath fading channels and provide guidelines in designing wireless communication systems. The PDPs chosen for FR1 are designed to exemplify environments with low, medium, and high r.m.s. DS of 30 ns, 100 ns, and 300 ns, respectively. The coherence bandwidth low, medium, and high r.m.s. DS are ~33.3 MHz, ~9.9 MHz, and ~3.3 MHz, respectively. The wireless channel modeled as frequency-selective fading when the coherence bandwidth is narrower than signal bandwidth and hence the channel impulse response has larger r.m.s. DS. The model parameters of low, medium, and high r.m.s. DS channels are listed in Table 13.2, and the TDL models, with 12 taps, are outlined in Tables 13.3–13.5. In non-line-of-sight environment, wireless channel is described by a Gaussian random variable using central limit theorem as the received signal is from the infinite number

Table 13.2 Delay profiles for NR channel models.

Model	Number of channel taps	Delay spread (r.m.s.) (ns)	Maximum excess tap delay (span) (ns)	Delay resolution (ns)
TDLA30	12	30	290	5
TDLB100	12	100	480	5
TDLC300	12	300	2595	5

Table 13.3 TDLA30 (DS = 30 ns).

Tap #	Delay (ns)	Power (dB)
1	0	−15.5
2	10	0
3	15	−5.1
4	20	−5.1
5	25	−9.6
6	50	−8.2
7	65	−13.1
8	75	−11.5
9	105	−11.0
10	135	−16.2
11	150	−16.6
12	290	−26.2

of scatters. Hence, the probability density function of the received signal in non-line-of-sight environment is modeled as Rayleigh distribution. Each tap in the PDPs is considered to be Rayleigh distributed. The NMSE in OFDM - SISO for TDL (30 ns, 100 ns, and 300 ns) NR channel models is shown in Figure 13.18. It is observed that for the three TDL NR channel models, both LS and MMSE channel estimator achieves similar performance. The pilot sequence observation is varied for $N_P = 8, 16$, and 32 and it is highlighted in Figure 13.18.

Monte-Carlo simulations are run with the parameters as in Table 13.6 to validate the effect of CU. Figure 13.19 depicts the uplink sum rate of a massive MIMO system with and without CU while considering power scaling of $P_u = \frac{E_u}{\sqrt{N_{BS}}}$. The number of BS antennas raises the sum rate of both systems. The massive MIMO system's uplink sum rate is significantly impacted by the CU. The sum rate for 400 antennas at the BS is 43.62 bits/s/Hz in the ideal scenario, compared

Table 13.4 TDLB100 (DS = 100 ns).

Tap #	Delay (ns)	Power (dB)
1	0	0
2	10	−2.2
3	20	−0.6
4	30	−0.6
5	35	−0.3
6	45	−1.2
7	55	−5.9
8	120	−2.2
9	170	−0.8
10	245	−6.3
11	330	−7.5
12	480	−7.1

Table 13.5 TDLC300 (DS = 300 ns).

Tap #	Delay (ns)	Power (dB)
1	0	−6.9
2	65	0
3	70	−7.7
4	190	−2.5
5	195	−2.4
6	200	−9.9
7	240	−8.0
8	325	−6.6
9	520	−7.1
10	1045	−13.0
11	1510	−14.2
12	2595	−16.0

to 31.78 bits/s/Hz in the CU scenario. The sum rate is decreased by ≈27.14% by the CU.

The uplink sum rate of a massive MIMO system with and without CU is shown in Figure 13.20 subject to power scaling of $P_u = \frac{E_u}{N_{\text{BS}}}$. As the number of

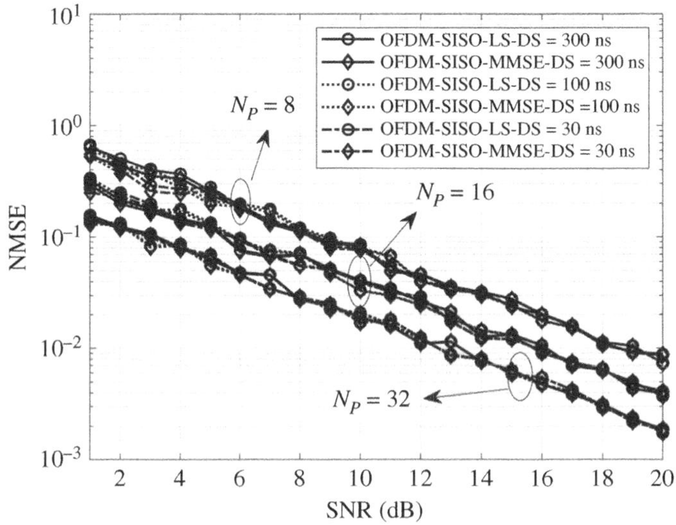

Figure 13.18 NMSE in OFDM - SISO for TDL (30 ns, 100 ns, and 300 ns) NR channel models.

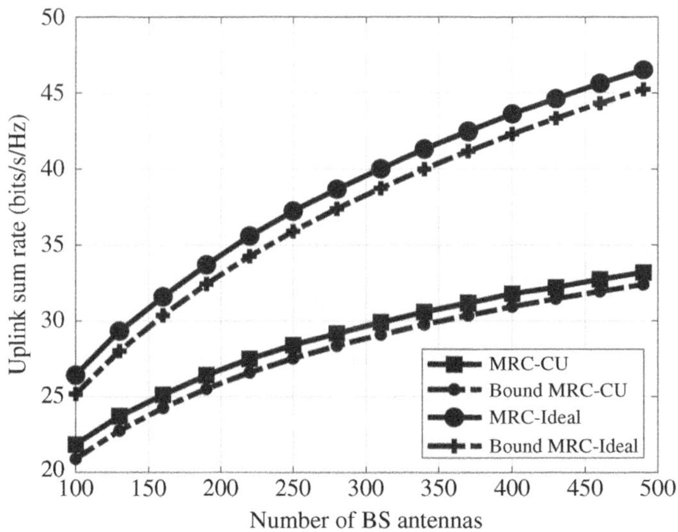

Figure 13.19 Uplink sum rate of massive MIMO system with and without CU considering power scaling of $P_u = \frac{E_u}{\sqrt{N_{BS}}}$.

Table 13.6 Simulation parameters.

Parameters	Values
Number of cell	1
Radius of cell	1000 m
Number of BS antennas (N_{BS})	100–500
Number of users	40
Fixed power per user (E_u)	10 dB
Reference distance (d_0)	100 m
Pathloss exponent	3.8
Mean of log-normal distribution	2
Number of iterations	5000

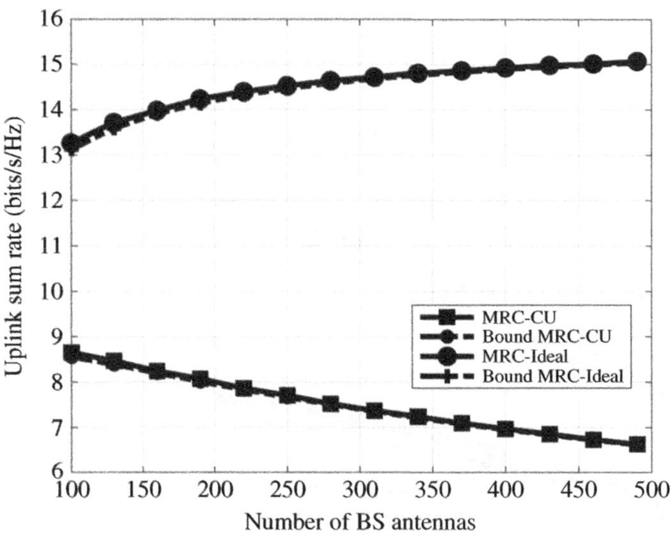

Figure 13.20 Uplink sum rate of massive MIMO system with and without CU considering power scaling of $P_u = \frac{E_u}{N_{BS}}$.

BS antennas increases, the sum rate of CU case drops. The sum rate approaches 0 when there are a large number of BS antennas, as pointed out in (13.54). The massive MIMO system's uplink sum rate is significantly affected by the CU. The total rate for 400 antennas at the BS is 14.93 bits/s/Hz in the ideal scenario, compared to 6.97 bits/s/Hz in the CU situation. The sum rate is decreased by ≈53.32% by the CU. Hence, the power scaling of $\frac{E_u}{\sqrt{N_{BS}}}$ is preferable over $\frac{E_u}{N_{BS}}$.

13.8 Conclusions

This chapter comprehensively reviews LS and MMSE-based channel estimation in various configurations of a wireless communication system in both narrowband (flat fading) and wideband (frequency-selective fading) environments. The impact of CU and pilot contamination on channel estimation performance in massive MIMO systems is analyzed in terms of NMSE and uplink sum rate respectively. The channel estimation framework for cell-free massive MIMO systems using CML and DML is presented. FL has great potential in addressing privacy policy in wireless communication. A brief overview of user data protection is presented for cell-free massive MIMO using FL. This chapter will assist learners in developing data sets for channel estimation in massive MIMO systems using FL. As future work, we plan to develop algorithms for user data protection in DML and communication overhead analysis of DML for cell-free massive MIMO systems.

References

1 Björnson, E., Hoydis, J., and Sanguinetti, L. (2017). Massive MIMO networks: spectral, energy, and hardware efficiency. *Foundations and Trends® in Signal Processing* 11 (3–4): 154–655.

2 Björnson, E., Chae, C.-B., Heath, R.W. Jr. et al. (2024). Towards 6G MIMO: massive spatial multiplexing, dense arrays, and interplay between electromagnetics and processing. *arXiv preprint arXiv:2401.02844.*

3 de Figueiredo, F.A.P. (2022). An overview of massive MIMO for 5G and 6G. *IEEE Latin America Transactions* 20 (6): 931–940.

4 Khansefid, A. and Minn, H. (2015). On channel estimation for massive MIMO with pilot contamination. *IEEE Communications Letters* 19 (9): 1660–1663.

5 Li, J., Zhang, Q., Xin, X. et al. (2019). Deep learning-based massive MIMO CSI feedback. *2019 18th International Conference on Optical Communications and Networks (ICOCN)*, 1–3. IEEE.

6 Dörner, S., Cammerer, S., Hoydis, J., and Ten Brink, S. (2017). Deep learning based communication over the air. *IEEE Journal of Selected Topics in Signal Processing* 12 (1): 132–143.

7 Hu, S., Chen, X., Ni, W. et al. (2021). Distributed machine learning for wireless communication networks: techniques, architectures, and applications. *IEEE Communications Surveys & Tutorials* 23 (3): 1458–1493.

8 Kim, M., Oh, I., Yim, K. et al. (2023). Security of 6G enabled vehicle-to-everything communication in emerging federated learning and blockchain technologies. *IEEE Access* 12: 33972–34001. https://doi.org/10.1109/ACCESS.2023.3348409.

9 Carrera, D.F., Vargas-Rosales, C., Azpilicueta, L., and Galaviz-Aguilar, J.A. (2020). Comparative study of channel estimators for massive MIMO 5G NR systems. *IET Communications* 14 (7): 1175–1184.

10 Ma, X., Yang, F., Liu, S. et al. (2017). Design and optimization on training sequence for mmWave communications: a new approach for sparse channel estimation in massive MIMO. *IEEE Journal on Selected Areas in Communications* 35 (7): 1486–1497.

11 Qin, Q., Gui, L., Gong, B., and Luo, S. (2018). Sparse channel estimation for massive MIMO-OFDM systems over time-varying channels. *IEEE Access* 6: 33740–33751.

12 Liu, L. and Yu, W. (2018). Massive connectivity with massive MIMO—Part I: Device activity detection and channel estimation. *IEEE Transactions on Signal Processing* 66 (11): 2933–2946.

13 Albataineh, Z., Hayajneh, K., Salameh, H.B. et al. (2020). Robust massive MIMO channel estimation for 5G networks using compressive sensing technique. *AEU-International Journal of Electronics and Communications* 120: 153197.

14 Sun, B., Zhou, Y., Yuan, J., and Shi, J. (2020). Interference cancellation based channel estimation for massive MIMO systems with time shifted pilots. *IEEE Transactions on Wireless Communications* 19 (10): 6826–6843.

15 De Figueiredo, F.A.P., Cardoso, F.A.C.M., Moerman, I., and Fraidenraich, G. (2017). Channel estimation for massive MIMO TDD systems assuming pilot contamination and frequency selective fading. *IEEE Access* 5: 17733–17741.

16 De Figueiredo, F.A.P., Cardoso, F.A.C.M., Moerman, I., and Fraidenraich, G. (2018). Channel estimation for massive MIMO TDD systems assuming pilot contamination and flat fading. *EURASIP Journal on wireless Communications and Networking* 2018 (1): 1–10.

17 Shen, K., Cheng, H.V., Chen, X. et al. (2020). Enhanced channel estimation in massive MIMO via coordinated pilot design. *IEEE Transactions on Communications* 68 (11): 6872–6885.

18 Han, Y., Jin, S., Wen, C.-K., and Ma, X. (2020). Channel estimation for extremely large-scale massive MIMO systems. *IEEE Wireless Communications Letters* 9 (5): 633–637.

19 He, H., Wen, C.-K., Jin, S., and Li, G.Y. (2018). Deep learning-based channel estimation for beamspace mmWave massive MIMO systems. *IEEE Wireless Communications Letters* 7 (5): 852–855.

20 Dong, P., Zhang, H., Li, G.Y. et al. (2019). Deep CNN-based channel estimation for mmWave massive MIMO systems. *IEEE Journal of Selected Topics in Signal Processing* 13 (5): 989–1000.

21 Yuan, J., Ngo, H.Q., and Matthaiou, M. (2019). Machine learning-based channel estimation in massive MIMO with channel aging. *2019 IEEE 20th International Workshop on Signal Processing Advances in Wireless Communications (SPAWC)*, 1–5. IEEE.

22 Demir, O.T. and Björnson, E. (2019). Channel estimation in massive MIMO under hardware non-linearities: Bayesian methods versus deep learning. *IEEE Open Journal of the Communications Society* 1: 109–124.

23 Chun, C.-J., Kang, J.-M., and Kim, I.-M. (2019). Deep learning-based channel estimation for massive MIMO systems. *IEEE Wireless Communications Letters* 8 (4): 1228–1231.

24 Hirose, H., Ohtsuki, T., and Gui, G. (2020). Deep learning-based channel estimation for massive MIMO systems with pilot contamination. *IEEE Open Journal of Vehicular Technology* 2: 67–77.

25 Sabeti, P., Farhang, A., Macaluso, I. et al. (2020). Blind channel estimation for massive MIMO: a deep learning assisted approach. *ICC 2020-2020 IEEE International Conference on Communications (ICC)*, 1–6. IEEE.

26 Wei, X., Hu, C., and Dai, L. (2020). Deep learning for beamspace channel estimation in millimeter-wave massive MIMO systems. *IEEE Transactions on Communications* 69 (1): 182–193.

27 Zhang, Y., Mu, Y., Liu, Y. et al. (2020). Deep learning-based beamspace channel estimation in mmWave massive MIMO systems. *IEEE Wireless Communications Letters* 9 (12): 2212–2215.

28 Jeon, Y.-S., Li, J., Tavangaran, N., and Poor, H.V. (2020). Data-aided channel estimator for MIMO systems via reinforcement learning. *ICC 2020-2020 IEEE International Conference on Communications (ICC)*, 1–6. IEEE.

29 Melgar, A., Fuente, A., Carro-Calvo, L. et al. (2022). Deep neural network: an alternative to traditional channel estimators in massive MIMO systems. *IEEE Transactions on Cognitive Communications and Networking* 8 (2): 657–671.

30 Vu, T.T., Ngo, H.Q., Dao, M.N. et al. (2022). Energy-efficient massive MIMO for federated learning: transmission designs and resource allocations. *IEEE Open Journal of the Communications Society* 3: 2329–2346.

31 Elbir, A.M. and Coleri, S. (2021). Federated learning for channel estimation in conventional and RIS-assisted massive MIMO. *IEEE transactions on wireless communications* 21 (6): 4255–4268.

32 Chen, S., Zhang, J., Björnson, E. et al. (2023). Energy-efficient cell-free massive MIMO through sparse large-scale fading processing. *IEEE Transactions on Wireless Communications* 22 (12): 9374–9389.

33 Femenias, G., Riera-Palou, F., and Björnson, E. (2023). Another twist to the scalability in cell-free massive MIMO networks. *IEEE Transactions on Communications* 71 (11): 6793–6804.

34 Maxwell, J.C. (2018). 5G NR User Equipment (UE) radio transmission and reception; Part 1: Range 1 standalone release 15 V. 15.2. 0 document 3GPP TS 38.101–1. In: *A Treatise on Electricity and Magnetism*, vol. 2, 68–73. Clarendon.

Index

Massive MIMO for Future Wireless Communication Systems: Technology and Applications,
First Edition. Edited by Webert Montlouis and Agbotiname Lucky Imoize.

t

Printed and bound by CPI Group (UK) Ltd, Croydon, CR0 4YY

17/09/2026

14964312-0006